# Neurophysiology of Consciousness

## Selected Papers and New Essays by Benjamin Libet

Ben Libet

CONTEMPORARY NEUROSCIENTISTS
Selected Papers of Leaders in Brain Research

# Neurophysiology of Consciousness

## Selected Papers and New Essays by Benjamin Libet

Springer Science+Business Media, LLC

Benjamin Libet
Department of Physiology and
Mount Zion Neurological Institute
University of California School of Medicine
San Francisco, CA 94143-0444

**Library of Congress Cataloging In-Publication Data**
Libet, Benjamin, 1916-
Neurophysiology of consciousness: selected papers and new essays of Benjamin Libet.
p. cm. -- (contemporary neuroscientists)
Collection of papers previously published.
Includes bibliographical references.

ISBN 978-1-4612-6722-5 ISBN 978-1-4612-0355-1 (eBook)
DOI 10.1007/978-1-4612-0355-1

1. Consciousness. 2. Neuropsychology I. Title. II. Series.
[DNLM: 1. Consciousness--physiology--collected works.
2. Neurophysiology--collected works. WL 705 L695n 1993]
QP411.L54 1993
612.8'2--dc20
DNLM/DLC
for Library of Congress 93-16896
CIP

*Birkhäuser*

Printed on acid-free paper

Originally published by Birkhäuser Boston 1993
Softcover reprint of the hardcover 1st edition 1993


Camera ready copy provided by the author.

9 8 7 6 5 4 3 2 1

*To my mother, Anna, and my wife, Fay*
*who both fostered my ability to carry out*
*these investigations*

# CONTENTS

# THE CEREBRAL NEUROPHYSIOLOGY OF CONSCIOUS EXPERIENCE

## Preface

*"All knowledge of reality starts from experience and ends in it."* Albert Einstein

*"The human mind is capable of such fantastic speculations. The great thing is that you think about it and also try to do measurements."* I.I. Rabi

Conscious, subjective experience is what is most important and meaningful to us as human beings. By this I mean our *awareness* of sensory inputs, of our will or intention to act or not to act, of our feelings and thoughts, indeed of our own mental existence. There is no doubt that the appearance of a conscious experience is intimately inter-related with neural functions of the brain. Consequently, it has long been recognized that a knowledge of the nature of that mind-brain inter-relationship presents perhaps the most significant challenge for science to achieve (for ex. Hook, 1960; Schrodinger, 1958; Eccles, 1979). There has been no shortage of theoretical speculation on the mind-brain relationship. But there has been surprisingly little direct experimental investigation of the way in which conscious subjective experiences are related to neural events in the brain.

A major difficulty in the experimental approach to this issue is the stubborn fact that conscious subjective experience is accessible only to the individual who is having the experience. This means that the only valid primary evidence of this private "inner" quality must come from an introspective report of it by the subject (see my paper nos. 5,7,15,16,18,19,20). Such a requirement is different from that in all other scientific investigations, which depend upon observations of "outer qualities" (Pepper, 1960) that are accessible to an external observer, the investigator. There has been a widespread reluctance and even antagonism to accept introspective reports as scientifically useful evidence, especially with the influence of positivism and behaviorism in psychology during most of this century. Admittedly, there is no fully "objective" way in which an external observer can be certain about the validity of the content of an introspective report (see paper nos. 9,15, 19). (Actually, quantum mechanics teaches us that even externally observable events at subatomic levels also suffer a lack of absolute specification.) But any discrepancies between what the subject experiences introspectively and what he/she reports to an external observer can be minimized and made insignificant in practice, by selecting for study simple kinds of experiences which are devoid of emotional content and which can be tested for reliability. A simple somatosensory "raw feel" fulfills these characteristics (see papers nos. 2,5). In any case, if we fail to find ways to use introspective reports in convincingly acceptable studies we would give up the ability to investigate the relation between conscious experience and neural activity, something warned against by William James (Krech, 1969).

Another factor in the dearth of direct experimental studies is, of course, the comparative inaccessibility of the human brain for such purposes. Meaningful investigations of the issue in question requires simultaneous study of brain events and introspective reports of experiences in an awake, cooperative human subject. Analysis by neuropsychologists of pathological lesions in the brain and the related disturbances of conscious functions have contributed much to mapping the possible representations of these functions. The non-invasive recording of electrical activity with electrodes on the scalp, starting from Berger's initial EEG recordings in 1929, has contributed much to the problems of states of consciousness and to various cognitive features associated with sensory inputs, but not as much to the specific issue of conscious experience. In more recent times, techniques have arisen for measuring localized or regional cerebral blood flow (r CBF), pioneered by Ingvar, Lassen, and Roland, and for localized measurements of changes in metabolic rate, using positron emission tomography or PET scans, pioneered by Sokoloff. These techniques have increasingly been utilized to study some conscious functions (e.g., Ingvar and Phillipson, 1977; Lassen et al. 1978; Roland and Friberg, 1985); but, again, the analysis is largely limited to identifying areas of representation; the methods still involve time resolutions too gross to analyze the time factors in most mind-brain functions.

Intracranial electrophysiological studies, with electrodes or other devices introduced subdurally or into the brain have offered the best chances for finer localizations of functions, and for resolution of the time factors in neural functions. Electrical stimulation, or local application of certain chemical agents, can also permit external manipulation of neuronal functions in a reversible manner; that capability is often essential in establishing causal relationships instead of merely correlative ones. The initial stimulations of the exposed human cerebral cortex by Cushing, and the more exhaustive studies by Penfield and his colleagues and by Foerster during neurosurgical procedures on epileptic patients, produced topographical mappings of conscious sensory and of motor functions for areas that gave such responses (chiefly the primary sensori motor areas). The discovery by Penfield, that electrical stimulation could also suppress or disrupt the normal functioning of a cortical area, made it possible to illuminate the functional representation of many areas of the cortex for which electrical stimulation does not elicit an overt subjective experience or a motor response. This feature has been exploited recently by Ojemann to analyze the representation for various features of language and speech functions; i.e. the ability of stimulation to stop or distort these activities suggests roles for the cortical area in these functions.

My interest has been a more physiological one. The general idea was to use stimulation and recording methods not for mapping representations, but for gaining evidence on the neural dynamics. The general question was — what kinds of neuronal activities mediate conscious experience, — rather than asking simply where are they located? My entry into this research area, and most of the subsequent experimental investigations, was made possible by my fortunate association with the late neurosurgeon Bertram Feinstein. When Feinstein converted himself into a neurosurgeon, by virtue of his training with Lars Leksell in Sweden in the early 1950s, he invited me to join him in exploiting opportunities for research that required intracranial studies in the unique human subject, and to suggest and devise specific physiological experiments to that end. This was all to be carried out in a newly constructed neurosurgery facility, completed in 1958 in the Mt. Zion Hospital and Medical Center in San Francisco; the operating room was arranged with suitable electromagnetic shielding and with an adjacent control room for equipment. Such studies were, of course, only to be done with the patients' informed consent and only when no significant risk was added by the research process. (I would note that in the hundreds of patients actually studied during a period of about 20 years there was not a single instance in which any irreversible deleterious effect was attributable to the research procedure.) I want also to acknowledge my special debt to the other two important contributors in the original team that began that work; they were, biophysicist W. Watson Alberts (who left us in 1972 to become a successful NIH administrator) and biomedical engineer Elwood ("Bob") W. Wright, who maintained his cheery collaboration until this work terminated at Mt. Zion in about 1982, several years after Feinstein's untimely death in 1978. Curtis A. Gleason took over Albert's biophysical functions. In about 1976, Dennis K. Pearl (now in the Dept. of Statistics at Ohio State University, Columbus) became an essential collaborator to help in statistical analysis and experimental design.

This book presents the more important of the original publications that stemmed from this experimental work. These include (I) complete experimental research papers (nos. 1,4,9,10,13,14,15,21), (II) review/analysis papers in which are also described some experimental findings *not* published in original form elsewhere (nos. 5,6,7,8), and (III) papers with only reviews/analysis (nos. 2,3,11,12,16,17,18,19,20,22). The publishing of some original experimental findings in group (II) which did not appear in group (I) papers occurred because of some extenuating circumstances. Chief among these was the prospective unavailability of suitable patient-subjects with whom a given experimental question could be pursued systematically, so as to provide sufficient numbers for a more satisfying statistical treatment. Rather than wait some indefinite time before publishing a fuller experimental paper on what were important issues in arguing a thesis, the available data that seemed qualitatively convincing was summarized and presented in a review type paper at the time. As it turned out, opportunities for further experiments on these issues did not appear in the ensuing years.

However, 1) a deficiency in experimental numbers was not a factor in the report of the qualities or modalities of the conscious somatosensory responses that could be elicited and altered by suitable electrical stimulation of the postcentral gyrus (primary somatosensory area of cerebral cortex); see paper nos. 5,6, 12) The study of event-related potentials (ERPs) recordable in the supplementary motor area (SMA), in response to sensory inputs of various modalities, was reported rather fully in the mixed review-data paper by Libet et al. 1975 (no. 6). Although the findings involved only one patient in whom the SMA was therapeutically accessible, that study was broadly based, solidly convincing and established the SMA as a source of most later ERP components recordable there. 3) Thirdly, the unique studies of direct cortical responses (DCRs), recorded adjacent to the cortical stimulating electrode, provided important insights into the kinds of neural activations related to a conscious sensory response but were only described in review-type articles (nos. 3,5,6). These extensive and important sets of data should have been published as separate experimental research papers; in retrospect, it was a mistake to include the original data from those studies simply within a chapter or review article.

We have now re-examined the original data for one of the critical issues, the production of retroactive enhancement of a skin sensation by a cortical stimulus that is delayed 200 to 500 msec after the skin stimulus. The quantitative analysis of those results is now presented in an additional paper [no. 10] inserted after the one by Libet, 1978 [no. 8] in which that phenomenon was first reported informally.

Finally, I would like to express my appreciation to George Adelman, for his encouragement and help in the development of this volume. We are of course indebted to the publishers of the original papers for permission to reprint them here.

Benjamin Libet
San Francisco

BIBLIOGRAPHY I

# Papers Reprinted in This Book


1 Libet B, Alberts WW, Wright Jr EW, Delattre LD, Levin G, Feinstein B (1964): Production of threshold levels of conscious sensation by electrical stimulation of human somatosensory cortex. *Journal of Neurophysiology* 27:546–578

2 Libet B (1965): Cortical activation in conscious and unconscious experience. *Perspectives in Biology and Medicine* 9:77–86

3 Libet B (1966): Brain stimulation and the threshold of conscious experience. In: *Brain and Conscious Experience*, Eccles JC, ed. New York: Springer-Verlag

4 Libet B, Alberts WW, Wright Jr EW, Feinstein B (1967): Responses of human somatosensory cortex to stimuli below threshold for conscious sensation. *Science* 158: 1597–1600

5 Libet B (1973): Electrical stimulation of cortex in human subjects and conscious sensory aspects. In: *Handbook of Sensory Physiology, Vol II: Somatosensory System*, Iggo A, ed. Berlin: Springer-Verlag

6 Libet B, Alberts WW, Wright Jr EW, Lewis M, Feinstein B (1975): Cortical representation of evoked potentials relative to conscious sensory responses, and of somatosensory qualities — in man. In: *The Somatosensory System*, Kornhuber HH, ed. Stuttgart: Georg Thieme Verlag

7 Libet B, Alberts WW, Wright Jr EW, Feinstein B (1972): Cortical and thalamic activation in conscious sensory experience. In: *Neurophysiology Studied in Man*, Somjen GG, ed. Amsterdam: Excerpta Medica

8 Libet B (1978): Neuronal vs. subjective timing for a conscious sensory experience. In: *Cerebral Correlates of Conscious Experience*, Buser PA, Rougeul-Buser A, eds. Amsterdam: Elsevier/North-Holland Biomedical Press

9 Libet B, Wright Jr EW, Feinstein B, Pearl DK (1979): Subjective referral of the timing for a conscious sensory experience: A functional role for the somatosensory specific projection system in man. *Brain: A Journal of Neurology* 102:193–224

10 Libet B, Wright Jr EW, Feinstein B (1992a): Retroactive enhancement of a skin sensation by a delayed cortical stimulus in man: Evidence for delay of a conscious sensory experience. *Consciousness and Cognition* 1:367–375

11 Libet B (1981): The experimental evidence for subjective referral of a sensory experience backwards in time: Reply to P.S. Churchland. *Philosophy of Science* 48:182–197

12 Libet B (1982a): Brain stimulation in the study of neuronal functions for conscious sensory experiences. *Human Neurobiology* 1:235–242

13 Libet B, Wright Jr EW, Gleason CA (1982b): Readiness-potentials preceding unrestricted "spontaneous" vs. pre-planned voluntary acts. *Electroencephalography and Clinical Neurophysiology* 54:322–335

14 Libet B, Wright Jr EW, Gleason CA (1983a): Preparation- or intention-to-act, in relation to pre-event potentials recorded at the vertex. *Electroencephalography and Clinical Neurophysiology* 56:367–372

15 Libet B, Gleason CA, Wright Jr EW, Pearl DK (1983b): Time of conscious intention to act in relation to onset of cerebral activity (readiness-potential): The unconscious initiation of a freely voluntary act. *Brain: A Journal of Neurology* 106:623–642

16 Libet B (1985): Unconscious cerebral initiative and the role of conscious will in voluntary action. *Behavioral and Brain Sciences* 8:529–566

17 Libet B (1987a): Are the mental experiences of will and self-control significant for the performance of a voluntary act? Response to commentaries. *Behavioral and Brain Sciences* 10:781–786

18 Libet B (1987b): Consciousness: Conscious, subjective experience. In: *Encyclopedia of Neuroscience*, Adelman G, ed. Boston: Birkhäuser Boston

19 Libet B (1989a): The timing of a subjective experience. Response to a commentary by D. Salter. *Behavioral and Brain Sciences* 12:181–187

20 Libet B (1989b): Conscious subjective experience vs. unconscious mental functions: A theory of the cerebral processes involved. In: *Models of Brain Function*, Cotterill RMJ, ed. New York: Cambridge University Press

21 Libet B, Pearl DK, Morledge DE, Gleason CA, Hosobuchi Y, Barbaro NM (1991): Control of the transition from sensory detection to sensory awareness in man by the duration of a thalamic stimulus: The cerebral "time-on" factor. *Brain: A Journal of Neurology* 114:1731–1757

22 Libet B (1992b): The neural time-factor in perception, volition and free will. *Revue de Métaphysique et de Morale* 97:255–272

# Bibliography II
# Selected Additional Papers by Libet on Brain Function Not Included in This Book


Libet B, Gerard RW (1939): Control of the potential rhythm of the isolated frog brain. *J Neurophysiology* 2:153–169

Gerard RW, Libet B (1940): The control of normal and "convulsive" brain potentials. *Amer J Psychiatry* 96:1125–1151

Libet B, Gerard RW (1941): Steady potential fields and neurone activity. *J Neurophysiology* 4:438–455

Libet B, Fazekas JF, Himwich HE (1941): The electrical response of the kitten and adult cat brain to cerebral anemia and analeptics. *Amer J Physiology* 132:232–238

Elliott KAC, Libet B (1942): Studies on the metabolism of brain suspensions I: Oxygen uptake. *J Biol Chem* 143:227–246

Elliott KAC, Scott DB, Libet B (1942): Studies on the metabolism of brain suspensions II: Carbohydrate utilization. *J Biol Chem* 146:251–269

Elliott KAC, Libet B (1944): Oxidation of phospholipid catalyzed by iron compounds with ascorbic acid. *J Biol Chem* 152:617–626

Libet B (1948): Adenosinetriphosphatase (ATP-ase) in nerve. *Fed Proc* 7:72

Libet B (1948): Enzyme localization in the giant nerve fiber of the squid. *Biological Bulletin* 95:277–278

Ralston HJ, Libet B (1953): The question of tonus in skeletal muscle. *Amer J Physical Med* 32:85–92

Libet B, Feinstein B, Wright Jr EW (1959): Tendon afferents in autogenic inhibition in man. *Electroenceph Clin Neurophysiol* 11:129–140

Eccles JC, Libet B, Young RR (1958): The behavior of chromatolyzed motoneurones studied by intracellular recording. *J Physiology (Lond)* 143:11–40

Libet B, Gerard RW (1962): An analysis of some correlates of steady potentials in mammalian cerebral cortex. *Electroenceph Clin Neurophysiol* 14:445–452

Libet B, Tosaka T (1970): Dopamine as a synaptic transmitter and modulator in sympathetic ganglia: A different mode of synaptic action. *Proc Nat Acad Sci* 67:667–673

Libet B, Kobayashi H, Tanaka T (1975): Synaptic coupling into the production and storage of a neuronal memory trace. *Nature* 258:155–157

Libet B, Gleason C, Wright Jr EW, Feinstein B (1977): Suppression of an epileptiform type of electro-cortical activity by stimulation in the vicinity of locus coeruleus. *Epilepsia* 18:451–462

Libet B (1978): Slow postsynaptic responses in sympathetic ganglion cells, as models for the slow potential changes in the brain. In: *Multidisciplinary Perspectives in Event-Related Brain Potential Research*, Otto D ed., EPA 600/9-77-043, Superintendent of Documents, Washington, D.C., pp 12–18

Libet B (1980): Mental phenomena and behavior. Commentary on Searle: Minds, brains and programs. *Behav and Brain Sci* 3:434

Libet B (1984): Heterosynaptic interaction at a sympathetic neurone as a model for induction and storage of a postsynaptic memory trace. In: *Neurobiology of Learning and Memory*, Lynch G, McGaugh JL, Weinberger NM, eds. New York: Guilford Press

Libet B (1985): Subjective antedating of a sensory experience and mind-brain theories: Reply to Ted Honderich. *J Theor Biol* 114:563–570

Libet B (1986): Nonclassical synaptic functions of transmitters. *Fed Proc* 45:2678–2686

Kaitin KI, Bliwise DL, Gleason C, Nino-Murcia G, Dement WC, Libet B (1986): Sleep disturbance produced by electrical stimulation of the locus coeruleus in a human subject. *Biol Psychiat* 21:710–716

Feinstein B, Gleason CA, Libet B (1989): Stimulation of locus coeruleus in man: Preliminary trials for spasticity and epilepsy. *Stereotact Funct Neurosurgery* 52:26–41

Libet B (1989): Neural destiny: Does the brain have a will of its own? *The Sciences* 29:32–35

Libet B (1990): Cerebral processes that distinguish conscious experience from unconscious mental functions. In: *The Principles of Design and Operation of the Brain*, Eccles JC, Creutzfeldt O, eds. Pontifical Academy Sci., Rome (also published by Springer-Verlag)

Libet B (1990): Attentional theories and conscious perception (Commentary on Naatanen: ERPs and attention). *Behav and Brain Sci* 13:247–248

Libet B (1990): Time delays in conscious processes (Commentary on Penrose: The Emperor's New Mind). *Behav and Brain Sci* 13:672

Libet B (1991): Conscious or unconscious. *Nature* 351:94–95

Libet B (1991): Conscious vs. neural time. *Nature* 352:27

Libet B (1991): Conscious functions and brain processes (Commentary on Velman: Is human information processing conscious?). *Behav and Brain Sci* 14:685–686

Libet B (1992): Voluntary acts and readiness potentials. *Electroenceph Clin Neurophysiol* 82:85–86

Libet B (1992): Models of conscious timing and the experimental evidence (Commentary on Dennett & Kinsbourne: "Time and the Observer"). *Behav and Brain Sci* 15:213–215

# Prologue:
# Introduction and Overview

I would like here to provide a framework and general view of the issues within which the reprinted papers may be viewed coherently by the reader. In doing so, I shall try to repeat as little as possible what is in the papers. Instead, I shall refer to the papers by their assigned numbers to indicate their relevance to particular points or issues within this overview.

## Principles for the Experimental Study of Brain and Conscious Experience

(1) An introspective report by our subject constitutes the operational definition of a conscious subjective experience. As a physiologist this seemed to me to be an obvious requirement from the start. I quickly found that some psychologists and philosophers, presumably schooled in behavioristic and logical positivist approaches, found this definition either intolerable or worrisome. This led me to discuss this issue repeatedly (paper nos. 1,2,3,5,7,9,11,12,16,18,20,22). Interestingly, physiologists generally found the definition acceptable without serious reservations. Of course, many psychologists and some philosophers also recognized the fundamental significance and validity of our operational definition as essential if one is to investigate conscious experience, a phenomenon that is outside the purview of classical behaviorist study.

(2) There are no *a priori* rules that can describe the relationsip between neural-cerebral events and subjective-mental events. The rules must be discovered by observing both of these phenomenologically independent categories of events simultaneously. It follows that even if one could achieve a complete knowledge of the pattern of neural-physical events this would not, in itself, produce a description of any correlated subjective experiences (see my paper nos. 9,11,12,15,16,18,20,22; see also Nagel, 1979).

Narrowing the experimental study down to the experience of a simple bodily (somatic) sensation allowed us to minimize and circumvent many of the difficulties in utilizing introspective reports. It also kept the psychological parameters simple and controllable and allowed us to focus analysis on the physiological ones (see papers nos. 1,3,4,6,9,10,11,21,22).

## Cerebral Activations Needed for a Conscious Sensory Experience

Electrical stimulation of primary sensory cortex, or of points in the cerebral ascending somatosensory projection to the cortex, enables one to control (albeit crudely) the nature of the neural input to sensory cortex at a level closer to cerebral production of the conscious sensation. By contrast, a peripheral sensory input can lead to patterns of multiple ascending neural activations the roles of which are not easily amenable to control and analysis in the human subject. Our initial study lay the groundwork for this approach by asking which stimulus parameters are significant to the eliciting of a conscious sensation, when stimulating postcentral gyrus or ventrobasal thalamus (nos. 1,4,5). The most striking requirement found was a surprisingly long train duration of repetitive stimulus pulses, i.e., a long period of repetitive inputs. The analysis of that time requirement for neural activation, and the pursuit of its implications, occupied a major portion of our efforts thereafter. Before pursuing that topic, some other related by-products of those cortical stimulation studies should be noted.

## Subjective Qualities of Sensation

Electrical stimulation of S-I cortex (postcentral gyrus) has generally produced sensations with a paresthesia-like quality (tingling, electric shock, pins and needles, etc.). Indeed, the group performing the most trials of such stimuli had concluded that the predominant or perhaps only function of S-I cortex lay in the discriminative aspects of sensation, i.e., spatial localization and sense of position and movement of limbs, rather than in mediating specific qualities of somatic sensation (e.g., Penfield and Jasper, 1954, pp. 68–69). We ourselves added a temporal function for S-I (somatosensory) cortex, to the already known spatial one (see below, and paper nos. 8,9,11,12,18,19,22). However, when we established the significance of applying stimuli with barely threshold (liminal) intensities and sufficiently long train durations (TD) of repetitive pulses, we found that more than 50% of reports described "natural-like" qualities of somatic sensation, such as touch, pressure, warmth, cold, etc., but not pain. The responses obtained in a total of 174 subjects, with many trials for each subject, were summarized and tabulated in full (paper nos. 5,6). When intensity was raised to 2 to 3 times the liminal level at the same electrode, previously natural-quality reports were consistently though reversibly converted to paresthesias. This would explain the predominance of paresthesia reports when stimulus intensities are not quantitatively kept to liminal levels.

Our findings supported the view that specificity not only for spatial localization but also for modality of peripheral sensory input is maintained even at the level of primary sensory cortex. They were also in general accord with a columnar representation for specific modalities (see Powell and Mountcastle, 1959), as discussed in paper nos. 5,6,12.

The particular subjective quality or modality of the conscious somatosensory experience appeared to be relatively independent of the nature of the stimulus to postcentral gyrus. This held for changes in pulse frequency, (except when dropping below 10 pps); changes in intensity, except that large increases produced only paresthesias; change in polarity; and even large changes in the area of electrode contact. The latter somewhat startling result is actually also in accord with the hypothesis of columnar representation (see paper nos. 1,3,5,6,12).

## Cerebral Time Required and the Delay for a Sensory Experience

*Minimum train-duration (TD).* When applying trains of pulses (15 to 120 per sec) to S-I cortex at the weakest effective (liminal) intensity for a conscious sensation, a minimum TD of about 0.5 sec was found necessary (nos. 1,3,4,5,9,12,21). With lesser TDs subjects reported feeling nothing, even though such pulses (at the same liminal I) could be shown to be eliciting considerable neuronal activity, as indicated by electrical recordings of the "direct cortical responses" (DCR) from nearby cortex (nos. 5,6,12). This minimum was termed the "utilization TD" or U-TD, by analogy with the "utilization time" for *pulse* durations when exciting a nerve fiber.

If intensity was raised, the effective TD (to elicit a conscious sensation) could be shortened within limits. The curves describing the relation between intensity and train duration (I-TD), of stimuli at thresholds for eliciting conscious sensation, were a bit choppy-looking for the few subjects with full series in paper no. 1. Putting together data from several additional subjects produced curves diagrammed in paper nos. 3,5,12 (and reproduced thereafter). The configuration of these curves, particularly in the TD range of 0.2 to 1 sec has been confirmed in many subjects since 1964 (most recently in Libet et al. 1991, no. 21).

The "abnormal" route of activation when applying stimuli at the outer (pia-arachnoid) surface of postcentral gyrus, does not account for the requirement of relatively long TDs with liminal I. Similar I-TD relationships were found with stimulation in subcortical white matter (just below postcentral cortex), in ventrobasal thalamus (whether n.VPL or nVPM) and in medial lemniscus. The requirement did not hold for stimulation at spinal dorsal columns (as observed by Nashold and Somjen) or, of course at skin or peripheral sensory nerves. In these, a single pulse is effective with intensities only slightly higher than for 1 sec trains (see paper no. 4).

The absolutely minimum effective TD when I is sufficiently raised at cerebral loci (S-I cortex, VB thalamus, medial lemniscus), is not unambiguously established. At S-I cortex, a strong single pulse can elicit some sensation, but it also can evoke a small muscular twitch that produces its own feedback of peripheral sensory input. Consequently, we re-checked this question with stimuli in n.VPL-thalamus, a purely sensory structure; additionally, we delivered the stimulus there via a bipolar coaxial electrode to minimize spread of current to surrounding structures. Under these conditions a single pulse was ineffective no matter how strong the stimulus; subject reported "feeling nothing" with a single pulse having a peak-current I up to 20 and even 40 times the level for liminal I (paper nos. 4,5). We concluded that a single pulse stimulus is simply unable to elicit conscious sensation, when applied in the cerebral somatosensory system. This has been confirmed by R.G. Grossman and by Tasker and colleagues (personal communications). Sensations have been reported with as little as 2 to 3 sufficiently strong pulses (paper nos. 4,5,6; also Tasker, personal communication). But with more modest and presumably closer to "normal" inputs, rather than the high intensity levels required for 2 to 3 pulses, it seems clear that substantial TDs up to 0.5 sec or even 1 sec are necessary to elicit conscious sensation (see papers nos. 4,5,21).

*What is the cerebral "time on" requirement for an effective single volley of peripheral sensory input?* If we hypothesize that to elicit even a normally induced conscious sensation requires substantial cerebral durations (hundreds of msec), how is this to be reconciled with the effectiveness of a single pulse at the peripheral sensory level? There are several lines of indirect evidence on this issue:

(1) *Evoked potentials (EP).* The EP recorded subdurally at postcentral gyrus, in response to a single pulse stimulus to skin of the contralateral hand, exhibits an initially surface-positive primary complex (latency about 15 msec) plus later components extending for 500 msec or more (see paper nos. 4,6). All of these components are already present when stimulus intensity (I) for the skin pulse is just at or above threshold for conscious sensation. With stimulus I just below threshold (subject never feeling any), the later components are lost but a small primary EP is still recordable (paper nos. 4,6,7,8). Neuronal activities represented by the later EPs seem to be associated with eliciting conscious sensation.

Additionally, the primary EP activities are neither necessary nor sufficient to elicit conscious sensation. That primary EPs are not necessary is illustrated by stimuli at the pial surface of S-I cortex. Each stimulus pulse, in a required 0.5 sec train of pulses, elicits a so-called *direct cortical response (DCR)*, recordable close to the stimulating electrode, (paper nos. 4,5,12). The DCR is quite different from a primary EP. The DCR is composed of a large surface negative component thought to be generated by apical dendrites of pyramidal cells; this may be followed by a slower, smaller surface positive component etc. (Ochs and Clark, 1968). It is clear that the surface cortical stimulus excites axons in superficial layers (paper no. 1) rather than those ascending axons of thalamic origin which terminate heavily in layer IV and elicit the primary EP response. Of further interest is our finding that even the neuronal activities represented by the DCR were *also* not necessary! DCR responses could be completely suppressed by a local surface application of GABA (gamma aminobutyric acid, a postsynaptic inhibitor); but this did not affect the ability of the surface cortical (or even of a peripheral skin) stimulus to elicit a conscious sensation (paper nos. 4,5,6). This suggested that deeper layers in S-I cortex and/or areas of cortex activated from

S-I are involved in mediating the conscious sensation.

That primary EPs are also not sufficient is already evident from the above noted observation of a primary EP in response to a skin pulse below threshold for eliciting any sensation (paper no. 4). However, this insufficiency is seen more strikingly with single pulse stimuli in n.VPL (thalamus) or medial lemniscus. In these structures, a strong single pulse produces no conscious sensation but does elicit a large primary EP response of S-I cortex (paper nos. 4,5,6). This large EP is expected, since these structures are part of the ascending specific projection pathway from skin to S-I cortex that normally elicits the primary EP; yet no conscious sensation appears. However, late EP components similar to those in the response to a skin stimulus are not elicited by a VPL stimulus (paper no. 4). EPs elicited from the skin evidently involve parallel, diffuse projection paths to cortex (Jasper, etc.) and, indeed, have a much wider topographical distribution than does the primary EP component (paper nos. 4,5,6; Goff et al., 1962).

Later EP components are widely distributed over the cortex, as recorded with scalp electrodes (e.g., Goff et al., 1962). They were also recorded by us, with primary EPs absent, by subdural electrodes located some several cm anterior and posterior to postcentral gyrus (paper nos. 4,5,6). In addition, strikingly large slow EP components were recorded by us in *supplementary motor cortex (SMA)*, in the mesial cortex of a patient suspected of, but actually found not to have, an epileptic focus in SMA (paper no. 6). EPs were elicited in SMA by somatic, visual and auditory stimuli. Furthermore, we were able to prove that several slow components were generated in SMA cortex itself and were not simply due to volume conductor spread from other areas.

(2) *Retroactive effects of a conditioning stimulus.* If cerebral neuronal responses to a skin stimulus must persist for up to about 500 msec, in order to become sufficient or adequate for conscious sensation, then it should be possible to modify the nature or content of that sensory experience by introducing other suitable neuronal activities during that substantial time period *after* a single pulse to the skin. "Backward masking" of a small weaker visual stimulus, when it is followed by a larger strong one applied after a delay of up to 100–200 msec, had already been described (the "Crawford effect", 1947). The locus of this delayed conditioning using peripheral stimuli is difficult to pinpoint in the visual system (see discussion in paper nos. 5,7). We applied a delayed or conditioning cortical stimulus (C) directly to postcentral gyrus and obtained retroactive effects on the subjective responses to a preceding single pulse stimulus at the skin (S). The C stimulus was applied via a large 10mm disc electrode. Retroactive masking or inhibition of the S responses could be obtained even when C began up to 200–500 msec after S. To be effective, C had to be a train of pulses, TD $\geq$ 100 msec, with intensities of 1.3–1.5 times the liminal I (i.e., 1.5 times the threshold I required by the cortical stimulus to elicit its own sensory response).

In another group of studies in which C was applied via a much smaller wire contact, retroactive *enhancement* of the S sensation was obtained (papers 8,10). The importance of this was in its elimination of the argument that could be made about retroactive inhibition, namely that the retroactive effects were simply due to disruption of the memory trace for S by the C conditioning stimulus; retroactive enhancement of course required that memory of S be retained. One concludes, therefore, that the conscious experience elicited by a single pulse peripheral stimulus pulse can be modified by other cortical actions initiated after a delayed period of at least 200 to 500 msec (and perhaps even later) following the pulse to the skin.

(3) *Reaction times, deliberately slowed.* In a study of reaction times, Jensen (1979) asked his subjects to try deliberately to slow these times gradually, from their usual values (of say 250 msec). To his surprise, he found that deliberate slowing to gradually increasing values could not be achieved. Instead, such slowed reaction times jumped discontinuously to values of about 600+ msec (see discussion in papers). Such a result was in fact explainable by our hypothesis: If one must first become consciously aware of the stimulus input in order deliberately to slow the response, then a substantial delay of some hundreds of msec would be expected.

## Neural Delay vs. Subjective Timing

Since there appears to be a substantial delay before "neuronal adequacy" becomes sufficient to elicit a conscious sensory experience, is there a corresponding delay in the *subjective* timing of that experience (following a brief peripheral sensory stimulus)? Intuitively, one does not normally discern delays between sensory inputs and our subjective perceptions of them. The question was put to an experimental test by matching the relative subjective timing for a skin-generated sensation with one felt in a nearby area of skin but generated by a cortical S-I stimulus (paper no. 9). We could objectively determine that the neural delay for appearance of the cortically induced sensation was at least 500 msec. If the sensation elicited by a threshold skin pulse involved a similar *cerebral* neural delay of about 500 msec, one might have expected the subject to report that a skin-induced sensation appears after a cortically-induced sensation, when the stimulus pulse to skin is delivered well after the onset of the cortical train of pulses, e.g., at +200 msec (see diagram in paper no. 9). But in fact, subjects consistently reported that such a skin sensation appeared before the cortically-induced sensation; this relative order of the experiences held until delivery of the skin pulse was delayed to the end of the 500 ms cortical stimulus train. Those results indicated that there was essentially no delay for the subjective timing of a skin stimulus, but that subjective timing of the cortically-induced sensation was indeed delayed by a time-interval corresponding to the duration of the required cortical stimulus, i.e., corresponding to the delay for achieving neuronal adequacy by that stimulus. How was one to reconcile a postulated substantial cerebral neural delay for achieving awareness of a peripheral sensory stimulus to skin, as supported by the indirect evidence, with the apparent absence of any appreciable delay in the subjective timing of the skin stimulus?

*Subjective referral backwards in time.* We realized that the *subjective* timing of an event need not be identical with the objective neural time at which neural activities become adequate for eliciting the sensory experience. (This is in accordance with "principle #2", for the study of brain processes and conscious experience — see above.) Our modified hypothesis to deal with this experimentally based dilemma contained two features (paper no. 9): 1) Although a sensory experience does not appear until some substantial though variable neural delay after arrival of the initial sensory signal at the cerebral cortex, the time of that experience is subjectively referred back to the time of the initial signal. 2) The initial signal utilized for this subjective referral consists of the neuronal processes represented by the primary EP, recordable directly at primary sensory cortex. This second feature explains why the sensation elicited by a stimulus at the surface of S-I cortex was *not* subjectively referred back to the time of the first cortical stimulus pulse; the cortical stimulus does not produce any primary EP like that elicited by a skin stimulus.

However, each stimulus pulse in n.VPL-thalamus or in medial lemniscus, in the ascending specific projection pathway from skin to cortex, does elicit a primary EP at S-I cortex (paper nos. 4,5,6). This provided us with the possibility of a direct experimental test of the modified hypothesis. Even though a liminal stimulus in these subcortical structures requires the same train duration (up to about 500 msec) as does one at S-I cortex, the production of the putative referral timing signal (primary EP) in the cortical response to each of these subcortical stimulus pulses should result in a subjective antedating of their sensory experiences, just as for a single pulse skin stimulus. That is, even though we can show objectively that thalamic stimulation must go on for at least 500 msec in order to elicit any conscious sensation, that sensation should appear subjectively to occur at the same time as that elicited by a single skin pulse that is delivered to coincide with the *start* of the thalamic stimulus train. These experimental predictions were in fact borne out (paper no. 9).

This apparently strange subjective referral of a sensory experience in time is actually related, in principle, to the long known phenomenon of subjective referral in space (see paper nos. 9,11,12,18,19,22). Both types of referral serve to "correct", subjectively, the distortion of the spatial and temporal representations of the sensory input by the responding neuronal systems of the brain.

## Conscious Intention in Voluntary Action

Does the principle of a minimum neural "time-on" and delay for eliciting a conscious experience extend beyond the case of a sensory experience? It became possible to test this for the initiation of a voluntary act, with the demonstration of a cerebral process (represented by the "readiness-potential," RP, recordable on the scalp at the vertex) which specifically precedes a self-paced act by 0.8 sec or more (Kornhuber and Deecke, 1965). Our question could then be framed as — does the *conscious* intention to perform this apparently voluntary act appear at the onset of (or before) the RP, or does it appear well after such RP onset? The latter would be in accord with the principle of a neural delay even for the awareness of wanting to move voluntarily.

*Free Volition.* Some limitations of free choice of when to act were present in studies of self-paced acts (by Kornhuber and Deecke; and others). In order to make our test more relevant also to the issue of free voluntary action or "free will," we first established that a fully spontaneous voluntary act (no restrictions on when to act) is also preceded by an "RP-II" with onset about 550 msec before activation of the muscles (paper no. 13). When subjects experienced some conscious preplanning of when to act, an "RP-I" with a much earlier onset similar to that reported for self-paced acts was recorded.

*Time of conscious intention.* To obtain this timing the subject fixed his/her gaze on a "clock," i.e., an oscilloscope spot revolving like a sweep second hand but about 25 times as fast (2.54 sec instead of 60 sec). The subject was asked to associate the first awareness of the urge or wish (W) to act with the clock position of the revolving spot. A control experiment to measure objectively the accuracy of this reporting procedure was added; in this, the subject was to wait for a weak stimulus to the hand delivered at random times, and then report the time of awareness of that sensation. We found that the conscious intention to act (W) appeared an average of 350 msec *after* the onset of the RP-II (onset of brain process) that preceded those same spontaneous voluntary acts (see paper nos. 15,16). This is in accord with a neural delay for the appearance of this kind of conscious event. It also meant that the cerebral process for this voluntary act was initiated *unconsciously*, before the subject was aware of the wish to move.

*Role of conscious will.* If the brain unconsciously initiates the specific process leading to a voluntary action, is there any active role for the conscious event which appears after a substantial delay? (See paper nos. 15,16,22.) It was important to note that awareness of the wish or decision to act does appear about 150 to 200 msec *before* actual action, even though it is delayed about 350 msec from onset of the brain process (RP-II). That timing provides an opportunity for the conscious function to affect the final outcome of the volitional process. The possible options were 1) one might consciously either permit the process to go on to produce the action or veto (abort) the process; or, 2) the conscious function might conceivably need to supply an active "trigger" in order to culminate in a voluntary action. The latter would seem to be a less attractive possibility, since spontaneous, endogenous acts can at times occur in the absence of any conscious awareness immediately *before* the action. Conscious intervention to cancel or veto the volitional process, and abort the motor act, is not only possible but was demonstrable by us (paper no. 14,16). Consequently, our findings do not rule out free will but do appear to impose constraints on how it may operate; free will would not initiate a voluntary act but it could select and control the outcome of the volitional process (paper nos. 14,15,16,18,22).

## Unconscious Mental Function: "Time-on" Theory

It is generally agreed that much if not most mental functions can proceed unconsciously. These are functions of which the individual is unaware; operationally one may define them as those for which the subject reports having no introspective experience. Unconscious mental functions include detection of sensory signals without awareness and making meaningful, purposeful responses to them (as in well-learned motor patterns or when driving a car while thinking of something else, etc.); they also include the *initiation* of voluntary actions (see above); and also intuitive thought patterns even when complex creative thinking and problem solving are involved (as attested to by great mathematicians, artists, etc.). On the other hand, one would not include functions which ordinarily do not rise to awareness, such as some postural reflexes and autonomic activities (e.g., regulation of heart rate, blood pressure, breathing, etc.). Unconscious *mental* functions are those which do have the potentiality of being brought into conscious awareness, under suitable conditions. How, then, do cerebral processes distinguish between conscious and unconscious mental function, and what controls the transition of an unconscious into a conscious function?

*"Time-on" theory.* (See paper nos. 2,20,21,22.) As indicated above, we already had substantial evidence for the view that a minimum duration (hundreds of msec), of appropriate cortical activities, is a unique requirement for eliciting a conscious sensory experience. To account for the difference between neural activities mediating conscious vs. unconscious mental functions, I proposed a more encompassing "time-on" theory (paper no. 20; Libet, 1990). This states that the transition from an unconscious to a conscious function can be mediated simply by changing the duration ("time-on") of the appropriate cerebral neural activities. That is, activities with durations less than those required for awareness can support an unconscious mental function; but the subjective awareness feature of the function can appear if and when the durations exceed certain minimum values. The theory does not exclude the possibility that there are other unique or controlling distinctions between processes mediating conscious vs. unconscious mental functions (paper nos. 20,21,22; Libet, 1990).

*Sensory detection with and without awareness, controlled by duration of cortical activations*: We were able to put the "time-on" theory to a direct experimental test which had the potentiality of either supporting it (as a causative, not merely correlative proposition) or falsifying it (paper no. 21). This was made possible by our earlier demonstration that stimuli in ventrobasal thalamus require utilization train-durations similar to the 500 msec ones for S-I cortex, in order to elicit a conscious sensation (paper nos. 4,5,7). We took advantage of the availability of ambulatory patients in whom electrodes had been chronically implanted in somatosensory thalamus for the therapeutic purpose of controlling intractable pain. In the experiment, durations of these thalamic stimuli were varied randomly over a wide range in successive trials. In each trial, subjects made a forced choice response on which of two intervals "contained" a stimulus, regardless of whether they subjectively felt anything; correct choices above the 50% chance level would indicate some *detection* of the stimulus. Subjects also indicated their level of awareness, if any, of any stimulus-induced sensation. This produced frequent reports of an uncertain sensation or of something different from nothing. The results showed statistically highly significant degrees of detection even with stimulus durations at which subjects reported feeling nothing; and also that a large additional duration of stimulus was required to go from mere detection without awareness to correct answers with awareness. The latter relationship held even when taking the "uncertain-different" report as the criterion for any awareness. It seems clear, then, that simply increasing the duration of the ascending activations to sensory cortex can control the transition between an unconscious mental function (detection without awareness) and a conscious one (detection with awareness).

## References for Prologue

*(other than those listed by number in this book)*

Eccles JC (1979): *The Human Mystery*. New York: Springer

Goff WR, Rosner BS, Allison T (1962): Distribution of cerebral somatosensory evoked responses in normal man. *Electroenceph Clin Neurophysiol* 14:697–713

Hook S (1960): *Dimensions of Mind*. Washington Square: New York University Press

Ingvar D, Phillipson L (1977): Distribution of cerebral blood flow in the dominant hemisphere during motor performance. *Ann Neurol* 2:230–237

Jensen AR (1979): "g": Outmoded theory or unconquered frontier? *Creative Sci & Technology* 11:16–29

Kornhuber HH, Deecke L (1965): Hirnpotential-anderung bei Willkurbewegungen und passiven Bewegungen des Menschen; Bereitschaftspotential und reafferente Potentiale. *Pflügers Archiv f ges Physiol* 284:1–17

Krech D (1969): Does behavior really need a brain? In: *William James: Unfinished Business*, McLeod RB, ed. American Psychological Association

Lassen NA, Ingvar DH, Skinhoj E (1978): Brain function and blood flow. *Sci American* 239:62–71

Libet B (1990): Cerebral processes that distinguish conscious experience from unconscious mental functions. In: *The Principles of Design and Operation of the Brain*, Eccles, JC, Creutzfeldt O, eds. Vatican: Pontifical Acad. Science (also by Berlin: Springer, vol. 21 in: *Exper Brain Res Series*)

Nagel T (1979): *Mortal Questions*. Cambridge University Press

Ochs S, Clark FJ (1968): Tetrodotoxin analysis of direct cortical responses. *Electroenceph Clin Neurophysiol* 24:101–107
Penrose R (1989): *The Emperor's New Mind.* New York: Oxford University Press
Pepper SW (1960): A neural-identity theory of mind. In: *Dimensions of Mind,* Hook S, ed. Washington Square: New York University Press
Roland PE, Friberg L (1985): Localization of cortical areas activated by thinking. *J Neurophysiol* 53:1219–1243
Schrödinger E (1958): *Mind and Matter.* London: Cambridge University Press

# PRODUCTION OF THRESHOLD LEVELS OF CONSCIOUS SENSATION BY ELECTRICAL STIMULATION OF HUMAN SOMATO-SENSORY CORTEX

BY

B. LIBET, W. W. ALBERTS, E. W. WRIGHT, JR., L. D. DELATTRE, G. LEVIN, AND B. FEINSTEIN

**Reprinted from**
*J. Neurophysiol. 1964, 27*
pp. 546-578

# PRODUCTION OF THRESHOLD LEVELS OF CONSCIOUS SENSATION BY ELECTRICAL STIMULATION OF HUMAN SOMATOSENSORY CORTEX[1]

B. LIBET, W. W. ALBERTS,[2] E. W. WRIGHT, JR., L. D. DELATTRE, G. LEVIN, AND B. FEINSTEIN

*Institute for the Study of Human Neurophysiology, Mt. Zion Hospital and Medical Center, and Department of Physiology, University of California School of Medicine, San Francisco, California*

(Received for publication August 22, 1963)

## INTRODUCTION

THE PRESENT WORK is part of a more general investigation into neurophysiological activities of the cerebral cortex which may be involved in the elaboration or mediation of conscious sensation. The primary concern in these studies is the dynamics of the processes involved, rather than their topographical representation.

In this study a number of parameters which affect the adequacy of direct cortical stimulation for the production of a conscious somatic sensation at the threshold level were investigated quantitatively.[3] The question of the specific modalities of sensation elicited in relation to the nature of the stimulus is to be considered more specifically in another paper; here we are concerned with the question of whether threshold has been achieved. The reasons for first developing this aspect were twofold. The longer range objective in these studies requires that one be able to examine responses just above and below the threshold for conscious awareness of a sensation which has been elicited by a direct cortical stimulus. Secondly, knowledge of the significant stimulus parameters, and of the parametric regions in which such stimuli are adequate, might itself lead to suggestions about the kinds of activities in primary sensory cortex which are involved in eliciting a subjective sensation. This hope of deriving from this work some clues to the underlying processes for conscious sensation was in fact realized.

Included in the set of parameters were peak current (I), polarity of pulses, pulse duration (PD), pulse repetition frequency (PF), train duration (TD), train repetition rate (TRR, i.e., interval between successive tests),

---

[1] This study was supported successively by the Air Force Office of Scientific Research, Contract AF 18(603)-48, Office of Naval Research Contract Nonr-2968(01), and Public Health Service Grant NB-01963; and by the Harry Freund Memorial Foundation.

[2] Supported by Research Career Program Award NB-K3 16,729 from the National Institute of Neurological Diseases and Blindness.

[3] Preliminary reports of this work have been presented (27, 28).

uni- and bipolar electrode arrangement, and electrode size or area. Since the parameters are interrelated in their effectiveness, one can draw from this study differing parametric points[4] for adequate stimuli that may be suitable for different experimental purposes; perhaps even more importantly, one finds that certain parametric regions may not be suitable at all for certain purposes. In addition, the influence of some aspects of the alerting procedures, and their correlated states of EEG-alpha rhythm, on cerebral and peripheral sensory thresholds was analyzed.

## METHODS

*Subjects and general procedure.* The subjects for this study were selected from among cooperative patients who were to undergo neurosurgical treatment for dyskinesias, mainly parkinsonism. Pneumoencephalography and trephination is carried out in a first stage under general anesthesia; the actual therapy is carried out at another time with the patient unanesthetized (13, 26) in order to permit appropriate activation and inactivation tests to be conducted on the responses of the target tissues (2). It was during these second stage operations that the present studies were carried out. The use of the Leksell stereotaxic instrument permits flexibility in choosing the cortical site of entry of the therapy electrode. A surprisingly high percentage of the patients was able to tolerate well the 30- to 90-min. period of cortical testing. A study period was terminated when the patient demonstrated reluctance or an unreliability of responses which often appeared in association with fatigue. The total number of such study sessions in the operating room which were successful in part or whole in contributing to this study was 99, carried out on 92 patients. Since each testing period could ordinarily contribute data significant to only one or a few of the points under investigation, the total number of subjects involved in attempting to establish individual relationships is in some instances not as large as one would like.

The general procedure during those second-stage operations which were used for investigative purposes was as follows. With the stereotaxic instrument in place and the available area of cerebral cortex exposed, the primary somatosensory area on the flat region of the postcentral gyrus was localized with an exploring stimulating electrode (usually with 0.5-msec. square pulses, alternating in polarity (1), 60 pulses/sec. trains of less than 1 sec. duration). Assurance that the area under investigation was in fact the postcentral gyrus was obtained in almost all cases by eliciting typical pyramidal-type motor responses (3) from the cortical gyrus just anterior to the sulcus bordering the gyrus in question. The cortical area of representation of sensation usually involved some portion of the contralateral hand or, less commonly, of the wrist or forearm. The appropriate stimulating electrode was then placed on that cortical point requiring the minimum current for eliciting this conscious sensory response. There was no particular locus on the postcentral gyrus, along its anteroposterior axis, which had the lowest threshold in the different subjects. Details of the stimulation technique, interrogation of the subject, and conditions in the environment are given below. Following the cortical testing period, the therapy electrode was inserted and the therapeutic procedure completed. Most patients received no special medication just before or during the second-stage operation; most of them also discontinued taking their anti-parkinsonism drugs the day before. They did, however, each receive approximately 40 ml. of 1% xylocaine, usually with 1:200,000 epinephrine, for local anesthesia.

*Electrodes and stimulation.* In order to eliminate electrode polarization, electrodes similar to those utilized by Lilly *et al.* (31) to stimulate motor cortex were used routinely, unless otherwise specified. These are Ag-AgCl electrodes with a large chlorided-metal surface area (ca. 3.5 cm.$^2$) inside a glass cylinder. The latter is filled with an 0.9% NaCl solution which

[4] Adopting the usage of Lilly *et al.* (31) for the following terms: A "parametric point" is defined as the group of simultaneously assigned single values, one for each parameter of a set. A "parametric line" is a set of parameters all of whose values are fixed except for one (e.g., peak current). A "parametric region" is determined by the simultaneous choice of an upper and lower limit (range) for the values of each parameter of a set.

makes electrical contact with the tissue at the tip. The flat circular surface of the tip ordinarily had a total diameter of 2 mm. with a central circular pore of 1 mm. diameter. This electrode was held at a right angle to the cortical surface in a clamp fastened to the stereotaxic frame. It was lowered onto the cortex by a rack-and-pinion control until there was gentle but distinct pressure against the pia-arachnoid membranes. The electrode slightly compressed the arachnoid, which is thicker in man than in laboratory animals. The cortex did not appear to be significantly compressed; there was no evidence of pial blanching adjacent to the electrode, and sensations elicited by stimulation through this electrode were located in body areas similar to those elicited earlier by stimulation through a lightly applied metal exploring electrode. In order to prevent the introduction of any air bubbles into the electrode's opening and to insure continuous electrical contact with the tissue, a constant inflow of saline from an intravenous drip bottle into the upper end of the electrode was maintained at about one to four drops/min. This also moistened the tissue around the electrode contact, but it may have carried some Ag ions to the pia-arachnoid surface. In addition, to eliminate drying and heating of the exposed cortex, the area surrounding the electrode contact was covered with strips of thin rubber sheeting and the operating lights were turned off. The cortical surface was carefully inspected, after removing the electrode at the end of the experiment, for evidence of any abnormalities in vascularity, color, or indentation (other than that of the arachnoid). Ordinarily a unipolar electrode arrangement was used. The reference electrode consisted of a large Ag foil over saline jelly fixed to the ipsilateral arm. No sensory or motor responses were elicited by this electrode, except on some rare occasions when very strong stimuli were employed.

For the experiments in which the effects of bipolar stimuli were studied, two separate glass cylinder electrodes were fused together along their lengths. This produced one rigid bipolar double-barreled unit containing two separate electrodes, each similar to the unipolar electrode. The cortical contact area of the double-barreled unit was similarly ground to a flat surface. The latter was ovoid in shape and contained two 0.5-mm. pores, in a line along the longer axis, separated by about 1.5 mm. of glass. A unipolar stimulus could be delivered for quantitative comparison without disturbing the electrode; this was done by employing a combination of one barrel and the usual indifferent electrode. Other electrodes were occasionally used for special purposes and these will be described in the appropriate sections below.

Stimuli consisted of rectangular pulses delivered from an AEL model 104A stimulator through an R.-F. isolation unit. The peak current of the pulses was monitored by measuring the voltage across a current indicating resistor in series with the stimulating electrode. This voltage was recorded with an oscilloscope which was isolated from the current indicating resistor by a General Radio transformer. The impedance of the electrode-subject circuit could be determined by matching a variable "dummy" resistor to this circuit. Actually, the approximately 8,000 ohms resistance in the typical electrode-subject circuit was almost entirely contributed by the 1-mm. pore of the electrode assembly itself. For each case, a plot of the measured peak current versus the voltage scale readings on the stimulator dial was made. The relationship was fairly linear over most of the range. Although unidirectional rectangular pulses were used in the experimental testing series, there was no visible superficial evidence of any local damage to the cortex, nor did any unexplained sensory aberrations or paresthesias develop during or after an experiment.

Thresholds were usually remarkably stable over testing periods of 15–30 min. Sharp rises in threshold occurred only occasionally and generally came on rather suddenly, late in the experiment. They were often accompanied by inconsistency of responses, and seemed to be correlated with fatigue of the patient and a concomitant flagging of interest in the testing. In the later periods of an experiment, however, there was often a progressive rise in threshold level. It is possible that some of this rise could have been due to a resting flow of a very small steady current which was presumably largely generated by the difference in EMFs of the stimulating electrodes themselves. Such a steady current, amounting to as much as 10 $\mu$A., probably produces negligible injury, however, when the rate of flow of charge is so very low.

*Environment and questioning of subject.* After the electrode had been fixed into position, the operating room was cleared of all personnel except for the surgeon and one nurse, who remained seated and unseen by the patient, and an observer who alone interrogated the patient. The observer was in communication with the control room, which contained all the stimulating and recording devices, by a telephone-operator type of headset. The patient could hear the remarks of the observer to the control room; these remarks thus had to be guarded in nature.

The subject was told that during the series of tests about to be conducted he might experience some sensations, occurring in the general area of those already experienced during the earlier localization of the somatosensory area. He was told that a number of the tests might not produce a sensation. Upon feeling a sensation, he was to say only "on" and "off" to designate its onset and cessation. This prevented long interruptions during a series of stimuli, which were to be delivered at set intervals of time. To learn the nature of the sensation, the subject was asked after an appropriate test to describe what "the sensation felt like to him" and precisely where in his body he felt it. In principle, the procedure was designed to encourage the patient to describe the sensation in his own words without suggestion from the observer. At times, additional information was elicited by asking the patient whether the sensation appeared to be located superficially, in the skin, or deeper, below the skin; or how a given sensation compared with one reported by him earlier when he used different descriptive terms; or what sort of sensation in his normal experience resembled the one he now reported. Occasionally toward the end of an experiment, after the subject had made reports which were unclear to the observer, outright questions of specific identity were asked, i.e., whether the sensation felt like an electric shock, tingle, pins and needles, etc.

To help maintain a relatively uniform state of alertness to the tests, the subject was asked to keep his eyes open (if he had a tendency to shut them), and, starting at approximately 5 sec. before the delivery of the stimulus, the observer, on cue from an operator in the control room, indicated verbally to the subject that there was about to be a test. The observer did this by calmly saying a few words, generally "ready now," or "ready Mr. ———," or "once again, Mr. ———." As will be reported below, however, it was found that a major change in this alerting procedure did not affect the threshold values of the adequate stimulus. Sensations at the threshold level, especially brief-lasting ones elicited by stimuli of short train duration, are sometimes confused by the subject with slight, spontaneously occurring parathesias, real or imaginary. To check the reliability of the subject's reports, dummy tests without any actual stimulus, but with the same alerting sequence, were occasionally interpolated. For the most part the consistency of the subject's positive and negative responses to stimuli delivered repeatedly just above and below threshold level was a sufficient indicator of reliability.

Usually a standard parametric point of reference was established initially in an experiment, for orientation to the quantities required in the particular case and for later checks of consistency. This parametric point was generally established by determining the threshold peak current value while fixing other parameters (cathodal polarity, 0.5-msec. PD, a PF of 30 or 60 pulses/sec., TD of 5 sec., and with the testing trains repeated at about 30-sec. intervals). Subsequently, one or more of the other parameters in the set were altered in value and then threshold I was redetermined with another series of tests along this new parametric line. Or, all parameters other than TD might be fixed and the TD varied, until the minimum along this line was discovered. It will be seen below that train durations considerably below 5 sec. may be used to establish minimum I levels, and that a TRR of 1/30 sec. is not slow enough for establishing appropriate thresholds under certain conditions.

The threshold point was defined as the least value of the parameter being varied, utilizing a fixed set of values for other parameters, at which the subject reported a subjective awareness of a somatic sensation. This held irrespective of the quality of the sensation or of its particular location in the body. The sensation did not have to be subjectively located in the identical area of a finger or hand in different tests at the threshold; in fact, the subjective locus did at times shift position in successive tests with the same threshold stimulus.

It should be noted that the question put to the subject and the nature of his response in this investigation differed from those used in certain other studies of sensory sensitivity in man. In the work of Eijkman and Vendrik (10) with stimulation of skin, for example, the subject was asked to indicate whether in his opinion a stimulus was delivered, or to choose which one of four stimuli (of different amplitudes) he thinks was given him, or to indicate in which one of four test periods he thinks an actual stimulus was given. None of these latter responses requires that the subject report a distinct subjective feeling or awareness of the sensation, such as was requested in the present work. The relative sharpness and consistency of thresholds in the present work, as opposed to the absence of a distinct threshold in the Eijkman and Vendrik experiments, may indeed indicate the existence of an important difference in the mechanisms underlying the different responses demanded in the two studies.

## RESULTS

### *Characterization of parametric regions*

A great range of parametric points can elicit threshold responses by stimulation of postcentral gyrus. It appears possible, however, to charac-

*Table 1. Summary of parametric regions for threshold stimuli*

| | Region A | Region B | Region C |
|---|---|---|---|
| I | Liminal | 2 or more times that in A | Intermediate range, between A and B |
| TD | >0.5–1 sec. | <0.1–0.3 sec., single pulses | 0.5 to 5 sec., at 8 pulses/sec. 0.1 to 0.5 sec. at PF ≥15 pulses/sec. |
| PF | ≥15 pulses/sec. (can get with 8 pulse/sec. if TD >5–10 sec.) | Any | Probably any, but best seen in low range (<15 pulses/sec.) |
| Facilitation evident with TRR =1 per 30 sec. | Usually none | Usually considerable | Usually present |
| Response at threshold | Purely somatosensory (no motor) | Observable muscular contraction | No observable contraction; sensation changes from that in A to one with aspect of motion (at <15 pulses/sec. often has slow pulsatile quality) |

I = peak current; TD = train duration; PF = pulse repetition frequency; TRR = train repetition rate.

terize three parametric regions according to certain features of both stimulus parameters and responses. This way of grouping the adequate stimuli, as it developed during this study, is presented at the outset in order to provide a reference framework to which the data described in the succeeding sections of RESULTS may be related. A summary of the features of the parametric regions is given in Table 1.

*Parametric region A* essentially encompasses all threshold stimuli in which trains of relatively long duration, i.e., >0.5–1.0 sec., and pulse frequencies of about 15 pulses/sec. and up are employed. As threshold I hardly changes, if at all, with increases in TD above 0.5–1 sec. in this range of PFs, the threshold currents are in fact the liminal I values as determined with any long train (usually 5-sec. TDs in this work) at each such pulse frequency. At the train repetition rate of 1/30 sec., there usually was no evidence of any facilitory effect of one test stimulus on the succeeding one. The threshold responses were all purely sensory in this parametric region. No movements or contractions could be seen or palpated.

Stimuli with pulse frequencies of 8 pulses/sec. and below, and with long TDs (5–15 sec.) appear to present a borderline problem of classification. In some of the tests with such threshold stimuli the sensation which was elicited resembled that reported with stimuli of higher PF in region A; in other tests slowly pulsatile sensations were elicited, like those obtained with shorter TDs at 8 pulses/sec. (see parametric region C).

*Parametric region B* includes those threshold parametric points in which the TD is very short ( <0.1–0.3 sec.) or in which the stimulus is a single pulse. These stimuli produce an observable muscular contraction at the threshold response. When using pulse frequencies of 30–60 pulses/sec., this motor threshold I level is on the order of two to three times the sensory liminal I values which are found with 0.5- to 5-sec. TDs. At the TRR of 1/30 sec., a facilitory effect on succeeding tests is usually observed in this parametric region. Our finding that single pulse stimuli apparently belong exclusively in this parametric region clarifies an observation that was reported by Foerster (14, 15). He found that galvanic stimuli (apparently brief) applied to the postcentral gyrus could elicit localized motor responses but were ineffective in producing sensation; faradic stimuli (i.e., repetitive pulses) were required by him to elicit sensation.

The sensation is usually in the nature of a "jerk" or "twitch," and is presumably elicited by peripheral sensory impulses, which are initiated by the observable muscular contraction. At times, however, subjects have described the sensation as a "tingle," etc., without a feeling of movement, or they have even reported an absence of any sensation, when a muscular contraction was in fact visible to the observer; it is evidently possible for the subject to be unaware of or to misjudge the nature of very fleeting and delicate muscular movements. On the other hand, there were occasions on which the subject reported feeling a twitch or jerk when no movement could be detected by the observer.

*Parametric region C* is a less certain and definable one than A and B, but we think it provides a useful category. It would include those threshold parametric points in which the TDs are intermediate between those for region A and those for region B. For PFs$\geq$15 pulses/sec., such TDs lie in the range between those for B (0.1–0.3 sec.) and those for A (0.5–1 sec.); for PFs <15 pulses/sec. these TDs are between about 1 and 10 sec. Threshold I levels with such intermediate TDs are generally markedly higher than the liminal I values required in region A, using the longer TDs. As in region B, a facilitory effect of one stimulus on the next is often apparent when using a TRR of 1/30 sec.

In region C, the sensory response was often different from that in region

A, often changing from the tingling, more commonly reported in region A, to one having an aspect of motion to it.[5] Feelings of a "quiver," "pull," or "drawing in the hand" have been reported (e.g., responses at points 4 and 5, of *subj. 78* in Fig. 2), and with pulse frequencies of 8 pulses/sec. and below sensations of a distinctly pulsatile nature, at about 1/sec., were often reported. No muscular contractions could ordinarily be detected by the observer with threshold stimuli in region C. Whether any undetected deep muscular contractions were in fact occurring generally and might have been responsible for the altered sensations in region C has not been adequately determined. It was not feasible to check this possibility by means of electromyogram recordings as the locus of the sensations often shifted about somewhat from test to test.

In connection with the difference in types of responses in the different parametric regions, it is of interest that Sigel (38) describes the threshold sensations in response to electrical stimulation of skin as predominantly "tapping," with pulse frequencies of 0.1 and 1 pulse/sec., "pulsating" at 1 or 10 pulses/sec., and "vibrating," etc. at high frequencies. Collins *et al.* (8) also reported threshold sensations of a tap, thump, etc. with single shocks to a sensory nerve, but those of electricity, tingling, etc. with pulse frequencies from 5 to 5,000 pulses/sec.

*Testing interval and facilitation*

Figure 1 illustrates some aspects of the question of the influence of previous tests on the responses to succeeding tests in a series, when using the usual TRR of 1/30 sec. In the first stimulus series, with 60 pulses/sec. and a 2-sec. TD (in this case, instead of the more usual 5 sec.), the threshold current for conscious sensation was achieved at 1.15 mA., the response being lost upon dropping I to 1.0 mA. A second test series with this same stimulus about 40 min. later (test series V) also produced a relatively sharp threshold I. This absence of any progressive lowering of threshold I within a given series of tests demonstrates that the 30-sec. test interval was adequately long for preventing any cumulative facilitory effect, when using near-threshold stimuli under these conditions.

Although the 30-sec. testing interval was usually adequate with stimuli in parametric region A, this was not the case in regions B and C. In Fig. 1, series II, for example, the threshold I was finally reached at 7.6 mA. after approaching it from below. Subsequent tests at lower currents, however, now were able to elicit responses. The responses finally began to fail at about 3.6 mA., but returned when the currents were again raised to levels of 4.5 or 5.5

[5] Unless a considerable pause, about 5 min., was allowed between the series of tests with higher pulse frequencies and the series at 8 pulses/sec., the latter often elicited the same type of sensation as the stimuli of higher pulse frequency. There appears then to be a long-lasting persistence or "perseveration" of a given quality of sensation. The possible dependence of this on a persistence of a facilitory effect on threshold current level (see below under *Testing interval*) is unclear, but such a relationship seems improbable at the present time.

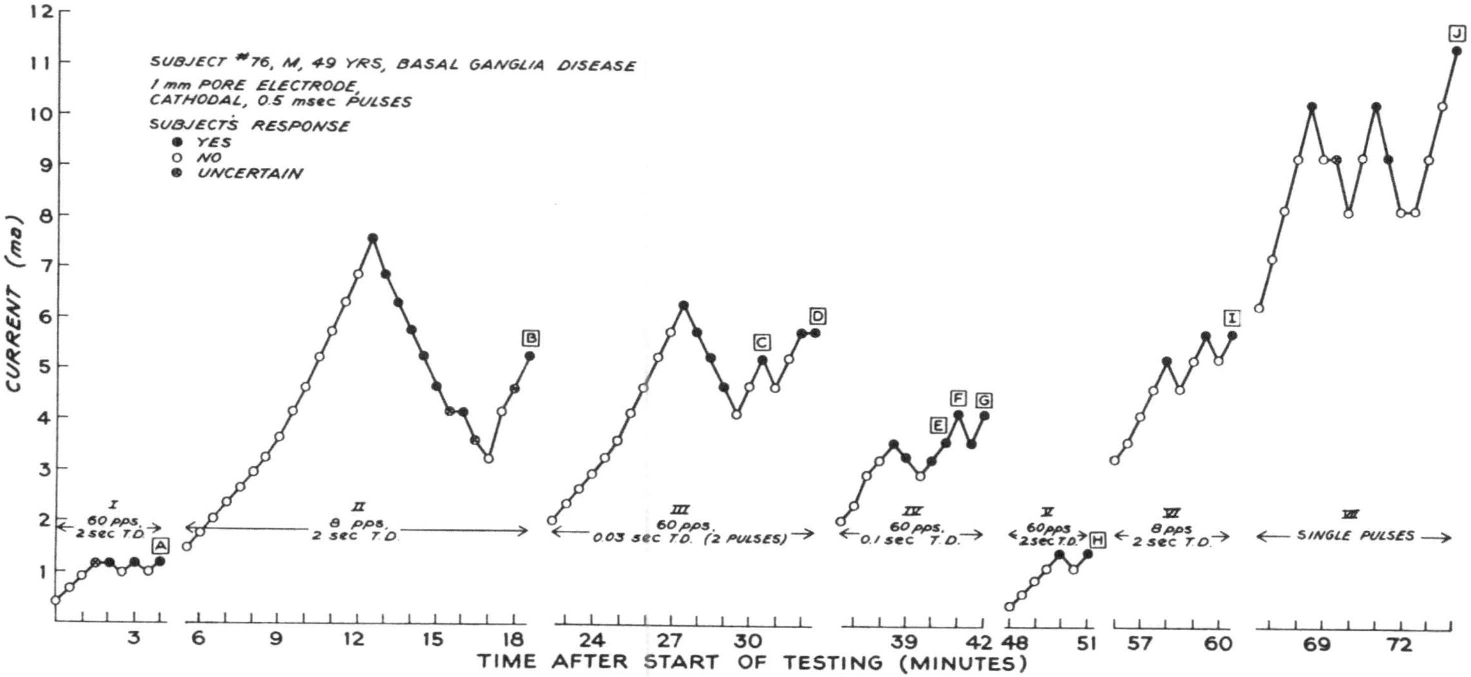


FIG. 1. Graphic record of values of current and responses for all test stimuli, in an experiment illustrating the sequential effects of stimulus trains in each series, at different pulse frequencies and train durations. Train repetition rate, or testing interval, was 1/30 sec. without interruption except as indicated after each series of tests. Following the number (roman numerals) of each series is given the pulse frequency and train duration used in that series. Capital letters over the positive responses at or near the end of each series refer to the lettered statements below, which summarize briefly the subject's report of sensations at these points along with the observer's notation as to the occurrence of any motor response. A: "a little numbness in the first three fingers." B: "same kind of numbness." C: "just a little twitch" (in thumb, and occasionally second and third fingers). D: subject reports same "twitch," with no other sensation such as numbness. Observer noted contraction of thumb adductor. E: Same "little twitch." Observer uncertain about contraction. F and G: observer confirms reported movement; just able to detect adductor contraction in thumb, which subject felt "in center of thumb," i.e., not in skin. H: "a little electricity going through—thumb and first finger—out toward the ends." Observer could not detect any movement or contraction. I: sensation similar to last one (H). Observer again detected no movement. J: "a little twitch." Observer detected twitch contraction in adductor of thumb and some other nearby muscles in the hand, in the region pointed to by the subject.

mA. Similar though less marked progressions of the threshold I values may be seen in series III, IV, and VII. In series of this kind then, there are two kinds of threshold I values; *a*) an initial high value, obtained upon approaching threshold I in successive tests from below, and *b*) a final low value, obtained upon reducing I in successive tests after having achieved a threshold or superthreshold response. This relationship, in which the threshold response level appears to lag behind the "cycling" of rising and falling applied I values in successive tests, bears an operational resemblance to the phenomenon of hysteresis in physical systems. Actually it appears to be due simply to persistence of a facilitory state for more than 30 sec. when using stimuli in these parametric regions (as will be seen further, below). It cannot be explained by an improvement in direction of the subject's attention after receiving some initial superthreshold stimuli, even though this can no doubt occur at times. One argument against the attention explanation stems from the prediction that this effect should then be observed with all kinds of stimuli; in fact it is rarely observed with series of stimuli that belong in parametric region A.

In addition, it was found to be possible to elicit a positive response simply by repetition of subthreshold stimuli in parametric region C at intervals of 30 sec. In an experiment using single pulse stimuli, for example, the peak current was held at 3.25 mA., below even the 3.8-mA. "final low" value (as in *b* above) which was obtained in a continuous series of tests; several repetitions at 1/30 sec., of these stimuli, at subthreshold I, was able to elicit a response.

The duration of the facilitory state which persists, following a threshold or moderately superthreshold stimulus that belongs in parametric regions B and C, may be estimated roughly to be about 1–4 min. After a series of tests such as series II in Fig. 1, an interval of several minutes appeared to be in general sufficient to raise the threshold I level at least part of the way back towards the "initial high" value (which had been obtained earlier when approaching threshold I from below). Evidently the duration of facilitory state following stimuli in parametric region A is ordinarily less than 30 sec., but no attempt was made to determine its value more precisely. It should also be noted that even with stimuli in regions B and C evidence of facilitation lasting longer than 30 sec. did not appear in all subjects. In addition, when evidence of it was present it was more likely to be evident in the first test series that utilized such stimuli (e.g., series II, Fig. 1) and less prominent in succeeding test series with such stimuli (e.g., series III, IV, and VII, Fig. 1).

The development of a long-persisting facilitory state was clearly not simply a function of using very low pulse frequencies, even though it appeared to be so early in the work when only stimuli of relatively long TDs were being tested. The effect can be exhibited readily at any pulse frequency if the TDs are sufficiently reduced. In order to elicit a threshold response the I level must then be raised well above that which obtains with long TDs; i.e., the combination of threshold parametric values must place the points in parametric regions B or C. Evidence of this is seen in stimulus series III and IV in Fig. 1.

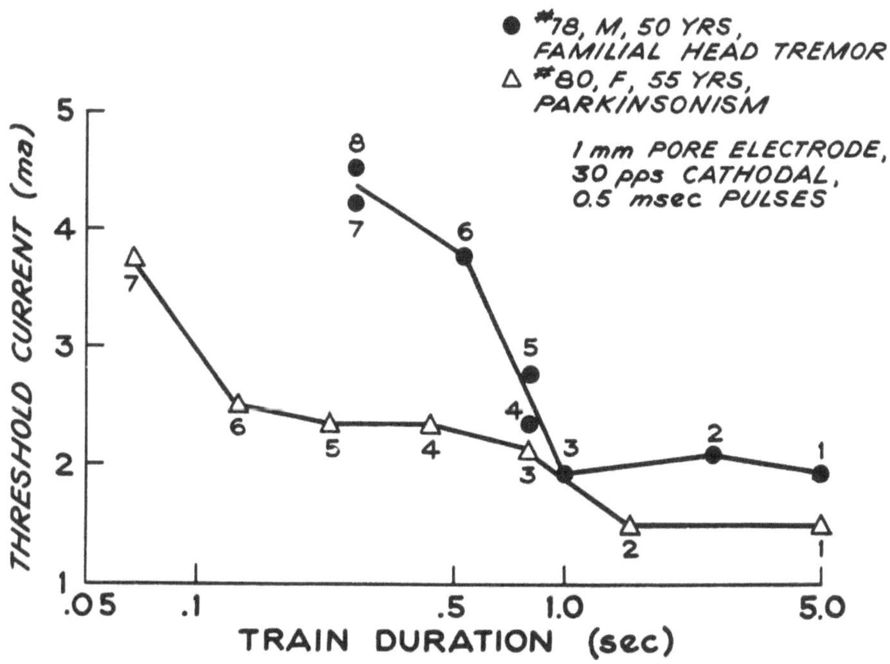


FIG. 2. Threshold peak currents at different train durations. The order of observations for each subject is given by the numbers of the points. To minimize the facilitory effect of a testing series at one TD on the threshold I of the next series at a new TD, pauses of 3 or 4 min. were introduced between tests for points 3 and 4, 5 and 6, 6 and 7, in *subject 78;* in *subject 80,* pauses of only 1 to 2 min. were allowed between successive determinations of I at each TD (except for a 4-min. pause between points 4 and 5). Threshold responses reported by *subject 78* may be summarized as follows: Points 1, 2 and 3—"tingling" in some surface skin areas of the contralateral hand. Points 4 and 5—"quiver" or "jerk" along the length of the last two fingers, with no tingling sensation. (No motion in the fingers was visible to the observer.) Point 6—"quiver" or "shiver" plus "tingling" in the last two fingers. (Some brief extension of these fingers accompanying each positive report by subject was seen by observer.) Point 8—"twitch" or "flick" of the end portion of little finger, seemingly sideways, with no tingling. (A tiny extension, rather than a lateral movement, of the terminal phalanx of the last digit was seen by the observer.) Sensations felt progressively shorter in duration as the TD was reduced. Threshold responses reported by *subject 80* are summarized as follows: Points 1 and 2—faint "shock of electricity" on portion of back of contralateral hand. Points 3, 4, 5, and 6—a vague sensation of "something" which moved along and "made a circle," mainly at the wrist, but also moving into the thumb with one of the positive responses at point 3, and back into the forearm at point 5. An aspect of distinct motion of the "something" was now present in contrast to sensations felt at points 1 and 2. (No movements were detected by the observer.) Point 7—something "deeper, running from wrist a little ways towards elbow." (In a few, but not all, of the positive responses here the observer detected a slight jerk motion of the forearm at the elbow and an uncertain jerk of a forearm tendon near the wrist.)

## *Train duration*

*Peak current versus train duration at threshold.* The two curves shown in Fig. 2 exhibit some differences in shape, but they tend in fact to overlap with the other available curves, not shown, in which points for smaller portions of the TD range were determined or in which the measurements were less acceptable (for reasons discussed below). Examining all available curves, certain important features of the threshold I versus TD relationship with 30 or 60 pulses/sec. stimuli were evident.

*a*) The curves were relatively flat between TDs of 5 sec. and about 0.5–1.0 sec. (at least with pulse frequencies of 15 pulses/sec. or greater). Actually, examination of all the data in this range of TDs indicates that on the average there probably was a slight rise in threshold I, when TD was reduced to about 0.5–1.0 sec. This rise is apparently only 5–10% above the threshold I found with the 5-sec. TD, however, and is thus virtually within the error of threshold I measurements. Subjectively, the quality and the intensity of the sensations did not appear to change within this range, although they did become shorter lasting with the shorter TDs in the range. Thus there appears to exist a minimum threshold I in a given subject below which one cannot go regardless of the train duration, at least up to 5 sec. This minimum I level for threshold responses is referred to as the "liminal I." Although liminal I is relatively independent of TD, when this is greater than 0.5–1 sec. (for pulse frequencies >15 pulses/sec.), it was affected by changes in other parametric values, such as pulse frequency, pulse duration, polarity, etc. It was not, however, affected by the TRR when this was 1/30 sec. or slower, as already described above.

*b*) When the TDs are reduced below the level of about 0.5–1.0 sec. the threshold I values usually begin to rise rather sharply (e.g., Fig. 2; but see below for 8 pulses/sec. stimuli). The TD at this point in the I-TD curve of threshold points will be referred to as the "utilization TD." The utilization TD, in other words, is the shortest TD at which a stimulus of liminal I intensity (to within ca. 10%) is still able to elicit a sensation; or, put in another way, train durations longer than utilization TD require the same or only a very slightly different threshold I. A summary and analysis of the available utilization TD values will be given below.

*c*) The actual shapes of the I-TD curves in the shorter range of TDs, below utilization TD, have been somewhat variable and difficult to characterize. Threshold I for a single pulse stimulus was greater than for a two-pulse train (at 30 or 60 pulses/sec.), although the difference in thresholds was in some instances relatively small. It is in this range of short TDs that special practical difficulties in conducting the experiment arise, and these probably account for at least some of the variability and for the questionable nature of some of the points in all the curves that were obtained. To avoid the facilitory effect, pauses of about 4 min. duration are required between successive test series when the stimuli are in the range of the shorter TDs; this makes the experiment a long and tiring one for the subject. In addition, the subject finds it more difficult, and apparently more fatiguing, to concentrate on and report the very fleeting and weak sensations which are produced by the shorter TDs at threshold levels.

*d*) The nature of the threshold subjective response, aside from its duration, changes at least once, and perhaps twice, as the TD is reduced in the range below the utilization TD. These changes have already been described above, in the characterization of the parametric regions.

*Utilization train durations.* These minimum TDs for a threshold response, when using liminal or near-liminal I strengths, mostly fell into the range of about 0.5–1.0 sec. (except for 8 pulses/sec. trains). Not enough measurements are available to draw very firm conclusions about the influence of various factors on utilization TD, but some observations appear to be significant. The utilization TDs measured with different types of electrodes overlap widely and all are apparently not distinctly affected by electrode differences. Utilization TDs were also similar at the different pulse frequencies tested, 15 pulses/sec. to 240 pulses/sec., though there tended to be more instances of lower TD values at the higher pulse frequencies. The mean values were: 0.9 sec. (SD $\pm 0.3$) at 15–20 pulses/sec., in 5 subjects; 0.8 sec. (SD $\pm 0.4$) at 30 pulses/sec., in 18 subjects; 0.6 sec. (SD $\pm 0.3$) at 60 pulses/sec., in 8 subjects. In one case at 120 pulses/sec. the utilization TD was about 0.5 sec. In another it was about 0.2 sec. at 120 and 240 pulses/sec.; but this unusually low value was obtained also at 60 and 30 pulses/sec. in this subject.

With 8 pulses/sec. stimuli, estimations of utilization TD were often subject to the additional variable of the facilitory effect of successive stimuli, when delivered at the usual 30-sec. intervals. Thus, no values for utilization TD can be given which would be acceptable under the criteria being used. It is, however, noteworthy that several of the values found were greater than 2 sec. Testing of 8 pulses/sec. stimuli with TDs of up to 15 sec. in three subjects indicated that at this pulse frequency the liminal I level is in fact not achieved unless TDs longer than 5 sec. are employed. The minimum TD required with liminal I stimulus at 8 pulses/sec. appears to be on the order of 10 sec.

When the postero-ventro-lateral nucleus (PVL) of the thalamus was stimulated, utilization TDs fell into the same general range of values found with cortical stimuli. Using a unipolar electrode with an exposed tip length of 2.5 mm., and 0.5-msec. pulses of alternating polarity at a frequency of 60 pulses/sec., the utilization TD was estimated to be 0.3–0.5 sec. for one subject and 0.5–2 sec. for the other. On the other hand, with electrical stimulation of the skin the utilization TDs were very short in comparison to those for the cerebral structures. Skin stimulation was carried out with a unipolar Ag mesh electrode, about 15 mm. square (Telectrode, Telemedics Co.) applied to the back of hand on an area free of palpable sensory nerve branches. Using 0.5-msec. cathodal pulses at 30 pulses/sec., in two parkinsonian patients, threshold I remained constant when the TD was reduced from 5 sec. to 0.13 sec. (i.e., 4 pulses); in one normal subject threshold I was unchanged at all TDs of 5 sec. and less, even with a single pulse; in another normal subject, threshold I rose about 10% when TD was reduced from 5 sec. to 0.07 sec. (i.e., 2 pulses).

*Pulse frequency*

The liminal current values are shown plotted against pulse frequency for a number of subjects in Fig. 3. The shapes of the curves are of more interest than the actual values, for purposes of comparison. In those cases (usually the 8 pulses/sec. stimuli) in which a facilitory effect was still evident after the 30-sec. testing intervals used, only the initial high threshold I value of any given series of tests was plotted; the low threshold value of such a series was presumed to reflect the effect of the accumulated temporal facilitation rather than merely the "resting" threshold.

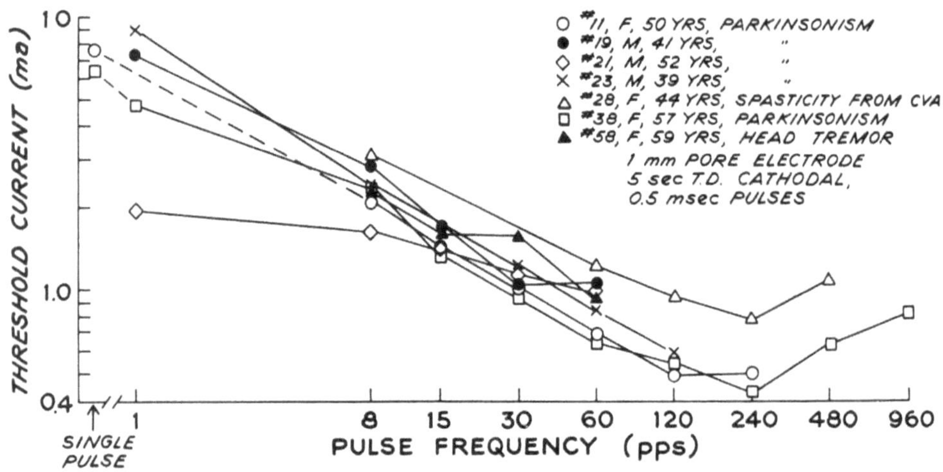


FIG. 3. Liminal I values (i.e., threshold peak currents required to elicit a sensation using 5-sec. train durations) at different pulse frequencies. Each curve represents a set of determinations made at one session for each of the seven subjects in the graph.

It may be seen from Fig. 3 that there was roughly an inverse linear relationship between the log of liminal I and the log of pulse frequency, in the range between 1 pulse/sec. and about 120–240 pulses/sec. The slope of this line is approximately −0.5 within this range; i.e., threshold I is roughly proportional to the pulse frequency raised to the power −0.5. In the pulse frequency range above 240 pulses/sec., liminal I rose instead of fell with an increase in frequency, in the two subjects tested. In the case of the subject whose curve shows values up to 960 pulses/sec. in Fig. 3, additional tests not shown on the graph were done with pulse frequencies up to 6,000 pulses/sec. Pulse duration, however, had to be reduced for the stimuli above 1,000 pulses/sec. to avoid a stimulus duty cycle greater than 50%. The results in this subject were as follows: At 2,000 pulses/sec. with pulse duration 0.2 msec., liminal I was 1.6 mA.; at 4,000 pulses/sec., 0.1-msec. PD, liminal I was 1.6 mA.; at 6,000 pulses/sec., 0.05-msec. PD, liminal I was 2.2 mA. These I values were all greater than the liminal I of 0.85 mA. found with 960 pulses/sec., 0.5-msec. PD, but the quantitative aspects of the comparison is obscure, as it is known (see below) that a reduction of pulse duration by itself can raise the threshold peak current level.

The inverse linear relationship of log values was retained into the lowest pulse frequency range, 8 pulses/sec. and 1 pulse/sec., even though these stimuli belong to parametric region C rather than A. As was pointed out above, TDs of about 10 sec. (rather than 5 sec.) are apparently required for 8 pulses/sec. stimuli, in order to obtain a liminal I value and response more related to parametric region A. Comparison with the available values for single pulse stimuli in these subjects indicates that the 5-sec. trains at 1 pulse/sec. required peak currents which were considerably closer to the single pulse than to the 8 pulses/sec. values.

In general, these relationships between liminal I and pulse frequency are similar to those found in the motor cortex by Mihailović and Delgado (33) and Lilly *et al.* (31), except that a relatively shallower slope in the low-frequency range was reported by the latter group.

For threshold motor responses with stimuli applied to postcentral gyrus the current (for 0.2-sec. TD) was found to be rather similar for PFs of 60 pulses/sec. and 120 pulses/sec., in the one case tested for this. Total production of motoneuronal excitatory postsynaptic potentials per unit time, however, has been found to be sensitive to change of pulse frequency in this range (9).

The relation of pulse frequency to utilization TD has already been treated above. A much greater effect of pulse frequency on the minimum TD for a threshold response was observed when a constant supraliminal I value was adopted for all tests; the minimum TD then decreased with increase in pulse frequency, according to a power function. The use of a constant I value might be expected to result in a clearer analysis of the dependence of facilitory effects (as manifested in minimum TD) upon pulse frequency. Practically, however, it was found that the frequency range is rather small in which this relationship can be meaningfully investigated. As the pulse frequency is raised the constant I level that is chosen becomes progressively supraliminal. This rapidly brings the threshold combination of I and TD out of parametric region A and into C or B; one may then be engaging components of the neural apparatus different from those involved in the threshold responses at the lower pulse frequency.

### *Polarity of stimulus pulses*

Table 2 summarizes the liminal current values (i.e., threshold I with 5-sec. TD) for cathodal (C) versus anodal (A) polarities of unipolar surface stimuli, at the different pulse frequencies. This shows that threshold I is consistently lower for C than for A, with C/A ratios between 0.6 and 0.9. This was true with all pulse frequencies used on the somatosensory cortex and included single shocks, even though threshold responses to the latter are generally frankly motor. The values given in Table 2 are the initial high threshold currents in those instances (especially at 8 pulses/sec.) in which there was evidence of facilitation lasting beyond the 30-sec. testing interval; but even if the lowest threshold values of such test series are examined, C retains its relative advantage over A. In one case the polarity was reversed for each successive pulse in the train. The liminal I value found was one between that for unidirectional cathodal and anodal pulse trains in this subject.

The stimulus polarity relation for motor cortex thresholds was tested

*Table 2. Polarity of pulses: liminal I values (threshold current in mA., using 5 sec. TD for the trains) for 0.5-msec. cathodal and anodal unipolar stimuli*

| Subj. No. | Single Pulse | | | 8 pulses/sec. | | | 30 pulses/sec. | | | 60 pulses/sec. | | |
|---|---|---|---|---|---|---|---|---|---|---|---|---|
| | C | A | C/A ratio | C | A | C/A ratio | C | A | C/A ratio | C | A | C/A ratio |
| | | | | | | *Sensory cortex (postcentral gyrus)* | | | | | | |
| 9 | 12 | 12 | 1.0 | | | | | | | | | |
| 10 | 13 | 19 | .7 | | | | 1.6 | 1.6 | 1.0 | .95 | 1.3 | .7 |
| 28 | | | | 3.1 | 4.0 | .8 | | | | 1.3 | 1.9 | .7 |
| 29 | | | | 2.5 | 3.6 | .7 | | | | .80 | 1.1 | .7 |
| 30 | | | | 3.8 | 5.2 | .7 | | | | 1.7 | 1.8 | .9 |
| 31 | | | | 1.3 | 2.1 | .6 | | | | .60 | .85 | .7 |
| 32 | | | | | | | | | | 2.8 | 4.4 | .6 |
| 33 | | | | | | | | | | 2.3 | 3.0 | .8 |
| 35 | | | | 1.2 | 1.6 | .8 | | | | .30 | .50 | .6 |
| 36 | | | | | | | | | | 1.6 | 2.2 | .7 |
| 37 | | | | | | | | | | .80 | 1.0 | .8 |
| 38 | | | | | | | | | | .65 | 1.1 | .6 |
| 43 | | | | | | | 1.9 | 2.7 | .7 | | | |
| 55 | | | | | | | | | | .60 | .85 | .7 |
| 74 | 3.2 | 3.5 | .9 | | | | | | | | | |
| 88 | | | | | | | 1.2 | 1.7 | .7 | | | |
| | | | | | | *Motor cortex (precentral gyrus)* | | | | | | |
| 61 | | | | | | | | | | 1.6 | 1.9 | .8 |
| 55 | | | | | | | | | | .95 | 1.1 | .9 |
| 87 | 11* | 12* | .9* | | | | 2.9† | 3.2† | .9† | | | |
| 89‡ | 22 | 25 | .9 | | | | | | | 13 | 14 | .9 |

* 1 pulse/sec., but each pulse in the train elicited a contraction. † 0.2 msec. PD.
‡ Patient under general anesthesia, 0.5-mm. pore electrode placed on intact dura over region giving lowest threshold for pyramidal type of motor response. The extra high threshold values may be largely due to both the extradural location of the electrode and to the general anesthesia.

only in a few subjects. A muscular contraction just detectable by vision or palpation by the observer served as the criterion of a threshold response. In one case the observer could see muscular "flicks" (of the lower eyelid) below the threshold for awareness of them by the subject. In others the thresholds for subject's awareness and observer's detection of the movements coincided. The exposed flat area of the precentral gyrus was stimulated in three unanesthetized subjects. The fourth subject was under general anesthesia (Fluothane, after induction with Pentothal sodium) and was stimulated extradurally at the point giving the lowest threshold for pyramidal-type muscular responses to cathodal stimuli. Cathodal stimuli had only slightly but consistently lower thresholds than anodal ones, whether 5-sec. trains at 30 pulses /sec. or 60 pulses/sec. were used (with which a just-visible sustained tetanic contraction appeared after some seconds of train duration) or whether single shocks which gave a flick or jerk were used. The differences between C and A thresholds were smaller than for sensations elicited at somatosensory cortex, with C/A ratio close to 0.9 for motor instead of the mean of 0.7 for sensory cortex. These findings on human precentral gyrus are in general agreement with those of Mihailović and Delgado (33) in the monkey (though apparently not with Lilly *et al.*, ref. 31) and with Hern, Landgren, and Phillips (22) in baboon. The latter authors point out that, when observable muscle flicks were taken to be the responses, the threshold currents were rather similar for anodal and cathodal pulses, with some advantage to each polarity in differing locations on the precentral gyrus. If the threshold response was taken as a detectable discharge in corticospinal fibers or as an excitatory postsynaptic potential in the appropriate lower motoneuron, however, they found that anodal thresholds were always lower than cathodal ones.

It may be noted additionally that the general range of threshold currents required for eliciting the respective threshold responses was similar for motor (precentral) and somatosensory (postcentral) areas. In the two subjects in whom both areas were accurately tested by the usual procedures, the thresholds were identical in one case (with 30 pulses/sec. stimuli) and somewhat greater for the motor cortex in the other (*subj. 55*, Table 2); no real significance can be attached to this variability, as movements of the electrode to different positions on the same gyrus in a given subject have produced differences in threshold of similar magnitudes. In a large number of other subjects the thresholds at both areas were estimated more crudely during the initial localization and exploration of the Rolandic region. In this group the threshold currents, for 30 or 60 pulses/sec. trains, were often somewhat lower for the precentral (motor) responses than for postcentral (sensory) responses.

### *Pulse duration*

The liminal I value decreased with increase in duration of the individual rectangular pulses used to make up the stimulus (Fig. 4). In the range of PDs tested, the steepest drop in threshold current relative to increase in PD took place between 0.1 and 0.5 msec. With a PD change between 0.01 msec. to 0.1 msec., even larger changes in threshold I for threshold subcortical activation of automatic responses have been reported (33). The employment of pulse durations shorter than 1 msec. permits a more meaningful analysis of effects of changes in pulse frequency, as single pulses of long duration (e.g., 5.0 msec.) may each produce a repetitive discharge in the responding units, as has been demonstrated for the corticofugal neurons of the motor cortex (24).

*Unipolar and bipolar stimulation*

Comparisons of liminal I values for unipolar and bipolar stimulation are presented for two subjects in Table 3. The unipolar thresholds were determined with each of the barrels so as to take into account any large differences between the cortical thresholds under the two-pore electrodes. In *subject 82*, the thresholds were stable throughout and were similar when either barrel of the double electrode was used in a unipolar manner. As was usual with unipolar stimuli, the anodal unipolar stimuli required a greater liminal I than the cathodal. The bipolar liminal I values were somewhat greater than the cathodal unipolar ones, but distinctly less than the anodal values. In *subject 79*, the results were more or less similar though they were complicated by a generally rising threshold during the experiment and also by a higher unipolar cathodal threshold at barrel 2 than at 1. If values of observations taken close together in time are compared for *subject 79* (e.g., observations 1 and 2, or 5 and 4, or 7 and 8) the bipolar liminal I values are almost the same as the unipolar values, with barrel 1 cathodal in all of these tests. A

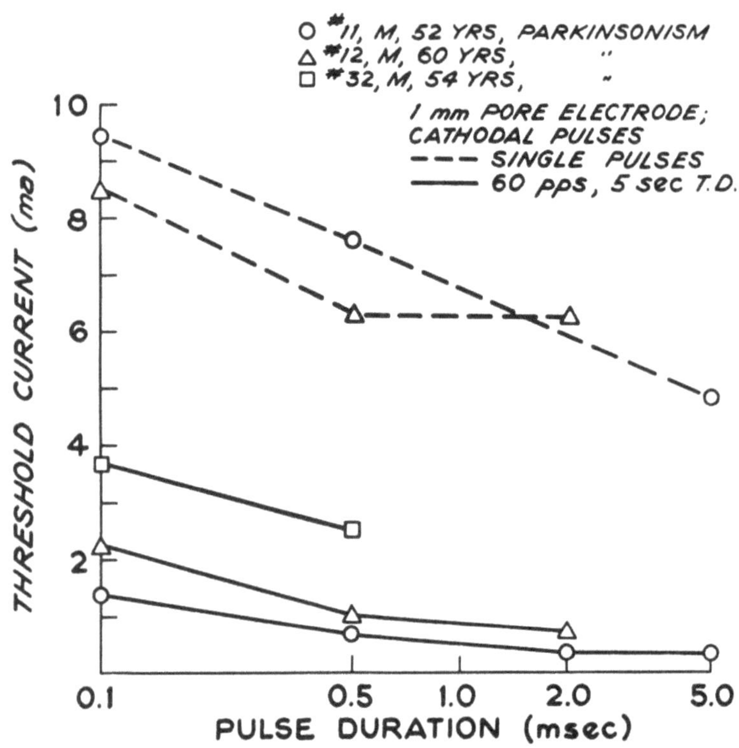


FIG. 4. Threshold peak currents at different pulse durations (PD).

similar but less satisfactory experiment in a third subject (using a 0.1 msec. pulse duration) also produced little difference between liminal I values determined with a unipolar cathode and with the bipolar arrangement (the same barrel remaining cathodal).

*Area of electrode contact with cortex*

When the 1-mm. pore electrode was compared with an electrode having very large contact surface, quite consistent differences appeared. For the first five subjects listed in Table 4, the large surface electrode consisted of a circular plate of Ag (with an AgCl surface) having a diameter of 10 mm. A thin layer of saline-soaked absorbent cotton of similar area provided an even, wet contact area under this electrode, which lay on the relatively flat surface of the postcentral gyrus. For *subjects 23* and *24* in Table 4, the large increase in contact surface was achieved by laying a pad of saline-soaked absorbent cotton, approximately 2 mm. thick and 5–10 mm. in diameter, on the gyrus and bringing the 1-mm. pore electrode into contact with the external surface of this pad. As the precise area of this pad was not measured, and the thickness of the pad was not uniform, values for current density are not calculated with this second method.

*Table 3. Unipolar and bipolar stimulation: liminal I values (30 pulses/sec., 0.5-msec. pulses), using double-barrel electrode throughout*

| Subj. No. | Unipolar | | | | Bipolar | |
|---|---|---|---|---|---|---|
| | Barrel 1 | | Barrel 2 | | Barrel 1 = cathode, mA. | Barrel 2 = cathode, mA. |
| | Cathode, mA. | Anode, mA. | Cathode, mA. | Anode, mA. | | |
| 82 (5-sec. TD) | 1.6 | 2.2 | 1.6 | 2.4 | 1.8 | 1.9 |
| 79 (2-sec. TD) | (1) 1.1<br>(5) 1.7<br>(7) 1.7 | (6) 2.1 | (9) 2.3 | | (2) 1.1<br>(4) 1.9<br>(8) 1.9 | (3) 2.0 |

Numbers in parentheses in front of the liminal I values for *subject 79* refer to the chronological order of the observations in this experiment (see further in text).

As seen in Table 4, the total liminal I with the 10-mm. plate electrode was generally several times as large as that with the 1-mm. pore electrode (range of ratios 1.5–8.7). This was also true for the cotton pad (*subjs. 23* and *24* in Table 4). When the current density (I, per $mm^2$) at the electrode contact area is calculated, however, it is largest for the 1-mm. pore electrode, smaller for the 2-mm. pore, and very much smaller for the large 10-mm. electrode. For liminal stimuli at 30 and 60 pulses/sec., the 1-mm. pore electrode passed an average of 1.7 $mA/mm^2$ of electrode contact area for the top five subjects in Table 4. The comparable value for the 10-mm. plate electrode averaged about 0.07 $mA/mm^2$. As calculated in this manner, the liminal current density of the large electrode is only 5% that of the 1-mm. pore electrode (avg. of the percentage comparisons for the individual subjects).

There was little change in the size or locus of the somatic area of the projected sensation, upon changing the electrode size from the 1-mm. pore to the 10-mm. circular plate, when using stimuli at the threshold currents for the respective electrodes. In two of the subjects there was no apparent change upon going to the larger electrode; another reported the same area

*Table 4. Effect of area of electrode on threshold current: liminal I values, using 5-sec. TD; 0.5-msec. cathodal pulses*

| Subj. No. | Pulse Repetition Frequency | 1-mm. Pore Electrode | | 10-mm. Plate Electrode | |
|---|---|---|---|---|---|
| | | Current, mA. | I/mm² | Current, mA. | I/mm² |
| 33 | 60 | 2.35 | 3.0 | 4.00 | .05 |
| 48 | 30 | 1.25 | 1.6 | 10.80 | .14 |
| 29 | 60 | .80 | 1.0 | 3.30 | .04 |
| | 8 | 2.45 | 3.1 | 3.75 | .05 |
| 28 | 60 | 1.25 | 1.6 | 5.40 | .07 |
| | 8 | 3.10 | 3.9 | 9.60 | .12 |
| 27 | 60 | 1.00 | 1.3 | 5.30 | .07 |
| | 8 | 1.85 | 2.4 | 7.20 | .09 |
| | | | | 1-mm. Pore on Cotton Pad | |
| 24 | 30 | .45 | .6 | 11.40 | |
| 23 | 60 | .85 | 1.1 | 7.90 | |
| | 8 | 2.35 | 3.0 | 8.40 | |
| | | | | 2-mm. Pore Electrode | |
| 19 | 60 | 1.05 | 1.3 | 1.80 | .6 |
| | 30 | 1.05 | 1.3 | 2.20 | .7 |
| | 8 | 2.55 | 3.2 | 5.15 | 1.6 |
| 18 | 30 | .65 | .8 | 1.00 | .3 |
| | 8 | 1.20 | 1.5 | 2.65 | .8 |

(hand or finger) or, in one response, a spread to the dorsum of the forearm; a fourth reported a complete shift of site from a forearm area near the wrist to a larger one around the elbow. In the two cases in which a saline cotton pad was interposed below the 1-mm. pore electrode, one showed no consistent changes in projected area and the other reported slightly larger areas and, in one response, a site changed from the fourth to the third finger.

Another remarkable finding with the large contact electrodes (10-mm. plate or cotton pad) was the relatively small effect of pulse frequency on liminal I, in contrast to the large increase in liminal I that occurs when changing from 60 pulses/sec. to 8 pulses/sec. while using the 1-mm. pore electrode. It is also seen that the increase in liminal I, going from the use of the small to the large electrode contact, was about 2.5-fold for 8 pulses/sec.

and about 5-fold for 60 pulses/sec. trains, for those subjects on whom both pulse frequencies were tested.

*Coulomb contents of threshold stimuli*

Analyses were made of the coulomb contents of the threshold stimulus trains along various parametric lines. When the TD was reduced within

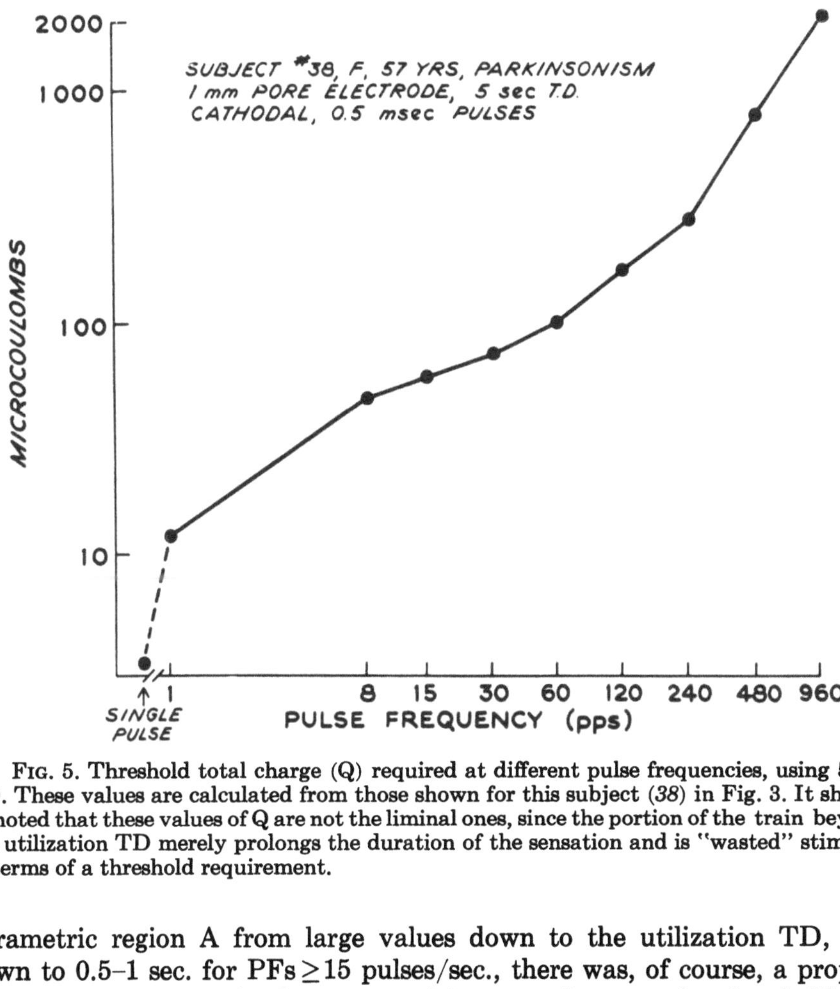


FIG. 5. Threshold total charge (Q) required at different pulse frequencies, using 5-sec TD. These values are calculated from those shown for this subject (*38*) in Fig. 3. It should be noted that these values of Q are not the liminal ones, since the portion of the train beyond the utilization TD merely prolongs the duration of the sensation and is "wasted" stimulus in terms of a threshold requirement.

parametric region A from large values down to the utilization TD, i.e., down to 0.5–1 sec. for PFs ≥ 15 pulses/sec., there was, of course, a proportional decrease in coulomb content; this occurs because the threshold I is relatively unchanged within this range. As TD was reduced further, i.e., within parametric regions C and B, the coulomb values for the subjects in Fig. 2 continued to decrease despite the rising values for the currents at threshold.

The effect of changing pulse frequency on the coulomb content of threshold stimulus trains is plotted for one subject in Fig. 5. This shows that the charge increases with increase in PF, even in the range in which liminal I progressively decreases with rise in PF (Fig. 3). It should be noted that the values in Fig. 5 include a large fraction of "wasted" coulombs, in that a threshold response is reached with TDs considerably less than the 5-sec. TDs employed there. On the assumption that the utilization TD in this subject would, as in general, have had a value of 0.5–1 sec. (at least for PFs $\geq$15 pulses/sec.), the "liminal" charge values would actually be about 10–20% of the microcoulomb values shown (at each such PF).

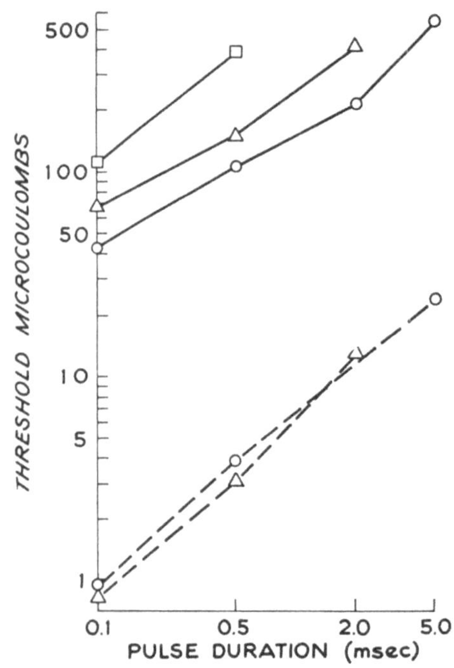


FIG. 6. Threshold total charge (Q) at different PDs. Calculated from data in Fig. 4. As in Fig. 5, it should be noted that these values of Q calculated for the 60 pulses/sec. trains are not actually the liminal ones, because of the use of longer than necessary 5-sec TDs; they are primarily of interest for the relative values at different PDs.

The coulomb values of the threshold stimuli with the different pulse durations are given in Fig. 6. It is seen that the charge required at threshold is much lower for stimuli using shorter pulse durations. The smaller requirement of total charge by stimuli with pulses of relatively short duration is in agreement with the findings of Lilly *et al.* (31) and of Mihailović and Delgado (33) for motor cortex responses. For threshold levels of self-stimulation of the septal region in rats, however, Ward (39) using a constant 0.2-sec. TD found a constant coulomb threshold regardless of the particular pulse duration, frequency, or intensity adopted. Use of a shorter pulse duration (e.g., 0.1 msec.) would presumably result in less tissue damage, in spite of the higher peak current and peak power requirements (30). The maximum peak power with stimulus trains, calculated from data in Fig. 4, and the total resistance in the electrode-subject circuit, came to about 0.1 W., using pulses of 0.1-msec. PD. All but a very small fraction even of this amount of power must be dissipated in the electrode portion of the circuit, which contains almost all the resistance, rather than in brain tissue. Injury from heating (30) therefore would be nonexistent with threshold stimulation of sensory cortex.

Rectangular pulses that alternately changed polarity had a threshold I between those for unidirectional cathodal and anodal pulses (at least in the one test of this point). The use of alternating polarity for rectangular or other shaped pulses should reduce any injurious effects (30), but one should note the possibility that different tissue elements may be responding to each of the pulses of opposite polarity in each pair.

It is generally agreed that injury tends to be proportional to the number of unidirectional coulombs passed (31, 32). In the present study minimum coulomb requirements of a threshold stimulus occurred with cathodal rather than anodal polarity, with the shortest of the pulse durations tested (0.1 msec.), and with the lowest pulse frequency and train duration that is usable (in terms of the type of response desired). This was true even though the peak currents at threshold were higher under some of these circumstances than in others. In a typical experiment in the present work a total of about 50 test stimuli were delivered, at intervals of 30 sec. or more. With each stimulus delivering on the order of 50 $\mu$coulombs, depending on the parametric values (Figs. 5 and 6), a total of around 2.5 mcoulombs of charge was passed in an experiment. This is only a small fraction of the amount found by Lilly *et al.* (31) to be injurious under comparable conditions. The total charge for stimuli in parametric region A can be reduced considerably below such values by restricting train durations to 1–2 sec. instead of 5 sec., and by using a pulse duration of 0.1 msec., instead of 0.5 msec.

*Alerting of subject: EEG-alpha rhythm, blocked versus present*

A quite different procedure for questioning the patient was compared with the alerting procedure ordinarily used throughout the above study. This procedure (*B*) differed from the usual one (*A*) as follows: The subject was asked to close his eyes and relax, rather than to keep his eyes open as in *A*. He was told that he might at times feel a sensation in the usual test area, but, that even when he did, he was to remain relaxed and not say anything about it until he was questioned by the observer. The observer gave no warning words to the subject that a stimulus was about to be delivered. Several seconds after the end of a stimulus the observer asked the subject whether he had just recently felt any test sensation. Unipolar scalp recordings of these subjects were made during the tests of the two procedures. Under the conditions of procedure *B*, the subjects exhibited resting EEG-alpha rhythms until the onset of the stimulus (when artifact made the EEG unreadable). In procedure *A*, the EEG showed a state of "alpha block," commonly associated with "arousal."

No significant difference in liminal I was found between the two procedures in five subjects. In two of these, this was tested with a pulse frequency of 30 pulses/sec., in one with 20 pulses/sec., in one with 8 pulses/sec., and in one with 8 pulses/sec. and with 60 pulses/sec. Nor was there any difference in the utilization TD between the two procedures; this was examined in one of the subjects tested with 30 pulses/sec. and in another with 8 pulses/sec.

In order to determine whether this lack of difference between the two procedures was unique to cerebral stimuli, similar comparisons were made with peripheral skin stimulation (procedure already described above, under *Utilization TD*). Neither liminal I, determined for 5-sec. TD, 30 pulses/sec. stimuli, nor the minimum number of pulses, 2–4, at liminal I strength, showed any significant difference as between the two alerting procedures. The threshold detection of tactile stimulation with graded von Frey hairs was also unaffected by the difference in alerting procedures in these subjects.

In view of the above it seems clear that any minor differences in the words used by the observer or in the degree of relaxation of the patient should have had no significant effects on the various threshold determinations made in this study. Distinct extraneous noises or the moving about of personnel in the operating room, however, did appear to be able to affect the consistency of the threshold determinations, especially with stimuli of short TD (below about 0.5 sec.), but this conclusion is based upon our impressions rather than on any careful comparison of conditions. The further significance of these findings to the problem of EEG-alpha state and "alertness" will be analyzed in the DISCUSSION.

## DISCUSSION

### *Significance of parametric regions*

A clear-cut difference was found in the type of response elicited by threshold stimuli in parametric region A as opposed to B, region C being less definitively different. This would suggest that different sets of neural components may be excited by stimuli in these different parametric regions. The components in the postcentral gyrus whose activity leads to a motor response appear to require a relatively higher current but much fewer repetitions of stimulus pulses for their adequate excitation than do those involved in producing a sensation. It was not clearly established whether any sensation could be directly elicited by stimuli with very short TDs which were at threshold for a motor response. It is quite clear, however, that responses elicited by stimulation of postcentral gyrus with the lowest current levels and sufficiently long TDs are purely somatic sensations. If one is to characterize the responsiveness of somatosensory cortex in terms of a meaningful response, particular attention obviously must be paid to the TD, PF, and I combinations of the stimuli being employed.

### *Motor responses from somatosensory cortex*

Apparently, stimulation of any point in the postcentral gyrus may elicit a motor response if the stimulus is characteristic of parametric region B.

In agreement with our findings is Foerster's (14, 15) observation, that the threshold galvanic current for eliciting a motor response by stimulating the postcentral gyrus was two or more times as great as that for the precentral gyrus, and the observations of similar ratios in the monkey's cortex (7, 15, 29). On the other hand, we found that liminal currents (parametric region A) for eliciting a sensation at postcentral gyrus were similar to those required of trains that elicited a threshold motor response at precentral gyrus. This similarity was also noted by Penfield and Boldrey (34).

The motor responses elicited by stimulation of the postcentral gryus showed a further characteristic different from those elicited at the precentral gyrus. A single pulse produced a twitchlike movement in either case, but stimulus trains did not. In the postcentral gyrus, a series of slowly repeating, weak, twitches was elicited during the delivery of a 30 pulses/sec. train (using a current level equal to that required of the threshold single pulse). Stimulus trains applied to the precentral gyrus, however, characteristically elicited motor responses that were smoothly tetanic and became progressively more powerful as the train went on. Trains of stimuli at the higher currents required of single pulses have not been tested on the precentral gyrus, because of the possibility of initiating a convulsion. In one subject a 4-pulse stimulus at 30 pulses/sec. was applied to the precentral gyrus and it produced a slight flexion twitch of the fingers, at a threshold I (1.7 mA.) about 150% of that required to produce a threshold contraction with a 5-sec. train. Using this higher I (1.7 mA.), a stimulus with a TD of 0.5 sec. produced a brief but smoothly sustained contraction, not a series of twitches. It seems unlikely that direct spread of current to the precentral gyrus was the cause of the threshold motor responses to postcentral stimulation, as observed in this work, partly in view of the different natures of the threshold motor responses found with stimulation of pre- and postcentral gyri. It should also be noted that postcentral motor responses were always contralateral and localized to the same approximate peripheral site in which stimuli of parametric region A, applied at the same cortical point, elicited a referred sensation.

There have been reports that stimulation of extensive areas of the lateral surface of the cortex outside the Rolandic region can elicit motor responses in the unanesthetized monkey (29). However, widespread motor responsiveness of the lateral surface of the cortex was not observed in unanesthetized man by Penfield and Rasmussen (36) or by Penfield and Jasper (35), when using stimuli at subconvulsive strengths. Foerster, too (14, 15), reported an absence of localized pyramidal-type motor responses outside the Rolandic region, although he did obtain mass body movements with strong faradic stimulation of the so-called extrapyramidal areas. Also in the present study, motor responses have not been observed upon exploratory stimulation of the available gyri surrounding the pre- and postcentral gyri, using stimuli in parametric region A.

The generally much higher threshold current for motor than for sensory responses, when stimulating postcentral gyrus in the present study, and the different nature of such threshold motor responses from those elicited at the precentral gyrus, supports the distinctiveness of the classical excitable motor and somatosensory areas of cortex, at least in man. These results and conclusion do not contradict the Hughlings Jackson postulate (14) of admixture of some types of motor and sensory representation, which can help explain the patterns of clinical seizures and the sequelae of lesions of the precentral cortex. The functioning of such mechanisms in the elaboration of motor patterns in normal man, however, remains an open question (see also 3, 14, 15, 16). Foerster's (15) analysis of the effects of stimulation and extirpation in man led him to suggest that the excitable motor regions outside, but acting in part through, the precentral gyrus may normally have supportive and integrative, rather than direct, motor functions.

It may be noted that threshold stimulation of the specific ascending pathway (thalamic PVL nucleus) in the present work elicited sensations without any visible motor responses, even when using pulse frequencies below 20 pulses/sec. (namely, 15 pulses/sec., 8 pulses/sec., and single pulses). Relatively greater increases in threshold I appeared to be necessary when changing to these lower pulse frequencies in PVL than was generally the case at the cortex.

### *Neural components activated for threshold sensation*

It may be presumed that surface cathodal stimuli excite relatively superficial structures, which may be presynaptic to other intracortical components, rather than deeper and perhaps corticofugal cells. Experimental evidence for such a presumption was obtained in the case of the motor cortex of the baboon, by Hern, Landgren, and Phillips (22). We have found a ratio of 0.7 for the cathodal/anodal currents, for threshold stimuli to the somatosensory cortex. The comparable ratio was higher for the motor cortex (precentral), about 0.9 in this work. These ratios indicate that the neural components, which are directly activated by a threshold postcentral stimulus which elicits sensation, lie in the more superficial rather than the deeper layers of the cortex.

Bipolar stimuli to the postcentral gyrus required only slightly greater liminal current to elicit a sensation than did unipolar cathodal stimuli, and slightly less current than unipolar anodal ones. This approximate similarity of thresholds was consistent in the relatively few tests made of this point. It is explained on the hypothesis that neural components relatively close to the surface of the cortex must be activated in the somatosensory response. This would be in accord with the inference already drawn above from the effect of change in polarity of the stimulus. It is, however, in contrast to the findings by Phillips and Porter (37), in the motor cortex of the baboon, that bipolar stimulation (using an interelectrode distance comparable to ours) generally required stronger currents than unipolar stimuli of either polarity. Further considerations of the paths of the current flow with the different types of electrodes, suggest that the relevant neural components are oriented vertically rather than horizontally. If the activated neural components were horizontal, one would have expected a considerably lower rather than similar threshold current with the bipolar electrodes, relative to the threshold for the unipolar arrangement. Activation predominantly by the vertical component of current flow would also fit better with the effectiveness of the large, 10-mm. electrode, which elicited a sensation at a much lower threshold current density than did the 1-mm. electrode. Current flow through most of the cortical area under the large electrode should be almost entirely vertical in direction.

*Temporal and spatial facilitation*

The subjective intensity, quality, and area of representation of sensations that are elicited by stimuli at the liminal I levels are apparently similar at all pulse frequencies, 15 pulses/sec. and up (27, 28). A comparison of liminal I values, therefore, should provide one index of the relative effectiveness of the different pulse repetition frequencies in achieving similar levels of excitability in the appropriate neural components. It should be recalled that this similar level of response to liminal I pulses is reached after about the first 0.5–1.0 sec. of the train duration, with most pulse frequencies. Furthermore, TDs greater than this utilization TD do not change the subjective intensity of the sensation; they merely lengthen its duration. This indicates that there is a net increment in excitatory state after each successive pulse in a train, but that this becomes zero after the utilization TD has been reached. The decay time of such excitatory increments after each pulse would appear to be rather long, on the order of 50–100 msec. This would permit stimuli with a PF as low as 15 pulses/sec. to achieve threshold effectiveness at a utilization TD that is not very different from that required with a higher PF (tested up to 120 pulses/sec.), but at a liminal I that is much greater than the one at the higher PF. The apparently striking increase in utilization TD at 8 pulses/sec. as compared even with 15 pulses/sec., on the other hand, indicates that this facilitory state falls to a rather small percentage of its peak value after intervals greater than about 50–100 msec. following each pulse.

It was pointed out that stimulus trains belonging in parametric regions B and C, even at threshold or somewhat subthreshold intensities, tended to be followed (at least upon retesting 30 sec. later) by a reduction of threshold current lasting for some minutes. Similar facilitation has been described for the motor cortex, lasting only some seconds in anesthetized monkeys (6), but up to 30 sec. in the unanesthetized monkey (29). Such long-lasting states of facilitation are presumably distinct in nature from the component of facilitation that occurs with repetition of pulses at liminal I strengths.

The liminal current density with the large, 10-mm. plate electrode was only about 5% that for the 1-mm. pore electrode, when calculated with respect to area of electrode surface. It seems clear, however, that the lateral spread of current through the pia-arachnoid membranes before reaching the cortex, as well as within the cortex, would involve only a much smaller fraction of the total current from the large plate electrode than from the 1-mm. pore electrode (see evidence on spread of effective current in motor cortex of cat, ref. 23). On this basis the current density calculated for the large plate electrode contact area might then be regarded as being much closer to the actual density of current which traverses the cortex below it. This liminal current density with the large plate electrode had a mean value of 0.07 $mA/mm^2$ (Table 4).

A comparison can be made of the approximate current densities in the cortex, as calculated for various horizontal distances from the centers of the 1-mm. and 10-mm. electrodes. From this it may be inferred that in the cortical region below the centers of these electrodes, liminal stimuli would produce more intense excitation of more neural elements under the smaller electrode than in any region under the large one; but in the surrounding regions, at some millimeters from the centers, the effectiveness of the small electrode would drop well below that of the large one. The ability of the large area electrode to elicit a threshold sensation, with this kind of distribution of liminal current densities relative to those under the small electrode, argues for a greater amount of spatial facilitation being involved in eliciting a response with the large electrode. The reduced effect of changes in pulse frequency on liminal I when using the large electrode suggests that spatial facilitation can to a considerable extent substitute for temporal facilitation. In addition, one may assume that the same or a similar sized group of neural components is appropriately activated by a threshold stimulus with the large or small electrode; this is indicated by the fact that both tend to elicit a sensation of the same intensity in the same area of the body. If so, the activation of such a group of neural elements would have been accomplished by the large electrode with the aid of extra spatial facilitation, originating from other neural components spread over a wider area than with the smaller electrode.

Such spatial facilitation within relatively large areas that are subjected to weaker stimulation may be related to similar types of phenomena observed by a number of investigators (35, pp. 240–241). More particularly, Brown (6) had found that a stimulus delivered at one point on the motor cortex could lower the threshold at another point about 5 mm. distant; he called this "secondary facilitation."

### *Period of activation required by liminal stimuli*

The minimum train durations required when using stimuli at the liminal current levels, i.e., the utilization TDs, turned out to be surprisingly long—about 0.5–1.0 sec. This range held for all pulse frequencies from 15 pulses/sec. to 120 pulses/sec. (although in some subjects it was down to 0.2–0.3 sec.); and it was not affected by using bipolar electrodes of various separations as opposed to unipolar. A long utilization TD also appears to obtain when the sensory cortex is activated via the ascending thalamic projection fibers. In the one subject in whom the subcortical white matter was stimulated by a unipolar wire electrode, inserted through the somatosensory cortex to a depth of about 5 mm., and in two other subjects, in whom the postero-ventro-lateral nucleus of the thalamus was stimulated, utilization TDs fell within the same range of values as those found with stimulation of somatosensory cortex. With stimulation of thalamic sensory regions, Ervin and Mark (11) reported that lengthening the TD above 5 sec. seemed to have no effect on threshold I and did not raise a subthreshold stimulus to threshold effectiveness. They did not, however, determine the minimum TD that was required by a stimulus of threshold intensity. The motor cortex, at least in the anesthetized cat, was found by Lilly *et al.* (31) to show no flattening in the curve for threshold current versus train duration, over a range of about 0.1–20 sec. in train durations, when using pulse frequencies of 20 pulses/sec. to 100 pulses/sec. (the curve was flat in this range of TDs for stimuli at 400–7,500 pulses/sec., but they note that at these frequencies there was only an "on," quick jerk, response). No attempt was made in the present study to determine this relationship for motor cortex in unanesthetized man.

The long utilization TDs, found with either direct surface cortical stimuli or with stimulation of the immediate afferent input to the cortex, are in marked contrast to those found with stimulation at the periphery, i.e., of skin or peripheral nerve. In the two parkinsonian and two normal subjects carefully tested with electrical stimulation of the skin the utilization TD was about 2–4 pulses long, i.e., about 0.1-sec. TD, with 30 pulses/sec. stimuli. The observation by Sigel (38), that electrical stimuli to the skin have approximately the same voltage threshold for eliciting sensation at pulse frequencies of 0.1 or 1 pulses/sec. as they do at 50 pulses/sec., is in agreement with this. The results of stimulation and electrical recording of exposed sural nerves in awake patients, reported by Collins, Nulsen, and Randt (8), are even more clear-cut in this respect. These authors found that, when stimulus intensity was just threshold for producing a conducted action potential (in the beta fibers, the largest ones present), a single shock was sufficient to elicit a sensation in the subject. At stimulus intensities below this level no sensations were reported. Even more strikingly, Hensel and Boman (21) have now shown that a mechanical stimulus to the skin which elicits only a single, nonrepetitive nerve impulse, as recorded from exposed nerve filaments, can be sufficient to elicit a sensation of touch in the human subject. It cannot be argued, therefore, that electrical stimuli of very short train durations or even single pulses, which are just strong enough to elicit a threshold sensation, are generating a lasting repetitive discharge in the afferent nerve fibers supplying the stimulated skin.

The very short utilization TDs with stimulation of the skin rule out the possibility that the relatively long utilization TDs required with direct cortical stimuli were simply a reflection of an arbitrary criterion of a positive sensation which the subjects might have set for themselves. There was, not uncommonly, some uncertainty on the part of subjects as to whether a sensation was experienced, at the near threshold current or train duration values, but this occurred over a range that was small compared to the total value of each such parameter.

The explanation of the long utilization TD could lie in the time required for sufficient temporal and spatial facilitation to be brought to bear on certain neural components, when stimulating with the different combinations of liminal intensity and pulse frequency. On the other hand, it may be that a certain minimum time period of activity in some neural components is required in order to elaborate a conscious sensation, regardless of the amount of temporal or spatial facilitation. An adequate test of these alternative hypotheses has not yet been carried out.

The demonstrations of evoked cortical afterwaves, which may go on for a considerable fraction of a second in unanesthetized animals following a single sensory stimulus (4, 17, 18), and the modification of such afterwaves (5, 17, 19) as well as of long-latency unit discharges (12) by the state of wakefulness, arousal, habituation, and conditioning, may be highly relevant to the requirement of 0.5- to 1-sec. utilization TDs for cortical stimuli that elicit sensation. Following single direct cortical stimuli to the postcentral gyrus no evidence of such repetitive afterwaves has been found in unanesthetized human subjects, at stimulus currents which can elicit sensation when delivered as a train of pulses (unpublished observations). This could explain why the repetitive activation for 0.5–1 sec. must be supplied by the stimulus in the case of direct cortical excitation. As an extension of a hypothesis that reverberating activity in neuronal circuits must go on for minutes or hours to fix experience in memory. Gerard (20) has suggested that circuits reverberating for seconds or fractions of a second may be necessary for the initial consciousness of an experience.

Whatever the mechanism is which brings about the necessity for such long utilization TDs, the fact of their requirement would seem to have considerable significance for an understanding of conscious responses to incoming somatosensory impulses. They indicate that not until 0.5 sec. or so after the arrival at the cortex of the initial impulses generated by a near-threshold sensory stimulus, will a subjective awareness of this stimulus take place. (This subjective awareness is of course one as defined here, i.e., one capable of being recalled and expressed within a few seconds after the stimulus.) Such lengthy "latent periods" for conscious sensations, at least at liminal levels of stimulation, introduce an interesting feature into the analysis of perception, mental processes in general, and behavior.

### *State of EEG-alpha rhythm and cortical responsiveness*

In the testing of the two different procedures for questioning the patient, a test of EEG-alpha state in relation to threshold values was also provided. The usual procedure (*A*) produces a state of alpha blockade before and presumably during each stimulus, while the other one (*B*) permits good alpha waves to be present before and probably during the test stimulus. (EEG recordings during the stimulus to the cortex were too obscured by artifact to be sure about alpha state during that time. Artifact-free recordings made during threshold electrical stimulation of the skin of the hand, however, exhibited no alteration in the alpha rhythm at the onset, during, or after the passage of a 1-sec. train at 30 pulses/sec., using procedure *B*.) It was somewhat surprising to find that there were no significant differences between the two procedures, in either the liminal currents or the utilization TDs (i.e., minimum train durations using liminal current) that were required to elicit a threshold-conscious sensation. The possibility existed that the influence of any arousal mechanism that may be associated with a state of alpha-blockade might be exerted at some level, in the neural apparatus of the somatosensory cortex, which is afferent to the portion that is engaged by the surface cortical stimulus. Stimulation of the cortical afferents subcortically, however, with an electrode in the white matter below the cortex of the postcentral gyrus, still produced no differences between the two alerting procedures. Even electrical stimulation of the skin produced the same threshold I and the same utilization TDs with both procedures. as tested in two parkinsonian patients and in two normal subjects.

These results do not, of course, contradict the findings of Lansing, Schwartz, and Lindsley (25), of a reduction in reaction time under certain conditions that produce alpha-blockade. The difference in responses of the subjects, i.e., leisurely recall and verbal reporting of a threshold sensation in our study versus speedy motor responses in the reaction time studies, could obviously involve mechanisms that are influenced differently by conditions associated with the states of alpha. Nor do the present studies necessarily imply that appropriate changes in attention state of the subject may not affect the threshold currents or utilization TDs required of stimuli at sensory cortex. The evidence would indicate, however, that processes involved in changing from alpha-present to alpha-blockade do not necessarily include the kind of arousal effects that might modify cortical excitability levels and attentive states, such as are relevant to the response studied here. A similar lack of any significant effect of the "EEG arousal" process on the responsiveness of the corticospinal system was found by Zanchetti and Brookhart (40) in cats with suitable mesencephalic lesions.

## Summary

The parameters of electrical stimuli which elicit a conscious sensation, when applied directly to a fixed area of the somatosensory cortex of awake human subjects, have been analyzed and characterized quantitatively. The threshold parametric points, i.e., the specific combinations of values of the set of parameters tested which are just adequate to elicit a threshold response, could be established within usable limits, with a consistency of about 5–10% in the value of the threshold current under suitable experimental conditions. The parameters studied, utilizing rectangular pulses, included peak current (I), polarity of pulses, pulse duration (PD), pulse repetition frequency (PF), train duration (TD), train repetition rate (TRR), uni- and bipolar electrode arrangement, and area of electrode contact.

Threshold parametric points can be grouped into three parametric regions. Parametric region A encompasses threshold stimuli with the trains of relatively long duration ( >0.5 sec.) and with pulse frequencies of about 15 pulses/sec. and up. The responses in this region are purely somatosensory. Parametric region B encompasses threshold stimuli requiring relatively high current and short train durations (including single pulse stimuli). The threshold response in region B was a motor one; this held true at almost all loci tested on the postcentral gyrus. Parametric region C, although less clearly characterizable, would encompass threshold stimuli and kinds of responses requiring intermediate current intensities and train durations, between those for regions A and B.

The motor responses obtained by stimulation of the postcentral gyrus differed from those elicited at the precentral gyrus, in their relative threshold currents and in the type of muscular activations elicited during a train of pulses.

Train duration may be reduced to about 0.5–1 sec. with very little or no rise in threshold current (with pulse frequencies of 15 pulses/sec. and up). This minimum TD that is required of stimulus trains having the "liminal" or lowest threshold intensity (other parameters remaining fixed), is named the "utilization TD."

While utilization TDs of about 0.5–1 sec. were characteristic of the somatosensory system at cerebral levels, they were found to be very short (2–4 pulses) with stimuli applied to skin. The relatively long cerebral utilization TD leads to inferences about a latent period for conscious awareness of sensory input, at least at the near-threshold levels, which may have important general implications.

The inverse effect of PF on liminal I, and the comparative absence of effect of PF from about 15 pulses/sec. up on utilization TD, suggest that an excitatory change is developed, by each stimulus pulse in a train, which has a decay time of 50–100 msec. In addition, stimuli in parametric regions B and C, though rarely in A, produce a longer lasting facilitory change, having a decay time greater than 30 sec.

Surface cathodal stimuli were consistently more effective than anodal ones, more so on somatosensory than on motor cortex. Bipolar stimuli, 2-mm. interelectrode distance, required threshold current levels between those for unipolar cathodal and anodal ones. These facts suggest that the relevant neural components lie relatively superficially in the cortex and are oriented perpendicularly to the surface.

An electrode with a cortical contact area of 10 mm. diameter can elicit a threshold-conscious sensation similar to that obtained with the 1-mm. electrode, but it does so with an apparently much lower current density (liminal intensity about 0.1 mA. peak current/mm$^2$). Spatial facilitation can thus be an important factor in eliciting threshold sensations.

The coulombic content of threshold stimulus trains is markedly different for differing parametric points. It tends to be lower the lower the pulse frequency, the shorter the pulse duration, and the shorter the train duration, even though the threshold current values change in the opposite direction with such changes in these variables.

A change in the alerting procedure from the usual one, in which a ready signal was given and EEG-alpha blockade obtained, to another, in which there was no ready signal and EEG-alpha was present, produced no significant difference in either the liminal I or the utilization TD. This was true for cortical as well as skin stimulation. Apparently, mechanisms that modify the state of EEG-alpha in the so-called arousal response do not necessarily modify the state of cortical responsiveness, or the attentive state, in so far as these states are relevant to the eliciting of conscious sensation.

ACKNOWLEDGMENT

We are pleased to express our gratitude to Dr. Alexander Riskin, Dr. Gerhardt von Bonin, Ann Hays, Elizabeth Leavy, Miriam Mueller, and David Swift for their assistance in various phases of this work.

## REFERENCES

1. ALBERTS, W. W. A stimulus pulse polarity reversal unit. *Electroenceph. clin. Neurophysiol.*, 1958, *10:* 172–173.
2. ALBERTS, W. W., WRIGHT, E. W., JR., LEVIN, G., FEINSTEIN, B., AND MUELLER, M. Threshold stimulation of the lateral thalamus and globus pallidus in the waking human. *Electroenceph. clin. Neurophysiol.*, 1961, *13:* 68–74.
3. BATES, J. A. V. Observations on the excitable cortex in man. *Lectures Sci. Basis Med.*, 1955–56, *5:* 333–347.
4. BRAZIER, M. A. B. *The Electrical Activity of the Nervous System* (2nd ed.). New York, Macmillan, 1960, pp. 178–179.
5. BREMER, F. Neurophysiological mechanisms in cerebral arousal. In: *The Nature of Sleep*. Ciba Foundation Symposium, edited by G. E. W. Wolstenholme and M. O'Connor. London, Churchill, 1961, pp. 30–50.
6. BROWN, T. G. Studies in the physiology of the nervous system. XXII: On the phenomena of facilitation. 1. Its occurrence in reactions induced by stimulation of the "motor" cortex of the cerebrum in monkeys. *Quart. J. exp. Physiol.*, 1915, *9:* 81–99.
7. BROWN, T. G. Studies in the physiology of the nervous system. XXVII: On the phenomena of facilitation. 6. The motor activation of parts of the cerebral cortex other than those included in the so-called "motor" areas in monkeys; with a note on the theory of cortical localization of function. *Quart. J. exp. Physiol.*, 1916, *10:* 103–143.
8. COLLINS, W. F., NULSEN, F. E., AND RANDT, C. T. Relation of peripheral nerve fiber size and sensation in man. *Arch. Neurol.*, 1960, *3:* 381–385.
9. ECCLES, J. C. *The Physiology of Synapses*. Berlin, Springer, 1964, pp. 83–86.
10. EIJKMAN, E. AND VENDRIK, A. J. H. Detection theory applied to the absolute sensitivity of sensory systems. *Biophys. J.*, 1963, *3:* 65–78.
11. ERVIN, F. R. AND MARK, V. H. Sterotactic thalamotomy in the human. Part II, Physiologic observations on the human thalamus. *Arch. Neurol.*, 1960, *3:* 368–380.
12. EVARTS, E. V. Photically evoked responses in visual cortex units during sleep and waking. *J. Neurophysiol.*, 1963, *26:* 228–248.
13. FEINSTEIN, B., ALBERTS, W. W., WRIGHT, E. W., JR., AND LEVIN, G. A stereotaxic technique in man allowing multiple spatial and temporal approaches to intracranial targets. *J. Neurosurg.*, 1960, *17:* 708–720.
14. FOERSTER, O. The motor cortex in man in the light of Hughlings Jackson's doctrines. *Brain*, 1936, *59:* 135–159.

15. FOERSTER, O. Motorische Felder und Bahnen. In: *Handbuch der Neurologie*, edited by O. Bumke and O. Foerster. Berlin, Springer, 1936, vol. 6, pp. 1–357.
16. FOERSTER, O. Sensible corticale felder. In: *Handbuch der Neurologie*, edited by O. Bumke and O. Foerster. Berlin, Springer, 1936, vol. 6., pp. 358–448.
17. FREEMAN, W. F. Alterations in prepyriform evoked potential in relation to stimulus intensity. *Exp. Neurol.*, 1962, *6:* 70–84.
18. GALAMBOS, R., AND SHEATZ, G. C. Evoked cortical auditory responses. *Fed. Proc.*, 1959, *18:* 49.
19. GALAMBOS, R., SHEATZ, G., AND VERNIER, V. G. Electrophysiological correlates of a conditioned response in cats. *Science*, 1956, *123:* 376–377.
20. GERARD, R. W. The biological roots of psychiatry. *Amer. J. Psychiat.*, 1955, *112:* 81–90.
21. HENSEL, H. AND BOMAN, K. K. A. Afferent impulses in cutaneous sensory nerves in human subjects. *J. Neurophysiol.*, 1960, *23:* 564–578.
22. HERN, J. E. C., LANDGREN, S., AND PHILLIPS, C. G. Selective excitation of corticofugal neurons by surface-anodal stimulation of the baboon's motor cortex. *J. Physiol.*, 1962, *161:* 73–90.
23. HERN, J. E. C., PHILLIPS, C. G., AND PORTER, R. Electrical thresholds of unimpaled corticospinal cells in the cat. *Quart. J. exp. Physiol.*, 1962, *47:* 134–140.
24. LANDGREN, S., PHILLIPS, C. G., AND PORTER, R. Minimal synaptic actions of pyramidal impulses on some alpha motoneurones of the baboon's hand and forearm. *J. Physiol.*, 1962, *161:* 91–111.
25. LANSING, R. W., SCHWARTZ, E., AND LINDSLEY, D. B. Reaction time and EEG activation under alerted and non-alerted conditions, *J. exp. Psychol.*, 1959, *58:* 1–6.
26. LEVIN, G., FEINSTEIN, B., KREUL, E. J., ALBERTS, W. W., AND WRIGHT, E. W., JR. Stereotaxic surgery for parkinsonism. *J. Neurosurg.*, 1961, *18:* 210–216.
27. LIBET, B., ALBERTS, W. W., WRIGHT, E. W., JR., LEVIN, G., AND FEINSTEIN, B. Sensory perception by direct stimulation of human cerebral cortex: stimulus parameters. *Fed. Proc.*, 1959, *18:* 92.
28. LIBET, B., ALBERTS, W. W., WRIGHT, E. W., JR., LEVIN, G., AND FEINSTEIN, B. Direct excitation of human somatosensory cortex. *Electroenceph. clin. Neurophysiol.*, 1961, *13:* 498.
29. LILLY, J. C. Correlations between neurophysiological activity in the cortex and short-term behavior in the monkey. In: *Biological and Biochemical Bases of Behavior*, edited by H. F. Harlow and E. N. Woolsey, Madison, Univ. of Wisconsin Press, 1958, pp. 83–100.
30. LILLY, J. C. Injury and excitation by electric currents. A. The balanced pulse-pair waveform. In: *Electrical Stimulation of the Brain*, edited by D. E. Sheer. Austin, Univ. of Texas Press, 1961, pp. 60–64.
31. LILLY, J. C., AUSTIN, G. M., AND CHAMBERS, W. Threshold movements produced by excitation of cerebral cortex and efferent fibers with some parametric regions of rectangular current pulses (cats and monkeys). *J. Neurophysiol.*, 1952, *15:* 319–341.
32. MACINTYRE, W. F., BIDDER, T. G., AND ROWLAND, V. The production of brain lesions with electric currents. *Proc. Nat. Biophys. Conf.*, 1st, 1957, (Pub. 1959) pp. 723–732.
33. MIHAILOVIĆ, L. AND DELGADO, J. M. R. Electrical stimulation of monkey brain with various frequencies and pulse durations. *J. Neurophysiol.*, 1956, *19:* 21–36.
34. PENFIELD, W. AND BOLDREY, E. Somatic motor and sensory representation in the cerebral cortex of man as studied by electrical stimulation. *Brain*, 1937, *60:* 389–443.
35. PENFIELD, W. AND JASPER, H. *Epilepsy and the Functional Anatomy of the Human Brain.* Boston, Little, Brown, 1954, pp. 41–106.
36. PENFIELD, W. AND RASMUSSEN, T. *The Cerebral Cortex of Man.* New York, Macmillan, 1952, pp. 47, 102–105, 185–196.
37. PHILLIPS, C. G. AND PORTER, R. Unifocal and bifocal stimulation of the motor cortex. *J. Physiol.*, 1962, *162:* 532–538.
38. SIGEL, H. Prick threshold stimulation with square-wave current; a new measure of skin sensibility. *Yale J. Biol. Med.*, 1953, *26:* 145–154.
39. WARD, H. P. Stimulus factors in septal self-stimulation. *Amer. J. Physiol.*, 1959, *196:* 779–782.
40. ZANCHETTI, A. AND BROOKHART, J. M. Cortico-spinal responsiveness during EEG arousal in the cat. *Amer. J. Physiol.*, 1958, *195:* 262–266.

# CORTICAL ACTIVATION IN CONSCIOUS AND UNCONSCIOUS EXPERIENCE

BENJAMIN LIBET

Reprinted for private circulation from
PERSPECTIVES IN BIOLOGY AND MEDICINE
Vol. IX, No. 1, Autumn 1965

# *CORTICAL ACTIVATION IN CONSCIOUS AND UNCONSCIOUS EXPERIENCE*

*BENJAMIN LIBET**

It has become generally accepted that a large, perhaps even a major part of our mental activities can take place without our being consciously aware of them. Though apparently unconscious, they are nevertheless a part of significant mental experience since there is evidence that such activities can participate in later mental and behavioral manifestations—cognitive, affective, or conative.

It is also generally assumed that specific temporo-spatial configurations of cerebral activity will be found to correspond to specific mental states or experiences. However, it is easier and at present more meaningful to look for the *differences* between conscious and unconscious experiences rather than to attempt a more complete specification of the cerebral processes of each mental state. Some clues to such differences stem from our investigations of those physiological parameters which are uniquely involved in eliciting conscious sensory experience by electrical stimulation of the human cerebral cortex [1, 2]. Cortical processes are studied which are accessible to measurement in awake human subjects during neurosurgical procedures. Conditions of cortical activation which are at the threshold level for eliciting a conscious sensory experience are compared to conditions of activation which are just below this threshold. ("Threshold" is considered here in the broad sense of adequacy in all parameters, not merely in intensity.) Since states of cortical activity which are just below the threshold for conscious sensory experience may represent some form

* Department of Physiology, University of California School of Medicine, San Francisco, and Mount Zion Neurological Institute, San Francisco. The unpublished experimental work included here was carried out in the Mount Zion Neurological Institute, San Francisco, in collaboration with Drs. W. W. Alberts, Bertram Feinstein, and Grant Levin, Mr. E. W. Wright, Jr., and Mrs. L. D. Delattre. It was partially supported by U.S. Public Health Service Grant NB 05061 from the National Institute of Neurological Diseases and Blindness.

of unconscious mental experiences, the differences between conditions at and below these thresholds may have a bearing on our general problem.

Before going on with the findings and hypotheses, I shall attempt to reformulate, out of the background of various notions that have been held by others, an appropriate usage of the terms "conscious" and "unconscious experience." (Incidentally, treating these mental phenomena as experiences gives them a more operational meaning than using the nouns "consciousness" and "the unconscious" would [3]). I begin with the premise that the subjective or introspective experience of awareness of something is the primary criterion of conscious experience. It would, then, be some clear indication of such an experience of awareness by the subject to an external observer that is operative in any investigation or analysis of conscious experience. "Awareness," as a subjective experience, is a "primitive (undefined) term" [4], with phenomenological rather than behavioral meaning. Although it is experienced and known privately by each individual, we are confident that others have experienced it too, and that they know what we mean by our questions about it [3, 5]. There is, however, the experimental problem of the suitability of the voluntary expression employed by the subject to indicate or report his conscious experience, which will be considered further below.

In addition, recallability, or at least recognizability, of the subjective experience is required in order for some report to be made. But this must be an essential requirement in any case if an experience is to supply content to the kind of subjective awareness which has continuity. There may well be an immediate but ephemeral kind of experience of awareness which is not retained for recall at conscious levels of experience. If such experiences exist, however, their content would have direct significance only in later unconscious mental processes, although, like other unconscious experiences, they might play an indirect role in later conscious ones. Dreaming and daydreaming, on the other hand, introduce borderline cases. Most of the dream experiences during natural sleep can hardly be recalled unless the individual is awakened within minutes of their occurrence [6]. The recall of experiences of awareness in the normal waking state is not subject to such limitations in any obligatory way. Nevertheless, investigation of these experiences is simplified by eliciting a report or indication of them from the subject shortly after their occurrence. The degree of recallability can, of course, be altered by some subsequent events or in certain psychological states, and this introduces additional issues [7, 8]. There is another possible difficulty which has been raised in accepting the subject's report. The subject himself may be setting some arbitrary subjective level of confidence of his awareness, below which level he reports that he is unaware. However, when a subject reports with a feeling of certainty that he has not experienced an awareness of the something in question, it would be an unwarranted distortion of the primary data to insist that he was in fact sufficiently aware of this something to consciously set a level for his response. That any such settings may go on unconsciously is, of course, not excluded.

A verbal report is usually employed as an acceptable indicator of conscious experiences to an external observer. Verbal descriptions of the actual content of an experience suffer, of course, from inarticulateness of a subject and from the inadequacy of language itself to describe perceptions, feelings, etc. This difficulty has been analyzed clearly by Eriksen [9] in relation to psychological experiments dealing with awareness. Non-verbal responses may thus be more appropriate indicators of awareness in some situations. For example, a lever may be pushed to indicate in a graded way the dimensions or intensities of a stimulus, or a picture of a visual experience may be drawn, etc. However, whether the most informative response happens to be verbal or non-verbal, or a combination of these, the instruction must make it quite clear to the subject that the response should represent only his subjective experience of awareness of the material—if it is to meet our criterion of conscious experience.

Lack of adherence to this criterion has, I believe, led to some confusion and irrelevancy in the attempts at experimental analysis of the problem of conscious and unconscious experience [4, 9]. The distinction between conscious and unconscious experience should not be made on the basis of the response or report being verbal versus non-verbal, or complex versus simple, or involving decisions and judgments or some degree of abstract mental processes. It should be made solely on the basis of an introspective experience of awareness which is recognized and reported as such in some suitable way by the subject. Techniques have been employed such that responses or reports about events are elicited regardless of the subject's introspective experiences of the events. For example, the subject may be required to choose among alternatives presented to him (forced-choice techniques), or he may be asked to guess about the occurrence of an event or validity of a choice. Such techniques have been shown to elicit correct responses at greater than chance levels even when the subject's introspective experience was not awareness of the events but simply guessing about them. To regard this kind of finding as evidence of a weaker or lower level of awareness which is qualitatively similar to one meeting our criterion would involve making a gratuitous assumption. On the other hand, when the objective is the study of responses which indicate that some detection of signals has occurred, irrespective of the introspective experiences of the subject, then various other techniques for obtaining and assessing responses are obviously suitable and desirable [10]. When one is applying experimentally the two different criteria of conscious experience, there are apparent differences in the range of threshold uncertainty and in the influence of alerting procedures related to block of cortical EEG-alpha rhythm [1, 2]. The primacy of introspective experience in defining conscious mental activity is a viewpoint shared with many scientists [3, 5, 11] and philosophers [12]. Some experimental psychologists seem to have difficulty accepting it in scientific investigations; but any purely behavioral evidence would require validation against the subjective experience of awareness.

Sensory experiences could be unconscious because either (*a*) attention was not sufficiently or selectively focused on them, or (*b*) the physical stimulus was so small that output from the sensory receptors was inadequate to elicit a conscious sensory experience, even with suitable attentiveness (subliminal sensory stimulation). The former of these alternatives is probably the more significant and interesting one. There appears to be no convincing psychological evidence of a superdiscriminating system which can respond at unconscious levels to subliminal stimuli [4, 9]. Indeed, there may be relatively little physiological potentiality for subliminal stimuli to have any significance. In the case of somatic sensation, at least, the minimum discharge from the sensory receptor which elicits a conscious sensory experience in an attentive subject can be at or close to the absolute physiological minimum, i.e., one nerve impulse [13]. This relationship is more difficult to assess, however, for other sensory modalities under conditions of stimulation in the normal environment. With alternative (*a*), on the other hand, there would seem to be more obvious potentialities. Attention is focused, if at all, on only small segments of the total sensory input at any moment. There could therefore be a great deal of input that continually escapes intrusion into introspective experiences of awareness even when it is above the liminal (potentially minimum) level for eliciting the latter. There is indeed good evidence that above-liminal inputs of which the subject is unaware can nevertheless be utilized in acquiring behavioral responses and perceptual changes [4, 9, 14] and as cues in making intelligent skilled responses. In fact, many of us have had the experience of driving a car on a thoroughly familiar road for a time while thinking of something else and then finding ourselves unable to recall any experience of awareness, not only of the sensory cues, but also of the traffic decisions and motor responses we made during that period.

A variety of unconscious experiences are apparently associated with "internal" cerebral processes, exclusive of any immediate sensory inputs. Freud was of course one of the chief originators of such hypotheses. One should not restrict these experiences to those related to affective states, for it appears that much of the purely intellectual activities, even of the highest order of complexity, may proceed without reportable conscious experience. Many instances have been documented [15–17] in which highly creative and elaborate "thinking" by scientists, poets, etc. has developed unconsciously, although interplay with conscious experiences is commonly involved. The impression is that not infrequently only the product—i.e., the new and illuminating idea or solution to a problem—springs into the light of conscious experience, and is then subjected to more deliberate conscious analysis. Probably all instances of intuitively generated ideas or artistic expressions share such processes. Perhaps even many ordinary mental operations, such as the occurrence of associational words or thoughts, are accomplished unconsciously, the person becoming introspectively aware of them only after they have been made.

In this broad view of unconscious experience, presumably a wide range of gradations and perhaps different kinds of mental processes are encompassed [17]. There will undoubtedly be disagreement over whether the term "unconscious" should be applied to all the processes which lack the criterion required here of a conscious experience. But the criterion does distinguish one group of experiences from the rest, and so one should not permit such a definition to obstruct investigation or analysis of an incisively significant aspect of mental existence.

Let us proceed to the evidence from brain stimulation and its possible bearing on the problem. As is now well known [18], conscious sensory experiences can be elicited by electrical stimuli applied directly to the surface of the somatosensory cortex (postcentral gyrus) in awake human subjects. Stimulation here has the advantage that one has more complete control of the parameters of direct cortical activation. By contrast, natural input from the sensory receptors can be subjected to complex alterations in the central nervous system before full delivery of its effects to the cortex. In our investigation of the nature of the stimuli to somatosensory cortex which were just adequate to elicit a conscious sensory experience, as opposed to those which were just inadequate, some interesting requirements were discovered [1]. Repetition of the individual (0.5 msec.) pulses of current is necessary to elicit a sensation. Indeed, when the intensity of a single pulse or of a very few pulses is raised sufficiently to produce some response, the threshold response so obtained by stimulation of the postcentral gyrus is a *motor* rather than a sensory one; i.e., a small muscular contraction can be observed in the appropriate region of the body. The intensity or current required of a stimulus to elicit a threshold sensation is lower with a longer train-duration of the repetitive pulses. If the duration of the train exceeds a certain minimum, however, there is no further appreciable drop in threshold intensity; i.e. a plot of threshold current vs. train-duration produces a relatively flat line for all points with train-durations greater than this minimum. We refer to this minimum threshold current level as the *liminal* intensity, and to the minimum duration required for a train of liminal pulses as the *utilization train-duration.* The values of the utilization train-durations mostly fall into the range of 0.5–1.0 sec.

Not only is repetition of liminal current pulses required, therefore, but such repetition must go on for the surprisingly long period of about 0.5 seconds or more in order to become effective in eliciting a conscious sensory experience. The utilization train-duration is relatively independent of the pulse repetition frequency in the stimulus trains, falling in the range of 0.5–1 sec. for frequencies of 15–60 per second. If the train is kept on longer than the utilization train-duration, the experience of conscious sensation lengthens in time correspondingly, but its subjective intensity remains stable at the same minimal level it had initially. The electrical responses to such stimulus pulses—i.e., the so-called direct cortical responses (DCR) recorded at the cortical surface—give no evidence of any progressive build-up in excitatory state during these stimulus trains. The DCR rises in amplitude with the first few pulse repetitions but then remains relatively stable throughout the remainder of the utilization train-duration and beyond (or it may decrease if a surface-negative shift in steady potential develops). In contrast to subliminal stimuli at the skin, cortical stimuli which are inadequate to elicit a conscious sensory experience nevertheless can evoke DCR's of considerable amplitude.

Stimulation at other cerebral levels (subcortical white matter; ventro-postero-lateral nucleus of thalamus) requires similarly long utilization train-durations. Liminal stimuli at the *peripheral sensory* level in the skin, on the other hand, are effective even with very short trains (a few pulses). But this does not mean that the cortical activation elicited by a brief cutaneous stimulus train also has such a short duration. (In this connection there is the insightful comment by Gerard [19,] that some period of reverberation of activity in neural circuits may be necessary not only for fixation of an experience in memory but also for the initial consciousness of an experience.) In fact, the primary evoked potential elicited in sensory cortex by a brief peripheral stimulus is often succeeded by a series of after-waves lasting 0.5 sec. or more, as has been shown by others in awake man, monkey, and cat. These after-waves may represent the neural elaboration of long-lasting activation similar to that found to be necessary with direct cerebral stimulation.

These findings lead me to formulate a general *hypothesis that a minimum period of suitable cortical activation, lasting 0.5–1 sec., is a necessary feature of any such activation* (at least when it is close to liminal level) *for eliciting any conscious experience. The corollary of this would be that shorter periods of such cortical activation may still elicit unconscious experiences.* (It is also possible that cortical activities which, for eliciting conscious experience, are subliminal in amount but not in duration may elicit unconscious experiences. And it seems likely that with stronger levels of cortical activation, well above the liminal one for conscious experience, the minimum duration might be considerably shorter than 0.5 sec.) This hypothesis is obviously not intended as an explanation of all significant differences between all conscious and unconscious experiences, but only of one difference that may be crucial in many circumstances. Let us now examine some of the possible implications of this hypothesis and also attempt to explain with it some of the differences between conscious and unconscious experiences.

The hypothesis proposes a quantitative variation in a physiological parameter to account for what is in introspective experience a qualitative difference. It thus makes it easier to deal with difficulties of borderline transitions between conscious and unconscious experience, and possibly also with gradations within each category. In addition, it offers a mechanism by which the *change-over* between the conscious and unconscious nature of an experience can be controlled by activities of the same or similar cerebral tissues and yet not require that certain localized structures must be the "centers" for specific experiences. This seems more attractive than having the controls separated widely (particularly if cortical versus subcortical). The unconscious processes in intuitive thinking which finally result in a conscious idea, for example, are much easier to begin to understand in the terms proposed.

The hypothesis also provides some understanding of how it is that, in complex, integrative and creative thinking, the play, interaction, and juxtaposition of mental events are often carried through unconsciously [15–17]. The requirement of relatively long duration, of some cortical activation at the near liminal level, for the onset of each increment in conscious experience obviously would impose a certain ponderousness on the thinking process. In contrast, if only short durations of liminal activities are needed in unconscious experiences, they would provide the kind of quick-acting only marginally intense nature that would facilitate complex interactions, rearrangements, and integrations of the type demanded.

If a rather long period of activation, e.g., 0.5–1 sec., is a requirement for conscious experiences at near liminal levels, this would constitute a "latent period" between the onset of activation and the "appearance" of the conscious experience. This would mean that one is not actually aware of a sensory stimulus (at least of a near-liminal one) for a period as long as 0.5 sec. or so after its occurrence. The suggestion that the time of initiating events in the nervous system is not identical with the time of conscious experience has also been made on the basis of other considerations by Lord Brain [20]. Until now it has been assumed that the only lag in any cerebral response to some stimulus would be the relatively short one involved in conduction time and delays in transmission at synaptic junctions in the pathways to the responding cerebral structure. A lag of conscious experiences behind the initiating events, which can be an order of magnitude greater than the delays involved in sensory and motor pathways, introduces a viewpoint about awareness which can have important psychological and philosophical implications. One of these is that unconscious experiences, if elicited by cerebral activations of shorter durations than conscious ones, would have a shorter latency for their development and be able to participate earlier in further cerebral activities.

Now we know that one can react to a sensory stimulus, for example, with reaction times as short as 0.05–0.1 sec. even when decisions are involved in making the responses. This could mean that such quick reactions, made in response to activation near the liminal level for conscious experience, would be made unconsciously. Only subsequent to the performance of such quick reactions would one become conscious or aware of the initiating stimulus as well as of the response; or one might remain unaware of these altogether depending upon the circumstances.

It is difficult to avoid taking a further step from this position and discussing its bearing on the problem of an individual making voluntary or "free" choices. (I do not mean to introduce here the question of the metaphysical nature of free will and the mind-body problem. If one accepts the phenomenon of the mental experience of free-choice-making, regardless of its nature, some aspects of its operational role are affected by the issues raised here.) The "exercise" of free will or voluntary choice is regarded as a function of conscious experience. Choices of alternative actions can undoubtedly be made unconsciously, but they would not be regarded as a demonstration of voluntary-choice-making or free will unless one was aware of what one was doing and could consciously exert some control over it. If, then, we maintain that much if not all of our quick motor reactions to stimuli (and perhaps to "mental patterns") are initially made unconsciously, these would not be tied directly to the operation of free will processes, i.e., they would not be subject to immediate conscious control. Processes involved in the operation of free will could only influence these actions retroactively, by attempting to change or undo the actions already promulgated, if these were in conflict with one's conscious choices. Unconscious controls or "censors" might of course exert more direct influence. The idea that conscious free choice operates better in deliberate consideration of alternatives than in quick actions or responses is, of course, not a new one. The possibility is developed here, however, that this may be related to the postulated difference in the latencies of conscious and unconscious responses and to the limitation this may impose on the potential influence of processes involved in the operation of free will.

The general hypothesis set forth in this paper admittedly makes a large speculative leap from the evidence thus far available. The hypothesis is, however, at least partly testable (we have begun an attempt in this direction), and it seems worthwhile to propose it as a working physiological hypothesis in an important problem area in which physiological hypotheses based upon any evidence are rather meager.

REFERENCES

1. B. Libet, W. W. Alberts, E. W. Wright, Jr., L. D. Delattre, G. Levin, and B. Feinstein. J. Neurophysiol., **27**:546, 1964.
2. B. Libet. *In:* J. C. Eccles (ed.). Brain and conscious experience. Berlin: Springer (in press).
3. J. C. Eccles (ed.). Brain and conscious experience.
4. J. K. Adams. Psychol. Bull., **54**:383, 1957.
5. C. J. Herrick. The evolution of human nature, pp. 288–9, 455. Austin: University of Texas Press, 1956.
6. J. Kamiya. *In:* D. W. Fiske and S. R. Maddi (eds.). The functions of varied experience, pp. 145–174. Homewood, Illinois: Dorsey Press, 1961.
7. R. W. Gerard. Sci. Amer., **189**:118, 1953.
8. R. W. Gerard. *In:* J. Field, H. W. Magoun, and V. E. Hall (eds.). Handbook of physiology: sec. 1, neurophysiology, vol. 3, pp. 1946–7. Washington, D.C.: American Physiological Society, 1960.
9. C. W. Eriksen. Psychol. Rev., **67**:279, 1960.
10. J. A. Swets (ed.). Signal detection and recognition by human observers. New York: Wiley, 1964.
11. J. C. Eccles. The neurophysiological basis of mind, pp. 262–4. London: Oxford University Press, 1953.
12. S. Hook (ed.). Dimensions of mind. New York: New York University Press, 1960.
13. H. Hensel and K. K. A. Boman. J. Neurophysiol., **23**:564, 1960.
14. G. J. W. Smith and M. Henriksson. Acta Psychol. (Amst.), **11**:346, 1955.
15. R. W. Gerard. Sci. Monthly, **62**:477, 1946.
16. A. Koestler. The act of creation. New York: Macmillan, 1964.
17. L. L. Whyte. The unconscious before Freud, p. 21. New York: Basic Books, 1960.
18. W. Penfield and T. Rasmussen. The cerebral cortex of man, pp. 21–44. New York: Macmillan, 1952.
19. R. W. Gerard. Amer. J. Psychiat., **112**:81, 1955.
20. Lord Brain. Brain, **86**:381, 1963.

**Reprinted from:**

# BRAIN *and* CONSCIOUS EXPERIENCE

*Study Week September 28 to October 4, 1964, of the Pontificia Academia Scientiarum*


Edited by

JOHN C. ECCLES

Institute for Biomedical Research, American Medical Association, Education and Research Foundation, Chicago, Ill.
formerly: The John Curtin School of Medical Research, The Australian National University, Canberra, Australia



SPRINGER-VERLAG NEW YORK INC.
175 Fifth Avenue
New York, N. Y. 10010

# 7

# Brain Stimulation and the Threshold of Conscious Experience

*by* B. Libet

*University of California Medical Center, San Francisco, California*

The subject of this paper stems out of a more general question which asks, what is the spatiotemporal configuration of neuronal activity which effectively elicits or is at least uniquely correlated with a conscious awareness of something? Now, to attempt a study of whole configurations of neuronal activities is obviously much too difficult and complicated. One would like to delimit severely the experimental approach to this question, and this we have done in two ways.

The first limitation consists in investigating only the simplest reportable elements of conscious experience, and for that we have chosen a primitive sensation or, as some philosophers (for example, Feigl) have termed it, a "raw feel." Lord Adrian has stated this position much more elegantly than I can. In his lecture in The Physical Basis of Mind series [1952] he said, "I think there is reasonable hope that we may be able to sort out the particular activities which coincide with quite simple mental processes like seeing or hearing. At all events that is the first thing a physiologist must do if he is trying to find out what happens when we think and how the mind is influenced by what goes on in the brain." In our work the simple sensations happen to be somesthetic ones, because of the regions of brain available to us for study.

The second limitation consists in studying the changes at the threshold level of awareness of a sensation. The general problem is then reformulated to ask, what are the critical events, within the whole necessary substratum of the dynamic and structural configuration, which are associated with bringing something into conscious experience? It is implied that there will be some unique general differences between the states that exist just below as opposed to just above this threshold level of awareness.

We must briefly define our usage of "threshold conscious experience of sensation." On the premise that the subjective or introspective feeling of a sensation is the primary criterion of conscious experience, it must be the report of such a feeling by the subject that is operational in such an investigation. Such a report, of course, includes processes of short memory, recall, and expression (verbal in our case) on the part of the subject. There may be a nonreportable type of immediate but ephemeral awareness, and I believe it probably does exist; to detect it for the purpose of study, however, is another matter. In any case, an ephemeral awareness would presumably not be of primary significance to the kind of human conscious experience which has continuity, although conceivably it would play a role in later unconscious processes and thus indirectly in conscious ones.

It should be noted that other criteria of threshold awareness of sensation have been employed by others in some psychophysical studies. These include (*a*) guessing by the subject as to whether he has or has not been stimulated, (*b*) an immediate unthinking motor reaction by the subject if he has any impression of a stimulus having been delivered, and (*c*) reaction times of such motor responses, taken as an index of the degree (or certainty) of such awareness. None of these involves a report of introspective awareness and may not represent the same phenomenon that we chose to study.

Indeed, significant differences in the behavior of the differently defined responses in certain experimental situations have already been observed. Using the report of subjective feeling as the criterion, we found relatively sharp thresholds for the intensity and the duration of the train of stimulus pulses, whether at the cortical somatosensory area or at the skin. There was at times uncertainty about the response in a narrow range of parametric values around the threshold level, but below this range level the subject never reported a sensation and above it he always did. In studies in which the subject was guessing as to whether he has been stimulated [for example, Eijkman and Vendrik, 1963], no definite thresholds were observable; instead the incidence of positive responses followed a probabilistic relationship over the whole range of parametric values. A second difference appears to be evident in the effects of a change in the procedure of alerting the subjects. Responsiveness may be studied in the condition of "EEG-arousal," when the alpha rhythm is "blocked" and replaced by a low voltage irregular activity, by asking the subject to keep his eyes open and by alerting him with a "ready" signal before the delivery of each stimulus. In the alternative procedure, alpha rhythm is present; the subject is asked to relax with the eyes closed, is given no "ready" signal and, in our study, was asked not to think about reporting any sensation that he might experience until requested to do so. These two different states of EEG activity have been thought to be associated with a difference in attentive responsiveness or alertness. Motor reaction times to a fixed stimulus have in fact been found to be considerably lower in the condition of blocked alpha rhythm which followed a ready signal [Lansing, Schwartz, and Lindsley, 1959]. Yet we found no significant differences between the two different alerting procedures, either in threshold intensities or train durations required of a stimulus, when using our criterion of conscious experience. This held true with stimuli at all points tested in the somatosensory pathway—skin, thalamic, and cortical.

It may be suggested, even more generally, that behavioral manifestations of attentiveness are not necessarily a reflection of responsiveness at the conscious, introspective level. A homely example may be cited in further support of this contention. You have no doubt had the experience of behaving as if you were listening to a conversation, even nodding correctly at appropriate cues in the conversation, and yet not consciously hearing a word of what is going on, as you are thinking of other things or of nothing at all. One should then be aware of a possible fundamental error in experimental approaches which assume that conscious experience of a reportable, continuous nature accompanies a particular motor response, even when the motor act is complicated and involves the making of decisions by the agent.

From this discussion of what it is we are attempting to study, we may proceed to our experimental questions. What is the unique nature of the input at cerebral levels which just reaches the threshold for eliciting a conscious sensory experience, as opposed to a subthreshold one which just fails to do so? Secondly, are there unique differences in the cerebral responses to these two classes of inputs? (We are only in the beginning stages of this second part of the study.) Peripheral sensory inputs do not appear to be very satisfactory for this experimental purpose, as the stimulus thresholds and the resulting discharges of sensory impulses are, in attentive subjects, generally rather close to the minimum physical or physiological quantities possible. Perhaps the most clear-cut experiment with somatic sensation in this regard has been carried out by Hensel and Boman [1960]. They exposed a small skin nerve in a human subject, and severed all the nerve fibers in it but one, whose electrical responses to mechanical stimulation of the skin innervated by it could be followed. The weakest stimulus that could be detected subjectively as a sensory experience (of touch) gave rise to one single conducted impulse in the one remaining nerve fiber. Such sensitivities mean that one can go from no input to some input at the threshold level, and thus the corresponding changes in cerebral activities may involve more than those which are unique to conscious experience.

Threshold stimuli applied to the somatosensory cortex are not, as we shall see further, of such a minimal character as are peripheral ones. There is the additional advantage that with stimulation at cerebral levels one has complete control of the immediate input here, instead of having to discover what complex alterations have been imposed on the nature of the original peripheral input by subcortical mechanisms. Much work has already shown that one can elicit simple sensations, usually paresthesia-like, by electrical stimulation of the postcentral gyrus in awake human subjects; Professor Penfield and his group [for example, Penfield and Rasmussen, 1950] have contributed greatly to this knowledge and have carried out elaborate topographical mapping of the representation of body sensibilities on the human cortex by this method. For our problem we have generally selected one responsive point in the postcentral gyrus and, with an electrode fixed at this position, have studied the nature of the adequate stimulus there for eliciting a conscious sensory experience. Such work is made possible by the cooperation of awake patients who are undergoing therapeutic surgical treatments of the brain which require that they be essentially non-medicated except for the local anesthetics injected into the scalp.

A number of parameters of the electrical stimuli applied to somatosensory cortex have potential physiological significance,[1] but I shall discuss chiefly those aspects which have turned out to be of special interest to the problem of conscious experience. The area of postcentral gyrus usually available to us produced sensation in the contralateral fingers or hand when suitably stimulated with rectangular current pulses, delivered generally through a unipolar surface electrode.

It is important to realize at the outset that the general nature of the threshold response to cortical stimuli can differ, depending on the combination of values for the different stimulus parameters, even though the electrode position is constant. The threshold cortical stimuli can thus be grouped into three parametric regions (Table 1). Those in parametric region A are characterized by the lowest or liminal threshold intensities, by relatively long durations of the train of repetitive pulses (0.5 sec or more), and by pulse repetition frequencies greater than about 10 per sec. These threshold stimuli elicit reports of the introspective experience of purely somatic sensations. The question of the control of the quality of the sensations by the input cannot be considered at length here; stimuli

[1] A fuller treatment of these may be found in the paper by Libet, Alberts, Wright, Delattre, Levin, and Feinstein [1964].

**Table 1**

Summary of Parametric Regions for Threshold Stimuli

| | Region A | Region B | Region C |
|---|---|---|---|
| Intensity (peak current) | Liminal | 2 or more times that in A | Intermediate range, between A and B |
| Train duration (T.D.) | >0.5–1 sec | <0.1–0.3 sec Single pulses | 0.5–5 sec, at 8 pps; 0.1–0.5 sec at pulse frequencies ≧15 pps |
| Pulse frequency | ≧15 pps (can get with 8 pps if T.D. >5–10 sec) | Any | Probably any, but best seen in low range (<15 pps) |
| Facilitation still evident after 30 sec or more | Usually none | Usually considerable | Usually present |
| Response at threshold | Purely somatosensory (no motor) | Observable muscular contraction | No observable contraction; sensation changes from that in A to one with aspect of motion (at <15 pps often has slow pulsatile quality) |

in this parametric region may produce virtually all types, both paresthesia-like and more normally specific types of qualities. On the other hand, with stimuli in parametric region B, having threshold intensities two or more times the liminal ones in region A and very short train durations (<0.1–0.2 sec), the response at threshold levels was a localized muscular contraction. The sensation of a movement was presumably due to returning sensory information of it from the periphery, and no purely somatic sensations independent of this were reported. Single pulses all belong in region B. Those threshold stimuli which fell between the characteristics of regions A and B seemed to comprise an intermediate class, region C. The latter produced no visible motor response but often did elicit a sensation with an aspect of motion in it. The difference from the responses in region A could be seen best when changing from stimuli in parametric region A to those in C. At pulse frequencies of 8 per sec or less, the sensation usually had a pulsatile character, with pulsations being felt at a frequency (about 1 per sec) which was independent of the pulse repetition frequency of the stimulus.

Another differential feature of the parametric regions which deserves consideration relates to the long-persisting effects following a stimulus. Threshold stimuli in parametric region A were not, in general, followed by any long-lasting facilitatory effects; they could be delivered at intervals of 30 sec with no effect on the thresholds of succeeding stimuli. In regions B and C, however, in which threshold responses are, respectively, frankly motor or have motion aspects to the sensations, a long-lasting facilitatory effect on the thresholds of succeeding stimuli was commonly seen, for as long as several minutes after a stimulus. This may be a function of the higher intensity levels characteristic of such stimuli.

Following stimuli in parametric region A, on the other hand, an interesting long-lasting effect on the quality of sensations in succeeding responses could be demonstrated. For example, a threshold stimulus with a pulse frequency of 60 per sec might elicit a tingling sensation. If this was followed after a 30-sec interval by a threshold stimulus now employing an 8 per sec pulse frequency, the subject tended to report the same type of tingling sensation. If one waited for about 5 min, however, the 8 per sec stimulus elicited a pulsatory sensation, as described earlier, instead of the tingling one. We have no very good explanation to suggest for this enduring effect on sensory quality, recalling that it is not accompanied by a similarly long-lasting effect on threshold requirements. It is possible that a persistence of facilitation occurs in those parts of the responding neural apparatus other than the ones that set the threshold requirements of the surface stimulus. What significance this phenomenon may have for the problems of long-persisting alterations of cerebral function such as memory remains to be seen.

We will not go into the question here of motor versus sensory specificity of responses to stimulation of the pre- and postcentral gyri in man, except to emphasize that at liminal intensity levels the responses were always purely sensory in postcentral gyrus and apparently purely motor in the precentral gyrus.

With respect to the question of which neural components in somatosensory cortex are initially activated at the threshold levels for conscious sensation, some general suggestions come out of this study of stimulus requirements. The analysis of the evidence on depth of responding elements in the motor cortex, as carried out by Hern, Landgren, and Phillips [1962], and Phillips and Porter [1962], encourages one to use a similar approach in sensory cortex. The critical neural components for threshold sensation (*a*) appear to lie in the more superficial layers of the postcentral cortex, and (*b*) are probably orientated vertically (that is, perpendicular to the surface) rather than horizontally. The evidence for (*a*) is, briefly, that unipolar cathodal pulses were found to require only about 70 per cent of the peak current required by anodal ones, in order to achieve threshold levels in each case. Suggestion (*a*) does not exclude the possibility that deeper neural components are also activated directly or transsynaptically by threshold stimuli, but it would indicate that it is the excitation of more superficial components that sets the threshold requirement. If suggestion (*b*) were wrong, that is, if neural components located in superficial layers were oriented horizontally, a bipolar stimulus should prove more effective than a unipolar one. Since, in fact, a unipolar cathodal stimulus was the somewhat more effective one, this supports the suggestion of a vertical orientation for such components. The even greater effectiveness, in terms of current density, of a large (10 mm diameter) unipolar electrode also supports this view.

The unipolar electrode with this large area could achieve threshold (using a 60 per sec pulse frequency) with a current density of less than 0.1 mA per $mm^2$, as compared with about 1 mA per $mm^2$ for the usual 1 mm electrode. It is of considerable interest that the threshold sensation with the large electrode was referred roughly to the same size area of the body and had the same quality associated with it as did the one elicited by the smaller electrode. This would indicate that spatial facilitation originating from a relatively large cortical mass can be an important factor in the effective excitation of a small group of neural elements that mediate conscious sensation.

I want to return now to a further consideration of what are probably the most interesting findings in these experiments with cerebral stimulation. These are the observations that, in order to elicit a somatic sensation not only is repetition of liminal stimulus pulses required but also such repetition must go on for about a half a second or so to become effective (parametric region A, above). This is in sharp contrast to peripheral input requirements, in which virtually a single liminal stimulus pulse (or indeed a single nerve impulse) can produce a conscious sensation. Figure 7.1 shows a graph of the minimum intensities required of the cortical stimuli for threshold responses when plotted against the duration of the repetitive train of pulses required at each such intensity. The curves are somewhat idealized, to represent our general experiences with this relationship, but are based upon the actual data found for several subjects. These curves resemble the intensity-duration curves for threshold single pulses that excite nerve fibers, but in the present case we are dealing with train duration, not pulse duration. Lengthening the train duration beyond a certain minimum value produces virtually no change in the minimum intensity required. This level, which we refer to as the liminal intensity, is lower for higher pulse repetition frequencies up to about 100–200 per sec, but it rises again when the pulse frequency goes above this. The mimimum train duration that is required when using a liminal intensity, that is, the point where further shortening of train duration be-

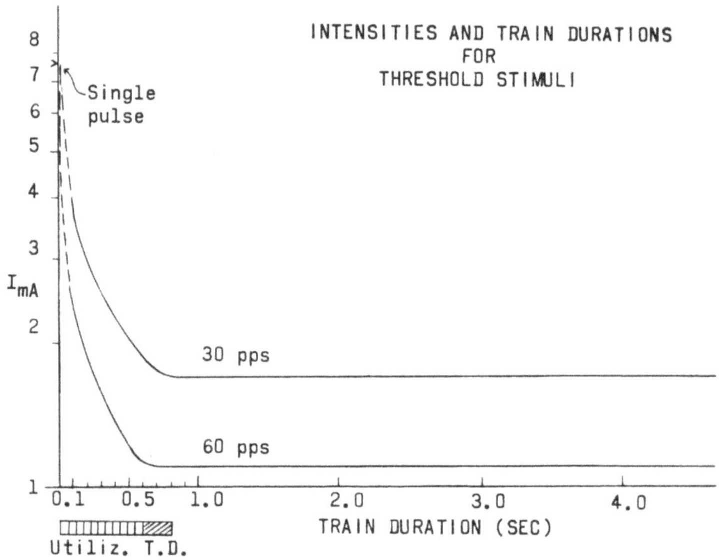


**Fig. 7.1.** Intensity-train duration combinations for stimuli (to postcentral gyrus) just adequate to elicit a threshold conscious experience of somatic sensation. Curves are presented for two different pulse repetition frequencies, employing rectangular pulses of 0.5 msec duration. "Utilization train duration" is discussed in the text.

gins to require distinctly greater intensities of stimulus, we refer to as the "utilization train duration." This is by analogy with the utilization time of the intensity-duration curve for axons. Most utilization train durations fall into the range of 0.5–1 sec for pulse frequencies from 15–120 per sec, although there have been occasional values as short as 0.2 sec and some as long as 1.5–2 sec. While the mean value of utilization train durations was somewhat lower when higher pulse frequencies were used, the data cannot be assessed for statistical significance of such differences; in any case, there have been subjects in whom there was no reduction of utilization train duration at the greater pulse frequencies (in the 15–60 per sec range).

It should also be noted that we found similar values for utilization train duration with liminal stimulation of the immediate afferent supply to the somatosensory cortex, that is, of its subcortical white matter or of the ventroposterolateral (VPL) nucleus of the thalamus. Such a requirement is not, therefore, something unique to the more abnormal mode of activation of the cortex by direct electrical stimulation.

This relationship of intensity to train duration lends a kind of all-or-nothing aspect to threshold awareness of a sensation. Below the liminal intensity level there is no reportable awareness of a sensation, in contrast to the findings using other criteria of sensory awareness. In addition, for any liminal stimulus with a train shorter than the utilization train duration, the subject reports that he does not feel anything.

One hypothesis to explain the utilization train duration would be that it represents the time required for sufficient summation of excitatory effects up to some critical level. It would also be inferred that the greater liminal intensity required with a lower pulse frequency approximately compensates for the longer pulse interval, in the development of excitatory level during the utilization period. This last could account for the absence of any large effects of change in pulse frequency on the utilization train duration. Another more intriguing hypothesis would be that it is some minimum period of time, *per se,* during which activation of certain neural components must proceed, which is essential for the elaboration of a reportable conscious sensation. This would not require a progressive buildup of excitatory level during the utilization period; rather, the suggestion is that a minimum "activation period" would be necessary even with a constant average level of excitatory actions during the time involved. The evidence we have available thus far appears to favor this second hypothesis, although more definitive types of experiments must still be done in order to settle the issue.

The evidence may be discussed by reference to the diagram in Fig. 7.2. The top line represents the 0.5 msec rectangular pulses at 20 per sec making up a stimulus train at liminal intensity for eliciting conscious sensation. The thresholds for eliciting the direct cortical electrical response (DCR), both its initial negative and the subsequent positive components as recorded from the adjacent cortical surface, were found to be well below those required for eliciting conscious sensory experience. Thus, as seen in line 2, Fig. 7.2, a usually submaximal DCR is already present with the first pulse. This gains in size with the first two or three repetitions of the stimulus pulse. The interesting point with respect to the hypothesis under consideration is that the DCR's remain fairly constant throughout the whole stimulus train, up to and beyond the utilization train duration (except for the rapid initial augmentation, and the decrease in amplitude later in the train in those cases in which a surface-negative shift in steady potential develops). At least for this type of response, then, there is no evidence of a progressive rise in excitatory level throughout the whole period of the utilization train duration.

The third line in Fig. 7.2 indicates that no sensation is elicited until the train has gone on for about 0.5 sec or more. If the train continues beyond the utilization train duration, however, the subjective intensity of the sensation does not increase progressively (except for some uncertainty in a relatively small range of time around the utilization point); the same threshold sensation merely has a longer duration with the longer train.

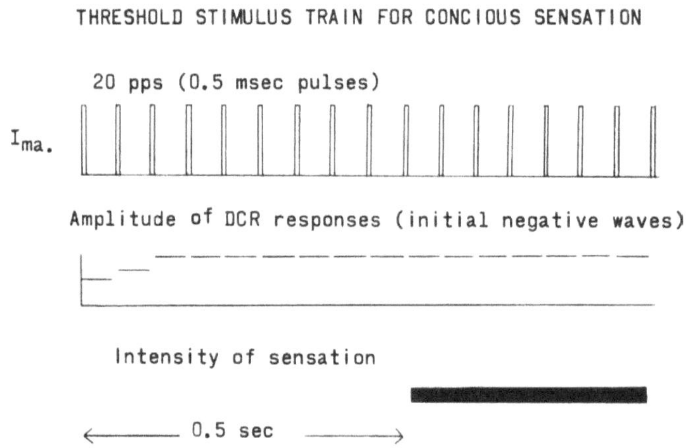


**Fig. 7.2.** Diagram of relationships between the train of 0.5 msec pulses at liminal intensity applied to postcentral gyrus, and the amplitudes of the direct cortical responses (DCR) recorded nearby. The third line indicates that no conscious sensory experience is elicited until approximately the initial 0.5 sec of events has elapsed, and that the just detectable sensation appearing after that period remains at the same subjective intensity while the stimulus train continues.

This is remarkable by contrast with the course of the motor response obtained when a similar type of stimulus is applied to the *precentral* gyrus, in the unanesthetized subject. A progressive increase in the motor response during a train of repetitive pulses is often observed even at strengths considerably below that required for a response to a single pulse; once the contraction begins it tends to become progressively more powerful as the train of similar pulses continues. Thus, excitatory processes in the somatosensory cortex, insofar as they are relevant to conscious sensory experience, appear to be maintained at a relatively constant level when the liminal stimulus train continues on beyond the utilization train duration. Such stability in the period following the utilization point does not tell us about the course of development of excitatory processes in the period before this point. It is at least compatible, however, with the suggestion that the relevant excitatory processes have ceased undergoing any further rise in level sometime before the end of the utilization train duration.

Some supporting evidence for the second hypothesis, which suggested a requirement of a minimum "activation period," comes from psychophysical experiments of the type initiated by Crawford [1947] and pursued by others. In these the visual threshold to a brief test flash of light was studied in relation to the timing of a second superthreshold flash [see Wagman and Battersby, 1964]. When the second ("conditioning") flash *followed* the first by as much as 200 msec, the first flash became invisible to the subject unless its intensity was raised considerably. Since the latency of the first evoked potential in the visual cortex is only about 35 msec, it may be presumed that the interaction is a central one. Such psychophysical findings thus indicate that awareness of a near-threshold visual stimulus does not occur for about 200 msec after the initial delivery of impulses to the cortex, and that such awareness can be prevented if there is sufficient interference at any time with the central processes that are going on during this relatively long period following the stimulus [see also Crawford, 1947].

Although the hypothesis of a minimum "activation period" itself has only a tenuous status at present, I will conclude by indulging in some interesting speculative inferences which can be developed from it. The first arises in response to the question of how one may explain the ability of a single peripheral nerve volley to elicit a conscious sensory experience? The answer could lie in the nature of the after-waves which follow the initial evoked potential. Such after-waves have been recorded from the sensory cortex of unanesthetized animals and human subjects by a number of investigators. The after-waves go on for some hundreds of milliseconds, are especially sensitive to conditions which modify the state of consciousness (anesthesia, sleep) and have been reported to change with shifts in attention and learning. They could represent or indicate activation processes which go on normally, in response to peripheral sensory input, during the requisite "activation period." We have not (even with averaging techniques) been able to detect any such after-waves in the DCR which is elicited in the human somatosensory cortex, in response to a stimulus at the liminal strength for eliciting a conscious sensory experience. One may suppose, therefore, that the equivalent of such after-waves must be supplied by the cortical stimulus itself, in the form of repetitive pulses during the activation period.

Secondly, one may speculate that an activation period could be involved in the elaboration or fixation of a memory trace, as we are dealing with a reportable conscious experience. Such a requirement is implicit in the more specific suggestion already proposed by Gerard [1955], that reverberation of activity in neural networks may be involved not only in the fixation of ordinary learned responses, but also in the development of the initial consciousness of an experience. The actual nature of the specific activity that may be critical in this latter connection is, of course, still to be developed, even if the general concept of an activation period is substantiated. It even seems worthwhile to consider the possibility that processes involved in producing memory of an awareness may have some unique differences from those involved in the retention of some types of learned behavioral responses.

Finally, the requirement of a utilization time of 0.5–1 sec indicates that awareness of a sensory experience does not take place immediately after the initial arrival of impulses at the cortex from the periphery. It should be recalled that this requirement was found even for stimuli applied to thalamic VPL nucleus at the liminal intensity for eliciting sensation, that is, when the input to the cortex is delivered via the specific projection afferent fibers. Such a requirement implies then that there is a latency of about 0.5–1 sec before the conscious experience of the sensation begins, at least with threshold levels of input. This inference would seem to hold regardless of what mechanism is postulated for the utilization train duration. Now, as we all know, one can react to a stimulus with a motor response much more quickly than in 0.5–1 sec, even when decisions are involved in making the response. Reaction times as low as 0.05 sec are demonstrable. This then can lead to the further speculative inference that the quick motor reactions which one makes in response to liminal stimuli in everyday life situations may not be mediated by processes at conscious levels, and that they may be confined to unconscious levels. Perhaps only subsequent to the performance of such a quick motor reaction does one become consciously aware of the sensation and response, at least in the sense of awareness having continuity and reportability; or one might remain unaware of it altogether depending upon the circumstances.

## ACKNOWLEDGMENTS

The experimental work reported here was carried out in collaboration with a group that included Dr. W. W. Alberts, biophysicist; E. W. Wright, Jr., electronics associate; L. D. Delattre, research assistant in psychology; and the neurosurgeons, Drs. Bertram Feinstein and Grant Levin. The work was done in the Mt. Zion Neurological Institute in San Francisco, and was partially supported by U. S. Public Health Service Grant NB 05061 from the National Institute of Neurological Diseases and Blindness.

## DISCUSSION

*Chairman:* PROFESSOR BREMER

PENFIELD: I have nothing to add to Dr. Libet's very careful, thoughtful analysis.

Our own stimulation of the human brain has been practical; it has always had an ulterior, a therapeutic, purpose of course. Experiment is never a primary purpose. We began in 1928 using the stimulation equipment that

Foerster was using in Breslau. We used a galvanic current for localization and added a faradic current with the greatest caution for fear that the faradic current would produce a major epileptic seizure.

After Dr. Herbert Jasper came to Montreal the apparatus was refined with his guidance. We now use a square wave generator and Dr. Theodore Rasmussen in our clinic has elaborated the parameters of stimulation much more completely than I have but without changing our general conclusions. Our purpose, from the outset, was to try to produce the very beginning of the epileptic seizure that we hoped to cure, without precipitating a major attack. We also used the electrode to establish the functional topography. Thus, as time passed, we stimulated all parts of the human brain. Scars or other lesions may well produce seizures anywhere in the cerebral cortex. In the early days, before we had electroencephalography we had to depend always on the patient to tell us that he felt as though an attack were coming and then we had to use the greatest caution to prevent producing a major convulsive seizure.

As Dr. Libet suggests, we have been impressed by the fact that the findings of Sherrington in cats and then in chimpanzees are borne out during exploration of the human cortex. Facilitation follows stimulation at a cortical point and that applies both to sensory and to motor and to psychical responding areas. There is evidence of fatigue, or inhibition, if one stimulates again too quickly. It is possible also to demonstrate displacement of response, as Sherrington first described it.

LIBET: In reply, I may say that your studies were of course invaluable to us as background in helping us to get into this area of study. Our work is just a beginning into these more quantitative analytical aspects, utilizing one locus. We did not study the sorts of things that you had. For example, we did not study displacement of sensation since we stayed in one place and simply studied the parameters at that point. I might add, however, that the area of projected sensation was not always constant, even though we stayed at one point; that is, even with a threshold stimulus one stimulus might produce a sensation in the tip of an index finger and the very next one with exactly the same parameters, 30 sec or more later, might produce a sensation in a somewhat adjacent area, not precisely the same one.

We did not find facilitation with sensory cortex stimuli lasting beyond the 30-sec interval, although we did find it if we used many stimuli at intensities somewhat above threshold level. So that there may be a difference there; if one stays at the liminal level using train durations of 5 sec or less, then I think you can get by with 30-sec intervals without producing appreciable facilitation on the next threshold point.

PENFIELD: In psychical responses, for example, the elicitation of previous experience, we found a little different kind of facilitation was present. If we produced an experience of one type, or an experience which came from a certain time in the individual's life, one was apt (from the region around within perhaps 3, 4, or 5 cm) to get another experience of the same type and from the same period. There was a certain tendency for the first experience thus to set the type of the later elicited experiences. That is a little different kind of facilitation. When, on the other hand, one jumped quite a long way across the cortex and set the electrode down on interpretive cortex at a distance, the experience summoned came from a different period of life or had different characteristics altogether.

LIBET: This is somewhat analogous to our requirement of a 5-min or so waiting period in order for us to change the modality of the sensation when stimulating at the same point on the sensory cortex; there is some persistence of a subjective quality once you elicit it at a certain point.

PENFIELD: If we went back to the same point with the lapse of a few seconds or a minute or so, we usually had the same experiential response. If, on the other hand, we waited a little longer time and then went back to the same point, another experience would appear.

GRANIT: I wanted to comment on the single fiber elementary sensation experiment. We have had so much information lately on small fibers that we should be very cautious in accepting evidence of that sort, because the small fibers may always contribute to these sensations. I was very impressed by the long time for utilization, but, of course, one can think of so many explanations for it that it is very difficult to put forward anything definitive. The fact in itself is most interesting, namely, that simple sensory developments take such a long time compared with what we know that the brain can do with fast reactions. So I want to compliment you on that finding, which seems to me of great importance.

LIBET: Thank you, Professor Granit. I agree that our experiments have not differentiated conclusively between the various alternative explanations of this long utilization train duration, although the fact is interesting. But we do intend to try to get further evidence on which of the possible hypotheses may be more probable.

MOUNTCASTLE: Dr. Libet, it is the duration in the hypothesis that you discussed at the end of your paper that disturbs me, because I believe that the human experiments which you refer to involve a decision and a discrimination of stimulus that can be made by humans with an intracortical time of 60 msec, as Grey Walter has recently told us. These are conscious reactions. And this does not fit very well with the idea that you need a time of a half second for perception. The difference, of course, may be that the cortex has a very difficult job in weeding out the conscious perception from the abnormal train of events set in motion by the electrical stimulus. I wonder whether one should really translate through to the process of perception what is going on when you pass the current through the cortex, because as Grey Walter showed especially in the learning situation, the human cortex can make a discrimination in a very few milliseconds.

LIBET: I have not seen the reports of the particular experiments you are talking about. One has to watch out for the distinction between making a decision response and then being consciously aware of it. In experiments of the kind you mention that I have looked at in this regard, they have not always been careful to make this distinction.* I am confident that one can

* (Note added later): The experiments reported by W. Grey Walter, referred to by Dr. Mountcastle, were evidently those described in a talk published in the *Ann. N. Y. Acad. Sci., 112*:1, 320–361 [1964]. Subjects could learn to turn off an

make a decision and act very rapidly, but this does not necessarily mean that awareness is involved immediately.

PHILLIPS: Could I ask you to say a little more about the qualities of the sensations that were evoked? I am interested in this partly from the point of view of the last question because, if the sensation is tingling or something abnormal, it may not be recognized as something normally significant in the environment. And the other question in which we are all interested is, how far does electrical stimulation evoke normal function in the cortex? If you ever get normal sensations, then this of course is a very important observation. I think you once told me that generally the sensations are abnormal: pins and needles or tingling or electricity or something of that sort. But if you ever get normal sensations, it is very important to know about those.

LIBET: Yes, that is certainly an important general question. One answer, to the suggestion that some of our findings may be a function of the very abnormal nature of electrical stimulation of the cortex, is that we have substantiated the long utilization train duration with stimuli other than those at the surface of the cortex. We have stimulated in some cases the VPL nucleus, the specific projection nucleus in the thalamus, and got again there at liminal intensity levels the same kind of requirement of a half a second or longer train duration; this, of course, involves stimulating the afferent pathways to the cortex. Stimulating, in one case at least, the subcortical white matter just below the somatosensory cortex gave the same kind of long utilization trains.

Now, as to the qualities of the sensations elicited by cortical stimuli: Although in the majority of cases the response was an abnormal or parasthesia kind of sensation, I would say that in about one third of the individuals a specific kind of quality was reported which was more akin to a natural one.

The most common quality of a more natural kind—this is at the threshold level of stimulation—has been one of a deep pressure sensation, a kind of wave moving about under the skin in the area that is involved. The less common ones have been those of a specific light touch, some reports of what appears to be pure motion without any movement detectable, and also reports of pure heat and pure cold with nothing else involved. So that at the threshold levels, specific qualities certainly do appear, and I am sure that Professor Penfield and others have also obtained reports of specific modalities. The reason for the specific sensations not appearing consistently is another matter, and this gets us into the general question of what is the nature of the activity that has to go on in the somatosensory cortex for the production of a specific kind of quality. I do not think we have time to go into that now, but we do have some evidence along this line to indicate what the reasons may be for our failure to elicit specific qualities consistently.

JASPER: I want to draw Dr. Libet out a little bit on some of these other findings that he did not have time to report. I first thought that perhaps your results might be explained in terms of the threshold required for conduction

unconditional stimulus with an apparent reaction time of 50 msec after its onset, when this stimulus followed a conditional one by a fixed interval (of 1 sec). The question of subjective experience of awareness of the unconditional stimulus, at the moment of response, was apparently not a specific consideration.

out of the cortex. This is known to be much higher when you study the pyramidal tract responses to direct cortical stimulation, where a much higher stimulus is required to get the impulse to conduct out than to excite the cortex superficially; I think Lord Adrian first showed this with the direct cortical response. And do you not find the same thing, that it requires many times the threshold of a direct cortical response to elicit sensation?

LIBET: Not usually. We did have to get above threshold for the direct cortical response in all cases in order to elicit a sensation. But the threshold for sensation could be as little as 10 per cent or so greater than that for the DCR, although it was relatively much greater in some instances. The other evidence that we have on this point is the fact that a distinctly weaker cathodal stimulus is required than an anodal one. This indicates that the response is generated at the superficial end of some neural elements.

JASPER: It does not require conduction out, you think?

LIBET: I do not think so. That is, there may eventually be conduction out, but I don't think we are directly stimulating efferent elements. The available evidence indicates that, at the threshold for sensation, we are stimulating something before that point.

PENFIELD: In regard to the response of the sense of movement. I am interested that it appeared so rarely in your liminal stimulation; I think we got it a little more frequently, probably because we were using just routinely one volt and perhaps were slightly above liminal strength. From the second sensory area of the cortex, which in man is the downward projection or continuation of the precentral gyrus into the fissure of Sylvius, we more frequently produced a sense of movement and often it was a more elaborate movement; just as from the supplementary motor area we sometimes got a sense of larger movements which seemed to include muscles and parts of both sides of the body.

CHAIRMAN: Before closing this interesting discussion, I want to ask Dr. Libet a question which may be answered "yes" or "no." Did you in your work combine a sensory stimulus with electrical stimulus in conditioning experiments, the sensory being the unconditioned stimulus? Such experiments might be interesting from the point of view of the organization of sensory modalities in the somatosensory cortex.

LIBET: I agree with you that this might be a very useful approach, and in fact we have intended to do experiments along such lines, but we have not done them yet.

## REFERENCES

ADRIAN, E. D. What happens when we think, in: *The Physical Basis of Mind* (ed. by P. Laslett). Oxford: Blackwell [1952].

CRAWFORD, B. H. Visual adaptation in relation to brief conditioning stimuli, *Proc. Roy. Soc. B* (London), *134*, 283–302 [1947].

EIJKMAN, E., and A. J. H. VENDRIK. Detection theory applied to the absolute sensitivity of sensory systems, *Biophys. J.*, *3*, 65–78 [1963].

GERARD, R. W. The biological roots of psychiatry, *Amer. J. Psychiat.*, *112*, 81–90 [1955].

HENSEL, H., and K. K. A. BOMAN. Afferent impulses in cutaneous sensory nerves in human subjects, *J. Neurophysiol.*, *23*, 564–578 [1960].

HERN, J. E. C., S. LANDGREN, and C. S. PHILLIPS. Selective excitation of corticofugal neurons by surface-anodal stimulation of the baboon's motor cortex, *J. Physiol.*, *161*, 73–90 [1962].

LANSING, A. W., E. SCHWARTZ, and D. B. LINDSLEY. Reaction time and EEG activation under alerted and non-alerted conditions, *J. Exp. Psychol.*, *58*, 1–6 [1959].

LIBET, B., W. W. ALBERTS, E. W. WRIGHT, Jr., L. D. DELATTRE, G. LEVIN, and B. FEINSTEIN. Production of threshold levels of conscious sensation by electrical stimulation of human somatosensory cortex, *J. Neurophysiol.*, *27*, 546–578 [1964].

PENFIELD, W., and T. RASMUSSEN. *The Cerebral Cortex of Man. A Clinical Study of Localization of Function.* New York: Macmillan [1950].

PHILLIPS, C. S., and R. PORTER. Unifocal and bifocal stimulation of the motor cortex, *J. Physiol.*, *162*, 532–538 [1962].

WAGMAN, I. H., and W. S. BATTERSBY. Neural limitations of visual excitability. V. Cerebral after-activity evoked by photic stimulation, *Vision Res.*, *4*, 193–208 [1964].

# Responses of Human Somatosensory Cortex to Stimuli below Threshold for Conscious Sensation

B. Libet

W. W. Alberts

E. W. Wright, Jr.

B. Feinstein

**Reprinted from Science, December 22, 1967, Vol. 158, No. 3808, pages 1597-1600**

## Responses of Human Somatosensory Cortex to Stimuli below Threshold for Conscious Sensation

Abstract. *Averaged evoked responses of somatosensory cortex, recorded subdurally, appeared with stimuli (skin, ventral posterolateral nucleus, cortex) which were subthreshold for sensation. Such responses were deficient in late components. Subthreshold stimuli could elicit sensation with suitable repetition. The primary evoked response was not sufficient for sensation. These facts bear on the problems of neurophysiological correlates of conscious and unconscious experience, and of "subliminal perception."*

Previous studies have indicated that the first appearance of any evoked potential in sensory cortex, elicited by a stimulus to skin or sensory nerve, coincides with the threshold for some report of subjective sensation by the human subject (*1*). A similar relationship was reported for the threshold of sensory discrimination in the cat, upon stimulation of a cutaneous nerve (*2*). Such conclusions were based upon recordings made with electrodes on the scalp or situated epidurally. This provides a relatively diffuse lead from unresponsive as well as responsive cortex (see, for example, *3*). It has been demonstrated in monkeys that localized responses recorded with cortical surface electrodes may be essentially indetectable with scalp electrodes (*4*); we have found this to be true in man (see also *3*). In the present work, the recording electrode is placed subdurally, directly on the pia-arachnoid surface of somatosensory cortex (postcentral gyrus). In addition, the stimulus to the skin or to ventral posterolateral (VPL) nucleus of thalamus is so located as to elicit a sensation within the same somatic area as that in which the sensation was subjectively felt when the recording site on somatosensory cortex was stimulated directly. With such relatively precise localization it has become quite evident that at least some components of the evoked potential are recordable in somatosensory cortex with stimulus levels which are distinctly below those required to elicit any conscious sensory experience. This was true whether stimuli were applied to the skin, the specific projection relay nucleus in the thalamus (VPL nucleus), or directly to somatosensory cortex.

Subjects were patients undergoing stereotaxic neurosurgical therapy for motor dyskinesias or intractable pain, who volunteered some study time during the operative stage in which they had to remain unanesthetized for purposes of therapy (*5*). Local anesthetic was injected into the scalp but generally no premedications were administered. Conditions of such experiments and criteria for conscious sensory experience have been described earlier (*6, 7*). The subject was alerted to attend to the stimulus and was asked to report (i) whether he subjectively experienced or "felt" a sensation even if it was very weak, (ii) whether he felt none at all, or (iii) if he was uncertain about having felt a sensation. With single-pulse stimuli the range of stimulus intensity of which the subject was uncertain was usually small (less than 5 percent for skin); false positive responses almost never occurred. (Throughout this report, the terms *threshold* or *subthreshold* refer exclusively to the ability or inability of stimuli to elicit conscious sensory experience, rather than to their ability to elicit electrophysiological responses. In order to keep this distinction clear to the reader, the term *threshold-c* will be used to describe threshold stimuli that can elicit conscious experience. We are not implying that a subthreshold stimulus as determined under the present conditions would necessarily remain subthreshold under all conditions of testing, for example, with extensive training of the subject.) All recordings were unipolar (*8*), with the three skull contacts of the stereotaxic frame generally serving as the reference or indifferent lead (except for d-c recordings). The stimuli were constant-current electrical pulses applied to the skin through a 5-mm disc, or to VPL by means of a coaxial needle electrode. To elicit direct cortical responses, stimuli were applied within 1 mm of the recording site by a twisted pair of metal wires.

1

Averaged evoked potentials elicited in somatosensory cortex by various strengths of skin stimuli are shown for two representative subjects in Fig. 1 (*9*). Threshold-c level (*T* in Fig. 1) refers to bare threshold intensity at which not all the stimuli in the series elicited a subjective sensory experience. All the components which are visible at much higher stimulus intensities (for example, at twice threshold-c) may already be present at threshold-c, though with smaller amplitudes. Intensity of the subthreshold-c stimulus (designated *SubT* in Fig. 1) was such that none of the stimuli gave rise to conscious sensation. Nevertheless, a distinct evoked potential was visible with such subthreshold-c stimuli. It should be noted that when a scalp lead was placed so as to lie over the postcentral gyrus, and was recording simultaneously with the subdural lead, it did not exhibit an evoked potential with subthreshold-c stimuli or generally even with threshold-c stimuli. This latter negative result was apparently a function of the small area of skin stimulated; when the median nerve was stimulated at the wrist, appearance of an evoked potential coincided roughly with threshold-c.

Although skin stimuli which were completely subthreshold-c could elicit distinct evoked potentials, recorded subdurally, the responses showed some differences from those obtained with threshold-c or stronger stimuli. The initial positive component and even more so the succeeding negative one were smaller than at threshold-c level, while the still-later positive and negative waves (which are smaller and longer lasting) were not detectable at all.

It might be argued that subthreshold-c skin stimuli which elicit evoked potentials are exciting a class of afferent fibers which can never elicit conscious sensory experience; indeed Swett and Bourassa (*2*) have found that cats were unable to discriminate afferent volleys in larger nerve fibers of deep nerves, even though certain of these (in the deep radial nerve) could evoke a large cortical response. In our studies, however, subthreshold-c stimulus pulses which produce an evoked potential could elicit a conscious sensory experience, that is, they could become threshold-c if delivered repetitively at higher frequencies (20 to 60 pulses per second) for a brief period. In fact, the minimum intensity of single-pulse stimuli which could evoke some detectable cortical potential appeared to coincide roughly with the minimum threshold-c current required by a train of such pulses to elicit a sensation. The minimum threshold-c current was achieved by trains with durations longer than a certain minimum, 0.05 to 0.1 second for stimulus pulses to skin. This minimum threshold-c current for such trains is referred to as the *liminal I* (see *6*, *7*). Liminal I was generally about 15 to 20 percent below the intensity required for threshold single pulses at the skin; it was also usually below the first completely subthreshold-c level for single pulses, as in subjects A and B in Fig. 1. The same afferent nerve fibers which were excited by subthreshold-c single pulses undoubtedly included those responsible for eliciting some conscious sensory experience when the subthreshold-c pulses were delivered at a suitable repetition rate; it is highly unlikely that afferent nerve fibers which are only excited by single pulses at threshold-c strength are recruited by temporal summation with subthreshold-c stimuli when pulse frequencies as low as 20 to 60 pulses per second are employed. That several cutaneous afferent nerve fibers from the finger must be excited to elicit a conscious experience, when single-pulse stimuli are used, has been found by Buchthal and Rosenfalck (*10*).

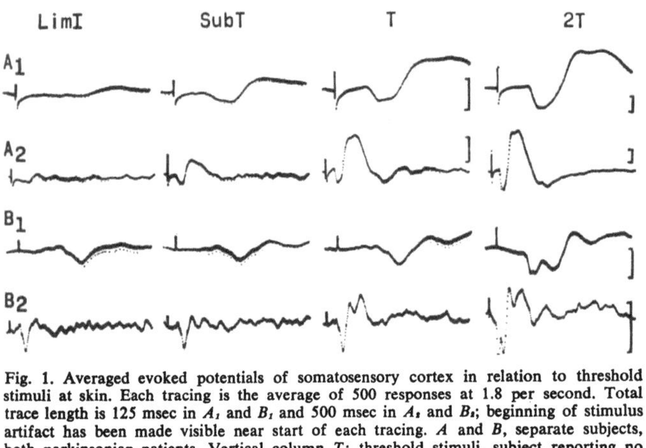


Fig. 1. Averaged evoked potentials of somatosensory cortex in relation to threshold stimuli at skin. Each tracing is the average of 500 responses at 1.8 per second. Total trace length is 125 msec in $A_1$ and $B_1$ and 500 msec in $A_2$ and $B_2$; beginning of stimulus artifact has been made visible near start of each tracing. *A* and *B*, separate subjects, both parkinsonian patients. Vertical column *T*: threshold stimuli, subject reporting no feeling of some of the 500 stimuli. Column 2*T*: stimuli at twice threshold current; all stimuli felt distinctly. Column *SubT*: subthreshold stimuli, none felt by subject; current about 15 percent below *T* in subject A, 25 percent below *T* in B. Column *LimI*: subthreshold stimuli at "liminal intensity" (see text), about 25 percent below *T* in subject A, about 35 to 40 percent below *T* in B. Polarity, positive downward in all figures. Vertical bars in *A*, under *T*, indicate 50 μvolt in $A_1$ and $A_2$ respectively, except for those in 2*T* as shown; for $B_1$ and $B_2$, 20-μvolt bars. (Calibration obtained by summating 500 sweeps of calibrating signal.)

The ability of subthreshold-c stimuli to elicit some components of the evoked potentials in sensory cortex is even more striking when localized stimuli are applied to the specific projection pathway in the thalamus, that is, to nucleus VPL. In VPL we have found that single stimulus pulses (or pulses repeated at 2 pulses per second) were completely inadequate to elicit a conscious sensory experience (or a motor response), even with peak currents which were as much as 20 times liminal I [liminal I being the minimum threshold-c current, with a train of pulses, usually at 60 pulses per second, lasting for 0.5 second or more, when stimulating VPL or somatosensory cortex directly; see Libet *et al.* (*6*)]. For example, in Fig. 2 the VPL stimulus pulses at 1.8 pulses per second and with a strength six times liminal I did not elicit any conscious experience. The primary (initial positive) evoked potential was nevertheless recorded from somatosensory cortex. To elicit a similar amplitude of primary evoked response to skin stimuli, the latter had to be raised to twice threshold-c current (*S* in Fig. 2). The difference between evoked responses to VPL and skin stimuli arises in components following the primary one. Ervin and Mark (*11*) have reported that the types of evoked potentials they recorded with scalp leads, in response to stimu-

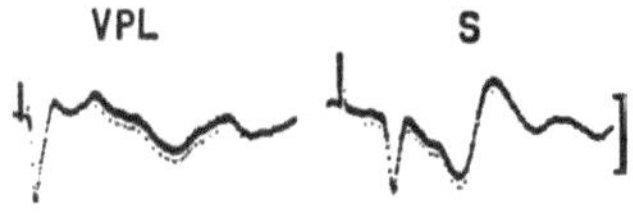


Fig. 2. Evoked potentials of somatosensory cortex in response to thalamic (*VPL*) and skin stimuli in the same subject (patient with heredofamilial tremor). Each tracing is the average of 250 responses at 1.8 per second; total trace length, 125 msec. *VPL*: stimuli in ventral posterolateral nucleus of thalamus; subject reported not feeling any of these stimuli, though current was 6 times liminal I for *VPL* electrode (liminal I being minimum current to elicit sensation with 60 pulses per second train of stimuli). *S*: stimuli at skin; current at twice threshold; all stimuli felt. Vertical bar, 50 μvolt. Note the shorter latency of the primary (positive) evoked response to *VPL* stimulus.

lation with a thalamic electrode, could also appear without any awareness of the stimulus by the patient.

Direct cortical responses (DCR's) to stimuli applied nearby directly on sensory cortex can also be elicited by stimuli which produce no conscious sensory experience. These responses are large enough to be clearly visible above the level of background activity without averaging. Here too, single-pulse stimuli (0.5 msec pulse duration) are relatively ineffective for eliciting sensation, the first response with a sufficiently strong single pulse being a muscular twitch (*6, 7*). Thus, strong single-pulse stimuli at intensities up to several times the liminal I can elicit large DCR's without any subjective experience (Fig. 3). Single pulses with strengths which are below even liminal I can still elicit DCR's (Fig. 3A). If a train of liminal I pulses is cut short below 0.5 second, one can observe the abbreviated train of relatively large DCR's, again with no conscious experience accompanying it. Indeed, no distinctive change in DCR responses other than in amplitude or in duration of repetition period could be ob. erved to accompany the transition between stimuli which were inadequate and those which were adequate for eliciting a sensory experience.

It appears evident, then, that neither the primary component of a single evoked potential nor a single DCR response complex represents or leads to the adequate cerebral condition which is associated with a conscious sensory experience, even in the awake, attentive human subject. Although the primary positive component of the evoked response was larger with threshold-c as opposed to subthreshold-c stimuli to skin, mere increase in amplitude of this response does not appear to provide the crucial difference for adequacy. This conclusion follows from the inability of strong single-pulse stimuli in VPL to elicit any sensation, even though these stimuli produced larger primary responses at the cortex than did threshold-c stimuli at skin. The evidence does not rule out some role for the primary response in sensation. However, the appearance of later components of the evoked potentials elicited by skin stimuli seem to be even better correlated with sensory awareness and may be equally if not more significant for conscious processes than the primary response (*6, 7*; see also *12*). The requirement of a minimum train duration of about 0.5 second for cortical or VPL stimuli at liminal I (minimum threshold-c current) may represent an activation period which substitutes for the normally occurring late components of the evoked potential elicited by single skin stimuli at threshold-c (*6, 7*). The further hypothesis has been proposed (*7*) that one physiological difference between conscious and unconscious experience in the awake and alert individual may lie in the duration of neuronal activation.

In contrast to earlier indications our results demonstrate that, when suitably recorded, cortical evoked potentials are detectable with sensory inputs below the adequate level for conscious sensation, even when the attention of the subject is directed to the stimulus. This fact may be taken to indicate that a possible physiological basis could exist for so-called "subliminal perception," at least under the conditions of stimulation employed (in contrast to the conclusion of Schwartz and Shagass, *13*), but this inference requires some clarification. Stimuli which were subthreshold-c, but could still elicit some evoked potential, could be made adequate for conscious sensation by simple repetition at a suitable frequency. It remains to be seen, then, whether responses evoked by subthreshold-c stimuli which are nonrepetitive or of low frequency can play any role in unconscious experience or in behavioral responses to sensory stimuli of which the subject is not aware.

B. Libet

*Mount Zion Neurological Institute* and *University of California School of Medicine, San Francisco*

W. W. Alberts

E. W. Wright, Jr.

B. Feinstein

*Mount Zion Neurological Institute, San Francisco 94115*

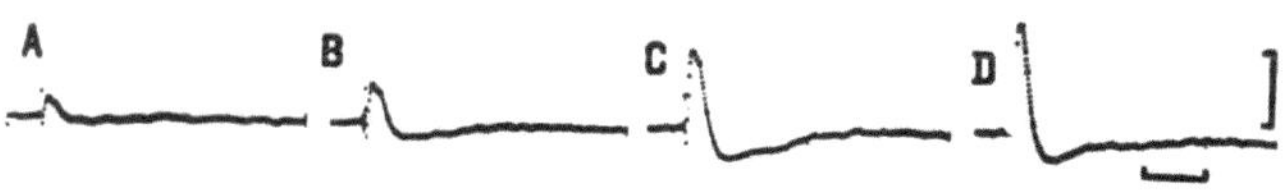


Fig. 3. Direct cortical responses (DCR) evoked in somatosensory cortex by adjacent direct stimuli (0.3-msec pulses). Each tracing is the average of 18 responses at 0.5 per second; horizontal bar in *D* indicates 100 msec. Subject is a parkinsonian patient. *A*, stimulus current 0.3 ma; *B*, 0.8 ma (equal to liminal I for trains of 20 pulses per second to elicit sensation in this subject); *C*, 1.7 ma; *D*, 5.0 ma (4 ma gave a similar response). Subject reported not feeling any of these stimuli, in *A* to *D*. Vertical bar, 200 μvolt.

#### References and Notes

1. S. Schwartz and C. Shagass, *J. Neuropsychiat.* **2**, 262 (1961); E. Domino, S. Matsuoka, J. Waltz, I. Cooper, *Science* **145**, 1199 (1964); J. Debecker and J. E. Desmedt, *Compt. Rend.* **260**, 687 (1965).
2. C. Bourassa and J. Swett, *J. Neurophysiol.* **30**, 515 (1967); J. Swett and C. Bourassa, *ibid.*, p. 530.
3. R. G. Heath and G. C. Galbraith, *Nature* **212**, 1535 (1966).
4. C. D. Geisler and G. L. Gerstein, *Electroencephalog. Clin. Neurophysiol.* **13**, 927 (1961).
5. B. Feinstein, W. Alberts, G. Levin, E. Wright, Jr., *Confinia Neurol.* **26**, 272 (1965).
6. B. Libet, W. Alberts, E. Wright, Jr., L. Delattre, G. Levin, B. Feinstein, *J. Neurophysiol.* **27**, 546 (1964); B. Libet, in *Brain and Conscious Experience*, J. C. Eccles, Ed. (Springer-Verlag, New York, 1966), p. 165.
7. B. Libet, *Perspectives Biol. Med.* **9**, 77 (1965).
8. For a row of multiple subdural electrodes a plate type of assembly designed by J. M. R. Delgado was used, similar to that described in *Electroencephalog. Clin. Neurophysiol.* **7**, 637 (1955).
9. Potentials evoked in other cortical areas were also studied but will be submitted for publication elsewhere.
10. F. Buchthal and A. Rosenfalck, *Brain Res.* **3**, 1 (1966).
11. F. Ervin and V. Mark, *Arch. Neurol.* **3**, 368 (1960).
12. I. Wagman and W. Battersby, *Vision Res.* **4**, 193 (1964); M. Haider, P. Spong, D. Lindsley, *Science* **145**, 180 (1964); H Davis, *ibid.*, p. 182; E. Donchin and L. Cohen, *Electroencephalog. Clin. Neurophysiol.* **22**, 537 (1967).
13. S. Schwartz and C. Shagass, *Science* **133**, 1017 (1961).
14. Supported by research grant NB-05061 and research career program award NB-K3 16,729 (W.W.A.), both from PHS. The present work is part of a larger study of cerebral mechanisms in conscious sensory experience.

7 November 1967

Reprint from

# Handbook of Sensory Physiology



Volume II

**Somatosensory System**

Edited by Ainsley Iggo

Springer-Verlag Berlin · Heidelberg · New York 1973
Printed in Germany

---

Not in Circulation

# Electrical Stimulation of Cortex in Human Subjects and Conscious Sensory Aspects

By

**B. Libet**

With 7 Figures

Chapter 19

# Electrical Stimulation of Cortex in Human Subjects, and Conscious Sensory Aspects

By

BENJAMIN LIBET, San Francisco (USA)

With 7 Figures

## Contents

# A. Introduction

Most problems in neurophysiology can be attacked more fruitfully in animals other than man, for the obvious reasons of controllability of conditions and of our moral restraints on the experimental procedures which are tolerable for human studies. But if one wants to investigate cerebral mechanisms underlying subjective experience (of sensation, in the present context), it should also be obvious that recourse must be had to human subjects for primary validation of the subjective phenomenon under study. Direct approaches to the brain of waking subjects are of course limited by compatibility with therapeutic procedures and by the patient's condition and informed consent. Electrical stimulation of (and, more recently, recording from) the cerebral cortex and deeper structures has provided one approach which, when suitably utilized, makes possible informative studies with no irreversible effects on the subject. The problems susceptible to investigation can be much broader than the initial classical one of the topographical relations of cortical sites to the body sites of the subjectively referred sensations, and some of these problems will be considered in this article.

# B. "Excitable" Somatosensory Cortex

## 1. Topography

Since the initial experiments by FRITSCH and HITZIG (1870) on electrical stimulation of the cerebral cortex of dogs, it has been recognized that only certain cortical areas would give rise to motor responses, when stimulated with subconvulsive strengths of current. Later, CUSHING (1909) demonstrated that sensory experiences without movement also could be elicited in conscious patients upon stimulation of postcentral gyrus[1]. The areas of "excitable" cortex in man, that is those areas where electrical stimulation gives rise to motor or conscious sensory responses, were subsequently enlarged (Fig. 1) especially by the extensive investigations of FOERSTER (1936a, b, c) and of PENFIELD and his collaborators (PENFIELD and BOLDREY, 1937; PENFIELD and RASMUSSEN, 1950; PENFIELD and JASPER, 1954; PENFIELD, 1958). The extent of such areas may vary with the condition of the subject, particularly as to whether his pathological disorder is one (like epilepsy) which affects cortical excitability; and it may vary with the stimulus parametric values employed, i.e., in relation to the range between threshold values (for motor, sensory or evoked electrical responses) and the values for producing seizure activity (either overt motor or sensory, or local electrical convulsive-type wave patterns).

Using stimulus trains at strengths below electrical seizure-wave production, the PENFIELD group (see Fig. 1) could consistently elicit motor responses in epileptic subjects only when applying such stimuli to the Rolandic region (mainly in precentral areas) and to supplementary motor area (superior mesial surface of cortex anterior to the precentral gyrus; see PENFIELD and Welch, 1951; BER-

1 A historical summary of the initial studies of motor and sensory responses to cortical stimulation in animals and man is given by PENFIELD and BOLDREY (1937), PENFIELD and RASMUSSEN (1950), and PENFIELD and JASPER (1954). Therefore, reference will be made here to the earlier works only incidentally, to help clarify points under discussion.

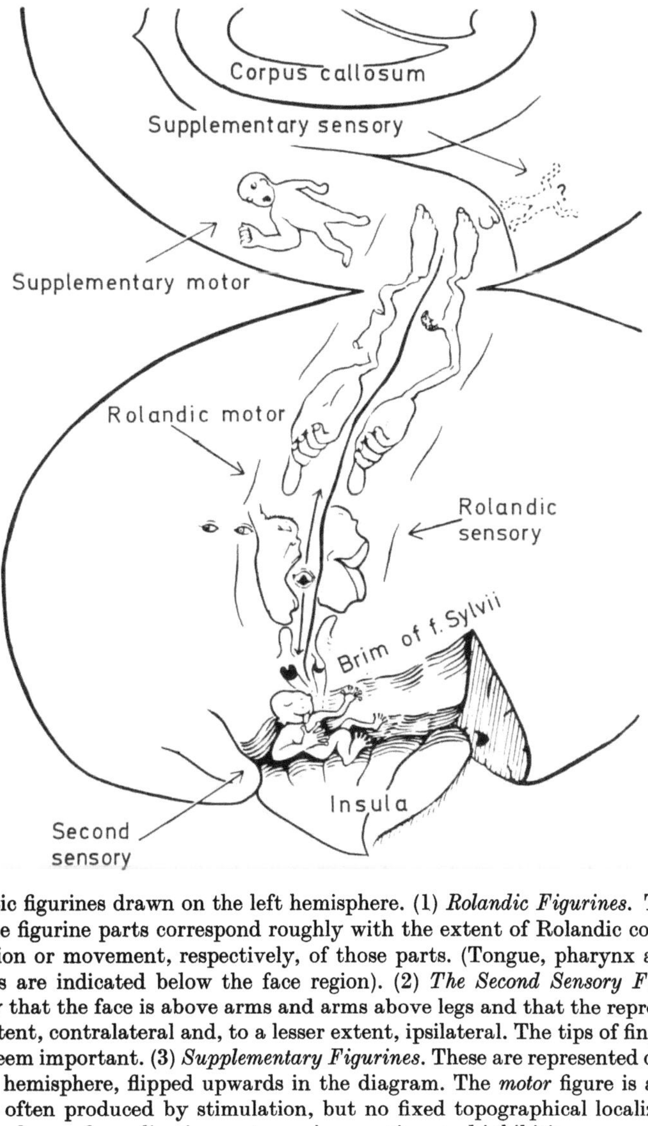


Fig. 1. Somatic figurines drawn on the left hemisphere. (1) *Rolandic Figurines.* The size and position of the figurine parts correspond roughly with the extent of Rolandic cortex devoted to the sensation or movement, respectively, of those parts. (Tongue, pharynx and intra-abdominal areas are indicated below the face region). (2) *The Second Sensory Figurine.* This indicates only that the face is above arms and arms above legs and that the representation is, to a larger extent, contralateral and, to a lesser extent, ipsilateral. The tips of fingers and toes are made to seem important. (3) *Supplementary Figurines.* These are represented on the medial aspect of the hemisphere, flipped upwards in the diagram. The *motor* figure is assuming the posture most often produced by stimulation, but no fixed topographical localization of the parts of the body or of vocalization, autonomic sensation, and inhibition were established by PENFIELD and JASPER. (See also TALAIRACH and BANCAUD, 1966). In regard to the *sensory* figure, PENFIELD and JASPER's observations were not yet sufficient to justify conclusion as to form or exact position. The positions of the parts of these figurines must not be considered topographically accurate in any precise way. The figurines are rough indicators of what has been generally observed. (From PENFIELD and JASPER, 1954, by permission of Little, Brown & Co.)

TRAND, 1956; TALAIRACH and BANCAUD, 1966). Occasionally stimulation of the second sensory area (lateral and posterior to postcentral gyrus) produced some trembling movements, and stimulation of an area of the mesial surface of the temporal lobe produced at times a stiffening of the limbs (PENFIELD, 1958, p. 15).

Results obtained from stimulation studies of the frontal and parietal areas of non-epileptic subjects, with peak currents limited to about 3 times the threshold level required at precentral gyrus (LIBET et al., 1964; ALBERTS et al., 1967) were in more or less agreement with those from PENFIELD's group, except that only a tremor-like movement could be obtained by stimulating postcentral gyrus (see below). With apparently stronger stimuli FOERSTER (1936a, c) had additionally produced rather generalized movements in epileptic patients by stimulation of rather large "extrapyramidal" areas, expecially in the superior frontal and parietal regions. In unanesthetized monkeys and cats, stimulation within wide areas of cortex is reportedly capable of eliciting movements, although often with threshold intensities greater than those required at the primary motor area (see BROWN, 1916; DOTY, 1969). This is in contrast to the relatively small extent of the excitable areas for motor responses in human cortex, when using moderate stimulation.

The extent of the "excitable" areas for eliciting conscious somatosensory responses in man is even more limited than for motor responses; it is essentially restricted to Rolandic cortex (FOERSTER, 1936b, c; PENFIELD and BOLDREY, 1937), perhaps to the postcentral gyrus portion of this, i.e. somatosensory area I (SS-I) in non-epileptics (LIBET et al., 1964; ALBERTS et al., 1967), and to somatosensory area II located inferiorly to SS-I, i.e. more laterally (PENFIELD and RASMUSSEN, 1950; PENFIELD and JASPER, 1954; see further below). FOERSTER (1936b) reported that strong stimulation of area 5, in the superior parietal region posterior to the upper part of postcentral gyrus, could also elicit sensations over the whole body, initially contralaterally; this area overlapped with that producing whole body extrapyramidal type movements. However, PENFIELD and BOLDREY (1937, p. 411) reported only an uncertain production of an epigastric aura for this area.

A relatively large proportion of human cerebral cortex is thus "inexcitable" or "silent". No subjective experiences or obvious behavioral changes of any sort have been elicited in these "inexcitable" areas, with some notable exceptions. The latter include the striking instances of various psychical responses (memories, hallucinations, etc.) aroused by stimulation of temporal lobe cortex in certain epileptic patients (PENFIELD and RASMUSSEN, 1950; PENFIELD and JASPER, 1954); and the negative or aphasic-like effects in certain association areas, such as the arrest of speech produced by stimulation at several regions (in Broca's area; in an inferior parietal area near the sensory area; and in an area in posterior temporal cortex — PENFIELD and RASMUSSEN, 1950). Much of the "silent" cortex is apparently also "dispensable", in the sense that excision or pathological destruction of individual portions of it does not produce defects in sensation or motor control, as occurs in the case of the Rolandic region (PENFIELD and RASMUSSEN, 1950). Of course, neither the "inexcitability" nor the "dispensability" as defined above, prove that "silent" cortex could not respond to suitable stimulation or does not have significant and important functions related to sensory input or motor output. Indeed it had been proposed by J. HUGHLINGS JACKSON (see FOERSTER, 1936c) and others that the whole cortex may be thought to contain nervous arrangements that mediate sensory impressions and movements, whether in a specific or more generalized manner.

Responses of "silent" cortex to stimulation, other than direct motor acts or reports of subjective experience, can in fact be detected. For example, local electrophysiological changes termed direct cortical responses (DCR) are elicited by relatively weak stimuli at all cortical areas in animals (PURPURA, 1959; OCHS, 1962), and they are recordable on human cortex as well (PURPURA et al., 1957b; GOLDRING et al., 1958, 1961; LIBET et al., 1967). Additionally, electrical stimuli applied to virtually any cortical area can be utilized successfully to establish conditioned behavioral responses in cats and monkeys (e.g., DOTY, 1969).

## 2. Adequacy and Limitations of Electrical Stimulation; Interference with Normal Functions

The "inexcitability" of extensive areas of the cerebral cortex is probably not due to an actual and special nonresponsiveness of these neurons to electrical stimulation. It is perhaps more appropriate to view such cortical "inexcitability" as a reflection of (a) the inadequacy of the stimuli employed and/or (b) the inadequacy of the kinds of observations employed to detect changes in behavioral or unconscious psychical processes.

Electrical stimuli, as ordinarily applied on the cortical surfaces, can only influence the mass of underlying neurons in a manner that is relatively non-selective and gross, compared to the organized spatial-temporal patterns of activities that presumably go on during normal functioning. There is, for example, evidence that different groups of cells in different cortical layers are selectively acted upon in certain sequences, by the arriving ascending impulses from natural peripheral sensory stimulation (TOWE et al., 1968; WHITEHORN and TOWE, 1968). The significant kinds of electrical changes may be difficult or impossible to produce with ordinary stimuli; for example, local postsynaptic potentials in dendrites etc. can be generated only indirectly by activating appropriate synaptic inputs rather than by direct electrical fields. The density of the particular units, that must be suitably affected in order to produce the response, may be too low in any given area of "silent" cortex for sufficient activation by the usual type of electrode. (An electrical converse of this latter possibility has been demonstrated in attempting to record ongoing activity instead of to elicit a response by a stimulus: A small electrode on the cortical surface could not record the dominant EEG rhythms that could be recorded with a scalp electrode, unless a number of the small suitably-placed cortical electrodes were connected together in parallel — DELUCCHI et al., 1962). In addition, the electrical stimulus may excite both excitatory and (when stronger) inhibitory presynaptic fibers to local or distant neurons (e.g., LI and CHOU, 1962; KRNJEVIĆ et al., 1966; PHILLIS and YORK, 1967; PRINCE and WILDER, 1967; OCHS and CLARK, 1968b), and may do so in a manner that is not conducive to developing a functional response.

Aside from these hypothetical possibilities for inadequacy, there is actual evidence that electrical stimuli can in fact interfere with or repress the normal function of a cortical area (PENFIELD, 1958). Stimulation of postcentral gyrus often produces a feeling of numbness in the referred bodily site (although this subjective response does not, by itself, necessarily indicate that an actual hypesthesia to peripheral stimulation has been produced). Stimulation of auditory cortex, i.e., on temporal lobe close to the sylvian fissure, has at times produced feelings of being deaf (PENFIELD and RASMUSSEN, 1950, Chapter VIII). PENFIELD (1958) states that stimulation of the calcarine cortex not only produces conscious sensory visions of lights, etc., but also makes the patient blind in that part of the visual field during stimulation. In this circumstance, then, the cortical stimulus elicits a positive sensory experience even when it is simultaneously interfering with sensation elicited by the peripheral sensory stimulus. An example of interference with normal motor functions is seen in the arrest of speech, without any direct motor responses to account for this arrest, when stimulating Broca's areas, etc. (PENFIELD and RASMUSSEN, 1950, Chapter V). It has been suggested (PENFIELD and RASMUSSEN, 1950) that this interference phenomenon may be related to the complete loss of normal sensibility to peripheral stimuli which occurs when there is seizure activity in the respective sensory cortex (i.e., somatic, visual, or auditory).

A more quantitative study of this interaction between cortical stimulation and responsiveness to peripheral input has now been carried out for the somatosensory system in non-epileptic subjects (LIBET et al., 1972a; see further below, under "retroactive masking"). The conditioning stimulus to postcentral gyrus consisted of a train of pulses; the test stimulus was a single pulse applied to the skin, within the referral area of sensation elicited by the cortical stimulus. When a threshold test stimulus to skin was delivered during the conditioning cortical stimulus, the conscious sensory experience otherwise elicited by the skin stimulus was absent. This suppression of the sensory response to threshold skin stimuli could occur even with strengths of cortical stimulus only slightly above the liminal intensity for cortical production of a conscious sensory response (see below, "parametric values"). The effect on skin threshold occurred not only when the cortical stimulus elicited a feeling of numbness but also when it elicited positive or even natural-like sensory experiences (e.g., "rolling"; "wave-like").

The mechanism of depression of peripheral sensibility by cortical stimuli remains an open question. Cortical efferent systems that are inhibitory to the subcortical sensory relay stations have been found in animals (HAGBARTH and KERR, 1954; see also WIESENDANGER, 1969) and probably function in man as well (HAGBARTH and HÖJEBERG, 1957; JOUVET et al., 1960). Production of local intracortical inhibitory actions by cortical stimuli has already been discussed above. Interference could also result from production of local cortical discharges which may disorganize neuronal activity or response patterns (e.g., PENFIELD, 1958; PHILLIPS, 1966). Such disorganizing effects would no doubt explain the loss of normal cortical functions during strong cortical stimulation or seizure activity. But their significance is less obvious when considering the interference with peripheral sensibility by cortical stimuli which are at only the liminal current strengths required for eliciting conscious sensory responses (see above; LIBET et al., 1972a). The interference is probably not due to a Leaõ-type spreading depression; the latter requires much stronger stimulation as well as other differences in experimental conditions. Recordings of the direct cortical responses, to trains of pulses near the liminal intensity for conscious sensory responses, have shown no changes in ongoing rhythms and no steady potentials shifts that characterize spreading depression (LIBET et al., 1972a). The possibility of producing spreading depression must be kept in mind, however, when rather long periods of apparent inhibition (some minutes) are produced by stimulation, especially of exposed cortex (see MARSHALL, 1959).

In view of the theoretically possible deficiencies of electrical stimulation at the surface of somatosensory cortex, as well as the actually observed depression of peripheral sensibility, it is perhaps surprising that any subjective sensory experiences can be elicited at all by such stimuli. The fact that responses are elicited may be a reflection of the relatively simple input-response relationship in a primary sensory area compared to that in "silent" cortex. For example, the stimulus might intracortically excite some of the neural elements normally involved in mediating the responses to the afferent projections in a sensory system (see further below). However, even for sensory cortex, the inability to elicit in any controllable manner each different quality of conscious somatic sensation (see below) could be a function of inadequately sophisticated spatial and temporal patterns of applied stimuli. Where more complex responses have been elicited, e.g. psychical responses to stimulation of temporal cortex in epileptics, it has been suggested that the stimulus may simply be triggering-off an established pattern in an abnormally irritable cortex (PENFIELD and RASMUSSEN, 1950). Probably this is true, in a sense, for responses elicited by stimuli at all points in the sensory system; that is, the stimuli may activate existing patterns of neuronal function by adequately exciting a suitable input at the site of stimulation. For example, direct surface stimulation of somatosensory cortex undoubtedly excites neural elements at a step beyond the ascending specific afferent projection fiber; even so, the requirements for adequate stimulation of either postcentral gyrus or of the specific afferent projection in n. VPL (thalamus) were found to be generally similar (LIBET et al., 1964, 1967; see below).

It has been argued that surface stimulation of "excitable" cortex never evokes natural functioning of the cortical neuronal apparatus, since the motor or sensory responses that are elicited have unnatural characteristics (e.g. PENFIELD, 1958; PHILLIPS, 1966). While this may be true in a strict sense, it may nevertheless be possible to approximate the evocation of natural functions to a greater degree than has been thought feasible. For example, most types of specific somatosensory qualities (rather than merely parasthesias) can in fact be elicited by stimulation of postcentral gyrus (see below).

## 3. Topographical Representation of Somatosensory Responses in SS-I Area

Upon stimulating postcentral gyrus of conscious human subjects, CUSHING (1909) obtained reports of parasthesia sensations (tingling, etc.) referred to the contralateral side, with no visible movements. FOERSTER (1936b) extended these observations to show that the representation, for body sites to which the sensations were referred, was similar to that for the motor responses from the precentral gyrus; i.e., leg area was near the longitudinal fissure on top, and head area below towards the Sylvian fissure. PENFIELD and his colleagues have made extensive and closely observed investigations of the responses to electrical stimuli that were near threshold or at least sufficiently moderate in intensities so as not to produce local seizure activity. (Fig. 1 shows their summary of the topographical representation for stimuli at different points on postcentral gyrus, in the form of a homunculus drawing; a more detailed homuncular drawing is given by PENFIELD and RASMUSSEN, 1950, p. 44). There is not any strict correspondence between this cortical sensory sequence and the sensory representation by spinal segment. Responses to stimulating postcentral gyrus are strictly contralateral, except for the face area where about 10 % of the responses were either bilateral or ipsilateral (e.g. PENFIELD and RASMUSSEN, 1950). It should be added that stimulation of the "bank" of the postcentral gyrus, down in the central sulcus, elicited responses with sites comparable to those obtained from the free surface; this could be accomplished in some patients after removal of the precentral gyrus. There were also no systematic differences in the referral sites of responses, or in the threshold intensities required, with stimulation of different points along the antero-posterior axis on the exposed face of the postcentral gyrus (PENFIELD and RASMUSSEN, 1950; LIBET et al., 1964).

## 4. Variation in Responses to Stimuli at a Given Cortical Point

The precise referral sites of the sensation elicited by stimulating a given point on the postcentral gyrus may vary, partly as a function of previous stimuli at this or at other points in the vicinity. (1) In successive tests with the same stimulus to the same cortical point, the somatic sites to which the sensation is subjectively referred can at times shift or change in size. This occurs even though the threshold intensity requirement for any response can remain remarkably stable (LIBET et al., 1964). Such changes perhaps reflect fluctuating degrees of spatial and temporal facilitation of the cortex in the vicinity of the electrode (see below for further quantitative aspects of such facilitations). (2) An initially subthreshold stimulus at a given cortical point can often elicit a response after it is repeated a number of times at sufficiently short intervals ("primary" facilitation; see TRR section below). (3) A type of "secondary" facilitation by a responsive cortical point (A) on another initially unresponsive point (B) has also been observed (PENFIELD and BOLDREY, 1937; PENFIELD and WELCH, 1949). When stimulation of a point (B) along the antero-posterior axis of the post-central gyrus gives no response with reasonable stimulus strengths (i.e. relative to the threshold values for points that do give responses), it may be induced to give responses if stimulated shortly after repeated stimulation of an adjacent point (A) which does give a response initially (see PENFIELD and BOLDREY, 1937, p. 402). In addition, when stimulation of a point (B) does elicit a sensory response, the somatic site to which the sensation is referred can be changed by prior stimulation of another point (A). For example, if stimulation of point B elicited a sensation in the thumb, while that of point A was referred to the lower lip, then stimulation of B following a series of stimuli at A could now elicit a sensation in the lower lip instead of the thumb (PENFIELD and WELCH, 1949). Such a shift in the referral site, produced by conditioning stimuli at another point, was termed "deviation of sensory response". Since the sensory response in the original referral site for point B is absent during such a "deviation" PENFIELD and WELCH (1949) suggested that the process includes a form of "sensory inhibition". (Such a cerebral inhibitory mechanism could be related to "surround inhibition" that was initially found with peripheral stimuli, e.g. TOWE and AMASSIAN, 1958; MOUNTCASTLE and POWELL, 1959b, and later demonstrated in cortical areas adjacent to a cortical stimulus by LI and CHOU, 1962; KRNJEVIĆ et al., 1966; PHILLIS and YORK, 1967; PRINCE and WILDER, 1967; OCHS and CLARK, 1968b). On the other hand, the change in referral site of the sensory response when stimulating point B after conditioning point A is presumably accomplished by a lasting state of facilitation at point A; i.e., weak presynaptic input to point A that can be initiated by stimulating at point B now becomes sufficient to generate the usual response at point A. Similar instabilities of threshold and site had been observed for motor responses to stimuli at points in precentral gyrus in man, by PENFIELD and BOLDREY (1937) and PENFIELD and WELCH (1949), and previously in the anthropoid brain by LEYTON and SHERRINGTON (1917). (4) Referral site may also be affected by the polarity and diameter of a surface unipolar electrode (LIBET et al., 1964). Shifting from cathodal to anodal polarity pulses can shift the site of referral. An increase in diameter of unipolar electrode, for example from 0.25 to 2 mm, tends to enlarge the referral site; but it was surprising to find how little actual change in referral site did occur, even with electrode diameter raised to as much as 10 mm, if stimulus strengths were carefully kept to the threshold or liminal level for each electrode (LIBET et al., 1964, and unpubl.; see further below).

It should be obvious from all this that the detailed representation of responses to stimuli is not a rigidly fixed condition but depends on past history of activity, area of electrode, polarity, etc. The site and total area of referral can fluctuate even with threshold level stimuli, and even when sufficiently long inter-stimulus time intervals are employed so that there is no evidence of any incidence of seizure-like activities. However, the quality of the sensation reported, in contrast to the referral site, tends to remain relatively constant for threshold stimuli applied to a given area of postcentral gyrus (see further below).

## 5. Uniqueness of Responses to Postcentral vs. Precentral Stimulation

Motor responses to stimulation of postcentral gyrus can be obtained (FOERSTER, 1936a; PENFIELD and BOLDREY, 1937) but they differ in stimulus requirements (FOERSTER, 1936b, c) and in character (LIBET et al., 1964) from those elicited at area 4 on precentral gyrus. Application of a single 0.5 msec constant current pulse of sufficient intensity to postcentral gyrus produces a twitch, but this is a very weak and restricted response compared with the twitch obtainable with a pulse at precentral gyrus. When trains of pulses (e.g. 60 pps) are applied to precentral gyrus, the response is a pyramidal-type, smoothly developing contraction with no intermittency, even when stimulus intensity is near threshold. With postcentral gyrus, trains of stimulus pulses generally elicit no motor responses when the intensity is near threshold level for conscious sensation or near the threshold level for precentral motor responses (LIBET et al., 1964). The latter difference, between pre- and postcentral thresholds, had been noted earlier, in apes (LEYTON and SHERRINGTON, 1917) and in man (FOERSTER, 1936a). Stimulating postcentral gyrus in man with a train of pulses at much higher intensities could elicit only a series of slowly repeating weak twitches, occurring in the same body region as the referred sensations produced by the lower intensity (Libet et al., 1964). (Not infrequently postcentral stimuli could initiate a somewhat irregular tremor, or exaggerate an ongoing one in Parkinsonians, often with stimulus intensities not far above threshold for sensation and in some cases at or below this threshold.) By contrast, Penfield and Rasmussen (1950, p. 211) stated that stimulation of postcentral gyrus (in epileptic subjects) produced movement that was similar in character to the response from stimulation of precentral gyrus. However, in describing an instance of a motor response to postcentral stimulation, Penfield and Boldrey (1937, p. 426) called it "up and down vibratory movements" associated with a tingling. Foerster (1936a, c) described only isolated focal twitches produced by galvanic (single pulse) stimulation of postcentral gyrus, when precentral pyramidal outflow was intact. He also noted that the motor part of seizures that originated in postcentral gyrus often began with a tremor, unlike precentral seizures.

On the possibility of eliciting *sensory* responses by stimulating *pre*central gyrus reports differ, perhaps in relation to whether epileptic subjects were tested. In non-epileptic patients, Libet et al. (1964) could not elicit any non-motion sensations that were unrelated to production of movement itself; no tingling etc. accompanied any motor response or appeared without the latter. The sensory responses to precentral stimulation that were reported by Penfield and Boldrey (1937) could not be explained as due to spread of stimulus or of neural excitation to the postcentral gyrus, for they could still be elicited after excision of the latter (Penfield and Rasmussen, 1950, p. 59). It should be noted further that somatosensory responses can be obtained with relatively strong stimulation of widely scattered points on the cortex of epileptics (Penfield and Rasmussen, 1950, p. 132) but not in nonepileptics (Libet et al., 1964; Alberts et al., 1967).

The pre- and postcentral areas adjoining the central fissure appeared to share functions sufficiently to be referred to as the "sensorimotor strip" by Penfield et al. (Penfield and Rasmussen, 1950; Penfield and Jasper, 1954) and by Foerster (1936b), although it was recognized that the primary (more indispensable) motor area was precentral and the primary somatosensory area was postcentral. However, on the basis of findings in non-epileptic subjects, and with close attention to threshold stimulus values (Libet et al., 1964), the functional distinction between these two areas in the adult human cortex appears to be even sharper than previously seemed to be the case. This is further supported by the results of excision, that of area 4 producing relatively little or no loss in sensibility (Foerster, 1936b; Penfield and Rasmussen, 1950) and that of areas 3-1-2 producing no motor losses except for some ataxia, an effect that may be assignable to the sensory loss, and some light spasticity (Foerster, 1936b). On the other hand, certain kinds of motor responses can be elicited by strong stimulation of postcentral gyrus even in non-epileptics (Libet et al., 1964), and generalized extrapyramidal-type movements under certain conditions (Foerster, 1936a, c). It has, of course, been known for some time that the postcentral gyrus contributes fibers to the pyramidal tract, although it now appears that these may be at least in part a descending inhibitory outflow to sensory input stations (see Wiesendanger, 1969); and that the gyrus has cortico-cortical U-fiber connections to precentral gyrus (e.g. Penfield and Jasper, 1954, p. 57). Also, group I afferents from muscle spindles project to the vicinity of somatosensory cortex, even though they appear to elicit no conscious sensory responses (see next section, below). There are therefore reasons for thinking that postcentral gyrus may have some functional roles in the organizing of motor patterns, as already envisioned by FOERSTER (1936a, c), and by PENFIELD (e.g. 1958), and even earlier by J. HUGHLINGS JACKSON (see FOERSTER, 1936c).

## 6. Relation of Sensory Responses to Representation of Evoked Potentials

Mapping of topographical representation of somatic sensory projections to the cortex can be carried out using, as the response indicators, the recordings of the early-latency or "primary" evoked potentials. This was initially carried out in animals including primates (see WOOLSEY, 1952), and then in man (WOOLSEY and ERICKSON, 1950; JASPER et al., 1960; KELLY et al., 1965). The topographical representation of the body parts on the postcentral gyrus produced by this method is in generally good accord with that obtained by the mappings of the body parts to which subjective sensory responses are referred, when the latter are elicited by stimulation of points on the postcentral gyrus (e.g. JASPER et al., 1960). The existence of another somatosensory area ("SS-II") was indicated by ADRIAN's (1941) observations of evoked potentials in a region below area SS-I in the cat, and it was then extensively mapped with this technique by WOOLSEY and collaborators (WOOLSEY, 1952). In SS-II the representation for head is oriented anteromedially and for legs posterolaterally, and evoked potentials are obtainable ipsilaterally and contralaterally. Stimulation of the apparently homologous cortical area in man (see Fig. 1) has elicited subjective sensory responses with a roughly appropriate representation, though only on the contralateral side of body (PENFIELD and RASMUSSEN, 1950; PENFIELD and JASPER, 1954). However excision of the regions lying in the presumed area SS-II did not produce any obvious loss of subjective sensibility (PENFIELD and RASMUSSEN, 1950), in contrast to the losses incurred when SS-I (postcentral gyrus) is removed in man.

While cortical evoked potentials to peripheral stimuli have been appropriately recordable wherever it has been possible to elicit conscious sensory responses by cortical stimuli, the converse is not true.That is, conscious sensory responses cannot be elicited at all the cortical areas that do respond with recordable evoked potentials. Evoked potentials, particularly later components with latencies greater than that of the specific primary response (about 20 msec $\pm$), can be recorded over regions outside areas SS-I and SS-II. This is true for both scalp and subdural electrodes (see DONCHIN and LINDSLEY, 1969; MCKAY, 1969) and for intracortical electrodes (WALTER, 1964); the scalp site at the vertex yields prominent evoked potentials with somatic (and other) peripheral sensory stimuli and is commonly employed especially in psychophysical studies (e. g. KATZMAN, 1964; COBB and MOROCUTTI, 1968; MACKAY, 1969; DONCHIN and LINDSLEY, 1969). Direct cortical stimulation of sites external to SS-I and SS-II has not elicited conscious sensory responses, as already noted above; this includes the cortical area in the vicinity of the vertex (e.g. PENFIELD and RASMUSSEN, 1950; LIBET et al., 1972b).

It has been argued that the evoked potentials elicited by somatic stimuli and recorded over non-sensory cortex actually originate only in the primary somatosensory areas (MACKAY, 1969, pp. 201–203; STOHR and GOLDRING, 1969; VAUG-

HAN, 1969). This question cannot be considered fully here. However, there now appears to be good evidence for at least some other cortical sources for somatosensory evoked potentials. Large evoked potentials are recordable at supplementary motor cortex in man and have been found to show a subcortical reversal of polarity, in accord with Stohr and Goldring's criterion (LIBET et al., 1972b). Indeed, the origination of evoked potentials by this area makes it a prime candidate for the actual source of at least some of the recorded evoked potentials at the vertex. Yet, no conscious sensory responses can be elicited by stimulation of supplementary motor cortex (PENFIELD and JASPER, 1954; TALAIRACH and BANCAUD, 1966; LIBET et al., 1972b). PENFIELD and RASMUSSEN, 1950, pp. 62–3, elicited some vague sensations there but these were similar to those experienced by the patients at the time of an epileptic attack. Evoked potentials can also originate in wide "polysensory" areas of the frontal lobe, in at least some primates (squirrel and rhesus monkeys), as has been shown by BIGNALL and IMBERT (1967); these responses also show reversal of polarity subcortically and they remain even when the primary sensory areas have been ablated. These frontally originating evoked potentials may be related to those recorded in man by WALTER (1964). There are further indicators of multiple generators of evoked potentials from the distributions of the recorded fields, especially for different components of the evoked potentials (VAUGHAN, 1969; BROUGHTON, 1969; LIBET et al., 1972b).

Finally, it must be noted that although some group I muscle afferent fiber inputs from the periphery have been found to elicit evoked responses in primary somatosensory or closely related cortex (AMASSIAN and BERLIN, 1958; OSCARSSON and ROSÉN, 1963, 1966; ALBE-FESSARD, 1967; SWETT and BOURASSA, 1967), these inputs do not elicit any subjective experience of motion, position, muscle length or other in man (BRINDLEY and MERTON, 1960; GELFAN and CARTER, 1967). The group I afferent inputs, in contrast to other somatic inputs, could not produce behavioral (GIAQUINTO et al., 1963) or conditional learned responses in the cat (SWETT and BOURASSA, 1967). The group I projection of muscle spindle afferent impulses to the cortex may therefore serve in the integration and organization of movements mediated by the cortex, without giving rise to any subjective sensory experience (e.g. OSCARSSON, 1965; ALBE-FESSARD, 1967). The actual subjective experience of position and movement would depend on other inputs, especially those originating in and about the joints (SKOGLUND, 1956; PROVINS, 1958; MOUNTCASTLE and POWELL, 1959a). OSCARSSON (1965) has suggested that the projection of *cutaneous* afferents to the primary *motor* area may also serve for motor rather than conscious sensory functions; this would fit with the inability to elicit conscious sensory responses by stimulation of precentral gyrus in nonepileptic human subjects (LIBET et al., 1964; LIBET, 1966). (See above.)

It should be realized, then, that it is possible for somatic sensory inputs to elicit, in the cerebral cortex, evoked potentials that represent processes whose functional roles are not necessarily involved in production of conscious sensory experience. MOUNTCASTLE (p. 228 in MACKAY, 1969) notes that the elucidation of these other functions provides a challenging task at present. Clearly, caution must be used when interpreting psychophysical correlations, between evoked potentials and the production of conscious sensory responses, in terms of cerebral mechanisms underlying subjective sensory experiences. The demonstrated experimental potentiality for error in such interpretations illustrates the point made below ("criteria for conscious sensory response") about the necessity for distinguishing experimentally among the different classes of detection indicators of sensory inputs.

## C. Significant Stimulus Parameters for Threshold Conscious Sensation

Clarification of the significant stimulus parameters and their values for producing conscious sensory responses is important for achieving adequate control and reproducibility of stimulation of sensory cortex for experimental and clinical purposes, and to avoid possible production of seizures or damage. There is the more interesting possibility, however, that an analysis of the "adequate" stimulus might be helpful in problems related to the dynamics rather than topography of cortical function (see LIBET et al., 1964; LIBET, 1965, 1966). For example, what can the differences between stimuli that are adequate and those that are inadequate, tell us about the requirements of sensory cortex for neural inputs that can elicit a conscious sensory experience? Can we elicit the different specific qualities of somatic sensation with uniquely different kinds of adequate stimuli? Although direct cortical stimulation has limitations, as already noted, it bypasses the unknown modifications imposed on the original peripheral sensory input by various subcortical mechanisms. It thus provides a possible route for analysing the more immediate requirements of cortical activation in the production of sensory experiences.

### 1. Criteria for Threshold Level of Conscious Sensory Response

This issue provides an exercise in the philosophy of scientific method, since it raises the problem of the validity of the evidence in relation to question under investigation. When the objective is to investigate conscious sensory responses to cortical (or other) stimuli, it should be clear that the indication to the experimental observer that the response has occurred must be based on the introspective experience of the subject (see ECCLES, 1966; LIBET, 1966). There are many possible response indicators that might show that a stimulus had affected the subject, i.e. that it had been "detected" in some way. Responses could range from internal and autonomic ones (such as cortical evoked potentials, changes in heart rate, galvanic skin responses etc.) to alterations in external behavior, either as overt actions or as covert modifications of the behavior that was expected without the cortical stimulus. But a clear distinction should be made between responses that indicate one of these kinds of detection of the stimulus, as utilized in detection theory studies (GREEN and SWETS, 1966), and those that indicate the specific kind of detection represented in the subject's awareness of a sensory experience. There is considerable evidence that behavioral responses can occur in normal subjects without awareness (ERIKSEN, 1956; ADAMS, 1957; GOLDIAMOND, 1958; RAAB, 1963; BEVAN, 1964; SHEVRIN and FRITZLER, 1968); there are also the automatisms and responses without awareness in certain types of epilepsies (e.g. PENFIELD and JASPER, 1954). Thus, many indicators of detection may or may not be accompan-

48*

ied by subjective experience of the sensory stimulus. (See also section above on "Relation of sensory responses to representation of evoked potentials.") The cerebral processes that mediate subjective experience may have some unique differences from those that mediate responses without such awareness, i.e. from those responses in which the subject is "unconscious" of the stimulus itself. It is precisely such possible differences that are of great interest (for further discussion of this issue see RAAB, 1963; LIBET, 1965, 1966; KRECH, 1969). Some of the physiological significance of the suggested distinction will become apparent in the discussion below (see also above, on evoked potentials). The failure to make the distinction can lead to a fundamental error in the validity of conclusions drawn from stimulus detection studies.

The distinction between these two classes of detection (conscious vs. other) has been made by psychophysicists on the basis of differences in "criterion" levels, allegedly adopted by the subjects for making a response at some point in what is regarded by those investigators as a stimulus response continuum (e.g. SWETS, 1961; EIJKMAN and VENDRIK, 1963; SUTTON, 1969). The threshold level for a report of conscious experience is considered by them to be an arbitrary one, with each subject setting his own "criterion" for the level at which he will report subjective experience (e.g. SUTTON, 1969; it should be noted, however, that there is no evidence for any such conscious deliberate action by the subject). On this view it would be argued that demonstrations of actual signal detection of some kind (e.g. by forced-choice responses or by the appearance of an evoked potential response, etc.) indicate that the subject really did have a conscious subjective experience even when the subject reports flatly and consistently that he had none. Such an argument would appear to constitute an unwarranted distortion of the primary evidence in order to make it fit a preconceived theoretical framework (LIBET, 1965, 1966).

In determining threshold levels on the basis of reports of subjective experience, a subject may be asked to report (a) whether he did feel or experience the sensation in question (no matter how weakly), or (b) whether he is uncertain about having felt it, or (c) whether he definitely did not feel it (LIBET et al., 1964). Speediness of report, for example by pressing a switch as soon as possible, is to be avoided as a potentially complicating factor; it is possible that speedy reports, at least at the instant they are made, could indicate detection without actual subjective awareness (see further below). Both the upper limit (a) and lower limit (c) can be approached by descending from a superthreshold value, or by ascending from a subthreshold value. The threshold level has often been specified as the level at which the subject makes the response in 50% of the tests, i.e. within the range of uncertainty.

In actual practice the threshold levels for conscious sensory responses can be very stable (5–10%) over periods of an hour or more (LIBET et al., 1964). There has usually been no important difference between the limiting values obtained by either ascending or descending approaches, except when using cortical stimuli with a low pulse frequency, $< 15$ per sec (LIBET et al., 1964). Re-establishing the threshold by presentation of stimuli with values more randomly distributed among the tests, and with false tests (no stimuli) thrown in at times, has usually not affected the values achieved with the method of approaching limits. The range of intensity for uncertainty of response, i.e. the difference between the upper limit (a, 100% positive) and the lower limit (c, virtually 100% negative), has turned out to be relatively small, i.e. about 10% of the stimulus intensity for the upper limit; and there remains a considerable range below the lower limit (c) in which the subject reports that he definitely feels nothing. This contrasts with the results obtained when a forced-choice answer is requested without any necessary regard to conscious sensory awareness (e.g. SWETS, 1961; EIJKMAN and VENDRIK, 1963; see also KIETZMAN and SUTTON, 1968).

The nature of the warning signal before a stimulus appears to have no major influence in a generally attentive and cooperative subject. However, it appears to be necessary that the subject focus his attention on the precise somatic site of the sensation and becomes familiarized in a few initial trials with the very weak sensory experiences generated by near-threshold stimuli. When the sensory experience is a very short-lasting one, e.g. with a single pulse to the skin, the degree of attention during the precise time period, in which the stimulus occurs, apparently becomes more significant. This could be seen when, instead of presenting the stimulus at some variable moment during a period of several seconds following the usual alerting auditory signal, the latter is followed at a fixed interval by a brief light. The light signal indicates the time period during which the stimulus is actually delivered (LIBET et al., 1972a). With this combination of indicator signals, the threshold values for brief skin stimuli could be reduced by some 10% (or more in some instances).

Somewhat surprisingly, a shift in the general state of "arousal" from that of EEG "alpha-blocked" (eyes open, alerting signals, etc.) to that with "alpha rhythm present" (eyes closed and no immediate alerting signal) had no significant effect on threshold values for stimulation of cortex or skin with pulse trains (LIBET et al., 1964); in the case of "alpha present" the subject was queried just after the stimulus was delivered so that any alerting could only be "retroactive". This kind of result provides another experimental difference between a conscious sensory response and other kinds of detection indicators. For example, reaction times have been found to change with such a shift in EEG "arousal" state (LANSING et al., 1959).

## 2. Definitions of Threshold Terminology

Different types of psychological and electrophysiological responses to stimuli in the sensory system (conscious sensory experience, forced choice detection, DCR's, evoked potentials, etc.) may have different stimulus threshold requirements. To avoid confusion and repeated qualifiers, the threshold levels for eliciting a conscious sensory experience will be referred to as *threshold-c*. Threshold-c level can refer to the middle of the usually small range of uncertainty, between the lower (0 response) and upper (100% responses) levels. However, it is useful for certain purposes to specify the threshold-c level as the lower or upper limit itself.

The term threshold as applied to the stimulus can and will be used in the broad sense of the just adequate levels of all significant parameters of a stimulus. For threshold-c, the significant parameters of electrical stimuli (rectangular wave pulses) applied to somatosensory cortex include not merely intensity, i.e. peak current (I), but also polarity (with unipolar surface electrode), pulse duration (PD), pulse frequency (PF, i.e. pulse repetition rate) for a stimulus made up of a train of pulses, train duration (TD), train repetition rate (TRR, the reciprocal of the interval between successive delivery of stimulus trains), and the contact area of the electrode (especially for surface unipolar electrodes). Since the threshold value for each parameter depends on the values selected for the other parameters in the set, it is desirable to adopt the usage of LILLY et al. (1952) which was applied to their study of threshold movements produced by stimulation of cortex. A

Table 1. *Summary of parametric regions for threshold stimuli*

| | Region A | Region B | Region C |
|---|---|---|---|
| I | Liminal | 2 or more times that in A | Intermediate range, between A and B |
| TD | > 0.5–1 sec | < 0.1–0.3 sec; or single pulse | 0.5 to 5 sec, at 8 pulses/sec<br>0.1 to 0.5 sec at PF ≥ 15 pulses/sec |
| PF | ≥ 15 pulses/sec (can get with 8 pulse/sec if TD > 5–10 sec) | Any | Probably any, but best seen in low range (< 15 pulses/sec) |
| Facilitation evident with TTR = 1 per 30 sec | Usually none | Usually considerable | Usually present |
| Response at threshold | Purely somatosensory (no motor) | Observable muscular contraction | No observable contraction; sensation often has aspect of motion (at 15 pulses/sec often has slow pulsatile quality) |

I = peak current; TD = train duration; PF = pulse repetition frequency; TTR = train repetition rate. (From LIBET et al., 1964, by permission of J. Neurophysiology).

"parametric point" is the group of simultaneously assigned single values, one for each parameter of a set. A "parametric line" is a set of parameters all of whose values are fixed except for one, e.g. peak current. (The threshold value of a given parameter is ordinarily determined with respect to a given parametric line.) A "parametric region" is determined by the simultaneous choices of an upper and lower limit (i.e., a range) for the values of each parameter of a set; as a consequence of such choices there is a "region" in which parametric points fall within the limits chosen for the various parameters.

## 3. Parametric Regions for Threshold Stimuli

It has been found useful to characterize threshold-c stimuli to somatosensory cortex into three parametric regions, according to some features of both the stimuli (especially their train duration) and the nature of the responses, as shown in Table 1 (LIBET et al., 1964). Parametric *region A* includes all threshold stimuli with the relatively longer TD's (> 0.5 sec), higher pulse frequencies (> 15 per sec) and lowest possible peak currents (i.e., liminal I; see below). Since the responses are all purely sensory, region A encompasses the kinds of stimuli which should generally be employed for somatosensory investigations. For stimuli in *region B*, the peripheral afferent input, generated by the observable muscular contraction that is produced at threshold, is probably responsible for any conscious sensory responses. This distinction of parametric regions clarifies an observation reported by FOERSTER (1936a, b, c,) that when brief galvanic (dc) stimuli of sufficient intensity were applied to postcentral gyrus in man they could only elicit localized motor responses; and that faradic stimuli (i.e. repetitive pulses) were required to elicit sensation. The conscious sensory responses elicited in *region C* often had an aspect of motion ("quiver", "pull", "drawing in") even when this quality was not elicited in *region A* in the same subject; with PF's of 8 per sec or lower, pulsatile sensations (having a subjective frequency of about 1 per sec irrespective of the actual PF) were commonly reported. However, no muscular contractions could ordinarily be detected by the external observer. The possibility that the sensations in region C may, at least in part, be due to undetected small but deep muscular responses has not been resolved. (The findings with short TD's of stimuli in the thalamus, see below, are relevant to this issue.) It should be clear, then, that particular attention must be paid to the TD, PF, and I combinations of the electrical stimuli employed, if one wants to relate mechanisms in somatosensory cortex to production of conscious sensory experience.

## 4. Train Duration (TD) and Intensity (I)

Holding other parametric values constant, the peak current values (I) for a threshold-c response can be determined for stimuli of different TD's. A plot of these parametric points (Fig. 2) shows that once the TD reaches a certain value, the threshold I requirement does not change with further increase in TD. This feature appears analogous to the rheobasic current level given by the curve for intensity vs. *pulse*-duration obtained for the threshold excitation of nerve or muscle fibers. The minimum intensity level in the I-TD curves for cortical stimuli has been termed "liminal I"; and the minimum TD that is required when using pulses with liminal I peak currents, has been termed the "utilization-TD" (LIBET et al., 1964; the significance of the utilization-TD is taken up below). With TD's less than utilization-TD, not only is threshold-I greater but the response is often different (already described above, under "parametric regions"). The unique nature of the I-TD relationship for threshold-c stimuli at somatosensory cortex may be compared to that for motor cortex and for two subcortical sites in the somatosensory pathway, namely the ventroposterolateral nucleus (n. VPL) in thalamus and the skin.

**a) Motor Cortex.** An I-TD curve has now been carefully determined for motor cortex in one unanesthetized human subject using cathodal 0.5 msec pulses at a PF of 60 pps (LIBET et al., 1972a). In contrast to sensory cortex, the peak current required at precentral gyrus for a threshold motor response was already liminal with very short TD's of about 0.05 sec (actually 4 pulses); i.e. longer TD's (up to 2 sec) required the same liminal I. Threshold I for a single pulse was greater than liminal I. Thus the "utilization-TD" was very short, resembling that for the skin (see below). On the other hand, for motor cortex in the anesthetized cat LILLY et al. (1952) found that the threshold I (for PF's of 20–100 pps) progressively decreased with increasing TD's over the range of 0.1 to 20 sec that was studied. In spite of the difference in utilization-TD's required for motor responses from precentral gyrus as compared to sensory responses from postcentral gyrus, which we found

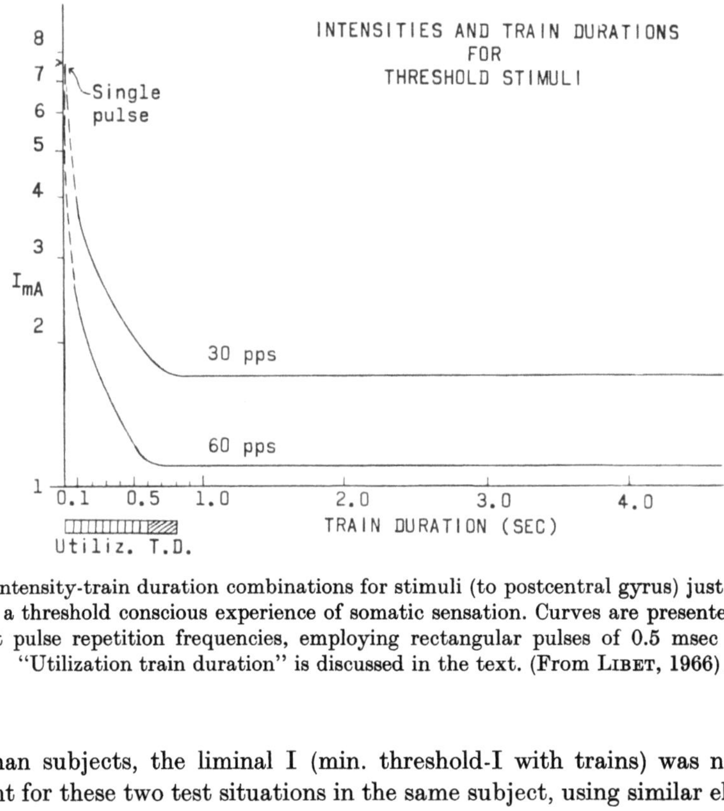


Fig. 2. Intensity-train duration combinations for stimuli (to postcentral gyrus) just adequate to elicit a threshold conscious experience of somatic sensation. Curves are presented for two different pulse repetition frequencies, employing rectangular pulses of 0.5 msec duration. "Utilization train duration" is discussed in the text. (From LIBET, 1966)

in human subjects, the liminal I (min. threshold-I with trains) was not much different for these two test situations in the same subject, using similar electrodes and stimulus parameters (LIBET et al., 1964; see also PENFIELD and BOLDREY, 1937). However, with brief galvanic current (i.e. single dc pulse) FOERSTER (1936a,c) observed that the threshold for eliciting a motor twitch response from postcentral gyrus was two or three times as great as that for precentral gyrus.

**b) Thalamus, n. VPL.** The I-TD relationship for threshold-c stimuli in n. VPL does show a leveling off of liminal I with TD's longer than about 0.5–1 sec. The utilization-TD is thus similar to that for somatosensory cortex. Stimuli with shorter TD's and higher threshold I's have often produced some motor response in addition to sensory ones, if unipolar or even bipolar electrodes with appreciable separation (e.g. 3 mm) were used (GUIOT et al., 1962; JOHANSSON, 1969). However, with a coaxial type electrode motor responses did not appear, indicating that with the other electrodes there was spread of stimulating current to internal capsule (LIBET et al., 1967; see also GUIOT et al., 1962). With the coaxial electrode, a single pulse stimulus did not elicit either a muscular twitch or any conscious sensory response, with peak currents even up to 20 or more times the liminal I level (LIBET et al., 1967, 1972a). [This finding is different from that for postcentral gyrus, where there are evidently some motor efferent elements that can respond to a single pulse; it supports the suggestion that any sensory response to a single pulse at postcentral gyrus is due purely to peripheral sensory impulses set up by the motor response, rather than to an activation of the somatosensory mechanism in the cortex.] On the other hand, stimulation of n.VPL with TD's as short as 0.05 sec (i.e. 3 pulses, at 60 pps) and sufficiently supraliminal I could produce conscious sensory responses, though also without any movement. The subjective

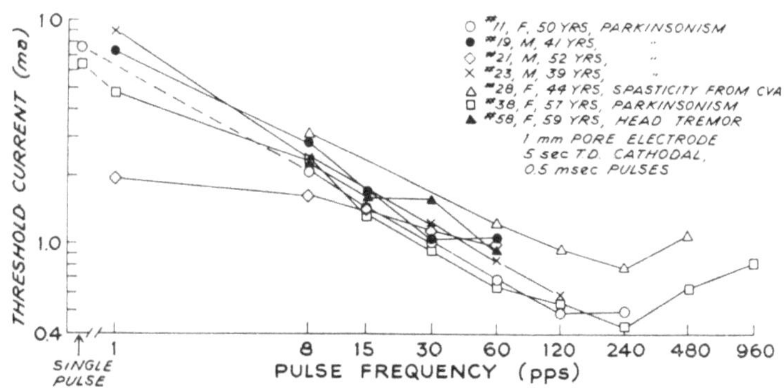


Fig. 3. Liminal I values (i.e., threshold peak currents required to elicit a sensation using 5-sec train durations in these tests) at different pulse frequencies (0.5 msec PD, 1 mm cathodal electrode). Each curve represents a set of determinations made at one session for each of the seven subjects in the graph. (From LIBET et al., 1964, by permission of J. Neurophysiology)

quality of the responses to these stimuli with short TD's (equivalent to parametric region C for cortex) was similar to that reported by the same subjects for stimuli with utilization-TD's or longer TD's and liminal I (equivalent to parametric region A); the sensory responses were all parasthesia-like in nature, i.e. tingling, numbness, or electricity (LIBET et al., 1972b). The findings with n.VPL showed that stimuli with TD's less than the utilization-TD could elicit pure sensory responses in the absence of any motor response (compare above, "parametric regions" for cortical stimuli).

**c) Skin.** The threshold-c peak currents for different TD's with electrical stimuli applied to the skin also level off to a minimum or liminal-I value, but this occurs at a relatively short utilization-TD of about 0.1 sec or less. That is, threshold-c I is the same for all TD's of about 0.1 second and longer (LIBET et al., 1964, 1967). Threshold-c I for a single pulse is generally only 110–130% that of liminal I. Thus while repetition of pulses does somewhat increase the effectiveness here, it is relatively unimportant when compared to the case of somatosensory cortex or n.VPL.

## 5. Pulse Frequency (PF) and Liminal I

Holding other parametric values constant (and using TD's longer than utilization-TD), the relationship between liminal I and PF for eliciting threshold-c responses is shown in Fig. 3 (LIBET et al., 1964). The relationship may reflect a temporal facilitation, such as has been seen with suitable intervals between two pulses (ROSENTHAL et al., 1967). Such a facilitation may be significant in at least the early portion of the stimulus train of pulses (see below, "direct cortical responses").

## 6. Pulse Duration (PD) and Liminal I

When PD for the individual pulses in the stimulus trains was changed from 0.1 msec to 0.5 msec liminal I decreased about 25–50% (LIBET et al., 1964). Increases in PD between 0.5 and 5.0 msec produced relatively little further decrease in liminal I. The longer PD's above 1 msec and especially at 5 msec or more, introduce the possibility of repetitive discharge of some neuronal elements with each pulse (e.g. PHILLIPS, 1969). This, in effect becomes equivalent to stimulating with a train of short pulses of unknown train duration. Therefore PD's for controlled stimuli with the lowest liminal I requirements should be in the vicinity of 0.5 msec. (See below for PD vs. total coulombs passed and possible damage.)

## 7. Polarity of Pulses and Liminal I

With unipolar electrodes on the exposed surface postcentral gyrus, surface cathodal polarity required uniformly lower liminal I values than did surface anodal polarity for threshold-c stimuli (LIBET et al., 1964). The mean ratio of cathodal/anodal values was about 0.7 (range 0.6–0.9), for 0.5 msec pulses and PF's of 8–60 pps. For threshold I's with single pulses (which produce a motor response) the mean ratio was about 0.9. On precentral gyrus, the mean ratio was also about 0.9 for threshold motor responses, whether stimulating with trains or single pulses. Cathodal pulses thus seem to be distinctly more effective than anodal ones for producing conscious sensory responses but not for motor responses. The near parity of cathode vs. anode for observable motor responses fits with the findings for precentral gyrus in monkey (MIHAILOVIĆ and DELGADO, 1956; but see LILLY et al., 1952) and in baboon (HERN et al., 1962; PHILLIPS, 1969). With more sensitive endpoints than an observable twitch (e.g. EPSP's in spinal motoneurones), or with certain positionings of the electrode, HERN et al. (1962) found the anode distinctly more effective.

## 8. Unipolar vs. Bipolar Electrode, and Liminal I

For threshold-c stimulus trains to somatosensory cortex, the liminal I requirement for a unipolar electrode (0.5 mm diameter) was roughly similar to that for bipolar electrodes (LIBET et al., 1964). This relationship was obtained with a 1.5 mm separation between the bipolar contracts and with the cathodal member of the pair employed as the surface cathode in the unipolar tests. For eliciting movements by stimulating the motor cortex of baboon, PHILLIPS and PORTER

(1962) found that a similar bipolar arrangement required stronger currents than a unipolar one of either polarity.

## 9. Area of Unipolar Electrode and Liminal I: Spatial Facilitation

An increase in diameter of the electrode contact from 1 mm increased the liminal I requirement (for a conscious sensory response) about twofold for a 2 mm electrode, and about fourfold for a 10 mm electrode (LIBET et al., 1964). However, the current density at the electrode, i.e. liminal I per $mm^2$ of contact area, was markedly lower with the larger electrodes; it was reduced to about 50% of the value for a 1 mm electrode when the latter was replaced by a 2 mm electrode, and down to about 4% when replaced by a 10 mm electrode. Because of lateral spread, the current density calculated for the contact area of a surface electrode would reflect the actual current density through the cortical tissue less accurately for a smaller than for a larger electrode. Nevertheless, it is safe to assume that the maximal current density through some area of cortex is considerably greater under a 1 mm than under a 10 mm electrode.

Although the liminal current density was lower with the larger electrode, the size of the body area to which the subjective sensation was referred did not decrease; if anything, there was a small increase of the referral area in some cases. The ability of the large electrode to produce a similar sensory response but with much lower current density would appear to represent a form of spatial facilitation. Under the large electrode there may be an initial activation of certain lower threshold neuronal elements, which are present in a lower density over the wider area of cortex. Excitation of these elements may induce the sufficient activation of certain neuronal elements that can lead eventually to a conscious sensory response.

## 10. Neural Elements Initially Excited

In an analysis of the latency and other characteristics of pyramidal cell responses to surface stimulation of motor cortex, PHILLIPS and his colleagues (HERN et al., 1962; PHILLIPS, 1969) concluded that a unipolar anode excited the axons of pyramidal cells directly, but that a unipolar cathode acted on them indirectly by exciting elements that lie more superficial and are presynaptic to the pyramidal cells. This appeared to hold at least for stimulus strengths that are near threshold levels. In producing conscious sensory responses by threshold stimulation of somatosensory cortex, the distinctly lower threshold currents required of surface cathodal as opposed to anodal pulse trains thus indicate that these responses involve the initial excitation of presynaptic fibers. Since the threshold intensities for sensory responses are similar when determined with bipolar or with unipolar cathodal stimuli (see above), the relevant presynaptic fibers initially excited probably lie very close to the cortical surface, i.e. in layer I. This conclusion would follow from the supposition that the fields of stimulus current would be very similar immediately below the cathodes of both a surface bipolar or unipolar electrode arrangement, but that the fields would be dissimilar elsewhere in the cortex (cf. GLEASON, 1970, for fields of applied currents). While the possibility of direct excitation of pyramidal cell axons cannot be excluded even with bipolar surface electrodes (e.g. Patton and Amassian, 1954; Rosenthal et al., 1967; Phillips, 1969), it would seem to be an improbable occurrence under the conditions of liminal stimulation required at sensory cortex (see Phillis and Ochs, 1971).

These parametric stimulus relationships, for cathode-anode polarity and for uni- vs. bipolarity would not distinguish between alternative anatomical orientations of the relevant presynaptic fibers in layer I, i.e. as to whether the orientation is tangential (horizontal) or vertical (radial or perpendicular to the surface). Fibers in layer I are predominantly tangential and each can spread for a distance of several mm in this direction (Colonnier, 1966), but they could be excited as they turned vertically to enter or leave layer I. Fibers below layer I are oriented predominantly in the vertical direction. However, there is little doubt that under the large, 10 mm electrode the relevant presynaptic fibers would be excited along a vertically oriented course, since the bulk of the lower density current at threshold-c must be taking a purely vertical path in the upper cortical layers directly below this electrode.

Axons in layer I are apparently derived from ascending intracortical axons originating mainly from cells in the deeper layers, V and VI (Colonnier, 1966). There may also be a contribution from axons of stellate cells particularly the fusiform type in layers II, III, and IV. Other vertically oriented fibers in primary sensory cortices would include the specific thalamo-cortical afferents, which appear to terminate on stellate cells and on the middle regions of apical dendrite shafts, most heavily in layer IV; nonspecific afferents from reticular core of brain stem and thalamus, ending diffusely on the entire apical shaft; and transcortical and transcallosal fibers terminating in all layers below I, but heavily in the lower layers (see Chow and Leiman, 1970). It would appear unlikely that surface stimuli at threshold-c intensities are exciting the specific afferents, because of the latter's depth and because of the form of the direct cortical response evoked by such stimuli (see below). The most likely elements initially excited by trains with liminal I strength would appear to be the axons of intracortical origin which enter into layer I. (See also Burns, 1958, and Ochs and Clark, 1968a, b, in relation to production of direct cortical responses by surface stimuli.) Excitation of other vertically oriented elements that rise through the upper layers to levels below layer I cannot, of course, be excluded, especially under the large 10 mm electrode. The location and nature of the *post*synaptic elements, that are innervated by the presynaptic fibers presumed to be excited initially, is considered below under "direct cortical responses".

## 11. Train Repetition Rate (TRR) and Liminal I: Temporal Facilitation

A 30 sec interval between successive stimulus trains is generally sufficient to prevent any progressive reduction in liminal I, with stimuli in parametric region A (Libet et al., 1964); the minimum interval for constancy of liminal I has not been established. However, with stimuli in parametric regions B and C repeated at a TRR of 2 per min there is often a progressive reduction in threshold-c I to a new level; also, repetition of stimuli with initially subthreshold-c I values at this TRR can achieve a threshold-c response without a change in I. Such changes are not observed when the TRR is slowed to 0.25–1 per min.

There are several points of interest in this post-stimulus facilitatory effect. (1) A sufficiently long stimulus interval, depending on the parametric region, must obviously be allowed if constancy of threshold is desired. (2) The use of peak currents greater than liminal I appeared to be more important for this effect than the total amount of activation. With stimulus currents initially at liminal I there was no evidence of any progressive change in threshold I at a TRR of 2 per min, whether the stimulus TD's were 0.5 sec or 5.0 sec. (3) A post-stimulus facilitatory effect with a duration of minutes must have a basis different from the type of facilitation that is a function of the pulse repetition rate during the stimulus train (see above under PF and liminal I). Whatever the synaptic mechanisms are for the former long-lasting type, they are probably related to the "primary" facilitation observed in motor and sensory cortex (BROWN, 1915; LEYTON and SHERRINGTON, 1917; LILLY, 1952; PENFIELD and WELCH, 1949). At motor cortex, repetition of such stimuli at shorter intervals can, of course, lead into a seizure pattern in unanesthetized individuals. (The possibility of a post-stimulus "surround inhibition" should also be borne in mind; see above under "Variation in responses etc.".)

## 12. Neuronal Injury and Parameters of Stimulation

It is generally agreed that irreversible injury due to passage of electrical current is chiefly a function of the number of unidirectional coulombs passed by the stimuli (McINTYRE et al., 1959). (This is apart from the possibility of injury from release of products at the metal-solution interface of the electrode, or from the generation of sufficient heat, as in the case of passage of currents in the radiofrequency range.) No injury has been found in brain when the total unidirectional charge passed is less than about 2.5 millicoulombs, with the ordinary macroscopic electrode exposure (McINTYRE et al., 1959). As pointed out by LILLY (1961), injury can be avoided by using a "balanced" diphasic pulse form with phases of opposite polarity separated by about 100 $\mu$sec, so that charges passed in both directions are equal. An interval as long as 50 msec between the alternate phases of bidirectional stimuli has been found to be effective for this purpose (ROWLAND et al., 1960). This permits one to adapt the usually available rectangular stimulus pulses by alternating the polarity of successive pulses in a train (ALBERTS, 1958). An alternating polarity of pulses could involve alternating shifts in the site of intracortical excitation with the successive pulses. This may not be a serious problem for many purposes; with surface stimulation of somatosensory cortex the values of stimulus parameters for threshold-c sensory responses (in parametric region A) have been found to fall into the same general range of values that are obtained when using unidirectional pulses.

The coulomb content of threshold-c stimuli was found to vary within the range of values for each parameter (LIBET et al., 1964). Coulombic content for threshold-c stimuli is lower (even though liminal I is greater) with lower pulse frequencies and with shorter pulse durations. Thus, the total amount of coulombs of unidirectional current could be minimized by proper selection of ranges used. In addition, coulombs passed may be minimized by using TD's that are no longer than utilization-TD (of 0.5–1 sec) whenever possible, since longer TD's merely give a longer-lasting sensory response with no decrease in liminal I (i.e. in the peak current required; see TD, above).

## D. Potentials Evoked in Somatosensory Cortex by Threshold-c Stimuli

The electrophysiological responses of somatosensory cortex may provide some further indications of the kinds of neuronal activity that are involved in mediating conscious sensory experience. The potentials elicited there by stimuli that are adequate for producing a conscious sensory response may be compared to those elicited by inadequate or subthreshold-c stimuli.

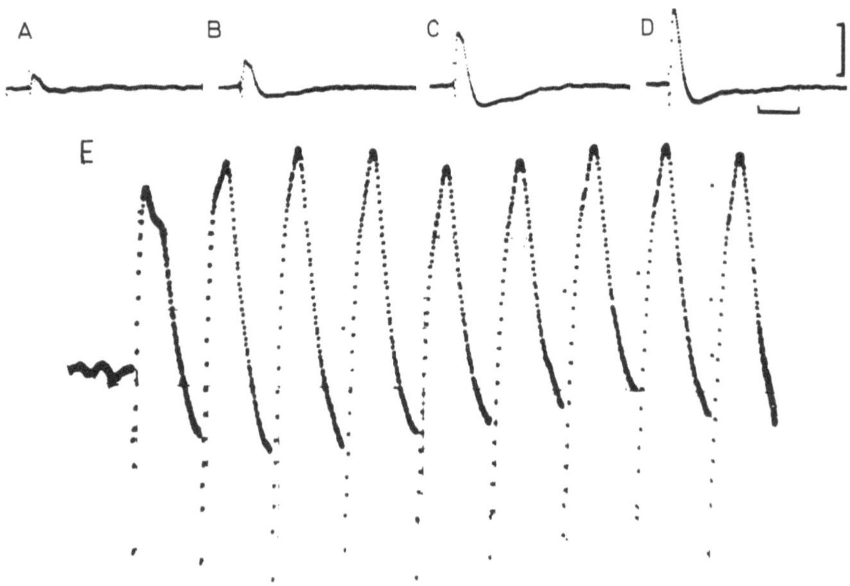


Fig. 4. Direct cortical responses (DCR) evoked in somatosensory cortex by adjacent direct stimuli (0.3-msec pulses). Subject is a parkinsonian patient, unanesthetized. Each tracing in A–D is the average of 18 responses at 0.5 per second; horizontal bar in D indicates 100 msec. A, stimulus current 0.3 ma; B, 0.8 ma (equal to liminal I, see E); C, 1.7 ma; D, 5.0 ma (4 ma gave a similar response). Subject reported not feeling any of these "single pulse" stimuli, in A to D. Vertical bar, 200 $\mu$volt. (From LIBET et al., 1967, by permission of Science). E, averaged response to 10 separate trains, 0.5 sec TD, PF of 20 pps and I of 0.8 ma peak currents; i.e. stimulus with utilization-TD at liminal I, each eliciting a conscious sensory response in the same subject as in A–D. Lines indicate time intervals of 55 msec. (From LIBET et al., 1972a)

### 1. Direct Cortical Responses; Postsynaptic Neuronal Elements Mediating Conscious Sensory Response

As found in animal studies (e.g. PURPURA, 1959; OCHS, 1962), the DCR recorded unipolarly on the surface of human cerebral cortex, adjacent to a bipolar stimulus, shows an initial surface negative component (N), followed by a smaller, longer-lasting positive (P) component and, with stronger stimuli, a slow negative wave (PURPURA et al., 1957b; GOLDRING et al., 1958, 1961). During a train of repetitive pulses, a slow negative shift builds up and the faster N response to each pulse tends to decrease concomitantly (GOLDRING et al., 1958, 1961). The N and P potentials are largely generated as a result of postsynaptic responses to the excitation of presynaptic fibers by the stimulus (ECCLES, 1951; PURPURA, 1959; OCHS and CLARK, 1968a). There is general agreement that the N wave represents a response of apical dendrites, with some additional evidence that this may include production of a decrementally propagating dendritic action potential or local response (OCHS and CLARK, 1968b; PHILLIS and OCHS, 1971). The P component apparently reflects a depolarizing response of pyramidal soma-dendritic regions in somewhat deeper cortical layers (IWAMA and JASPER, 1957; BINDMAN et al., 1962; SUZUKI and OCHS, 1964; LANDAU, 1967) and perhaps IPSP responses in the apical shafts as well (PURPURA, 1959; LI and CHOU, 1962; CREUTZFELDT et al., 1966; OCHS and CLARK, 1968b).

The threshold-I levels needed for eliciting DCR-N and P responses were found to be below the liminal I values that were required to elicit conscious sensory responses by cortical stimulus trains of 20 or 30 pps in parametric region A, as in Fig. 4 (LIBET et al., 1967). On the other hand, liminal I for sensory responses was found to be below the strength required for producing maximal N and P waves, and it was below the threshold I for producing any slow negativity in single DCR responses. When stimulating with a 20–30 pps train of liminal I pulses, which become adequate for conscious sensation after a utilization-train duration of about 0.5 sec (as in Fig. 4), there were no striking changes in the DCR responses during or at the end of such a train (LIBET, 1965; LIBET et al., 1972a). In addition, no after-waves could be detected after such stimulus trains. Nor was there any correlation between threshold-c stimulus adequacy and the appearance of a negative steady potential shift during such trains. There is, therefore, no evidence in the surface recorded DCR's of any unique neuronal response that appears to be correlated with the attainment of adequacy, by the cortical stimulus, for eliciting subjective sensory experience (see further discussion of this below, in section on utilization-TD).

One approach to discovering which components of a compound neuronal response may at least be necessary, for the production of the conscious sensory response, lies in applying selective blocking agents to the tissue. Gamma-aminobutyric acid (GABA) appears to mimic or in fact be an inhibitory synaptic transmitter in the cerebral cortex (KRNJEVIĆ and SCHWARTZ, 1967; CURTIS et al., 1970); it hyperpolarizes and increases ionic conductance of postsynaptic membranes and should, therefore, depress the excitability of postsynaptic structures as well as the amplitudes of postsynaptic potentials. The application of GABA to the cortical surface can produce a suppression of the DCR-N wave, or a reversal of its polarity to surface positivity (PURPURA et al., 1957a; IWAMA and JASPER, 1957; OCHS, 1962); surface negative components of evoked potentials responding to peripheral sensory stimuli, and of other evoked or spontaneous potentials also are depressed or inverted to positivity. The best explanation of these effects of GABA is that, when applied to the cortical surface, it penetrates only into the uppermost layers and depresses postsynaptic responses there. The unaffected postsynaptic responses of the deeper layers still contribute to the surface recordings and thus account for the change in surface polarity (IWAMA and JASPER, 1957; BINDMAN et al., 1962;

OCHS and CLARK, 1968a). GABA does not interfere with excitation of superficial fibers and transmission of their presynaptic impulses to adjacent areas, where they can still elicit normal DCR responses (IWAMA and JASPER, 1957; OCHS and CLARK, 1968a).

GABA has now been applied to somatosensory cortex of human subjects, in a manner sufficient to abolish the DCR-N wave or invert the polarity of this response (LIBET et al., 1972b). Under such conditions, no significant changes were produced in the ability of surface cortical stimuli to elicit conscious sensory responses; i.e. there was no effect on the liminal I or the utilization-TD required for threshold-c trains, or on the quality of the conscious sensory responses to such cortical stimuli. Nor were any changes observed by the subjects in their peripheral somatic sensibility (e.g. there were no reports of "numbness"), and preliminary tests in a few cases for thresholds with von Frey hairs and for 2-point discrimination showed no obvious changes.

These findings indicate that postsynaptic responses of apical dendrites and other neuronal elements in the superficial cortical layers of sensory cortex (including the neuropil of the molecular layer) are not necessary for the production of a subjective sensory experience. Instead, it would seem that postsynaptic responses of neurons or parts thereof that lie below the upper few layers are probably more directly involved in mediating the conscious sensory response. Such neuronal elements could presumably be acted upon by presynaptic fibers that are initially excited by the surface stimuli (see above), e.g. by fibers in the molecular layer that dip down vertically to end on deeper neuronal elements (OCHS and CLARK, 1968a, b) or by radially ascending fibers. The findings with GABA might be taken to support PENFIELD's contention (e.g. PENFIELD, 1958) that, to elicit a sensory response, the cortical stimulus directly excites corticofugal fibers, and that the primary sensory cortex should be regarded as a simple relay station. However, the other available evidence, from stimulus parameter requirements and from analyses of responses in cortex of lower animals, points to a presynaptic site for the initial excitatory action of the cortical stimulus (see above). Additional evidence is accumulating which suggests (e.g. TOWE et al., 1968; WHITEHORN and TOWE, 1968) that the primary sensory cortex has a more elaborate function than that of a simple distributory relay station. On the other hand, any such functions of sensory cortex are quite compatible with a requirement for eventual efferent discharge, from sensory cortex to other cerebral areas, as part of the processes mediating conscious sensory experience.

## 2. Evoked Potentials, with Stimulation of Skin or n. VPL

When single pulse stimuli are applied to the skin (or to a peripheral nerve), evoked potentials can be detected with electrodes on the scalp with the aid of averaging techniques (e.g. KATZMAN, 1964; COBB and MOROCUTTI, 1968; DONCHIN and LINDSLEY, 1969). With suitable placement of electrodes, even the short latency (20 $\pm$ msec) relatively localized, surface-positive component can be detected (e.g. GIBLIN, 1964; DEBECKER et al., 1965). The thresholds for scalp-recorded evoked potentials have been reported to coincide reasonably well with those for eliciting a conscious sensory response (SHAGASS and SCHWARTZ, 1961);

evoked potentials could also be detected in the "uncertainty" range of stimulus intensities, in which some but not all the stimuli are felt by the subject (DEBECKER et al., 1965), and even at about the "absolute threshold for psychophysical detection" (ROSNER and GOFF, 1967). But it was also reported that scalp-evoked potentials could be reduced to the vanishing level by inhalation of cyclopropane without markedly depressing the psychophysical detection of the stimulus by the subjects (CLARK et al., 1969).

However, recordings of evoked potentials on the scalp are considerably attenuated as compared to those recordable on the cortex directly (e.g. HEATH and GALBRAITH, 1966; LIBET et al., 1967; BROUGHTON, 1969); this is especially the case for the early, more localized components (see also GEISLER and GERSTEIN,

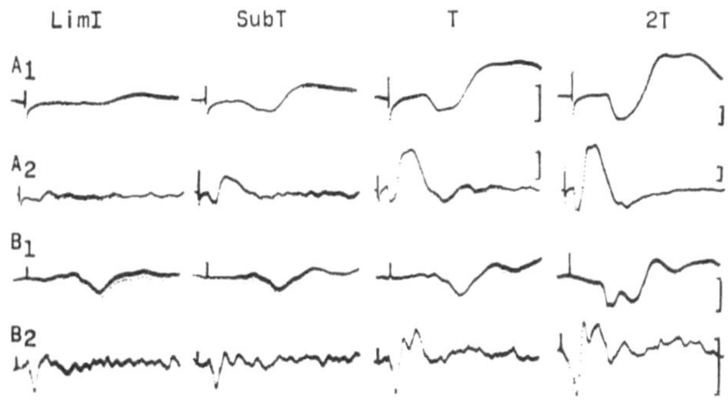


Fig. 5. Averaged evoked potentials of somatosensory cortex in relation to threshold stimuli at skin. Each tracing is the average of 500 responses at 1.8 per second. Total trace length is 125 msec in $A_1$ and $B_1$ and 500 msec in $A_2$ and $B_2$; beginning of stimulus artifact has been made visible near start of each tracing. A and B, separate subjects, both parkinsonian patients. Vertical column T: threshold stimuli, subjects reporting not feeling some of the 500 stimuli. Column 2T: stimuli at twice threshold current; all stimuli felt distinctly. Column SubT: subthreshold stimuli, none felt by subject; current about 15 percent below T in subject A, 25% below T in B. Column LimI: subthreshold stimuli at "liminal intensity" (see text), about 25% below T in subject A, about 35 to 40% below T in B. Polarity, positive downward in all figures. Vertical bars in A, under T, indicate 50 $\mu$volt in $A_1$ and $A_2$ respectively, but gains are different in 2T as shown; for $B_1$ and $B_2$, 20-$\mu$volt bars. (Calibration obtained by summating 500 sweeps of calibrating signal.) (From LIBET et al., 1967, by permission of Science)

1961; CELESIA et al., 1968; STOHR and GOLDRING, 1969; VAUGHAN, 1969). Scalp recordings may also be contaminated with potentials originating extracranially (BICKFORD et al., 1964; CELESIA et al., 1968; VAUGHAN, 1969). When recordings are made subdurally the position of the electrode on the skin can be matched with the referred location of the conscious sensory response obtainable by directly stimulating the cortical electrode; this can improve the cortical recording of highly localized responses to skin stimulation (see also JASPER et al., 1960). Subdural recording can also obviate the problem of extracranial potentials.

With subdural recordings at a point on postcentral gyrus and with the locus for the skin stimulus matched to the referral from this cortical point, it was found that an averaged evoked potential could be detected with stimulus strengths that were distinctly below the lower limit for threshold-c level (see Fig. 5). That is, evoked potentials could be detected with skin stimuli which were well below the range of uncertainty and which never elicited any conscious sensory responses (LIBET et al., 1967). The evoked potentials produced by such single subthreshold-c pulses to skin consist mainly of an initial or primary surface positive component; the later components only appear at about the threshold-c level (though the first negative component that follows the primary positive one has at times also been evident with subthreshold-c strengths). This finding indicates (a) that the neuronal activity represented by the primary component of the evoked potential is not sufficient for initiating the cerebral events involved in the appearance of the conscious sensory response (see further below, under n.VPL); and it suggests (b) that the later components represent activity that may be necessary for such a response, as has also been suggested more indirectly from other types of findings (DAVIS, 1964; HAIDER et al., 1964; WAGMAN and BATTERSBY, 1964; DONCHIN and COHEN, 1967).

The skin threshold strength for detection of some cortical evoked potentials did coincide roughly with the liminal I value for conscious sensory response, as seen in Fig. 5 (LIBET et al., 1967); liminal I is obtained when the stimulus is a brief train of pulses at 60 pps instead of a single pulse. This indicated that a single pulse (i.e. repeated at 1 pps or less), which has a somewhat higher threshold-c I than liminal I for the train, has to excite more than one cutaneous sensory nerve fiber in order to elicit a conscious sensory response, even in subjects who are closely attentive to the stimulus (LIBET et al., 1967). Such a conclusion is in accord with the quantitative peripheral nerve studies of BUCHTHAL and ROSENFALCK (1966). With recordings from an exposed cutaneous nerve bundle in trained human subjects, HENSEL and BOMAN (1960) concluded that one nerve impulse in one sensory fiber might be sufficient to produce a conscious sensation, although they admit the possibility that additionally excited fibers may not have been included in the recorded bundle.

In *n.VPL*, a single pulse of relatively high current is ineffective for eliciting a conscious sensory experience (see above section on "train duration"). Nevertheless, such a pulse can produce a large evoked potential in the appropriate area of human somatosensory cortex (LIBET et al., 1967), as was to be expected from studies on animals (DEMPSEY and MORISON, 1943; ANDERSEN and ANDERSSON, 1968). The primary surface-positive component of these evoked potentials is presumably elicited in the same cortical elements and by the same specific projection fibers to the cortex as is the primary component elicited by a skin stimulus. The primary evoked potential elicited by single pulse-stimuli to n.VPL can attain an amplitude that is greater than the one elicited by a skin stimulus that is well above threshold-c level for skin, as seen in Fig. 6 (LIBET et al., 1967). The absence of a conscious sensory response to a single pulse in n.VPL, therefore, establishes even more firmly the conclusion that the primary component is not sufficient to initiate subjective experience (see above, "evoked potentials with the skin stimuli").

It follows from these considerations, that the presence or amplitude of an evoked potential, particularly of the primary component, cannot be assumed to be an indicator of the occurrence or intensity of subjective sensory experience without other validation under the conditions of study (see also, section above on "relation of sensory responses to representation of evoked potentials").

## E. Significance of Utilization-TD

The utilization-TD, i.e. the minimum train duration required for cortical stimuli at liminal I strength (parametric region A) to elicit a conscious sensory response, appears to be a physiologically significant phenomenon and quantity. Values for utilization-TD have fallen mostly in the range of 0.4–1.0 sec (clustering at 0.5–0.6 sec) for all subjects. The value is relatively constant for a given subject. Changes in other stimulus parameters have little or no effect on utilization-TD, even when they markedly affect the liminal I requirements. The value for utiliza-

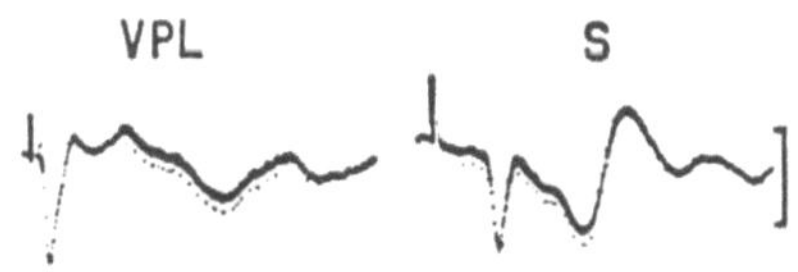


Fig. 6. Evoked potentials of somatosensory cortex in response to thalamic (VPL) and skin stimuli in the same subject (patient with heredofamilial tremor). Each tracing is the average of 250 responses at 1.8 per second; total trace length, 125 msec. VPL: stimuli in ventral posterolateral nucleus of thalamus; subject reported not feeling any of these stimuli, though current was 6 times liminal I for VPL electrode (liminal I being minimum current to elicit sensation with 60 pulses per second trains of stimuli). S: stimuli at skin; current at twice threshold, all stimuli felt. Vertical bar, 50 $\mu$volt. Note the shorter latency of the primary (positive) evoked response to VPL stimulus. (From LIBET et al., 1967, by permission of Science)

tion-TD is relatively independent of PF's (except for the range below 10 pps, where it seems to be distinctly longer, — see LIBET et al., 1964). Values for utilization-TD were found to be in the usual range whether the stimulus pulses were cathodal or anodal in surface polarity (unidirectional pulses), alternating in polarity, had pulse durations varying from 0.1–0.5 msec, or were applied via bipolar instead of unipolar electrodes, or by larger (up to 10 mm) electrodes instead of smaller ones (down to 0.25 mm).

The relatively large utilization-TD (about 0.5 sec) for eliciting conscious sensory responses with stimuli at sensory cortex, as compared to skin (0.1 sec), is not a peculiarity due to "abnormal" routes of input that may be excited by the surface cortical electrode; similarly long values were found with stimulation of the ascending projection system, both in n.VPL and in the subcortical white matter below somatosensory cortex (LIBET et al., 1964, 1967). Incidentally, the finding that the same subject shows a large difference, between the utilization-TD's for skin and cortical stimuli, rules out a possible contention that the longer utilization-TD is due to an arbitrary criterion level, for some minimal duration of the sensation, before the subjects are willing to report the experience of a conscious sensory response.

The longer utilization-TD for stimuli in the cerebral portions of the somatosensory system is not a peculiar requirement for all types of stimulus-response processes at these levels. For, the motor cortex required only a very brief utilization-TD for liminal stimuli in the unanesthetized human subject, and it showed no definable utilization-TD in the anesthetized cat (see above). In addition, the effect of prolonging the stimulus TD beyond 0.5–1 sec is strikingly different for sensory and motor areas in man. For somatosensory cortex, stimuli with TD's longer than the utilization-TD, but at the same liminal I strength, simply elicit longer-lasting conscious sensory experiences without any change in subjective intensity. For motor cortex, stimuli kept at a liminal-I value elicit motor responses that progressively increase in intensity and in extent of responding body musculature as the TD is lengthened. Indeed, motor responses to longer trains of even relatively weak stimuli tend to progress into seizures, even in non-epileptic patients.

A utilization-TD of about 0.5 sec for eliciting a conscious sensory experience would indicate a requirement of a surprisingly long period of "activation" of somatosensory cortex, at least for near-liminal inputs. It has been postulated that such a period of activation, of the cortical areas that may be involved, is a fundamental physiological requirement for the mediation of conscious sensory and other subjective experiences (LIBET et al., 1964; LIBET, 1965, 1966). If one accepts the relatively long period of activation as a requirement for conscious experience of at least near-threshold sensory stimuli, then a number of interesting psychological and philosophical inferences arise (LIBET, 1965, 1966). One of these is that the conscious sensory response at threshold has a kind of all-or-nothing aspect to it; "activation" below the minimum, either in duration or intensity, would not give rise to any conscious experience, as defined here, although they are generating demonstrable neural responses in the cortex.

Evidence for such a minimum time requirement for the experience also of peripheral sensory inputs is presented below ("Retroactive masking"). Peripheral stimuli to skin require only a very short utilization-TD (0.1 sec) and a single pulse stimulus can easily be made adequate for a sensory response, but this does not exclude the postulated cortical requirement. Peripheral sensory nerve impulses may, by various routes, initiate cortical activations that outlast the initial short-latency cortical response (e.g. WAGMAN and BATTERSBY, 1964; see also KATZMAN, 1964; BERGAMINI and BERGAMASCO, 1967; COBB and MOROCUTTI, 1968; DONCHIN and LINDSLEY, 1969; MACKAY, 1969). Even a single pulse stimulus to skin, at or just above threshold-c intensity, has been shown to elicit at the cortex a sequence of evoked potential components that may have a total duration of several 100 msec or more (LIBET et al., 1967, and 1972b; see also GIBLIN, 1964, DEBECKER et al., 1965, and others, for scalp recordings on this point).

At least two alternative possibilities exist to describe the nature of the physiological "coding" of the conscious sensory experience, when a relatively long utilization time of about 0.5 sec is required: (a) Some special neuronal event is triggered at the end of the adequate period; or (b), the occurrence of a series of certain similar neuronal events for a given period itself constitutes the code. With stimulus sites at sensory cortex or n.VPL, where the actual utilizations-TD's are long, one can have more direct control of the period of activation. As already indicated, no distinctive differences can be seen between the DCR's elicited by trains of liminal I pulses that are shorter, equal to, or somewhat longer than utilization TD. Nor have any unique new potentials or after-waves been detected following completion of a utilization-TD, at least in surface recordings from somatosensory or other cortical areas (LIBET, 1966; LIBET et al., 1967, and 1972a). This absence of distinctive changes in the recorded electrical responses during and after the utilization-TD is in accord with the absence of any progressive build-up in subjective intensity of the sensory experience when cortical stimulus trains longer than the utilization-TD are delivered. (The relationships between stimulus train, DCR's and "appearance" of conscious sensory response are diagrammed in Fig. 7).

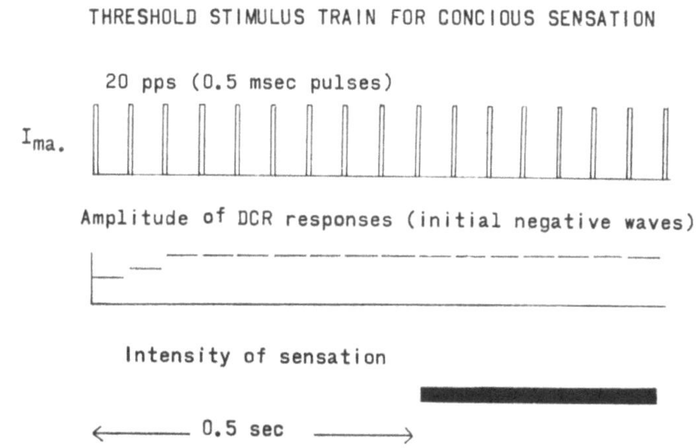


Fig. 7. Diagram of relationships between the train of 0.5 msec pulses at liminal intensity applied to postcentral gyrus, and the amplitudes of the direct cortical responses (DCR) recorded nearby. (A train of actual DCR's elicited by an adequate stimulus train is seen in Fig. 4E). The third line indicates that no conscious sensory experience is elicited until approximately the initial 0.5 sec of events has elapsed, and that the just detectable sensation appearing after that period remains at the same subjective intensity while the stimulus train continues. (From LIBET, 1966)

Similarly, the evoked potentials which are elicited at sensory cortex by a utilization-TD stimulus to n.VPL show no unique components at the end or after the train (LIBET et al., 1972a, b). Such evidence appears to favor alternative (b) for the "coding mechanisms", but this can only be regarded as indicative rather than conclusive. Obviously, DCR's and evoked potentials are incomplete representations of the neuronal events that may be occurring in the somatosensory cortex or elsewhere in the brain at such times.

## 1. Retroactive or Backward Masking, by a Second Stimulus, of the Conscious Sensory Response to a Preceding Stimulus

It has been demonstrated that a second but stronger peripheral stimulus ($S_2$) can in various ways mask or completely blank out the awareness of a *preceding* relatively weak and shorter-lasting peripheral stimulus ($S_1$) (see review by RAAB,

1963). With visual stimuli the backward masking effect can be obtained with the conditioning stimulus $S_2$ following the test stimulus $S_1$ by about 100 to 200 msec or more, in the so-called Crawford effect (CRAWFORD, 1947; see also WAGMAN and BATTERSBY, 1964). Backward masking effects with auditory stimuli have generally been shown for intervals of about 50–100 msec, but some effects have been reported for intervals as long as 1000 msec (see RAAB, 1963; BÉKÉSY, 1971). Cutaneous backward masking has been less frequently studied but has been demonstrated for intervals of about 50 to 100 msec (see RAAB, 1963; MELZACK and WALL, 1963). The question of the nature and location of the neural elements and processes involved in retroactive masking is complicated by the fact that there are both peripheral and central contributions possible (see CRAWFORD, 1947; RAAB, 1963). This complication could be avoided if backward masking could be demonstrated when the conditioning or second stimulus $S_2$ is applied directly to a cortical site, in a known temporal relationship to the initial arrival at the cortex of the afferent projection of impulses initiated by $S_1$.

## 2. Retroactive Masking by a Cortical Stimulus

Positive results with such a retroactive masking paradigm have now been achieved (LIBET et al., 1972a). The conditioning or delayed stimulus $S_2$ was applied directly to somatosensory cortex (postcentral gyrus) in human subjects. It consisted of a brief train (usually of 0.5 sec TD) of pulses (60 pps), with peak currents at about 1.3–1.5 times the liminal I that was required by this cortical stimulus for eliciting a conscious sensory response. The test or first stimulus $S_1$ was a single pulse at threshold-c intensity; it was applied to the skin inside the body area to which was referred the sensory response elicited by the cortical stimulus. The sensations generated by $S_1$ and $S_2$ separately could usually be distinguished readily, by their different qualities and extent of somatic area. The $S_1$–$S_2$ interval in this procedure refers to the time between $S_1$ (single pulse) and the *beginning* of the $S_2$ train of pulses.

The conscious sensory response obtained with $S_1$ alone completely vanished when $S_1$ and $S_2$ overlapped in time (see above, "Adequacy of electrical stimulation . . ."). Response to $S_1$ continued to be masked, however, when the $S_1$–$S_2$ interval was increased up to 125–200 msec for most subjects, and up to 500 msec in one subject; i.e. at these intervals only a single sensory response was reported which was identical with that elicited by $S_2$ alone. When the interval was greater, or when the strength of $S_1$ was raised sufficiently, the subjects experienced both of the sensations in the same temporal order as the responsible stimuli. When $S_2$ was a single pulse stimulus to the cortex it appeared to be ineffective for masking; the minimum $S_2$ train duration that is needed in order to achieve retroactive masking, and therefore the total effective interval for masking, is yet to be determined. The latency to the beginning of the primary evoked potential produced at somatosensory cortex by $S_1$ (when applied to the hand) is about 20 msec, and this should be subtracted from the total effective interval. Fortunately, the latent periods of the early evoked potential components do not vary significantly with the strength of peripheral somatic stimuli (GIBLIN, 1964; LIBET et al., 1967), and are not increased even for skin stimuli at threshold-c levels or below (LIBET et al., 1967; In prep. b); this is unlike the increase in latencies of the visual evoked response with decreasing intensities of light flash (e.g. VAUGHAN et al., 1966).

Retroactive masking of the peripheral ($S_1$) sensation by a later stimulus ($S_2$) directly to somatosensory cortex, with $S_1$–$S_2$ intervals of up to 200 msec or more, could only be due to interference with some late components of the brain responses to $S_1$ that are necessary for the mediation of a conscious sensory response. The mechanism of such interference is an open question. It seems probable that the maximum potential $S_1$–$S_2$ interval for effective backward masking is greater than that achieved, as it is unlikely that all of the potentially effective cortex was stimulated by the $S_2$ stimulus that was applied. In any case, the extent of retroactive masking that could be demonstrated provides powerful support for the hypothesis that a relatively long period of suitable cerebral activations is a necessary feature of the processes mediating conscious sensory experiences (LIBET et al., 1964; LIBET, 1965, 1966).

## F. Qualities of Sensory Responses to Cortical Stimuli

The discussion thus far has considered the relationship between stimulation of somatosensory cortex and the production of some conscious sensory experience, but without any special consideration of the way in which specific subjective qualities of sensory experiences may be related to such stimulation. The latter consideration relates to the question of what the specific spatial-temporal configurations of neuronal activity are that provide the coding for the subjective experiences of different sensory qualities in general. For the differences among the more general modalities (i.e. somatic, visual, auditory, olfactory, gustatory), a modified version of Johannes Müller's Doctrine of Specific Nerve Energies would still appear to satisfy the findings obtained by electrical stimulation of the cortex. MÜLLER (1843) had observed that the general quality of a sensation depended on the specific nerve that was excited rather than on the nature of the stimulus, e.g. a visual experience resulted whether the eye was stimulated by light or mechanically. He concluded that: "Sensation consists in the sensorium receiving through the medium of the nerves, and as the result of the action of an external cause, a knowledge of certain qualities or conditions, not of external bodies, but of the nerves of sense themselves; and these qualities of the nerves of sense are in all different, the nerve of each sense having its own peculiar quality or energy" (Book VI, p. 712). He added with some perspicacity that "it is not known whether the essential cause of the peculiar 'energy' of each nerve or sense is seated in the nerve itself, or in the parts of the brain and spinal cord with which it is connected" (p. 714). The unique relationship between the subjective experience of a given modality and an afferent pathway appears to hold, at least crudely, for the primary sensory receiving areas of the cerebral cortex. Direct electrical stimulation of each such cortical area not only elicits subjective sensory experiences but does so only within the one general sensory modality associated with the specific afferent projection to that area (e.g. PENFIELD and RASMUSSEN, 1950). Convergence of different sensory inputs has been found in wide areas of cortex outside the primary sensory areas (e.g. BUSER and BIGNALL, 1967; BIGNALL and IMBERT, 1969), but the relation of such polysensory projections to subjective modalities is yet to be determined. (See also section above, on "Relation of sensory responses to representation of evoked potentials.") The questions about (a) whether a given peripheral afferent fiber or a given ascending fiber in the CNS can be involved in the mediation of more than one quality of sensory experience, and (b) the nature and extent of the interaction of effects among different inputs and ascending pathways in relation to modalities, have been considered elsewhere in this volume.

It is theoretically possible that suitably sophisticated stimulation of the "silent" cortical areas would elicit subjective experiences that could include sensory ones in some form, as already indicated above (section on "Limitations of electrical stimulation"). However, the present results of cortical stimulation and excision (or pathological damage) would support a re-stated version of Müller's Doctrine — namely, that the quality of a sensation depends on the activation of specific cerebral areas by the sensory nerve fibers mediating that quality. Such a view would not exclude the possibility that temporal and/or spatial patterns of the inputs to such cerebral areas play a significant role in determining quality. Nor would it rule out the possibility that cerebral areas other than the primary sensory areas of cortex play a role, whether activated through a primary area or otherwise. But such a view does continue the Müllerian idea that the site(s) of the cerebral neurones activated is (are) of critical significance in mediating quality; in this view, any spatio-temporal patterns of inputs that might be specifically involved in mediating a given sensory quality would be effective only if imposed on certain specified groups of cerebral neurones, not on any nonspecified groups anywhere in the cortex or subcortex.

The question to be considered more fully in the present context is that of the production, by stimulation of somatosensory cortex, of subjective experiences of the subqualities within the modality of somatic sensation (i.e. touch, pressure, motion, heat, cold, pain, etc.). Earlier investigators have emphasized the generally non-specific qualities of the responses, i.e. that the experiences were usually of parasthesia-like character (tingling, electric shock, etc.), although other more natural-like qualities were elicited by them at times (e.g. FOERSTER, 1936b; PENFIELD and RASMUSSEN, 1950). PENFIELD and BOLDREY (1937) summarized the qualities described by their subjects in a total of 426 responses. (Since their compilation evidently included many stimulus responses at various cortical sites in a given individual, the total number of individual subjects was presumably much smaller than 426). Of the 426 responses, 335 responses were called tingling or electricity or numbness (131 being numbness). Of the remaining 91 responses that did not have a purely parasthesia-like quality, 49 were described as a "sense of movement" when no change in position was observed; 13 were called "coldness", most of these in the face region (though PENFIELD and BOLDREY suggested that it seemed possible that tingling may have been called or taken for coldness); 2 were called heat; 10 were described as a "sense of blood rushing"; 6 were termed thickness or swelling, all in connection with the tongue. For 11 responses the word "pain" was used, but the "pain" was never severe and never caused the patient to object; they concluded that pain has "little if any true cortical representation". (A subjective "desire-to-move", without resultant movement was reported a number of times with stimuli on the precentral cortex.) Seizures involving the

Table 2.[a] *Incidence of responses among subjects, to stimulation of postcentral gyrus*[b]

| Type of response | (X) Stimulus parametric region A[c] | | (Y) Subdural testing[d] | | (Z) Exploring electrode[e] | |
|---|---|---|---|---|---|---|
| A. "Parasthesia-like" (totals) | 41 | | 47 | | 40 | |
| B. "Natural-like" (totals) | 64 | | 51 | | 18 | |
| a) something "moving inside" | | 17 | | 15 | | 5 |
| b) feeling of movement of part | | 15 | | 17 | | 3 |
| c) deep pressures | | 12 | | 8 | | 7 |
| d) surface mechano-type | | 6 | | 6 | | 2 |
| e) vibration | | 4 | | 2 | | – |
| f) warmth | | 8 | | 3 | | 1 |
| g) coldness | | 2 | | – | | – |
| *Actual number of subjects,* in the stimulus and response categories | | | | | | |
| Totals (164) | 60 | | 64 | | 50 | |
| I. Subjects reporting parasthesias *only* | | 12 | | 24 | | 33 |
| II. Subjects reporting "natural-like" qualities *only*[f] | | 14 | | 21 | | 10 |
| III. Subjects reporting parasthesias plus any non-parasthesias[g] | | 34 | | 19 | | 7 |
| III'. Subjects reporting parasthesias plus only *one* non-parasthesia[h] | | (17)[8] | | (15)[8] | | |

[a] Data from LIBET et al. (1972b).

[b] Each type of response is counted only once for a given subject, even if it was elicited repeatedly by multiple tests in that subject. However, if a given subject experienced more than one type of response he was listed in each appropriate category. The actual numbers of subjects in each general category are given below. Referral sites for almost all responses were in the upper extremity, mostly in the regions of the hand and fingers.

[c] Stimulus values were all at threshold-c in parametric region A, i.e. at liminal I carefully determined for trains with PF's $>$ 15 pps and train durations $>$ 0.5 sec. All tests in a given subject were usually applied at the same cortical site with the electrode stationary.

[d] Stimuli were applied via a subdural Delgado-type electrode, usually as part of procedure of localizing area SS-I for evoked potential studies (LIBET et al., 1967); stimulus values were in parametric region A except that intensity may have been greater than liminal I by some unknown but usually not large fraction.

[e] Stimuli applied via a stigmatic hand-held electrode to points on postcentral gyrus; trains of 60 pps with variable durations and with intensities generally fixed at reasonable values that were usually in the range of being superthreshold, though not strongly so.

[f] In these subjects, with sensory responses that did not include any parasthesias, the qualities reported were almost all of the types in B-a and/or B-c.

[g] In subjects listed as reporting both parasthesias and natural-type (non-parasthesia) qualities for their responses the two different types of qualities often were elicited independently, by separate stimulus tests with a given subject, as well as jointly by the same test (i.e. as one response with several qualities distinguishable in it).

[h] This group is, of course, a subgroup of group III and is already included in the total given for group III.

Rolandic area, pre- and postcentral gyri, give rise to sensory experiences that are not far different in their distribution of qualities from those obtained by subconvulsive cortical stimulation in epileptics (PENFIELD and JASPER, 1954; PENFIELD and KRISTIANSEN, 1951). Most of the experiences reported with seizure activity were parasthesia-like, i.e. tingling, numbness, pins and needles. However a small minority of the seizure experiences were not of this type (PENFIELD and KRISTIANSEN, 1951), and included "throbbing", "sensation of movement", and temperature sensations (although the 4 subjects reporting the latter all had lesions in the temporo-parietal region).

The relative incidence and variety of natural-like qualities have since been found to be considerably greater by LIBET et al. (1972b; see also LIBET, 1959). In their studies, parametric values of the stimuli to somatosensory cortex were carefully controlled for proximity to liminal levels, and the subjects were not

Table 3. *Kinds of descriptions included in each category of qualities*[a]

A. *Parasthesia-like:* tingling; electric shock; pins and needles; prickling; numbness.

B. *"Natural-like" sensory responses*
- a) Something "moving inside": wave moving along inside through the affected part; or wavy-like feeling inside; or wavy "like a snake back's" motion; rolling or flowing motion inside; moving back and forth inside; circular motion inside; crawling under the skin, or more deeply.
- b) Feeling of movement of the part (but with no actual motion observable to outsider): quiver; trembling; shaking; flutter; twitching; jumping; rotating; jerking; pushing; pulling; straightening; floating, or sensation of hand raising up, or lifting.
- c) Deep pressures: throb; pulsing; swelling; squeeze; tightening.
- d) Surface mechano-type: touch; tapping; hairs moving; rolling (a ball etc.) over surface; water running over surface; talcum powder sprinkling on; light brushing of skin; holding a ball of cotton; rubbing something between thumb and index finger.
- e) Vibration: vibration, buzzing (distinct from tingling, etc.).
- f) Warmth, or warming.
- g) Coldness.

[a] Data from LIBET et al. (1972b).

epileptics. Table 2 presents a summary of the reports of qualities in these studies. It may be seen that, for the 124 subjects with stimulation in parametric region A, at or not far above the liminal levels, the incidence of natural-like qualities was in fact greater than that of parasthesia-like ones. For most of the subjects both categories of responses were elicited, even though the stimulus site was usually stationary. In a minority of cases only parasthesia-like responses were reported, but an approximately equal minority reported only natural-like responses. For subjects in the last column (Z) of Table 2, the manner of stimulating postcentral gyrus more or less resembled that usually employed by other investigators; in this column, the relative incidence of parasthesia-like and natural-like qualities also more closely resembled that given for all responses by PENFIELD and BOLDREY (1937).

The large variety of the sensory descriptions, on which the sub-categories of natural-like stimuli under B of Table 2 are based, is indicated in Table 3. PEN-

FIELD and RASMUSSEN (1950) had concluded that "the tactile and proprioceptive sensation is detailed only in regard to the location of the part of the body represented (p. 158)", and they stated that "the patient never suggested that something rough or smooth or warm or cold had actually touched the part (p. 217)". By contrast, descriptions of detailed sensory qualities, in addition to their precise locations, were obtained among the responses summarized in Table 2, columns X and Y. For example, descriptions included statements such as "like talcum powder being sprinkled" on the index finger; or "like rolling a deodorant jar lightly over the surface" of the base of the last three fingers; or "crinkling touch, like picking up a paper thing" with the whole hand (see further examples in Table 3). The experiences of "warmth" and "coldness", although relatively infrequently obtained, appeared to the subjects to be qualitatively similar to their natural sensations. (It should be noted that FOERSTER, 1936b, reported obtaining descriptions of vibration, waves, tickle, burning and cold, though he did not specify their rate of incidence among the total number, which were "mostly tingling, etc.".) However, in agreement with other investigators (PENFIELD group; FOERSTER, 1936b), no definite sensations of pain were ever elicited by stimulating postcentral gyrus. (Stimulation of n.VPL in thalamus similarly never elicits pain in "normal" subjects, i.e. who are not suffering from intractable pain — ALBE-FESSARD and BOWSHER, 1968; LIBET et al., 1972a). A large fraction of the natural-like sensory responses involved some sort of motion quality (B–a and B–b in Table 2), although only about one half of these involved any kind of feeling of actual movement of a part (B–b), such as might be related to the "sense of movement" that predominated among the non-parasthesia-like responses reported by the PENFIELD group (e.g. PENFIELD, 1958).

It should be made clear that the so-called "natural-like" sensory responses were usually peculiar and different from all of the normal sensory experiences of the subject, even though they resembled the specific normal qualities and were clearly in a different category from what are defined here as parasthesia-like qualities (LIBET et al., 1972b). The "natural-like" qualities were often difficult for the subjects to describe except by indicating similarities to or analogies with naturally occurring ones, and the subjects often insisted that these descriptions were really inadequate. This situation provides a fascinating illustration of the breakdown in the ability of one person to communicate the content of his primary subjective experience to another person, when both individuals have not experienced similar situations in common. In spite of these difficulties, many subjects used similar verbal expressions to describe their sensory responses; for example, an experience of "a wave moving inside", from one site to another in the arm or hand, was one of the more commonly reported ones.

It appears, then, that most of the various qualities that are at least related to natural somatosensory experiences (except for pain) can be elicited by liminal stimulation of somatosensory cortex. The large fraction of responses to cortical stimuli that are parasthesia-like, and clearly un-natural in quality, may be a result of inadequately sophisticated stimulation (see below) rather than due to an actual lack of representation of the various specific qualities in postcentral gyrus. This conclusion about representation of qualities would suggest that the function of the primary somatosensory cortex is not limited to that proposed by PENFIELD and colleagues (e.g. PENFIELD and JASPER, 1954, pp. 68–69), i.e. to the discriminative aspects of sensation (localization and sense of position and movement) that are involved in perceiving the form of objects. It is true that the long-term defect that remains following ablation of postcentral gyrus is primarily an astereognosis of the affected part of the body, involving loss of two-point discrimination and of sense of position and movement (e.g. PENFIELD and RASMUSSEN, 1950, p. 184). But FOERSTER (1936b) emphasized the fact that there are losses of all somatic sensory qualities, including those of warm, cold and even pain, for some variable time after ablation. He regarded this as understandable in terms of some kind of representation of all somatosensory qualities in the postcentral gyrus, although he noted that nothing was then known about how the various qualities were arranged in the gyrus (FOERSTER, 1936b, p. 431). Representation of specific somatosensory qualities in the SS-I area has also been demonstrated by recording of evoked potentials in response to some individual natural sensory stimuli at the periphery (e.g. MOUNTCASTLE, 1967), including cortical evoked responses to radiant thermal stimulation of the skin (MARTIN and MANNING, 1969). However, as already noted above, the appearance of evoked potentials is not necessarily matched by an ability to elicit conscious sensory experience.

## 1. Cortical Mechanisms Mediating Specific Qualities

If there is a representation of the various specific somatic sensory qualities in postcentral gyrus, what is the nature of this representation and why has it not been possible by electrical stimulation of the gyrus to elicit specific natural-like qualities more regularly and predictably? Two alternative hypotheses have been proposed in relation to cortical mechanisms for specific somatosensory qualities: (a) Each of several "basic" qualities is represented in a separate type of vertical (radial) column of cells in the cortex. Such a functional columnar arrangement was demonstrated by POWELL and MOUNTCASTLE (1959), for at least the different types of mechanoreceptors in the skin, by single unit electrical recordings of cortical responses to peripheral stimuli. Columnar organization is also discernable histologically, with an apparent columnar width (horizontally) of the order of 100 microns (e.g. COLONNIER, 1966; BONIN and MEHLER, 1970), but with impreciseness of borders as seen in the extension of horizontally oriented elements through more than one column. This hypothesis would extend Müller's Doctrine to a uniqueness of different microscopic sites, within the somatosensory cortex, for the different qualities of somatic sensation. The hypothesis also has a further basis in the present evidence for specificity of first order afferent fibers in relation to the different natural stimuli (IGGO, 1965; BURGESS et al., 1968; BESSOU and PERL, 1969). (b) In the second hypothesis, the production of each specific quality of somatic sensation depends on an appropriate patterning of incoming impulses for each quality, rather than on activation of a specific site. The patterning could involve specific configurations of temporal and/or spatial distributions. Such a viewpoint has been advocated by WEDDELL (1961) and others for the coding of the afferent discharges in first order afferent fibers, each of which was hypothesized to respond nonspecifically to different physical stimuli at the periphery; the present evidence, however, favors the concept of specificity, rather than non-specificity, of function for individual first order afferent fibers (Iggo, 1965; Perl, 1968). With either hypothesis (a) or (b), one is referring to those activities occurring only at the level of somatosensory cortex, which enable this area to participate suitably in the overall cerebral processes that mediate the subjective experience of a given quality.

If specific somatosensory qualities are represented separately in different vertical columns of cortex, i.e. on hypothesis (a) above, it ought to be possible to elicit some specific quality (not a parasthesia) in all tests by adequately localizing the stimulus. When the exposure of a unipolar electrode on the pia-arachnoid surface was reduced from 1 mm to 0.25 mm in diameter, there was in fact no significant difference in the incidence of parasthesia-like qualities of the responses to stimuli in parametric region A (Libet et al., 1972b). However, even the smaller surface electrode probably excites neural elements in many vertical columns, even in areas outside the area of electrode contact (see Landgren et al., 1962, and Stoney et al., 1968, for examples of such spread in motor cortex). Adequate localization of stimulating current to one vertical column would undoubtedly require the use of intracortical microelectrodes, as has been achieved for motor cortex (Asanuma and Sakata, 1967; Stoney et al., 1968). Stimulation with microelectrodes has been employed in n.VPL of the thalamus in waking human subjects, by Marg and Dierssen (1966). They obtained mostly reports of parasthesia-like qualities for superficial cutaneous responses although rare reports of some more natural-like ones also occurred (e.g. tickle, light touch, warmth); they also obtained some deep pressure responses. Macroelectrode stimulation of n.VPL, with trains carefully adjusted to liminal I levels, also elicited chiefly parasthesia-like sensations, though there were also rare reports of warmth or position change (Alberts et al., 1966; Libet et al., 1972b). It may be that the much closer packing of neurones of different types, in the thalamic nucleus as opposed to cortical area SS-I, does not permit a sufficiently selective stimulation even with microelectrodes. On the other hand, it is possible that the specific projection nucleus in thalamus transmits information only of a localizing, epicritic nature (as suggested by Marg and Dierssen, 1966), but that somatosensory cortex receives a greater variety of inputs that encompasses information about the various qualities of somatic sensation.

Although reducing the number of excited vertical columns in the cortex is technically difficult, one should be able easily to increase the number of responding columns by raising the intensity of the stimulus. If one makes the assumption that stimulation of a larger group of presumably mixed types of columns would be more likely to elicit a parasthesia-like experience, then the latter should appear more readily when intensity of stimulation is raised to supraliminal levels. This assumption would be in accord with the common experience of tingling or electric shock, etc. when a mixed peripheral nerve is stimulated, especially with a train of pulses, (e.g. Collins et al., 1960); it contrasts with responses to stimulation of the skin, where presumably more selectively uniform fibers are excited by near-threshold stimuli and give rise to more natural-like sensations (e.g. Sigel, 1953). When stimulating somatosensory cortex it was indeed found that raising the intensity of a stimulus train to 1.5–2 times liminal I did change the quality of the sensory response to a parasthesia-like one, when the response at liminal I strength had a specific or natural-like quality (LIBET et al., 1972b). A less direct demonstration of this may also be deduced from Table 2; for the group in column (Z), in which the stimulus intensities were often likely to be distinctly supraliminal, the incidence of parasthesia-like qualities predominates (cf. columns X and Y). The effects of raised stimulus intensities therefore do provide some indirect support for the columnar specificity hypothesis (a), above.

If the different somatosensory qualities depend upon specific temporal patternings of neuronal activation in somatosensory cortex and perhaps elsewhere (hypothesis b, above), rather than on the sites of activation, there are at least two kinds of direct cortical tests that might be applied. (1) It ought to be possible to alter the quality elicited by a peripheral sensory stimulus by simultaneously introducing inputs via cortical electrical stimulation which would change any naturally specific input patterns. This test is difficult to apply, since cortical stimuli tend to mask or inhibit the conscious sensory response to a peripheral stimulus (see above). (2) It should also be possible to elicit different specific qualities of responses to cortical electrical stimuli by changing the various parametric values of the stimulus in some appropriate manner. Test number (2) has been applied in a limited way by LIBET et al. (1972b), by investigating the influences of changes in various parametric values of the cortical stimuli on the quality of the conscious sensory responses. With stimuli in parametric region A, at their respective liminal I values, differences in pulse frequency over a large range (15 to 120 pps), train duration, polarity, etc. did not result in changes in quality. When different qualities of responses were reported by a given subject in different tests with a stationary electrode, there was no apparent correlation of the different quality reports with any changes in stimulus parameters, except for two limited alterations with two parametric changes. One of these is the conversion of a specific quality to one of a parasthesia by raising intensity above liminal I, as already discussed above. Another effective change in parametric values was found to be the lowering of pulse frequency to the range below approximately 10 pps (part of parametric region C). Stimuli with PF's of 8, 6, or 1 pps, using 5 sec TD's and peak currents near the liminal I for such PF's, elicited sensory responses which often had a subjective quality of pulsation; no actual movement in the part was apparent (LIBET et al., 1964). This was often described as an internal arterial-like beat, with a frequency of about 1 per sec regardless of the stimulus PF. It remains to be determined whether this sensation was simply elicited indirectly by actual deep small muscle contractions which could not be detected by the observer, rather than by direct activation of the sensory cortex. The pulsation frequency of about 1 per sec does in fact resemble that of the series of observable muscular twitches which are elicited by a train of higher PF's (e.g. 60 pps) when the peak current is raised to that of the threshold-I for a single pulse (see section on stimulus parameters above).

It would appear, then, that the particular subjective quality of a conscious sensory experience is relatively independent of the nature of the stimulus to somatosensory cortex, at least within the limits of the variations in parametric values tested thus far (LIBET et al., 1959; and 1972b). If this is correct it would favor hypothesis (a) above, as opposed to (b). However, it would be desirable to extend the presently limited studies to include further kinds of stimulus patterning, for example by temporal groupings of pulses into bursts or modulations of intervals between pulses or bursts within a given stimulus train. This will become more feasible as clues to possible significant patternings of afferent inputs become available (e.g. IGGO, 1964; BULLOCK, 1968). It is important to note, however, that even if specific patterns of sensory nerve discharge were found to be characteristically correlated with different types of natural peripheral stimuli, it would not necessarily follow that such patterns were meaningful for CNS function (e.g. BULLOCK, 1968). This caution would apply particularly to the production of the different subjective qualities of sensory experience; significance for subjective experience would have to be established by appropriate studies in conscious human subjects. Particularly instructive on this point is the ability of an actual cooling of the skin to elicit subjective sensory experiences of pressure. This phenomenon is known as "Weber's deception" and is discussed by HENSEL and ZOTTERMAN (1951) and HENSEL and BOMAN (1960), as it relates to their findings that mechanoreceptor fibers can often also respond to thermal stimuli though with a lower degree of sensitivity (see also IGGO, 1965). They conclude (HENSEL and BOMAN, 1960) that certain mechanoreceptor fibers arouse a subjective sensory response of touch or pressure whether these fibers are excited either by pressure or by cooling! Their observation incidentally also provides an elegant demonstration of Müller's Doctrine at the level of a single sensory fiber.

## 2. Perseveration of Quality

A surprising requirement for the production of a pulsatile quality, in the response to a stimulus at low pulse frequency, was found to be an interval of at least 4–5 min after any preceding test with a higher PF (60 pps) stimulus (LIBET et al., 1964; this feature has already been referred to above, in the section on TRR and temporal facilitation). If a shorter period of time was allowed, the same quality (e.g. tingling) that was elicited by a 60 pps train at its liminal I was also elicited by a succeeding 8 pps train at its liminal I. (The absence of changes in the subjective quality of sensory responses with changes in stimulus values within parametric region A, as described above, was not due to a similar perseveration; it was found to hold even in those instances when intervals of 4 min or more between tests did occur.) This perseveration or persistence of the given subjective quality occurs without any concomitant changes in the respective threshold-c requirements of the different stimuli; temporal facilitation as evidenced by a change in liminal I following a stimulus in parametric region A persists for less than 0.5 min, and for about 2–3 min after a stimulus in parametric region C (see TRR section). (However, it seems likely that the persistence of a given quality applies only for the change to a pulsatile quality, when shifting to a low pulse frequency, and not to changes among other qualities in the range of stimulus pulse frequencies of 15–120 pps, — LIBET et al., 1972b). An analogous perseveration of subjective experiences was described by PENFIELD (1958, p. 37; see also in Discussion after the paper by LIBET, 1966), for "psychic" responses to stimulation of the temporal lobe cortex. These responses could include experiences of elaborate memory "flash backs" as well as a variety of types of illusions. PENFIELD reported that, after a given subjective experience was produced by stimulation at a given site, the same experience was often produced again from the same point or even from other points unless there was a considerable lapse of time. The mechanism providing for cortical changes lasting many minutes could have a possible general significance. In any case, the perseveration of changes over long inter-stimulus intervals is a variable that must be considered when studying responses to cortical stimulation.

## Acknowledgements

The author is indebted to his colleagues in the Mount Zion Neurological Institute, W. Watson Alberts, Bertram Feinstein, Mary Lewis and Elwood W. Wright, Jr., for permission to include much of our as yet unpublished work and also expresses his gratitude to them and to an additional colleague, Curtis Gleason, for helpful discussion of issues presented in this article. This work was supported in part by U.S.P.H.S. Research Grant NS0601 from the National Institute for Neural Diseases and Stroke and by National Science Foundation Grant GB-30552X1.

# References

Adams, J.K.: Laboratory studies of behavior without awareness. Psychol. Bull. **54**, 383–405 (1957).

Adrian, E.D.: Afferent discharges to the cerebral cortex from peripheral sense organs. J. Physiol. (Lond.) **100**, 159–191 (1941).

Albe-Fessard, D.: Organization of somatic central projections. In: Contributions to sensory physiology, **2**, 101—167. Ed. by W.D. Neff. New York: Academic Press 1967.

Albe-Fessard, D., Bowsher, D.: Central pathways for painful messages. Proc. Int. Congr. Physiol. Sci. XXIV, Washington, D.C. **6**, 241–242 (1968).

Alberts, W.W.: A stimulus pulse polarity reversal unit. Electroenceph. clin. Neurophysiol. **10**, 172–173 (1958).

Alberts, W.W., Feinstein, B., Libet, B.: Electrical stimulation of "silent" cortex in conscious man. (abs) Electroenceph. clin. Neurophysiol. **22**, 293 (1967).

Alberts, W.W., Feinstein, B., Levin, G., Wright, E.W., Jr.: Electrical stimulation of therapeutic targets in waking dyskinetic patients. Electroenceph. clin. Neurophysiol. **20**, 559–566 (1966).

Amassian, V.E., Berlin, L.: Early cortical projection of Group I afferents in forelimb muscle nerves of cat. J. Physiol. (Lond.) **143**, 61P (1958).

Andersen, P., Andersson, S.A.: Physiological basis of the alpha rhythm. New York: Appleton-Century-Crofts 1968.

Asanuma, H., Sakata, H.: Functional organization of a cortical efferent examined with focal depth stimulation in cats. J. Neurophysiol. **30**, 35–54 (1967).

Békésy, G. von: Auditory backward inhibition in concert halls. Science **171**, 529–536 (1971).

Bergamini, L., Bergamasco, B.: Cortical evoked potentials in man. Springfield: Charles C. Thomas 1967.

Bertrand, G.: Spinal efferent pathways from the supplementary motor area. Brain **79**, 461–473 (1956).

Bessou, P., Perl, E.R.: Response of cutaneous sensory units with unmyelinated fibers to noxious stimuli. J. Neurophysiol. **32**, 1025–1043 (1969).

Bevan, W.: Subliminal stimulation: a pervasive problem for psychology. Psychol. Bull. **61**, 81–99 (1964).

Bickford, R.G., Jacobson, J.L., Cody, D.T.R.: Nature of average evoked potentials to sound and other stimuli in man. In: Sensory evoked response in man. Ed. by R. Katzman. Ann. N. Y. Acad. Sci. **112**, 205–223 (1964).

Bignall, K.E., Imbert, M.: Polysensory and cortico-cortical projections to frontal lobe of squirrel and rhesus monkeys. Electroenceph. clin. Neurophysiol. **26**, 206–215 (1969).

Bindman, L.J., Lippold, O.C.J., Redfearn, J.W.T.: The non-selective blocking action of $\gamma$-aminobutyric acid on the sensory cerebral cortex of the rat. J. Physiol. (Lond.) **162**, 105–120 (1962).

BONIN, G. von, MEHLER, W.R.: In: The structural and functional organization of the neocortex. Ed. by K.L. CHOW and A.L. LEIMAN. Neurosci. Res. Prog. Bull. 8, 174–175 (1970).

BRINDLEY, G.S., MERTON, P.A.: The absence of position sense in the human eye. J. Physiol. (Lond.) 153, 127-130 (1960).

BROUGHTON, R.J.: In: Averaged evoked potentials. Methods, results, and evaluation. Ed. by E. DONCHIN and D.B. LINDSLEY. Washington, D.C.: NASA SP-191 (1969) pp. 79–84.

BROWN, T.G.: Studies in the physiology of the nervous system. XXII: On the phenomena of facilitation. I. Its occurrence in reactions induced by stimulation of the "motor" cortex of the cerebrum in monkeys. Quart. J. exp. Physiol. 9, 81–99 (1915).

BROWN, T.G.: Studies in the physiology of the nervous system. XXVII: On the phenomena of facilitation. 6. The motor activation of parts of the cerebral cortex other than those included in the so-called "motor" areas in monkeys; with a note on the theory of cortical localization of function. Quart. J. exp. Physiol. 10, 103–143 (1916).

BUCHTHAL, F., ROSENFALCK, A.: Evoked action potentials and conduction velocity in human sensory nerves. Brain Res. 3, 1–402 (1966–1967).

BULLOCK, T.H.: Representation of information in neurons and sites for molecular participation. Proc. nat. Acad. Sci. (Wash.) 60, 1058–1068 (1968).

BURGESS, P.R., PETIT, D., WARREN, R.W.: Receptor types in cat hairy skin supplied by myelinated fibers. J. Neurophysiol. 31, 833–848 (1968).

BURNS, B.D.: The mammalian cerebral cortex. London: Arnold 1958.

BUSER, P., BIGNALL, K.E.: Nonprimary sensory projections on cat neocortex. Int. Rev. Neurobiol. 10, 111–165 (1967).

CELESIA, G.C., BROUGHTON, R.J., RASMUSSEN, T., BRANCH, C.: Auditory evoked responses from the exposed human cortex. Electroenceph. clin. Neurophysiol. 24, 458–466 (1968).

CHOW, K.L., LEIMAN, A.L., eds.: The structural and functional organization of the neocortex. Neurosci. Res. Prog. Bull. 8, 153–220 (1970).

CLARK, D.L., BUTLER, R.A., ROSNER, B.S.: Dissociation of sensation and evoked responses by a general anesthetic in man. J. comp. physiol. Psychol. 68, 315–319 (1969).

COBB, W., MOROCUTTI, C., eds.: The evoked potentials. Electroenceph. clin. Neurophysiol. Suppl. No. 26. Amsterdam: Elsevier 1968.

COLLINS, W.F., NULSEN, F.E., RANDT, C.T.: Relation of peripheral nerve fiber size and sensation in man. Arch. Neurol. (Chic.) 3, 381–385 (1960).

COLONNIER, M.L.: The structural design of the neocortex. In: Brain and conscious experience, pp. 1–23. Ed. by J.C. ECCLES. Berlin-Heidelberg-New York: Springer 1966.

CRAWFORD, B.H.: Visual adaptation in relation to brief conditioning stimuli. Proc. roy. Soc. B. 134, 283-302 (1947).

CREUTZFELDT, O.D., WATANABE, S., LUX, H.D.: Relations between EEG phenomena and potentials of single cortical cells. I. Evoked responses after thalamic and epicortical stimulation. Electroenceph. clin. Neurophysiol. 20, 1–18 (1966).

CURTIS, D.R., DUGGAN, A.W., FELIX, D., JOHNSTON, G.A.R.: GABA, bicuculline and central inhibition. Nature (Lond.) 226, 1222–1224 (1970).

CUSHING, H.: A note upon the faradic stimulation of the postcentral gyrus in conscious patients. Brain 32, 44–54 (1909).

DAVIS, H.: Enhancement of evoked cortical potentials in humans related to a task requiring a decision. Science 145, 182–183 (1964).

DEBECKER, J., DESMEDT, J.E., MANIL, J.: Sur la relation entre le seuil de perception tactile et les potentiels évoqués de l'écore cérébrale somato-sensible chez l'homme. C.R. Acad. Sci. (Paris) 260, 687–689 (1965).

DELUCCHI, M.R., GAROUTTE, B., AIRD, R.B.: The scalp as an electroencephalographic averager. Electroenceph. clin. Neurophysiol. 14, 191–196 (1962).

DEMPSEY, E.W., MORISON, R.S.: The electrical activity of a thalamocortical relay system. Amer. J. Physiol. 138, 283–296 (1943).

DONCHIN, E., COHEN, L.: Averaged evoked potentials and intramodality selective attention. Electroenceph. clin. Neurophysiol. 22, 537–546 (1967).

DONCHIN, E., LINDSLEY, D.B.: Averaged evoked potentials. Methods, results, and evaluation. Washington, D.C.: NASA, SP-191 (1969).

Doty, R.W.: Electrical stimulation of the brain in behavioral context. Ann. Rev. Physiol. **20**, 289–320 (1969).

Eccles, J.C.: Interpretation of action potentials evoked in the cerebral cortex. Electroenceph. clin. Neurophysiol. **3**, 449–464 (1951).

Eccles, J.C.: Conscious experience and memory. In: Brain and conscious experience, pp. 314–344. Ed. by J.C. Eccles. Berlin-Heidelberg-New York: Springer 1966.

Eijkman, E., Vendrik, A.J.H.: Detection theory applied to the absolute sensitivity of sensory systems. Biophys. J. **3**, 65–78 (1963).

Eriksen, C.W.: Discrimination and learning without awareness. Psychol. Rev. **67**, 279–300 (1956).

Foerster, O.: Motorische Felder und Bahnen. In: Handbuch der Neurologie, vol. 6, pp. 1–357. Ed. by O. Bumke und O. Foerster. Berlin: Springer 1936a.

Foerster, O.: Sensible coıticale Felder. In: Handbuch der Neurologie, vol. 6, pp. 358–448. Ed. by O. Bumke and O. Foerster. Berlin: Springer 1936b.

Foerster, O.: The motor cortex in man in the light of Hughling Jackson's doctrines. Brain **59** (2), 135–159 (1936c).

Fritsch, R., Hitzig, E.: Über die elektrische Erregbarkeit des Großhirns. Arch. f. Anat., Physiol. u. wissensch. Med. **37**, 300–332 (1870).

Geisler, C.D., Gerstein, G.L.: The surface EEG in relation to its sources. Electroenceph. clin. Neurophysiol. **13**, 927–934 (1961).

Gelfan, S., Carter, S.: Muscle sense in man. Exp. Neurol. **18**, 469–473 (1967).

Giaquinto, S., Pompeiano, O., Swett, J.E.: EEG and behavioral affects of fore- and hindlimb muscular afferent volleys in unrestrained cats. Arch. ital. Biol. **101**, 133–148 (1963).

Giblin, D.: Somatosensory evoked potentials in healthy subjects and in patients with lesions of the nervous system. In: Sensory evoked response in man. Ed. by R. Katzman. Ann. N.Y. Acad. Sci. **112**, 93–142 (1964).

Gleason, C.: The use of applied DC fields in the analysis of interictal epileptiform discharges. In: The nervous system and electric currents, pp. 93–98. Ed. by N.L. Wulfson and A. Sances, Jr. New York: Plenum Press 1970.

Goldiamond, I.: Indicators of perception: I. Subliminal perception, subception, unconscious perception. An analysis in terms of psychophysical indicator methodology. Psychol. Bull. **55**, 373–411 (1958).

Goldring, S., O'Leary, J.L., King, R.B.: Singly and repetitively evoked potentials in human cerebral cortex with D.C. changes. Electroenceph. clin. Neurophysiol. **10**, 233–240 (1958).

Goldring, S., Jerva, M.J., Holmes, T.G., O'Leary, J.L., Shields, J.R.: Direct response of human cerebral cortex. Arch. Neurol. (Chic.) **4**, 590–598 (1961).

Green, D.M., Swets, J.A.: Signal detection theory and psychophysics. New York: Wiley 1966.

Guiot, G., Albe-Fessard, D., Arfel, G., Hertzog, E., Vourc'h, G., Hard, Y., Derome, P., Aleonard, P.: Interpretation of the effects of thalamus stimulation in man by isolated shocks. C.R. Acad. Sci. (Paris) **254**, 3581–3583 (1962).

Hagbarth, K.-E., Höjeberg, S.: Evidence for subcortical regulation of the afferent discharge to the somatic sensory cortex in man. Nature (Lond.) **179**, 526–527 (1957).

Hagbarth, K.-E., Kerr, D.I.B.: Central influences on spinal afferent conduction. J. Neurophysiol. **17**, 295–307 (1954).

Haider, M., Spong, P., Lindsley, D.B.: Attention, vigilance, and cortical evoked-potentials in humans. Science **145**, 180–182 (1964).

Heath, R.G., Galbraith, G.C.: Sensory evoked responses recorded simultaneously from human cortex and scalp. Nature (Lond.) **212**, 1535–1537 (1966).

Hensel, H., Boman, K.K.A.: Afferent impulses in cutaneous sensory nerves in human subjects. J. Neurophysiol. **23**, 564–578 (1960).

Hensel, H., Zotterman, Y.: The response of mechanoreceptors to thermal stimulation. J. Physiol. (Lond.) **115**, 16–24 (1951).

Hern, J.E.C., Landgren, S., Phillips, C.G., Porter, R.: Selective excitation of corticofugal neurons by surface-anodal stimulation of the baboon's motor cortex. J. Physiol. (Lond.) **161**, 73–90 (1962).

Iggo, A.: Temperature discrimination in the skin. Nature (Lond.) **204**, 481–483 (1964).

Iggo, A.: The peripheral mechanisms of cutaneous sensation. In: Studies in Physiology, pp. 92–100. Ed. by D.R. Curtis and A.K. McIntyre. Berlin-Heidelberg-New York: Springer 1965.

Iwama, K., Jasper, H.: The action of gamma aminobutyric acid upon cortical electrical activity in the cat. J. Physiol. (Lond.) **138**, 365–380 (1957).

Jasper, H., Lende, R., Rasmussen, T.: Evoked potentials from the exposed somatosensory cortex in man. J. nerv. ment. Dis. **130**, 526–537 (1960).

Johansson, G.C.: Electrical stimulation of a human ventrolateral-subventrolateral thalamic target area, I–IV. Acta physiol. scand. **75**, 433–475 (1969).

Jouvet, M., Lapras, C., Tunisi, G., Wertheimer, P.: Mise en evidence chez l'homme au cours d'enregistrements stereotaxiques thalamiques d'un contrôle central des afférences somesthésiques. Acta neurochir. (Wien) **8**, 287–292 (1960).

Katzman, R., ed.: Sensory evoked responses in man. Ann. N.Y. Acad. Sci. **112**, 1–546 (1964).

Kelly, D.L., Goldring, S., O'Leary, J.L.: Averaged evoked somatosensory responses from exposed cortex of man. Arch. Neurol. (Chic.) **13**, 1—9 (1965).

Kietzman, M.L., Sutton, S.: The interpretation of two-pulse measures of temporal resolution in vision. Vision Res. **8**, 287–302 (1968).

Krech, D.: Does behavior really need a brain? In: Wm. James: Unfinished business. Ed. by R.B. McLeod. Washington: Amer. Psychol. Assoc. 1969.

Krnjević, K., Randić, M., Straughan, D.W.: An inhibitory process in the cerebral cortex. J. Physiol. (Lond.) **184**, 16–48 (1966).

Krnjević, K., Schwartz, S.: The action of gamma-aminobutyric acid on cortical neurones. Exp. Brain Res. **3**, 320–326 (1967).

Landau, W.M.: Evoked potentials. In: The neurosciences, pp. 409–482. Ed. by G.C. Quarton, T. Melnechuk and F.O. Schmitt. New York: Rockefeller Univ. Press 1967.

Landgren, S., Phillips, C.G., Porter, R.: Cortical fields of origin of the monosynaptic pyramidal pathways to some alpha motoneurones of the baboon's hand and forearm. J. Physiol. (Lond.) **161**, 112–125 (1962).

Lansing, R.W., Schwartz, E., Lindsley, D.B.: Reaction time and EEG activation under alerted and non-alerted conditions. J. exp. Psychol. **58**, 1–6 (1959).

Leyton, A.S.F., Sherrington, C.S.: Observations on the excitable cortex of the chimpanzee, orang-utan, and gorilla. Quart. J. exp. Physiol. **11**, 135–222 (1917).

Li, C.-L., Chou, S.N.: Cortical intracellular synaptic potentials and direct cortical stimulation. J. cell. comp. Physiol. **60**, 1–16 (1962).

Libet, B.: Cortical activation in conscious and unconscious experience. Perspect. Biol. Med. **9**, 77–86 (1965).

Libet, B.: Brain stimulation and the threshold of conscious experience. In: Brain and conscious experience, pp. 165–181. Ed. by J.C. Eccles. Berlin-Heidelberg-New York: Springer 1966.

Libet, B., Alberts, W.W., Wright, E.W., Jr., Delattre, L., Levin, G., Feinstein, B.: Production of threshold levels of conscious sensation by electrical stimulation of human somatosensory cortex. J. Neurophysiol. **27**, 546–578 (1964).

Libet, B., Alberts, W.W., Wright, E.W., Jr., Levin, G., Feinstein, B.: Sensory perception by direct stimulation of human cerebral cortex: stimulus parameters. Fed. Proc. **18**, 92 (1959).

Libet, B., Alberts, W.W., Wright, E.W., Jr., Feinstein, B.: Responses of human somatosensory cortex to stimuli below threshold for conscious sensation. Science **158**, 1597–1600 (1967).

Libet, B., Alberts, W.W., Wright, E.W., Jr., Feinstein, B.: Cortical and thalamic activation in conscious sensory experience. In: Neurophysiology Studied in Man, pp. 157–168. Ed. by G.G. Somjen. Amsterdam: Excerpta Medica 1972a.

Libet, B., Alberts, W.W., Wright, E.W., Jr., Lewis, M., Feinstein, B.: Some cortical mechanisms mediating conscious sensory responses and the somatosensory qualities in man. In: Somatosensory System. Ed. by H.H. Kornhuber. Stuttgart: Georg Thieme (In press, 1972b).

LILLY, J.C.: Injury and excitation by electric currents. A. The balanced pulse-pair waveform. In: Electrical stimulation of the brain, pp. 60–64. Ed. by D.E. SHEER. Austin: Univ. of Texas Press 1961.

LILLY, J.C., AUSTIN, G.M., CHAMBERS, W.W.: Threshold movements produced by excitation of cerebral cortex and efferent fibers with some parametric regions of rectangular current pulses (cats and monkeys). J. Neurophysiol. **15**, 319–341 (1952).

MACINTYRE, W.J., BIDDER, T.G., ROWLAND, V.: The production of brain lesions with electric currents. Proc. Nat. Biophys. Conf. 1st, 1957 (pub. 1959) pp. 723–732.

MACKAY, D.M.: Evoked brain potentials as indicators of sensory information processing. Neurosci. Res. Prog. Bull. **7**, 184–276 (1969).

MARG, E., DIERSSEN, G.: Somatosensory reports from electrical stimulation of the brain during therapeutic surgery. Nature (Lond.) **212**, 188–189 (1966).

MARSHALL, W.: Spreading cortical depression of Leaõ. Physiol. Rev. **39**, 239–279 (1959).

MARTIN, H.F., III, MANNING, J.W.: Peripheral nerve and cortical responses to radiant thermal stimulation of skin fields. Fed. Proc. **28**, 458 (1969).

MELZACK, R., WALL, P.D.: Masking and metacontrast phenomena in the skin sensory system. Exp. Neurol. **8**, 35–46 (1963).

MIHAILOVIĆ, L., DELGADO, J.M.R.: Electrical stimulation of monkey brain with various frequencies and pulse durations. J. Neurophysiol. **19**, 21–36 (1956).

MOUNTCASTLE, V.B.: The problem of sensing and the neural coding of sensory events. In: The neurosciences, pp. 393–408. Ed. by G.C. QUARTON, T. MELNECHUK and F.O. SCHMITT. New York: Rockefeller Press 1967.

MOUNTCASTLE, V.B., POWELL, T.P.S.: Central nerve mechanisms subserving position-sense and kinesthesis. Bull. Johns Hopk. Hosp. **105**, 173–200 (1959a).

MOUNTCASTLE, V.B., POWELL, T.P.S.: Neural mechanisms subserving cutaneous sensibility, with special reference to the role of afferent inhibition in sensory perception and discrimination. Bull. Johns Hopk. Hosp. **105**, 201–232 (1959b).

MÜLLER, J.: Elements of physiology., transl. by W. BALY, LEA, and BLANCHARD. Book VI, Of the senses (1843).

OCHS, S.: Analysis of cellular mechanisms of direct cortical responses. Fed. Proc. **21**, 642–647 (1962).

OCHS, S., CLARK, F.J.: Tetrodotoxin analysis of direct cortical responses. Electroenceph. clin. Neurophysiol. **24**, 101–107 (1968a).

OCHS, S., CLARK, F.J.: Interaction of direct cortical responses — a possible dendritic site of inhibition. Electroenceph. clin. Neurophysiol. **24**, 108–115 (1968b).

OSCARSSON, O.: Proprioceptive and exteroceptive projections to the pericruciate cortex of the cat. In: Studies in physiology, pp. 221–226. Ed. by D.R. CURTIS and A.K. MCINTYRE. Berlin-Heidelberg-New York: Springer 1965.

OSCARSSON, O., ROSÉN, I.: Projection to cerebral cortex of large muscle-spindle afferents in the forelimb nerves of the cat. J. Physiol. (Lond.) **169**, 924–945 (1963).

OSCARSSON, O., ROSÉN, I.: Short latency projections to the cat's cerebral cortex from skin and muscle afferents in the contralateral forelimb. J. Physiol. (Lond.) **182**, 164–184 (1966).

PATTON, H.D., AMASSIAN, V.E.: Single- and multiple-unit analysis of cortical stage of pyramidal tract activation. J. Neurophysiol. **17**, 345–363 (1954).

PENFIELD, W.: The excitable cortex in conscious man. Liverpool: Liverpool Univ. Press 1958.

PENFIELD, W., BOLDREY, E.: Somatic motor and sensory representation in the cerebral cortex of man as studied by electrical stimulation. Brain **60**, 389–443 (1937).

PENFIELD, W., JASPER, H.: Epilepsy and the functional anatomy of the human brain. Boston: Little, Brown & Co. 1954.

PENFIELD, W., KRISTIANSEN, K.: Epileptic seizure patterns. Springfield: Thomas 1951.

PENFIELD, W., RASMUSSEN, T.: The cerebral cortex of man. New York: Macmillan 1950.

PENFIELD, W., WELCH, K.: Instability of response to stimulation of the sensorimotor cortex of man. J. Physiol. (Lond.) **109**, 358–365 (1949).

PENFIELD, W., WELCH, K.: The supplementary motor area of the cerebral cortex. Arch. Neurol. Psychiat. (Chic.) **66**, 289–317 (1951).

PERL, E.R.: Relation of cutaneous receptors to pain. Int. Union Physiol. Sci., 24th, Washington, D.C. **6**, 235–236 (1968).

PHILLIPS, C.G.: Changing concepts of the precentral motor area. In: Brain and conscious experience, pp. 389–421. Ed. by J.C. ECCLES. Berlin-Heidelberg-New York: Springer 1966.

PHILLIPS, C.G.: Motor apparatus of the baboon's hand. Proc. roy. Soc. B. **173**, 141–174 (1969).

PHILLIPS, C.G., PORTER, R.: Unifocal and bifocal stimulation of the motor cortex. J. Physiol. (Lond.) **162**, 532–538 (1962).

PHILLIS, J.W., OCHS, S.: Occlusive behavior of the negative wave direct cortical response (DCR) and single cells in the cortex. J. Neurophysiol. **34**, 374–388 (1971).

PHILLIS, J.W., YORK, D.H.: Cholinergic inhibition in the cerebral cortex. Brain Res. **5**, 517–520 (1967).

POWELL, T.P.S., MOUNTCASTLE, V.B.: Some aspects of the functional organization of the cortex of the postcentral gyrus of the monkey: a correlation of findings obtained in a single unit analysis with cytoarchitecture. Bull. Johns Hopk. Hosp. **105**, 133–162 (1959).

PRINCE, D.A., WILDER, B.J.: Control mechanisms in cortical epileptogenic foci; "surround" inhibition. Arch. Neurol. (Chic.) **16**, 194–202 (1967).

PROVINS, K.A.: The effect of peripheral nerve block on the appreciation and execution of finger movements. J. Physiol. (Lond.) **143**, 55–67 (1958).

PURPURA, D.P.: Nature of electrocortical potentials and synaptic organizations in cerebral and cerebellar cortex. Int. Rev. Neurobiol. **1**, 47–163 (1959).

PURPURA, D.P., GIRADO, M., GRUNDFEST, H.: Selective blockade of excitatory synapses in the cat brain by $\gamma$-aminobutyric acid. Science **125**, 1200–1202 (1957a).

PURPURA, D.P., POOL, J.L., RANSOHOFF, J., FRUMIN, M.J., HOUSEPIAN, E.M.: Observations on evoked dendritic potentials of human cortex. Electroenceph. clin. Neurophysiol. **9**, 453–459 (1957b).

RAAB, D.: Backward masking. Psychol. Bull. **60**, 118–129 (1963).

ROSENTHAL, J., WALLER, H.J., AMASSIAN, V.E.: An analysis of the activation of motor cor tical neurons by surface stimulation. J. Neurophysiol. **30**, 844–858 (1967).

ROSNER, B.S., GOFF, W.R.: Electrical responses of the nervous system and subjective scales of intensity. In: Contributions to sensory physiology, 2, pp. 169–221. Ed. by W.D. NEFF. New York: Academic Press 1967.

ROWLAND, V., MACINTYRE, W.J., BIDDER, T.G.: The production of brain lesions with electric currents. II. Bidirectional currents. J. Neurosurg. **17**, 55–69 (1960).

SHAGASS, C., SCHWARTZ, M.: Evoked cortical potentials and sensation in man. J. Neuropsychiat. **2**, 262–270 (1961).

SHEVRIN, H., FRITZLER, D.E.: Visual evoked response correlates of unconscious mental processes. Science **161**, 295–298 (1968).

SIGEL, N.: Prick threshold stimulation with square wave current: a new measure of skin sensibility. Yale J. Biol. Med. **26**, 145–154 (1953).

SKOGLUND, S.: Anatomical and physiological studies of knee joint innervation in the cat. Acta physiol. scand. **36**, Suppl. 124, 1—101 (1956).

STOHR, P.E., GOLDRING, S.: Origin of somatosensory evoked scalp responses in man. J. Neurosurg. **31**, 117–127 (1969).

STONEY, S.D., JR., THOMPSON, W.D., ASANUMA, H.: Excitation of pyramidal tract cells by intracortical microstimulation: effective extent of stimulating current. J. Neurophysiol. **31**, 659–669 (1968).

SUGAYA, E., GOLDRING, S., O'LEARY, J.L.: Intracellular potentials associated with direct cortical response and seizure discharge in cat. Electroenceph. clin. Neurophysiol. **17**, 661–669 (1964).

SUTTON, S.: The specification of psychological variables in an average evoked potential experiment. In: Averaged evoked potentials. Methods, results, and evaluation, pp. 237–297. Ed. by E. DONCHIN and D.B. LINDSLEY. Washington, D.C.: NASA, SP-191 1969.

SUZUKI, H., OCHS, S.: Laminar stimulation for direct cortical responses from intact and chronically isolated cortex. Electroenceph. clin. Neurophysiol. **17**, 405–413 (1964).

SWETS, J.A.: Is there a sensory threshold? Science **134**, 168–177 (1961).

SWETT, J.E., BOURASSA, C.M.: Comparison of sensory discrimination thresholds with muscle and cutaneous nerve volleys in the cat. J. Neurophysiol. **30**, 530–545 (1967).

Talairach, J., Bancaud, J.: The supplementary motor area in man. Int. J. Neurol. (Montevideo) **5**, 330–347 (1966).

Towe, A.L., Amassian, V.E.: Patterns of activity in single cortical units following stimulation of the digits in monkeys. J. Neurophysiol. **21**, 292–311 (1958).

Towe, A.L., Whitehorn, D., Nyquist, J.K.: Differential activity among wide-field neurons of the cat postcruciate cerebral cortex. Exp. Neurol. **20**, 497–521 (1968).

Vaughan, H.G., Jr.: The relationship of brain activity to scalp recordings of event-related potentials. In: Averaged evoked potentials. Methods, results, and evaluation, pp. 45–94. Ed. by E. Donchin and D.B. Lindsley. Washington, D.C.: NASA, SP-191 1969.

Vaughan, H.G., Jr., Costa, L.D., Gilden, L.: The functional relation of visual evoked response and reaction time to stimulus intensity. Vision Res. **6**, 645–656 (1966).

Wagman, I., Battersby, W.S.: Neural limitations of visual excitability. V. Cerebral afteractivity evoked by photic stimulation. Vision Res. **4**, 193–208 (1964).

Walter, W.G.: The convergence and interaction of visual, auditory, and tactile responses in human nonspecific cortex. Ann. N.Y. Acad. Sci. **112**, 320–361 (1964).

Weddell, G.: Receptors for somatic sensation. In: Brain and Behavior, vol. I, pp. 13–48. Ed. by M.A.B. Brazier. Washington, D.C.: Amer. Inst. Biol. Sci. 1961.

Whitehorn, D., Towe, A.L.: Postsynaptic potential patterns evoked upon cells in sensorimotor cortex of cat by stimulation at the periphery. Exp. Neurol. **22**, 222–242 (1968).

Wiesendanger, M.: The pyramidal tract. Recent investigations on its morphology and function. Ergebn. Physiol. **61**, 72–136 (1969).

Woolsey, C.N.: Patterns of localization in sensory and motor areas of the cerebral cortex. In: The biology of mental health and disease, Chap. 14. New York: Hoeber 1952.

Woolsey, C.N., Erickson, T.C.: Study of the postcentral gyrus of man by the evoked potential technique. Trans. Amer. neurol. Ass. **75**, 50–52 (1950).

# The Somatosensory System

## Edited by H. H. Kornhuber

With Contributions by

D. Albé-Fessard
W. W. Alberts
E. R. Arbuthnott
J. C. Aschoff
E. A. Asratyaw
S. F. Atweh
J. M. Besson
D. Bowsher
I. A. Boyd
G. Broggi
A. G. Brown
M. G. Cesa-Bianchi
A. M. Dart
I. Darian-Smith
M. S. Dash
S. S. Deshpande
J. C. Eccles
B. Feinstein
H. Goodman
G. Gordon
Y. Hassmannova
T. Hongo
J. Hyvärinen
A. Iggo
Y. Iwamura
S. J. Jabbur
K. O. Johnson
Y. Jokinen
K. U. Kalu
O. Keller
D. R. Kenshalo
H. Koike
H. H. Kornhuber
C. La Motte
M. Lewis
B. Libet
I. Linnankoski
A. K. McIntyre
M. Mancia
M. Manfredi
M. Margnelli
M. Merzenich
R. Näätänen
K. Nier
R. L. Paul
A. Poranen
D. A. Poulos
T. A. Quilliam
P. Rudomin
N. H. Sabah
H. Sakata
M. Sassen
R. F. Schmidt
M. L. Sotgiu
I. Spector
W. A. Spencer
H. Táborková
Ch. J. Vierck Jr.
L. Vyklicky
S. Watanabe
W. V. Weber
E. W. Wright Jr.
D. W. Young
M. Zimmermann

181 Figures, 10 Tables



Georg Thieme Publishers Stuttgart 1975

# Cortical Representation of Evoked Potentials Relative to Conscious Sensory Responses, and of Somatosensory Qualities – in Man

B. LIBET, W. W. ALBERTS, E. W. WRIGHT, Jr., M. LEWIS and B. FEINSTEIN

The study of cortical evoked potential and neuronal unit responses to sensory stimuli can give us essential information about the locations and types of projections to the cortex and about the physiological mechanisms activated by such projections. By themselves, however, such studies cannot establish the relationship between neuronal activities and the production of conscious sensory experiences. Obviously this can be done only by direct validations of the electrophysiological actions against the other independent variable, the subjective experience of the human subject. Indeed, since sensory projections and their cortical responses would be expected to subserve other functions in addition to that of conscious sensory experience, it should be no surprise to find dissociations between the occurrence of certain evoked electrophysiological responses to stimuli and the eliciting of a subjective response. The clearest example of this is seen in the responses to certain muscle afferents. Group I muscle afferents have been found to elicit evoked responses in primary somatosensory or closely related cortex (Amassian and Berlin, 1958; Oscarsson and Rosen, 1963, 1966; Albe-Fessard, 1967; Swett and Bourassa, 1967). Yet these inputs do not elicit any subjective experience of motion, position, muscle length, or any other sensation in man (Brindley and Merton, 1960; Gelfan and Carter, 1967). Also, in contrast to other somatic inputs, the group I afferent inputs could not produce behavioral (Giaquinto, Pompeiano, and Swett, 1963) or conditional learned responses in the cat (Swett and Bourassa, 1967). The projection to the cortex of group I muscle-spindle afferent impulses, as well as of other somatic afferents, may of course serve in the integration and organization of movements mediated by the cortex, without giving rise to any subjective sensory experience (e.g. Oscarsson, 1965; Albe-Fessard, 1967; Kornhuber, 1971). The actual subjective experience of position and movement would depend on other inputs, especially those originating in and about the joints (Skoglund, 1956; Provins, 1958, Mountcastle and Powell, 1959).

In the present paper we shall describe relationships or the absence of them between electrophysiological responses of cortical areas to peripheral or cerebral stimuli, on the one hand, and the ability of such stimuli to elicit conscious sensory responses, on the other. In addition, the problem of the cortical representation of mechanisms for the various specific subjective qualities of somatic sensation will be considered, on the basis of evidence from studies of the different somatosensory qualities elicited by various modes of stimulation of somatosensory (SS-I) cortex in the postcentral gyrus.

## "Excitable" Somatosensory Cortex

The extent of the "excitable" areas of the human cerebral cortex, in which electrical stimulation gives rise to conscious somatosensory experiences, may vary with the condition of the subject, expecially with regard to whether his pathological disorder is one (like epilepsy) that affects the excitability of the cortex. It can also vary with the stimulus values used, i.e., how they relate to the range between those for threshold responses and those for producing seizure activity, including local electrical convulsive-type waves. Earlier studies have used predominantly epileptic subjects (Foerster, 1936; Penfield and Boldrey, 1937; Penfield and Rasmussen, 1950; Penfield and Jasper, 1954). In such subjects and with the use of stimulus strengths below levels for producing electrical seizure wave patterns, the ability to elicit conscious somatosensory responses was essentially restricted to Rolandic areas and to somatosensory area II (Penfield and Rasmussen, 1950; Penfield and Jasper, 1954). Foerster (1936) reported that strong stimulation of area 5, in the superior parietal region posterior to the upper part of the postcentral gyrus, could also elicit sensations over the whole body, initially contralaterally; this area overlapped with that producing whole body extrapyramidal type movements. However, Penfield and Boldrey (1937, p. 411) reported only an uncertain production of an epigastric aura for this area.

Even in the Rolandic region, stimulation of the *pre*central gyrus could not elicit any non-motion sensations that were unrelated to the production of movement itself, when this was done in nonepileptic subjects with currents not much above threshold for eliciting motor responses by trains of pulses (Libet et al., 1964). This would be in accord with the suggestion of Oscarsson (1965) that the projection of *cutaneous* afferents to the primary *motor* area may serve motor rather than conscious sensory functions (see also Kornhuber and Aschoff, 1964, and Kornhuber, 1971). Stimulation of the *post*central gyrus could elicit motor responses, but these were twitchlike in nature, appearing repetitively at about 1 per sec when stimulus trains of sufficient intensity were used; these twitches were produced only at levels of current several times greater than the liminal one for sensory responses (Libet et al., 1964; Libet, 1973). A smoother tremor at about 5 per sec could often be initiated or augmented in parkinsonian subjects with currents not far from the threshold for sensory responses. The electrical "inexcitability" of an area for eliciting a given response does not in itself exclude the possibility of the area's having a role in mediating that response, but it at least indicates a real difference between this area and those that are "excitable." The functional distinction between pre- and postcentral gyri in the adult human cortex may, therefore, be sharper (Libet et al., 1964; Libet, 1973) than was previously thought by those who referred to the combined Rolandic region as a "sensorimotor strip" (e.g. Penfield and Rasmussen, 1950; Penfield and Jasper, 1954; compare also Glees, 1961, p. 240, and Kornhuber, 1971).

## Evoked Potentials and Conscious Somatosensory Responses

### Relation of "excitable" Somatosensory Cortex to Distribution of Somatic Evoked Potentials

When the early or "primary" (surface positive and succeeding negative) evoked potentials (EP) are used as the indicators of somatic sensory projections to the cortex, there is generally good accord between their representation and the distribution of excitable cortex (e.g. Jasper, Lende, and Rasmussen, 1960; Hirsch et al., 1961; Kelly, Goldring, and O'Leary, 1965). One exception to this is the generation of a primary EP response by the precentral gyrus even though stimulation there apparently does not directly elicit a conscious sensory response in nonepileptic subjects (see above). However, EPs with later components, especially in the latency ranges of 100–300msec, can be recorded widely over areas outside areas SS-I and SS-II, not only contralaterally but also ipsilaterally (e.g., Katzman, 1964; Cobb and Morocutti, 1967; MacKay, 1969; Donchin and Lindsley, 1969).

It has been argued that the evoked potentials elicited by somatic stimuli but recorded over cortical areas other than the primary somatosensory cortex actually originate only in the primary cortex in man (Stohr and Goldring, 1969; Vaughan, 1969; MacKay, 1969, pp. 201–203). For the criterion that a recorded EP is generated locally, rather than sampling the field of one generated elsewhere, Stohr and Goldring (1969) have insisted that the evoked potential waves recorded at the cortical surface must have polarities opposite to those recorded subcortically at the same site. On this basis, they reported that EPs elicited by stimulation of the contralateral median nerve could be found to originate only in the primary somatosensory hand area (postcentral gyrus) and occasionally in the motor or precentral area. This conclusion has been difficult to accept in view of the large differences among the relative distributions of the recorded fields for different components of the evoked potentials (e.g. Gastaut et al., 1967; Vaughan, 1969; Broughton, 1969; Libet 1973). It would be especially difficult to explain why our subdural recordings of the primary, initially surface positive component (latency about 20msec) should be highly localized to the Rolandic cortex, whereas later components can be recorded additionally at more remote sites and also may have highly differing wave forms when recorded simultaneously at various sites. Furthermore, there is now evidence that even meets the criterion of Stohr and Goldring (1969) for cortical origins other than primary sensory cortex. In at least some primates (squirrel and rhesus monkeys) EPs can originate in wide "polysensory" areas of the frontal lobe (e.g. Bignall and Imbert, 1967); these EPs may be related to those recorded in the human frontal lobe by Walter (1964). Additionally, we have now found that the supplementary motor area of the human subject can itself generate large late EP components in response to various sensory stimuli.

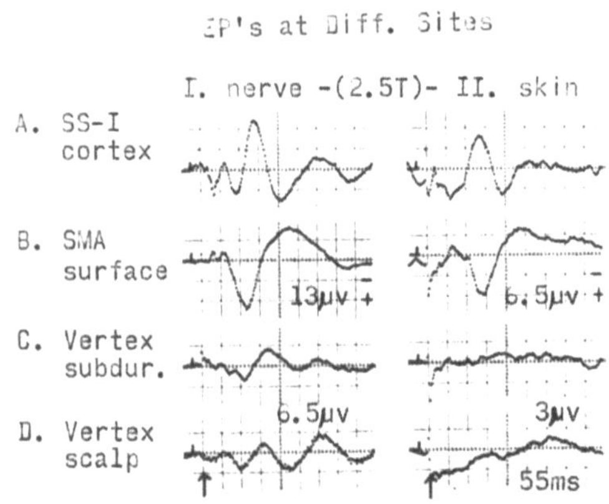


Figure 1. Evoked potentials at various cortical sites in response to somatic stimulation of peripheral nerve and skin. Subject had grand mal epilepsy with a suspected focus in left supplementary motor area. *Stimuli.* Column I: right median nerve; II: "skin" (large unipolar electrode on ventral surface of right forearm which elicited only a local sensation). Stimulation with 0.5msec single pulses at 1 per sec, with peak currents for each electrode at 2.5 times the threshold intensity required for eliciting a conscious sensory response with that electrode. *Recordings. A:* Contralateral (left) somatosensory cortex (postcentral gyrus), 1mm subdural surface electrode; stimulation at this site elicited tingling sensations whose referral area overlapped with that obtained with the skin electrode on the forearm but not with that for median nerve. *B:* Contralateral (left) supplementary motor area, with 1mm tip of Schryver-made electrode, inserted from frontolateral burr hole until contact was made with the falx; stimulation at this electrode elicited a generalized movement of the right shoulder and arm with rotation inwards, but with no conscious sensory responses independent of the sensations generated by the movement. *C:* 1mm surface electrode inserted subdurally, on contralateral (left) side, until it reached almost to midline at the vertex. *D:* Scalp needle at the vertex. (The large initial shift after the skin stimulus was, of course, a stimulus artifact.) All recordings unipolar; reference electrodes on both ear lobes connected together; grounding band on the arm, central to the peripheral stimulating electrodes. Each tracing is average of 256 responses. Stimulus pulse time indicated by arrow (preceding pip is a calibration signal). Sweep time = 55msec per division, giving a 500msec analysis period after the stimulus. Vertical calibrations of 13μv. per division in I B applies to I A–C, 6.5μv. in II B to II A–C; vertex scalp gains in D are about twice those in the respective A–C recordings. Positivity of active electrode is downwards in all figures

*Responses at supplementary motor area (SMA) and vertex.* Stimulation of SMA cortex in man (on the frontal mesial surface above the cingulate gyrus) gives rise to generalized movements of contralateral portions of the body (Penfield and Rasmussen, 1958; Talairach and Bancaud, 1966), but no conscious sensory experiences are elicited (aside from those of the movements). Stimulation of the cortical surface near the vertex has elicited neither movements nor sensations, in our experience. EPs can be recorded at these sites (as well as on the postcentral gyrus) in response to somatic, auditory, and visual sensory stimuli (figs. 1, 2). We have been able to show that certain later components of the EP responses of SMA cortex to all these stimulus modalities are generated locally there, on the criterion of their reversal of polarity when recorded by a deeper electrode at the same site; this conclusion appears to be quite clear even though the findings are limited at present to one subject. In figure 3-*I, II* it may be seen that reversal of polarities in the case of somatic stimuli applies chiefly to the dominant surface positive wave with a peak latency of about 125msec for median nerve stimuli and about 150msec for skin; in addition, there appeared to be reversal of a small preceding surface negative component and of at least the first portion of a succeeding surface negative wave, making up a triphasic complex that reversed in the period between about 55msec and 275msec after the stimulus. For visually evoked responses (fig. 3-*IV*), reversal appeared to start only with the surface positive wave having a peak at about 165msec; the surface negative peak at about 260msec is also clearly reversed, and many of the succeeding components may also be. For audio stimuli (fig. 3-*III*), the reversal of the surface positive component having a peak at about 135msec was not quite as clear, perhaps because of the 100msec duration of the tone stimulus itself; but the surface negative peak at about 220msec is clearly reversed, and many of the later components probably show reversal. With all stimuli, there were earlier evoked components that were not reversed; these include a small surface negative component with a starting latency of about 22msec after a somatic stimulus (fig. 3-*I*), which corresponds to the primary, surface positive component recorded at SS-I cortex (see also fig. 1-*I A* and *B*). The nonreversed components are presumably volume pickups from generators located elsewhere, though it is possible that some local reversals could have been missed with the particular electrode locations employed. One may speculate that the function of the various sensory projections to SMA cortex is concerned with their roles in the elaboration of motor responses, perhaps in orienting movements, etc., rather than with the mediation of conscious sensory experience in any direct way.

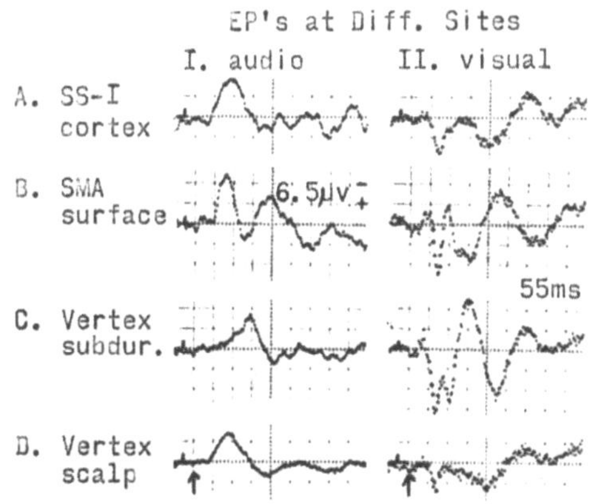


**Figure 2.** Evoked potentials at various cortical sites in response to audio and visual stimuli. Subject and recordings same as in fig. 1 (but vertical calib. of 6.5μv./div. applies to all tracings, and each tracing is average of 128 responses). Stimuli: Column I-audio tone, 1.2KHz and 100msec duration: reference recording electrode on nasium and ground on the arm. Column II-light flasch from Grass photic stimulator; ref. electrode on left mastoid and ground on the nasium. Intensities of audio and visual stimuli were well above threshold levels

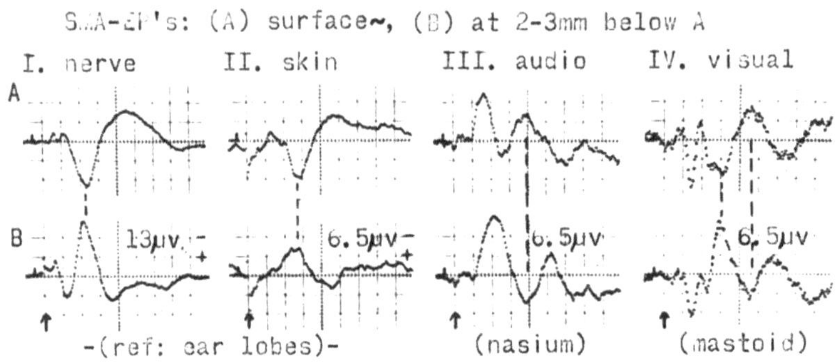


**Figure 3. Evoked potentials at supplementary motor cortex, surface vs. depth recordings. Subject, stimuli, and recordings for columns I—II same as in fig. 1, and for columns III—IV as in fig. 2. *A:* Same surface electrode as figs. 1 *B* and 2 *B*. *B,* second 1mm contact on the Schryver-made electrode; center of this contact was 3mm deeper than *A*. *A* and *B* were recorded simultaneously. Dashed vertical lines indicate the largest components of those that show inversion of polarity; see text**

Whether the cortex near the vertex also generates some of the EP components recorded there cannot be told in the absence of a subcortical lead. It is quite clear, however, that the vertex EPs (at least for somatic stimuli) are *not* all simply volume conducted potentials generated at the primary sensory cortex, as has been argued for example by Vaughan (1969); Kooi, Tipton, and Marshall (1971) have recently reported evidence from scalp recordings that also contradicts Vaughan's proposal. Although Stohr and Goldring (1969) found that vertex scalp EPs in response to median nerve stimuli disappeared after ablation of the hand area of the contralateral postcentral gyrus, they noted that this proved only that the sensory cortex was necessary for these (extracranially recorded) EPs; it did not necessarily mean that the vertex EPs were actually generated by the primary sensory cortex. Our data bear on this point as follows: (1) With median nerve stimuli, the positive peak at about 120msec latency in the vertex EPs (fig. 1-I *C, D*) resembles the main one that was found to be generated by SMA cortex (fig. 3 1-I *B;* 3-I *A, B*). (2) Even more striking, on this point, are our preliminary findings that the vertex recordings, whether subdural or scalp, exhibit relatively small or no EPs in response to electrical stimulation of *skin,* in contrast to median nerve. (Selective stimulation of the "skin," i.e., cutaneous afferent fibers in a restricted area, was accomplished by using close bipolar electrodes, which were oriented at right angles to the axis of known nerves in the region; the sensory response was then subjectively felt only under or very close to the electrodes.) This difference in vertex EPs was true for strengths of stimuli to the skin that elicited large EPs at the primary sensory cortex and at SMA (fig. 2-II *A, C, D*). The large difference between median nerve and skin stimuli for eliciting vertex scalp EPs was also seen in several normal subjects tested. This finding indicates that the vertex recordings are not simply volume conducted reflections of EPs generated at either the postcentral gyrus or SMA, and also that the vertex EPs that can be elicited by median *nerve* stimuli (fig. 1-I *C, D*) are probably generated at sites other than SS-I or SMA, perhaps by the cortex at the vertex itself. With visual stimuli, however, there was a remarkable similarity between all of the wave forms in the subdural vertex record (fig. 2-II *C*) and those in the SMA deep cortical record (fig. 3-IV *B*).

The relative absence of vertex EPs with electrical stimuli to the skin, at strengths that are comparable (in relation to the respective thresholds) to those which produce substantial EPs when a median nerve is stimulated, suggests further interesting possibilities. It may be that cutaneous afferent projections are relatively poorly represented in the cortical area that does generate the vertex EPs, in contrast to afferents from deeper somatic strucutres such as the joints, muscles, etc. On the other hand, some types and intensities of natural cutaneous stimuli are evidently capable of eliciting vertex EPs (Klinke, Fruhstorfer, and Finkenzeller, 1968; Fruhstorfer and Kenshalo, personal communication); it would thus be desirable to compare the effects of natural and electrical stimuli in the same subject. While it might be expected from various other reports that peripheral natural stimuli would be more effective than electrical ones for eliciting central responses, the large difference between the vertex responses to stimulation of median nerve vs. skin when *both* stimuli are electrical would still suggest that there is a significant difference between the potentials evocable by the different sources of afferent fibers. Indeed, a similar difference, though of much less magnitude, appears to hold even for EPs recorded subdurally at the postcentral gyrus. (This difference is seen in figs. 1-*II A*, in spite of the fact that the referral area with stimulation at the cortical electrode was closer to that for the skin electrode than for median nerve stimulation. A similar difference between nerve and skin EPs appeared to prevail in two additional subjects suitably tested.) The attenuation of EP responses when recorded at the overlying scalp (e.g. Libet et al., 1967; Broughton, 1969) could thus make relatively negligible the contribution (to the parietal scalp-recorded EP) from cutaneous afferents that are contained in a median nerve. This might be especially true when the nerve is electrically stimulated at levels near the threshold for conscious sensory responses, when the detectable scalp EP may be virtually entirely elicited by simultaneously excited large afferent fibers from deeper structures. It should be noted that almost all reports dealing with vertex EP responses to somatic stimuli have involved stimulation of a peripheral nerve, and that there probably are no purely "cutaneous" nerves (Mountcastle and Powell, 1959).

One may note our additional observation that median nerve stimuli could elicit substantial EPs *ipsi*laterally at SS-I, SMA, and vertex, all recorded with subdural electrodes. These EPs were all much smaller than the responses to the contralateral nerve and were lacking in early components (latencies shorter than 100msec); the largest wave in the SS-I and vertex records was one with a negative peak at 175–200msec, and at SMA a positive peak at about 150msec. However, skin stimuli again elicited little or no detectable EPs at any of these ipsilateral sites (except for SMA, and then only at 4 times threshold).

It should be evident, from these various findings with stimulation and recording, that considerable caution should be exercised when one attempts to interpret psychophysical studies of EPs in relation to subjective sensory experience and specific perceptual processes; this is particularly true when the EPs are recorded extracranially and especially at the vertex in response to electrical stimulation of mixed somatic nerves.

### Relation of EP Components to Conscious Sensory Responses

If EPs are at least crude indicators of neuronal population responses when generated in relevant areas, they might provide clues to some of the neuronal processes mediating the functions of those areas (e.g. MacKay, 1969; Donchin and Lindsley, 1969). A single pulse electrical stimulus to skin elicits both the primary and later EP components in SS-I cortex when the stimulus is at or above the threshold for producing a conscious sensory response (fig. 4). At subthreshold levels for the latter response, single pulse stimuli elicit a small primary EP (surface positive, and/or the succeeding surface negative wave at times), but virtually no later EP components (Libet et al., 1967). Incidentally, brief periods of repetition (at 60 pps for $< 0.1$sec) of subthreshold single pulses can elicit a conscious sensory response. The somewhat lower threshold intensity for eliciting, with repetitive pulses, a sensory response that has a subjective quality similar to that elicited by the single pulse, indicates that a single pulse normally must excite more than one peripheral sensory nerve fiber in order to elicit a conscious sensory response (cf. Buchtal and Rosenfalck, 1966–7; Hensel

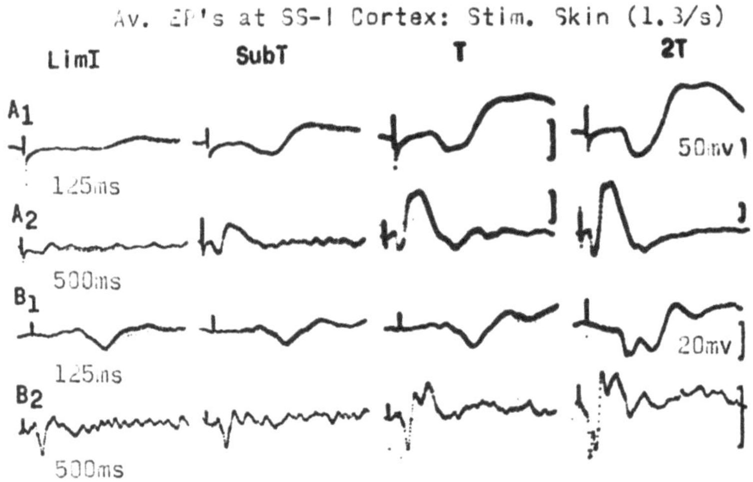


Figure 4. Averaged evoked potentials of somatosensory cortex in relation to threshold stimuli at skin. Each tracing is the average of 500 responses at 1.8 per sec. Total trace length is 125msec in $A_1$ and $B_1$ and 500msec in $A_2$ and $B_2$; beginning of stimulus artifact has been made visible near start of each tracing. A and B: separate subjects, both parkinsonian patients. Vertical column T: threshold stimuli, subject reporting not feeling some of the 500 stimuli. Column 2T: stimuli at twice threshold current; all stimuli felt distinctly. Column SubT: subthreshold stimuli, none felt by subject; current about 15% below T in subject A, 25% below T in B. Column Lim-I: subthreshold stimuli at "liminal intensity" (see text), about 25% below T in subject A, about 35 to 40% below T in B. Polarity, positive downward in all figures. Vertical bars in A, under T, indicate 50$\mu$v. in $A_1$ and $A_2$ respectively, but gains are different in 2T as shown; for $B_1$ and $B_2$, 20$\mu$v. bars. (Calibration obtained by summating 500 sweeps of calibrating signal.) (From Libet *et al.*, 1967, by permission of *Science*.)

and Bowman, 1960; Shagass and Schwartz, 1961). In many subjects, however, the quality changed from a "tap" or "twinge" with a threshold single pulse to skin, to a more piercing, burning quality with the brief train of pulses. When this occurred, the liminal intensity for the repetitive pulses could be considerably lower (down by 25–50%) than the threshold-I for a single pulse. Perhaps in this situation the threshold single pulse excites more than one type of afferent fiber, but the type giving rise to the piercing, burning kind of sensory quality needs to be excited repetitively in order to elicit this sensory response. This latter type of afferent fiber could be in the unmyelinated C fiber group and/or the small myelinated delta group from noxious receptors (e.g. Bessou and Perl, 1969; Burgess, Petit, and Warren, 1968); the paradoxical requirement of a lower threshold-I to excite such fibers in these cases could be explained by their being more optimally located relative to the stimulus current, in the skin under the electrode. (The latter point was made by A. Iggo in discussion.)

Single-pulse stimuli applied with a close bipolar (coaxial) electrode in n.VPL of the thalamus are completely ineffective for producing any conscious sensory responses; this inadequacy holds even at very high intensities (20–40 fold) relative to the peak current for pulses in a train that can elicit a response (Libet et al., 1967). One should recall our finding that not only is repetition of pulses necessary for production of conscious sensation by stimuli at SS-I cortex, n.VPL and medial lemniscus, but also that there is a minimum (or "utilization") train duration of about 0.5sec required of such repetition when liminal intensities are used, regardless of changes in pulse frequency from 15 to 120 pps (Libet et al., 1964; Libet, 1973). The single pulse stimuli in the thalamus, though ineffective for conscious sensory responses, do elicit a large primary EP in SS-I cortex (see fig. 5), as expected from animal studies. Consequently, it is quite clear that the process represented by the primary EP are not sufficient for mediating a conscious sensory response, even within the set of functional conditions

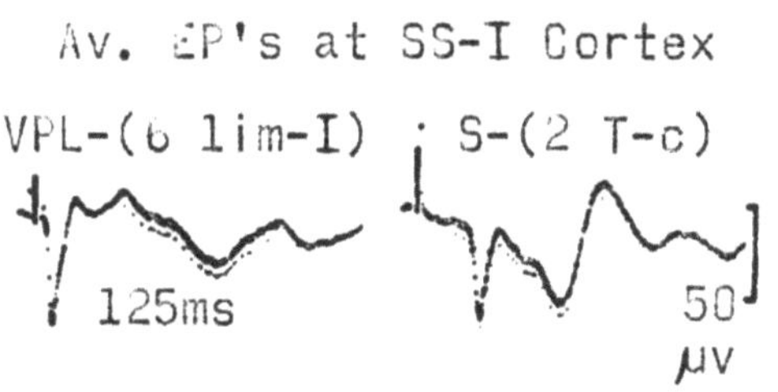


Evoked potentials of somatosensory cortex in response to thalamic (VPL) and skin stimuli in the same subject (patient with heredofamilial tremor). Each tracing is the average of 250 responses at 1.8 per sec; total trace length, 125msec. VPL: stimuli in ventral posterolateral nucleus of thalamus; subject reported not feeling any of these stimuli, though current was 6 times liminal-I for VPL electrode (liminal-I being minimum current to elicit sensation with 60 pulses per sec train of stimuli). S: stimuli at skin; current at twice threshold, all stimuli felt. Vertical bar, 50µv. Note the shorter latency of the primary (positive) evoked response to VPL stimulus. (From Libet *et al.*, 1967, by permission of *Science*)

prevailing in the brain of an awake, attentive individual. A further test, of the necessity of the later evoked activities for the awareness even of a peripheral stimulus, is described by Libet et al. (1972); this involves retroactive masking by a cortical stimulus applied 200msec or more *after* a threshold peripheral stimulus. (The proposition that activities represented by both the early and late EPs are not necessary for conscious somatosensory experience has been suggested by Clark, Butler, and Rosner [1969], but reservations about the interpretation of extracranial EP recordings, made with scalp electrodes make this suggestion inconclusive at present-see Libet et al. [1972].

The surface application of GABA can alter EP components and depress postsynaptic responses in the upper layers of the cortex (Bindman, Lippold, and Redfearn, 1962; Iwama and Jasper, 1957; Ochs and Clark, 1968). However, the surface application of GABA to SS-I cortex did not alter the ability of direct cortical or peripheral stimuli to elicit conscious sensory responses (see Libet, 1973). This finding indicates that postsynaptic responses of neuronal elements that lie below the upper few layers are more directly involved in mediating the conscious sensory responses; it would also appear to rule out, at least for primary sensory cortex, earlier speculations that postsynaptic activity in the neuropil of the upper (outer) layers of the cortex may be the mediator of conscious experience (e.g. Papez, 1956).

## Somatosensory Qualities

The question of what the cortical mechanisms are that mediate the production of the different subjective qualities of somatic sensation can, also, only be settled by validation of any differences in neuronal responses to the different types of sensory inputs against the conscious experiences of human subjects. We have approached this problem with two specific questions: (1) Can suitable stimulation of the somatosensory cortex elicit the various somatic qualities of touch, pressure, motion, heat, cold, pain, etc.? (2) If so, are there specific modes of cortical stimulation that are effective in eliciting each such quality? An answer to the second question could shed light on the neuronal processes involved in their selective mediation.

**Qualities of Responses to Liminal Stimulation of Somatosensory Cortex**

Earlier investigators have emphasized the generally nonspecific qualities of the responses to cortical stimulation in epileptic subjects, i.e., that the experiences were usually of a paresthesialike character (tingling, electric shock, etc.), although other more "natural-like" qualities were elicited at times (e.g. Foerster, 1936; Penfield and Boldrey, 1937; Penfield and Rasmussen, 1950). The relative incidence and variety of natural-like qualities that can be elicited by cortical stimuli have since been found to be considerably greater than was previously reported. These findings have been described in preliminary form by Libet (1966, 1973) and by Libet et al. (1959). In our studies parametric values of the stimuli to the somatosensory cortex were carefully controlled for proximity to liminal levels, and the subjects were not epileptics. Table 1 presents a summary of the reports of qualities in these studies. It may be seen that, for the 124 subjects with stimulation at or not far above the liminal levels, the incidence of natural-like qualities was in fact greater than that of paresthesialike ones. For most of the subjects, both categories of responses were elicited, even though the stimulus site was usually stationary. In a minority of cases only paresthesia-like responses were reported, but an approximately equal minority reported only natural-like responses. For subjects in the last column (Z) of table 1, the manner of stimulating the postcentral gyrus more or less resembled that usually employed by earlier investigators; in this column, the relative incidence of paresthesialike natural-like qualities also more closely resembled that given for all responses by Penfield and Boldrey (1937).

The large variety of the sensory descriptions, on which the subcategories of natural-like stimuli under B of table 1 are based, is indicated in table 2. Penfield and Rasmussen (1950) had concluded that "the tactile and proprioceptive sensation is detailed only in regard to the location of the part of the body represented" (p. 158), and they stated that "the patient never suggested that something rough or smooth or warm or cold had actually touched the part" (p. 217). In contrast, descriptions of relatively detailed sensory qualities, in addition to their precise locations, were obtained among the responses summarized in table 1, columns X and Y. For example, descriptions included statements such as "like talcum powder being sprinkled" on the index finger; or "like rolling a deodorant jar lightly over the surface" of the base of the last three fingers; or "crinkling touch, like picking up a paper thing" with the whole hand (see further examples in table 2). The experiences of "warmth" and "coldness", although relatively infrequently obtained, appeared to the subjects to be qualitatively similar to their natural sensations. (It should be noted that Foerster [1936] reported obtaining descriptions of vibration, waves, tickle, burning and cold, though he did not specify their rate of incidence among the total number, which were "mostly tingling, etc.") A large fraction of the natural-like sensory responses involved some sort of motion quality (B-a and B-b in table 2), although only about one-half of these involved any kind of feeling of actual movement of a part (B-b), such as might be related to the "sense of movement" that predominated among the nonparesthesialike responses reported by the Penfield group (e.g. Penfield, 1958).

Sensations of pain were virtually never elicited by us with stimulation of the postcentral gyrus; this is in agreement with reports by previous investigators (Penfield group [see References]; Foerster, 1936). The absence of reports of painful sensory responses held for all values of the stimulus parameters tested, including stimuli with intensities of three to four times the liminal current level that was required to elicit some sensory response; it is, of course, possible that a type of stimulus pattern not tested by us is required for eliciting a painful sensory response. Several subjects did initially report a sensory response as being painful. But further questioning revealed that these qualities were in the nature of unpleasant prickling or electric paresthesias, especially when a supraliminal stimulus was delivered, rather than those of natural pain associated with noxious stimuli. In almost every case, the subject did not object to further applications of the stimulus, having become

Table 1. Incidence of responses among subjects to stimulation of postcentral gyrus.[1]

| | (X) Stimulus parametric region A[2] | (Y) Subdural testing[3] | (Z) Exploring electrode[4] |
|---|---|---|---|
| *Type of response* | | | |
| A. "Paresthesialike" (totals) | 41 | 47 | 40 |
| B. "Natural-like" (totals) | 64 | 51 | 18 |
| a. something "moving inside" | 17 | 15 | 5 |
| b. feeling of movement of part | 15 | 17 | 3 |
| c. deep pressures | 12 | 8 | 7 |
| d. surface mechano-type | 6 | 6 | 2 |
| e. vibration | 4 | 2 | – |
| f. warmth | 8 | 3 | 1 |
| g. coldness | 2 | – | – |
| *Actual number of subjects,* in the stimulus and response categories. Totals (174) | 60 | 64 | 50 |
| I. Subjects reporting paresthesias *only* | 12 | 24 | 33 |
| II. Subjects reporting "natural-like" qualities *only*[5] | 14 | 21 | 10 |
| III. Subjects reporting paresthesias plus any nonparesthesias[6] | 34 | 19 | 7 |
| III'. Subjects reporting paresthesias plus only *one* nonparesthesia[6] | (17)[7] | (15)[7] | |

[1] Each type of response is counted only once for a given subject, even if it was elicited repeatedly by multiple tests in that subject. However, if a given subject experienced more than one type of response, he was listed in each appropriate category. The actual numbers of subjects in each general category are given below. Referral sites for almost all responses were in the upper extremity, mostly in the regions of the hand and fingers.

[2] Stimulus values were all at threshold-c in parametric region A (Libet et al., 1964), i.e., at liminal-Is carefully determined for trains with pulse frequencies > 15pps and train durations > 0.5sec. All tests in a given subject were usually applied at the same cortical site, with the electrode stationary.

[3] Stimuli were applied via a subdural Delgado-type electrode, usually as part of the procedure of localizing area SS-I for evoked potential studies (Libet et al., 1967); stimulus values were in parametric region A except that intensity may have been greater than liminal-I by some unknown but usually not large fraction.

[4] Stimuli applied via a stigmatic hand-held electrode to points on postcentral gyrus; trains of 60pps with variable durations and with intensities generally fixed at reasonable values that were usually in the range of being superthreshold, though not strongly so.

[5] In these subjects, with sensory responses that did not include any paresthesias, the qualities reported were almost all of the types in B-a and/or B-c.

[6] In subjects listed as reporting both paresthesias and natural-like (nonparesthesia) qualities for their responses, the two different types of qualities often were elicited independently, by separate stimulus tests with a given subject, as well as jointly by the same test (i.e., as one response with several qualities distinguishable in it).

[7] This group is, of course, a subgroup of group III and is already included in the total given for group III.

Table 2. Kinds of descriptions included in each category of qualities

A. *Paresthesialike sensory responses:*
tingling; electric shocks; pins and needles; prickling; numbness.

B. *"Natural-like" sensory responses:*
a. Something "moving inside": wave moving along inside through the affected part; or wavy-like feeling inside; or wavy "like a snake back's" motion; rolling or flowing motion inside; moving back and forth inside; circular motion inside; crawling under the skin, or more deeply.
b. Feeling of movement of the part (but with no actual motion observable to outsider): quiver; trembling; shaking; flutter; twitching; jumping; rotating; jerking; pushing; pulling; straightening; floating; or sensation of hand raising up, or lifting.
c. Deep pressures: throbbing; pulsing; swelling; squeezing; tightening.
d. Surface mechano-type: touch; tapping; hairs moving; rolling (a ball etc.) over surface; water running over surface; talcum powder sprinkling on; light brushing of skin; holding a ball of cotton; rubbing something between thumb and index finger.
e. Vibration: vibration, buzzing (distinct from tingling, etc.).
f. Warmth, or warming.
g. Coldness

familiar with what to expect. A "natural" and almost intolerable painful sensation in the head could be elicited readily by us when the plastic strip electrode carrier (Delgado-type) was inserted into the subdural space upside down, as sometimes occurred inadvertently, so that the exposed contacts faced the dura instead of the pial surface. The need for fuller clarification by the subject of what he is calling "painful," as well as the question of localization of the stimulus to the neuronal group allegedly stimulated, has also been emphasized by Bates (1972), in view of discrepancies among the reports of various investigators as to their ability to elicit pain by stimulation of the VP region of the human thalamus (e.g. Albe-Fessard and Bowsher, 1968; reports by Bates, Bertrand, Donaldson, McComas and Halliday in the symposium edited by Somjen, 1972; and Hassler, this symposium).

It should be made clear that the so-called natural-like sensory responses were usually peculiar and were different from all of the normal sensory experiences of the subject, even though they resembled the specific normal qualities and were clearly in a different category from what are defined here as paresthesialike qualities. The natural-like qualities were often difficult for the subjects to describe except by indicating similarities to or analogies with naturally occurring ones, and the subjects often insisted that these descriptions were really inadequate. This situation provides a fascinating illustration of the breakdown in the ability of one person to communicate the content of his primary subjective experience to another person, when the two individuals have not experienced similar situations in common. In spite of these difficulties, many subjects used similar verbal expressions to describe their sensory responses; for example an experience of "a wave moving inside," from one site to another in the arm or hand, was one of the more commonly reported ones.

It appears, then, that most of the various qualities that are at least related to natural somatosensory experiences (except for pain) can be elicited by liminal stimulation of the somatosensory cortex. The large fraction of responses to cortical stimuli that are paresthesialike and clearly unnatural in quality may be a result of inadequately sophisticated stimulation (see below) rather than being due to an actual lack of representation of the various specific qualities in the postcentral gyrus. This conclusion about representation of qualities would suggest that the function of the primary somatosensory cortex is not limited to that proposed by Penfield and his colleagues (e.g. Penfield and Jasper, 1954, pp. 68–69), i.e., to the discriminative aspects of sensation (localization and sense of position and movement) that are involved in perceiving the form of objects. It is true that the long-term defect that remains following ablation of the postcentral gyrus is primarily an astereognosis of the affected part of the body, involving loss of two-point discrimination and of sense of position and movement (e.g. Penfield and Rasmussen, 1950, p. 184). But Foerster (1936) emphasized the fact that there are losses of all somatic sensory qualities, including those of warmth, cold, and even pain, for some variable time after ablation. He regarded this as understandable in terms of some kind of representation of all somatosensory qualities in the postcentral gyrus, although he noted that nothing was then known about how the various qualities were arranged in the gyrus (Foerster, 1936, p. 431). Representation of specific somatosensory qualities in the SS-I area has also been demonstrated by recording of evoked potentials in response to some individual natural sensory stimuli at the periphery (e.g. Mountcastle, 1967), including cortical evoked responses to radiant thermal stimulation of the skin (Martin and Manning, 1969). However, as already noted above, the appearance of evoked potentials is not necessarily matched by an ability to elicit conscious sensory experience.

### Stimulus Parameters and the Neuronal Mechanisms for Specific Qualities

This question may be considered in relation to two alternative hypotheses that have been proposed for the mediation of specific somatosensory qualities: (a) Each of several "basic" qualities is represented in a separate type of vertical (radial) column of cells in the cortex. Such a functional columnar arrangement was demonstrated by Powell and Mountcastle (1959), for at least the different types of mechanoreceptors in the skin, by single unit electrical recordings of cortical responses to peripheral stimuli. Additional selective activation of units by light touch as opposed to firm touch or joint movement has been reported (Mountjoy and Baker, 1971). This hypothesis also has a further basis in the present evidence for specificity of first order afferent fibers in relation to the different natural stimuli (Iggo, 1965; Burgess et al., 1968; Bessou and Perl, 1969). (b) In the second hypothesis, the production of each specific quality of somatic sensation depends on an appropriate patterning of incoming impulses for each quality, rather than on activation of a specific site. The patterning could involve specific configurations of temporal and/or spatial distributions. Such a viewpoint has been advocated by Weddell (1961) and others for the coding of the afferent discharges in first order afferent fibers, each of which was hypothesized to respond nonspecifically to different physical stimuli at the periphery; the present evidence, however, favors the concept of specificity, rather than nonspecificity, of function for individual first order afferent fibers (Iggo, 1966; Perl, 1968). With either hypothesis (a) or (b), one is referring to those activities occurring only at the level of somatosensory cortex, which enable this area to participate suitably in the overall cerebral processes that mediate the subjective experience of a given quality.

#### 1. Area of Stimulating Electrode

If specific somatosensory qualities are represented separately in different vertical columns of the cortex, i.e., on hypothesis (a) above, it ought to be possible to elicit some specific quality (not a paresthesia) in all tests by adequately localizing the stimulus. When the exposure of a unipolar electrode on the pia-arachnoid surface was reduced from 1mm to 0.25mm in diameter, there was, in fact, no significant difference in the incidence of paresthesialike qualities of the responses to stimuli in parametric region A (see table 1, footnote 2). However, even the smaller surface electrode probably excites neural elements in many vertical columns, even in areas outside the area of electrode contact (see Landgren, Phillips, and Porter [1962] and Stoney, Thompson, and Asanuma [1968], for examples of such spread in motor cortex). Adequate localization of stimulating current to one vertical column, as has been achieved for motor cortex, would undoubtedly require the use of intracortical microelectrodes (Asanuma and Sakata, 1967; Stoney et al., 1968).

*Increasing the area* of the unipolar electrode might be expected to increase the number of neuronal elements or columns excited, and thereby possibly to change the quality of the sensation. However, a change from the usual 1mm contact to a 10mm disc (in five different subjects) produced no definite or consistent changes in the quality of the sensory responses to stimulus trains having liminal intensities appropriate for the respective electrode. This result is not too surprising when one recalls that the liminal current density (I per $mm^2$ of contact) is much smaller for the larger electrode, indicating that the larger electrode can activate neuronal elements by means of a greater degree of spatial facilitation (Libet et al., 1964; Libet, 1973). The bodily sites for referral of the sensation were overlapping or adjacent to one another, and in most cases somewhat larger with the larger electrode. However, small shifts in referral sites without changes in quality of sensory response have often been observed in successive tests even with the same liminal stimulus applied at the same cortical site through a stationary electrode (Libet et al., 1964; Libet, 1973). In terms of the columnar hypothesis, this may mean that with the conditions in a given individual, columns for a given type of quality tend to be activated more readily by the cortical stimulus than those for other types, even when they represent different topographical sites. An alternative possibility could be that a column for a given quality receives convergent input from a receptive field that is considerably larger than the sites that can be discriminated from each other position-wise, as would appear to be the case for the columnar fields mapped by Powell and Mountcastle (1959); and that finer specification of the locality of the site itself is a function of other neuronal elements or columns that are not directly concerned with quality. In this alternative explanation, the identical "quality" (column(s) would have been activated by the different cortical stimuli even with some shifting of the referral site.

2. Intensity of Stimulus

By raising the intensity of the stimulus to the postcentral gyrus above the liminal current level (for a given electrode, pulse frequency, etc.), one should be able to increase the number and types of neuronal elements or columns effectively excited, though there is also the possibility of producing more effective inhibition of some elements during the train of stronger stimulus pulses. In the subject most extensively tested on this point, liminal-I stimuli at pulse frequencies of 120, 60, 30, 15, and 8pps all elicited a variety of surface tactile quality ("soft, gentle water" or "camels hair brush" or "wire brush" moving lightly over skin). Raising the peak current values to 3 times liminal-I in each case converted these specific or natural-like qualities into a paresthesialike one (tingling). In several trials the current values were returned to liminal-I in succeeding stimuli; the natural-like quality without the tingling was restored. Additional subjects, in whom liminal-I stimuli elicited sensory responses with natural-like qualities, were tested more incidentally with stimulus trains at about 2 times liminal-I and some of these also reported changes to tingling or electric shock qualities with these stronger stimuli. (A less direct indication that paresthesia-like qualities tend to predominate when using supraliminal stimuli may be deduced from table 1, comparing the responses for the group in column Z with those for the groups in X and Y.) In other subjects the sensory response to the supraliminal stimulus had a proprioceptive or movement quality (e.g. rotation of finger joint, tightening of tendons, fingers closing into a claw or "wanting to rise and open," or movement back and forth, – with no actual motion observed – as in table 1, B-b). In two of these cases the sensory quality with liminal intensity was of a surface mechano-type, in two others it was of a paresthesia-type, and in one other it was already one of movement. Aside from any changes in quality, sensory responses were generally subjectively stronger with the higher intensity stimuli, although in a few they were also, surprisingly, reported by the subject to be shorter lasting; the latter change may indicate an increased inhibitory development during the stimulus train. In only some cases was the somatic area for referral of the sensation enlarged (see also Libet et al., 1964). (Incidentally, the fact that changes in quality could be brought about deliberately, by raising the intensity of the stimulus, argues against the possibility that psychological fixation or set of some sort was the cause of the limitation to one or two different qualities for the responses in many individual subjects to different stimuli – see table 1).

The conversion of a natural-like quality to a paresthesialike one by a rise in intensity could be explained on the assumption that stimulation of a larger group of presumably mixed types of cortical columns would be more likely to elicit a paresthesialike experience. This assumption would be in accord with the common experience of tingling or electric shock, etc., when a mixed peripheral nerve is stimulated, especially with a train of pulses (e.g. Collins, Nulsen, and Randt, 1960); it contrasts with responses to stimulation of the skin, where presumably more uniform fibers are selectively excited by near-threshold stimuli and give rise to more natural-like sensations (e.g. Sigel, 1953). The production of a quality involving a feeling of movement of a part in a number of tests with supraliminal stimulus intensity could mean that cortical columns for this modality have a higher threshold and/or that cortical inhibitory or mutually masking interactions are more effective for other types of columns. In any case, the tendency for this kind of quality to appear when higher intensity stimuli are used could help to explain why reports of "a sense of movement" constituted a larger percentage (of those responses that did have a natural-like quality) in the studies of the Penfield group (e.g. Penfield and Boldrey, 1937; Penfield, 1966) than they did in our studies with liminal-I stimuli (see table 1, columns X and Y.

### 3. Polarity of Unipolar Stimuli

Unipolar stimulation at the surface of the postcentral gyrus had been found to be somewhat more effective for eliciting conscious sensory responses when cathodal rather than anodal (Libet et al., 1964; Libet, 1973). If the mediation of different qualities were to involve activation at different depths in the cortex, some differential effect on the quality of the response might be expected for the two different polarities of stimulus pulses. However, no consistent differences were found. Observations were made in 5 subjects, with stimulus trains of pulses at 60, 30, and 8pps at the respective liminal-I levels for each set of parametric values. In most cases the subjective qualities of the responses, whether initially natural-like or paresthesialike, were similar for a given stimulus whether cathodal or anodal pulses were used. In one subject there was a consistent change to a quality of "warmth" when polarity was changed from cathodal to anodal. However, in another subject there was a change from a feeling of "cramps" (in the hand) with anodal pulses, to one of "warmth and pulsation" with cathodal ones; and in two others warmth was reported with both cathodal and anodal stimuli.

### 4. Pulse Frequency

If the different somatosensory qualities depend upon specific temporal patternings of neuronal activation in the somatosensory cortex and perhaps elsewhere (hypothesis (b), above), rather than on the sites of activation, it should be possible to elicit different specific qualities of responses to cortical electrical stimuli by changing the temporal parametric values of the stimulus in some appropriate manner. This test has been applied in a limited way by investigating the influence of changes in the pulse frequencies (PFs) of the cortical stimuli on the quality of the conscious sensory responses; such changes were tested extensively and deliberately in 10 subjects, but additional, more incidental, observations were made in many others. With stimulus trains of 0.5sec or longer, at their respective liminal-I values, differences in PF over a large range (15 to 120pps) did not result in changes in quality. When different qualities of responses were reported by a given subject in different tests with a stationary electrode, there was no apparent correlation of the different qualities with the changes in pulse frequency over this range.

However, when the PF was reduced to below approximately 10pps (i.e., to PFs of 8, 6, or 1pps, using 5sec TDs, and using the higher peak currents required for the liminal-I with such PFs), the sensory responses often had a subjective quality of pulsation without any actually observable movement in the part (Libet et al., 1964; see below for effect of interstimulus interval on incidence of this pulsatile quality). This sensation was described as an internal artery-like beat, with a subjective frequency of about 1 or 2 per sec that was independent of the stimulus PF. It remains to be determined whether this sensation was simply elicited indirectly by actual deep small muscle contractions, which could not be detected by the observer, rather than by direct activation of the sensory cortex. The pulsation frequency of about 1 per sec did in fact resemble that of the series of observable muscular twitches that can be elicited by a train of higher PFs (e.g. 60pps) when the peak current is raised to that of the threshold-I for a single pulse (Libet et al., 1964). Liminal or threshold-I levels for eliciting a conscious sensory response with low PF trains were often not much different from the threshold-Is for eliciting observable movements.

### 5. Interstimulus Interval and Changes in Quality

The change to a pulsatile quality, when the stimulus pulse frequency was changed from the usual PFs of 30–60pps to low PFs of 8pps or less, appeared to require the elapse of a relatively long time interval, about 4min or more, between the two kinds of stimuli. If the interval were shorter than this, the low PF stimulus usally elicited a sensory response similar in quality to that obtained with the preceding higher PF stimulus (Libet et al., 1964; Libet, 1973). However, this point was difficult to establish in any quantitative manner since, with the brief experimental periods available only during the surgical procedure, it was usually not feasible to repeat tests deliberately with long and short interstimulus intervals. In two subjects in whom direct comparisons were made, the pulsatile quality was elicited by 8pps stimuli when tested 5 or more min after the last stimulus at 60pps, but not after short intervals (one of those subjects appeared to require > 9min). In other subjects, tested with either a short or a long interstimulus interval (between 60pps and 8pps), 7 or 8 of the 15 subjects with long intervals did report a change to a form of pulsatile quality at 8pps, while only 1 or 2 of the 7 subjects with short intervals (about 2min) did so. (On the question of stimulus interval effects when changing the PF in the reverse direction, from a PF of 8pps to one in the 15–60pps range, only a small number of subjects who reported pulsation at 8pps were tested; two of these switched back to the nonpulsatile quality characteristic of the responses to the higher PFs with only a 1–3min interval after the last 8pps stimulus, while two others continued to report a pulsatile response to the higher PF stimulus even after a 4–5min interval following the last 8pps test.)

The effect of the interstimulus interval (i.e. the inverse of train repetition rate) on changes in the quality of the responses within the higher range of PFs (15pps to 120pps) was not analyzed as fully as the change from higher PFs to those of less than 10pps; we have not made a full accounting of all the recorded data on this point, and did not carry out deliberate testing of different intervals in a given subject. However, our impressions were that duration of the interstimulus interval did not affect the quality of the response within the higher PF range, and it was certainly not a controlling factor (although it can effect the liminal-I value; Libet et al., 1964). In subjects with implanted electrodes, it was not uncommon for the same quality to be elicited by near-liminal stimuli separated in time by very long periods, of hours or even days. On the other hand, there were numerous examples of changes in the quality of the sensory response to different stimuli delivered within relatively short intervals of one another i.e., with interstimulus intervals of about 1–2min. Many of these occurred when the intensity of successive stimuli was changed (see above), as well as at other times for no obvious or consistently predictable reason. If it is true that the long interstimulus interval is a requirement only for the change to a pulsatile quality, i.e., when changing from a higher to a low (< 10pps) PF, then the effect may not be one of facilitatory perseveration of any given quality; rather, it may be a function of some long lasting inhibition of the neuronal elements involved in producing the pulsatile quality. Penfield and his group (e.g. Penfield and Jasper, 1954; Penfield, 1966) have reported that there tends to be a perseveration of the same hallucinatory memory elicited by stimulation of the temporal lobe unless a substantial time period is allowed to elapse between the successive stimuli. It is, of course, possible that their effect could be either one of an enduring facilitation of a given pattern or one of inhibition of other patterns of neuronal responses.

### 6. General Conclusion on Stimulus Parameters

It would appear, then that the particular subjective quality of a conscious sensory experience is relatively independent of the nature of the stimulus to the somatosensory cortex, at least within the limits of the variations in parametric values tested thus far (see also Libet et al., 1959). The chief exceptions to this have been the effects of a change to a strongly supraliminal-I, explainable on hypothesis (a) above, and of a change to a low pulse frequency. However, it would be desirable to extend the presently limited studies to include further kinds of stimulus patterning, for example by temporal groupings of pulses into bursts, or modulations of intervals between pulses or bursts within a given stimulus train. This will become more feasible as clues to possible significant patternings of afferent inputs become available (e.g. Iggo, 1964; Bullock, 1968). It is important to note, however, that even if specific patterns of afferent nerve discharge were found to be characteristically correlated with different types of natural peripheral stimuli, it would not necessarily follow that such patterns were meaningful for CNS function (e.g. Bullock, 1968); or, in particular, for conscious sensory qualities (Libet, 1973).

## Acknowledgement

This work was supported in part by PHS Research Grant No. NS 05061 from the National Institute of Neurological Diseases and Stroke.

## References

Albe-Fessard, D.: Organization of somatic central projections. In: Ed. by W.D. Neff, Contributions to sensory physiology 2 New York: Akademie Press, 1967

Albe-Fessard, D., D. Bowsher: Central pathways for painful messages. Proc. Int. Congr. Physiol. Sci. XXIV, Washington, D. C. 6: 241–242, 1968

Amassian, V.E., L. Berlin: Early cortical projection of Group I afferents in forelimb muscle nerves of cat. J. Physiol. (London) 143: 61 P, 1958

Asanuma, H., H. Sakata: Functional organization of a cortical efferent examined with focal depth stimulation in cats. J. Neurophysiol. 30: 35–54, 1967

Bates, J.A.V.: Some practical aspects of electrical stimulation of the human diencephalon. In: Edited by G.G. Somjen, Neurophysiology studied in man. Amsterdam: Excerpta Medica, 1972

Bessou, P., E.R. Perl: Response of cutaneous sensory units with unmyelinated fibers to noxious stimuli. J. Neurophysiol. 32: 1025–1043, 1969

Bignall, K.E., M. Ibert: Polysensory and cortico-cortical projections to frontal lobe of squirrel and rhesus monkeys. Electroenceph. Clin. Neurophysiol. 26: 206–215, 1969

Bindman, L. J., O.C.J. Lippold, J.W.T. Redfearn: The non-selective blocking action of alpha-aminobutyric acid on the sensory cerebral cortex of the rat. J. Physiol. 162: 105–120, 1962

Brindley, G.S., P.A. Merton: The absence of position sense in the human eye. J. Physiol. (London) 153: 127–130, 1960

Broughton, R.J. Edited by E. Donchin and D.B. Lindsley. In: Averaged Evoked Potentials. Methods, Results, and Evaluation. NASA SP191, Washington, D. C., 1969, pp 79–84

Buchtal, F., A. Rosenfalck: Evoked action potentials and conduction velocity in human sensory nerves. Brain Res. 3: 1–402, 1966–67

Bullock, T.H.: Representation of information in neurons and sites for molecular participation. Proc. Nat. Acad. Sci. (U.S.) 60: 1058–1068, 1968

Burgess, P.R., D. Petit, R.W. Warren: Receptor types in cat hairy skin supplied by myelinated fibers. J. Neurophysiol. 31: 833–848, 1968

Clark, D.L., R.A. Butler, B.S. Rosner: Dissociation of sensation and evoked responses by a general anesthetic in man. J. Comp. Physiol. Psychol. 68: 315–319, 1969

Cobb, W., C. Morocutti, eds.: The Evoked Potentials. Electroenceph. Clin. Neurophysiol. Suppl. 26. Amsterdam: Elsevier, 1967

Collins, W.F., F.E. Nulsen, C.T. Randt: Relation of peripheral nerve fiber size and sensation in man. Arch. Neurol. 3: 381–385, 1960

Donchin, E., D.B. Lindsley, eds.: Averaged Evoked Potentials. Methods, Results and Evaluation. NASA, SP191, Washington, D. C., 1969

Foerster, O.: Sensible corticale Felder. Edited by O. Bumke and O. Foerster. In: Handb. d. Neurol. Berlin: Springer-Verlag, 1936

Gastaut, H., H. Regis, S. Lyagoubi, T. Mano, L. Simon: Comparison of the potentials recorded from the occipital, temporal & central regions of the human scalp, evoked by visual, auditory & somato-sensory stimuli. Edited by W. Cobb and C. Morocutti. In: The Evoked Potentials. Electroenceph. Clin. Neurophysiol. Suppl. 26. Amsterdam: Elsevier, 1966

Gelfan, S., S. Carter: Muscle sense in man. Exp. Neurol. 18: 469–473, 1967

Giaquinto, S., O. Pompeiano, J.E. Swett: EEG and behavioral affects of fore- and hindlimb muscular afferent volleys in unrestrained cats. Arch. Ital. Biol. 101: 133–148, 1963

Glees, P.: Experimental Neurology. Oxford: Clarendon Press, 1961, pp 239–240

Hassler, R.: Interaction between cortex-dependent and cortex-independent thalamic pain representation in the human and baboon. Communicated at Symposium "The Somatosensory System."

Hensel, H., K.K.A. Boman: Afferent impulses in cutaneous sensory nerves in human subjects. J. Neurophysiol. 23: 564–578, 1960

Hirsch, J.F., B. Pertuiset, J. Calvert, J. Buisson-Ferey, H. Fischgold, J. Scherrer: Etude des responses electrocorticales obtenues chez l'homme par des stimulations somesthesiques et visuelles. Electroenceph. Clin. Neurophysiol. 13: 411–424, 1961

Iggo, A.: Temperature discrimination in the skin. Nature 204: 481–483, 1964

Iggo, A.: The peripheral mechanisms of cutaneous sensation. Edited by D.R. Curtis and A.K. Mc Intyre. In: Studies in Physiology. Heidelberg: Springer-Verlag, 1965

Iwama, K., H. Jasper: The action of gamma aminobutyric acid upon cortical electrical activity in the cat. J. Physiol. (London) 138: 365--380, 1957

Jasper, H., R. Lende, T. Rasmussen: Evoked potentials from the exposed somatosensory cortex in man. J. Nerv. Ment. Dis. 130: 526–537, 1960

Katzman, R., ed.: Sensory responses in man. Ann. N. Y. Acad. Sci. 112: 1–546, 1964

Kelly, D.L., S. Goldring, J.L. O'Leary: Averaged evoked somatosensory responses from exposed cortex of man. Arch. Neurol. 13: 1–9, 1965

Klinke, R., H. Fruhstorfer, P. Finkenzeller: Evoked responses as a function of external and stored information. Electroenceph. Clin. Neurophysiol. 25: 119–122, 1968

Kooi, K.A., A.C. Tipton, R.E. Marshall: Polarities and field configurations of the vertex components of the human auditory evoked response: A reinterpretation. Electroenceph. Clin. Neurophysiol. 31: 166–169, 1971

Kornhuber, H.H.: Motor functions of cerebellum and basal ganglia: The cerebellocortical saccadic (ballistic) clock, the cerebellonuclear hold generator, and the basal ganglia ramp (voluntary speed movement) generator. Kybernetik 8: 157–162, 1971

Kornhuber, H.H., J.C. Aschoff: Somatisch-vestibuläre Integration am Neuronen des motorischen Cortex. Naturwissenschaften 51: 62–63, 1964

Landgren, S., C.G. Phillips, R. Porter: Cortical fields of origin of the monosynaptic pyramidal pathways to some alpha motoneurons of the baboon's hand and forearm. J. Physiol. 161: 112–125, 1962

Libet, B.: Brain stimulation and the threshold of conscious experience. Edited by J.C. Eccles. In: Brain and Conscious Experience. New York: Springer-Verlag, 1966

Libet, B.: Electrical stimulation of cortex in human subjects, and conscious sensory aspects. Edited by A. Iggo. In: Handbook of Sensory Physiology. Vol. 2, Somatosensory Systems. Heidelberg: Springer-Verlag, 1973

Libet, B., W.W. Alberts, E.W. Wright, Jr., L. Delattre, G. Levin, B. Feinstein: Production of threshold levels of conscious sensation by electrical stimulation of human somatosensory cortex. J. Neurophysiol. 27: 546–578, 1964

Libet, B., W.W. Alberts, E.W. Wright, Jr., G. Levin, B. Feinstein: Sensory perception by direct stimulation of human cerebral cortex: Stimulus Parameters. Fed. Proc. 18: 92, 1959

Libet, B., W.W. Alberts, E.W. Wright, Jr., B. Feinstein: Responses of human somatosensory cortex to stimuli below threshold for conscious sensation. Science 158: 1597–1600, 1967

Libet, B., W.W. Alberts, E.W. Wright, Jr., B. Feinstein: Cortical and thalamic activation in conscious sensory experience. In: Ed. by G.G. Somjen, Neurophysiology studied in man. Amsterdam: Excerpta Medica, 1972

MacKay, D.M.: Evoked brain potentials as indicators of sensory information processing. Neurosci. Res. Prog. Bull. 7: 184–276, 1969

Martin, H.F., J.W. Manning: Peripheral nerve and cortical responses to radiant thermal stimulation of skin fields. Fed. Proc. 28: 458, 1969

Mountcastle, V.B.: The problem of sensing and the neural coding of sensory events. Edited by G.C. Quarton, T. Melnechuk, F.O. Schmitt. In: The Neurosciences. New York: Rockefeller University Press, 1967

Mountcastle, V.B., T.P.S. Powell: Central nerve mechanisms subserving position-sense and kinesthesis. Bull. Johns Hopkins Hosp. 105: 173–200, 1959

Mountjoy, D.G., M.A. Baker: Single unit activity in somatosensory cortex of waking monkeys. Physiologist 14 (3): 199, 1971

Ochs, S., F.J. Clark: Tetrodotoxin analysis of direct cortical responses. Electroenceph. Clin. Neurophysiol. 24: 101–107, 1968

Oscarsson, O.: Proprioceptive and exteroceptive projections to the pericruciate cortex of the cat. Edited by D.R. Curtis and A.K. McIntyre. In: Studies in Physiology. Heidelberg: Springer-Verlag, 1965

Oscarsson, O., I. Rosén: Projection to cerebral cortex of large muscle-spindle afferents in the forelimb nerves of the cat. J. Physiol. (London) 169: 924–945, 1963

Oscarsson, O., I. Rosén: Short latency projections to the cat's cerebral cortex from skin and muscle afferents in the contralateral forelimb. J. Physiol. 182: 164–184, 1966

Papez, J.W.: Central reticular path to intralaminar and reticular nuclei of thalamus for activating EEG related to consciousness. Electroenceph. Clin. Neurophysiol. 8: 117–128, 1956

Penfield, W.: The Excitable Cortex in Conscious Man. Liverpool: Liverpool University Press, 1958

Penfield, W.: In Discussion of Libet. Edited by J.C. Eccles. In: Brain and Conscious Experience. New York: Springer-Verlag, 1966, pp 176–180

Penfield, W., E. Boldrey: Somatic motor and sensory representation in the cerebral cortex of man as studied by electrical stimulation. Brain 60: 389–443, 1937

Penfield, W., H. Jasper: Epilepsy and the Functional Anatomy of the Human Brain. Boston: Little, Brown & Co., 1954

Penfield, W., T. Rasmussen: The Cerebral Cortex of Man. New York: Macmillan, 1950

Powell, T.P.S., V.B. Mountcastle: Some aspects of the functional organization of the cortex of the postcentral gyrus of the monkey: A correlation of findings obtained in a single unit analysis with cytoarchitecture. Bull. Johns Hopkins Hosp. 105: 133–162, 1959

Provins, K.A.: The effect of peripheral nerve block on appreciation and execution of finger movements. J. Physiol. 143: 55–67, 1958

Shagass, C., S. Schwartz: Evoked cortical potentials and sensation in man. J. Neuropsychiat. 2: 262–270, 1961

Sigel, N.: Prick threshold stimulation with square wave current: A new measure of skin sensibility. Yale J. Biol. Med. 26: 145–154, 1953

Skoglund, S.: Anatomical and physiological studies of knee joint innervation in the cat. Acta Physiol. Scand 36 (Suppl. 124): 1–101, 1956

Somjen, G.G ed.: Neurophysiology studied in man. Amsterdam: Excerpta Medica, 1972

Stohr, P.E., S. Goldring: Origin of somatosensory evoked scalp responses in man. J. Neurosurg. 31: 117–127, 1969

Stoney, S.D., Jr., W.D. Thompson, H. Asanuma: Excitation of pyramidal tract cells by intracortical microstimulation: Effective extent of stimulating current. J. Neurophysiol. 31: 659–669, 1968

Swett, J.E., C.M. Bourassa: Comparison of sensory discrimination thresholds with muscle and cutaneous nerve volleys in the cat. J. Neurophysiol. 30: 530–545, 1967

Talairach, J., J. Bancaud: The supplementary motor area in man. Int. J. Neurol. 5: 330–347, 1966

Vaughan, H.G., Jr.: The relationship of brain activity to scalp recordings of event-related potentials. Edited by E. Donchin and D.B. Lindsley. In: Averaged Evoked Potentials, Methods, Results, and Evaluation. NASA, SP191, Washington, D. C., 1969

Walter, W.G.: The convergence and interaction of visual, auditory, and tactile responses in human nonspecific cortex. Ann. N. Y. Acad. Sci. 112: 320–361, 1964

Weddell, G.: Receptors for somatic sensation. Edited by M.A.B. Brazier. In: Brain and Behavior. Vol. I. Washington, D.C.: Amer. Inst. Biol. Sci., 1961

*Reprinted from International Congress Series No. 253*
NEUROPHYSIOLOGY STUDIED IN MAN
Proceedings of a Symposium held in Paris at the Faculté des Sciences, 20-22 july 1971
*Excerpta Medica, Amsterdam* **ISBN 90 219 0184 6**

# Cortical and thalamic activation in conscious sensory experience*

B. LIBET, W. W. ALBERTS, E. W. WRIGHT Jr and B. FEINSTEIN

Neurological Institute, Mt. Zion Hospital and Medical Center, and Department of Physiology, University of California School of Medicine, San Francisco, Calif., U.S.A.

Our work has been aimed directly at the question of the nature of the spatio-temporal configurations of cerebral neural activities which elicit, or are at least uniquely correlated with, the conscious awareness of a somatosensory experience in the awake and alert individual. We have focused the study on the differences between cerebral functions associated with stimuli that are at or just above threshold level for eliciting a conscious sensory experience, as opposed to those with stimuli that are below such a threshold. The implication is that there are some unique differences between physiological cerebral states above and below this threshold level. The term threshold is used by us in the broad sense of the just adequate levels for all significant parameters of a stimulus (electrical in our studies) not merely that for intensity (Libet *et al.*, 1964; Libet, 1972). (To avoid confusion with threshold levels for various other types of responses to stimuli, particularly for evoked potentials (EP's), the threshold levels for eliciting a conscious sensory response will be referred to as ,threshold-c.')

*Conscious sensory response*

The definition of conscious sensory experience and the operational criteria used to characterize it are, of course, a crucial issue in any such investigation (Libet *et al.*, 1964, 1967; Libet, 1965, 1966). A clear distinction must be made between the kind of detection of the stimulus that is specified by the subject's introspective awareness of a sensory experience, and other kinds of detection in which the response does not specify such an awareness. The latter may include a variety of behavioral responses of the skeletal or autonomic systems, electrophysiological responses (*e.g.* EP's in the brain), detection of a stimulus by forced-choice or reaction-time methods in psychophysical studies, *etc.* The criterion for the conscious sensory response must be based on a report by the subject that he is aware of having had a subjective sensory experience. The appearance of other types of detector-responses cannot be used as indicators of the conscious sensory experience without being validated by the subjective reports. There is in fact considerable evidence that many of the other types of signal detection can occur without awareness (*e.g.* Adams, 1957; Bevan, 1964; Raab, 1963; Shevrin and Fritzler, 1968; Libet, 1972).

In determining threshold levels on the basis of reports of subjective experience, a subject may be asked to report (*a*) whether he did feel or experience the sensation in question (no

* This work was supported in part by U.S. Public Health Service grant NS-05061 from the National Institute for Neurological Diseases and Stroke and by National Science Foundation Grant GB-30552X.

matter how weakly), or (*b*) whether he is uncertain about having felt it, or (*c*) whether he definitely did not feel it (Libet *et al.*, 1964). Speediness of report, for example by pressing a switch as soon as possible, is to be avoided as a potentially complicating factor; it is possible that speedy reports, at least at the instant they are made, could indicate detection without actual subjective awareness. The nature of the warning signal before a stimulus appears to have no major influence in a generally attentive and cooperative subject. However, it appears to be necessary that the subject focus his attention on the precise somatic site of the sensation and become familiarized in a few initial trials with the very weak sensory experiences generated by near-threshold stimuli. When the sensory experience is of very short duration, *e.g.* with a single pulse to the skin, the degree of attention during the precise time period in which the stimulus occurs apparently becomes more significant. This could be seen when a rigidly timed warning replaced the cruder one of a verbal 'ready' followed by the delivery of the stimulus at some variable moment during the next second or two. In the rigidly timed procedure an alerting tone signal was followed at a fixed interval by a light signal lasting several tenths of a second. The light signal indicated the time period during which the stimulus is actually delivered. With the latter combination of indicator signals, the threshold values for brief skin stimuli could be reduced by some 10% (or more in some instances). It should be noted that our primary concern is not to establish the absolutely minimum stimulus level that could give rise to a conscious sensory response (*i.e.* with special training and other techniques); it is rather to arrange conditions of stability so that the subject can reliably know and report whether he has (or has uncertainly) experienced subjectively a sensation, even though a very weak one.

One general experimental approach to investigating the cerebral physiology of conscious sensory experience, in the framework outlined above, has been to determine the characteristics of the threshold-c electrical stimuli, when these are applied to the skin, or to VPLN of the thalamus, or to the somatosensory cortex (SS-I) on postcentral gyrus (Libet *et al.*, 1964, 1967; Libet, 1972). Stimulation at the two cerebral levels in the sensory projection system enables one to bypass the unknown modulations imposed on the original peripheral input by various subcortical mechanisms. It could, therefore, generate clues as to the kinds of activations of the primary sensory cortex that are directly required in order for it to mediate the production of a conscious sensory experience. Another general experimental approach has consisted of studying the electrophysiological responses elicited by threshold-c versus subthreshold-c stimuli applied to the various sites in the sensory system; thus far we have utilized only macroelectrode recordings of the evoked electrical responses (Libet *et al.*, 1967; Libet, 1972).

*Threshold-c stimuli; utilization train duration (TD)*

The most interesting stimulus parameter, of those that are significant for threshold-c stimulation of somatosensory cortex and VPLN-thalamus, has turned out to be the train duration of repetitive pulses. (See Libet *et al.*, 1964 and Libet, 1972 for analyses of other significant parameters.) The general relation between intensity (peak currents) and TD for threshold-c stimuli to SS-I cortex is given in Figure 1. The minimum effective intensity for long trains of a given pulse frequency is called the liminal I. The minimum I does not increase with decrease in TD until the TD is shortened to about 0.5 sec or less. There is thus a minimum effective TD when using liminal I pulses, and this minimum TD is referred to as the utilization-TD (Libet *et al.*, 1964). The utilization-TD appears to be a physiologically significant phenomenon and quantity. The values have fallen in the range of about 0.4-1.0 sec for all subjects, clustering at about 0.5 sec. The value is relatively constant for a given subject. Changes in other stimulus parameters have little or no effect on utilization-TD, even when they markedly affect the liminal I requirements. The value for utilization-TD is

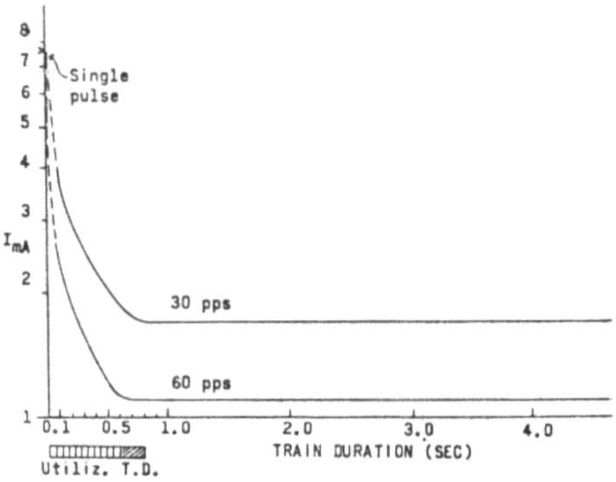


*Fig. 1* Intensity-train duration combinations for stimuli (to postcentral gyrus) just adequate to elicit a threshold conscious experience of somatic sensation. Curves are presented for two different pulse repetition frequencies, employing rectangular pulses of 0.5 msec duration. (From Libet, 1966 in: J. C. Eccles (ed.): *Brain and Conscious Experience;* courtesy of the publishers.)

relatively independent of pulse frequency (except for the range below 10 p.p.s., where it seems to be distinctly longer – see Libet *et al.*, 1964). Values for utilization-TD were found to be in the usual range whether the stimulus pulses were cathodal or anodal in surface polarity (unidirectional pulses), alternating in polarity, had pulse durations varying from 0.1-0.5 msec, or were applied via bipolar instead of unipolar electrodes, or by larger (up to 10 mm) electrodes instead of smaller ones (down to 0.25 mm).

The relatively long utilization-TD (about 0.5 sec) for eliciting conscious sensory responses with stimuli at sensory cortex, as compared to skin (0.1 sec or less), is not a peculiarity due to 'abnormal' routes of the input via the surface cortical electrode. Long values were similarly found with stimulation of the ascending projection system, both in VPLN and in the sub-cortical white matter below somatosensory cortex (Libet *et al.*, 1964, 1967) and recently also in the midbrain portion of the medial lemniscus (Libet *et al.*, unpublished data).

*Stimulation of VPLN* Results with localized stimulation of VPLN are particularly interesting. The form of the I-TD relationship is similar to that for SS-I cortex (and for medial lemniscus). But, if care is taken to increase the localization of the stimulus by using a coaxial electrode instead of a wider bipolar separation (of 3 mm for example), a single pulse could not elicit any response, sensory or motor, even with peak currents of up to 20 or more times the liminal I value (Libet *et al.*, 1967; *cf.* Guiot *et al.*, 1962). This provides clearer evidence of the need for repetition to produce a conscious sensory response, than does stimulation of postcentral gyrus; in the latter, the presence of some motor outflow which can respond to a single pulse (Libet *et al.*, 1964) makes it difficult to be sure that no sensation can be elicited by stimulating non-repetitively. On the other hand, stimulation of VPLN with trains shorter than utilization-TD (*i.e.* with trains as brief as 0.05 sec at 60 p.p.s., at threshold-c intensities that are of course above liminal I for the longer TD) could produce conscious sensory responses without any movement. This indicates that repetitive activations lasting much less than 0.5 sec may if intense enough be effective for eliciting conscious sensory experience. However, it is also possible that the requirement for about 0.5 sec of actual activation holds even for these supraliminal, brief train inputs; the more intense though briefer repetitive input could conceivably give rise to some appropriate after-activations which continue for some tenths of a second after the end of the stimulus train. As yet we have no clear answer on this latter possibility.

*Stimulation of motor cortex* For motor cortex an I-TD curve has now been carefully determined in an unanesthetized human subject using cathodal 0.5 msec pulses at 60 p.p.s. (Libet *et al.*, unpublished data). In contrast to sensory cortex, the peak current required at precentral gyrus for a threshold *motor* response was already liminal with very short TD's of about 0.05 sec (actually 4 pulses), *i.e.* longer TD's (up to 2 sec) required the same liminal I. Threshold I for a single pulse was greater than liminal I. Thus the utilization-TD was very short, resembling that for the skin (see below). (Single pulses to motor cortex can of course produce a twitch, and do so at a lower intensity than at postcentral gyrus, see Foerster, 1936.) In addition, the effect of prolonging the stimulus TD beyond 0.5-1 sec is strikingly different for the sensory and motor areas in man. For somatosensory cortex, stimuli with 5 sec TD's but at the liminal I strength simply elicit longer-lasting conscious sensory experiences than do those with utilization-TD, without any subjective change in intensity of the sensation during the 5 sec period. For motor cortex, stimuli kept at liminal I value elicit motor responses that progressively increase in intensity and in the extent of responding body musculature as the TD is lengthened. Indeed, motor responses to longer trains of even relatively weak stimuli tend to progress into seizures, even in non-epileptic patients.

*Stimulation of the skin* With stimulation of the skin (or of a peripheral nerve) it is well known that a single pulse can effectively elicit a conscious sensory response. Even here, it can be shown that threshold-c peak current does decrease with repetition of pulses and that the I-TD curve levels off to a minimum or liminal I value (Libet *et al.*, 1964, 1967). But this levelling occurs at a relatively short utilization-TD of about 0.1 sec or less, *i.e.*, threshold-c I is the same for all TD's of about 0.1 sec and longer (Libet *et al.*, 1964, 1967). Threshold-c I for a single pulse is generally only 110-130% that of liminal I. Thus, repetition of pulses is relatively unimportant when compared to the cases of somatosensory cortex, VPLN and medical lemniscus. However, the somewhat lower threshold-c intensity for repetitive pulses does imply that a single pulse normally must excite more than one peripheral sensory nerve fiber in order to elicit a conscious sensory experience (Libet *et al.*, 1967). This implication is borne out by the studies of Buchthal and Rosenfalck (1966-7) in peripheral nerve action potentials (though compare Hensel and Boman, 1960 and Shagass and Schwartz, 1961).

The surprisingly long utilization-TD of about 0.5 sec, that is uniquely characteristic for stimulation at cerebral levels, led to the hypothesis that such a relatively long period of 'activation' of the appropriate cortical elements is a fundamental physiological requirement for eliciting conscious sensory experience, at least for near-liminal inputs (Libet *et al.*, 1964; Libet, 1965, 1966). ('Activation' is used in the broad sense, as suggested by Jasper (1963), of integrated patterns of neural activity that may be brought into play, rather than of simple uncomplicated excitation.) This hypothesis can lead to interesting related inferences of psychological and philosophical significance (Libet, 1965, 1966). In physiological sensory terms such a temporal requirement could mean that briefer activations, even if they are part of other significant processes, will not clutter up conscious experience, *i.e.* it could act as a kind of temporal filter in conscious processes. The appropriate duration of cortical activations may develop only with adequate peripheral inputs and with cerebral conditions prepared to elaborate the inputs into an effectively long duration of cortical function. Experimental tests of the hypothesis, as it applies to natural or peripheral sensory stimulation, are very difficult to devise; a chief difficulty lies in the inability to specify the moment at which subjective awareness of a sensory stimulus takes place, as distinguished from the moment of a behavioral response to a stimulus (see Libet, 1966). However, it has been possible to make some experimental tests and relevant observations, as will appear below.

*Retroactive or backward masking of the conscious experience of a peripheral sensory stimulus*

One kind of test of the above hypothesis, as it applies to development of a conscious response to a brief peripheral sensory input, would lie in the ability of other inputs to interfere with that conscious experience when they are applied during the 0.5 sec *after* the peripheral stimulus. The ability of a second but stronger peripheral stimulus ($S_2$) to mask the awareness of a preceding relatively weak and briefer peripheral stimulus ($S_1$), has in fact already been demonstrated in a variety of situations (see review by Raab, 1963). With visual stimuli the backward masking effect can be obtained with the conditioning stimulus $S_2$ following the test stimulus $S_1$ by about 100 to 200 msec or more, in the so-called Crawford effect (Crawford, 1947; see also Wagman and Battersby, 1964). Backward masking effects with auditory stimuli have generally been shown for intervals of about 50-100 msec, but some effects have been reported for intervals as long as 1000 msec (see Raab, 1963; Von Bekesy, 1971). Cutaneous backward masking has been less frequently studied but has been demonstrated for intervals of about 50 to 100 msec (see Raab, 1963; Melzack and Wall, 1963). The question of the nature and location of the neural elements and processes involved in retroactive masking is complicated by the fact that there are both peripheral and central contributions possible (see Crawford, 1947; Raab, 1963; Galbraith and Heath, 1970). This complication could be avoided if backward masking could be demonstrated when the conditioning or second stimulus $S_2$ is applied directly to a cortical site, in a known temporal relationship to the initial arrival at the cortex of the afferent projection of impulses initiated by $S_1$.

*Retroactive masking by a cortical stimulus*, of the conscious sensory response to a preceding skin stimulus, has now been achieved (Libet *et al.*, 1972*a*; see Libet, 1972). The conditioning or delayed stimulus $S_2$ was applied directly to somatosensory cortex (postcentral gyrus) in human subjects. It consisted of a brief train (usually of 0.5 sec TD) of pulses (60 p.p.s.), with peak currents at about 1.3 to 1.5 times the liminal I that was required by this cortical stimulus for eliciting a conscious sensory response. The test or first stimulus $S_1$ was a single pulse at threshold-c intensity; it was applied to the skin within the body area to which was referred the sensory response elicited by the cortical stimulus. The sensations generated by $S_1$ and $S_2$ separately could usually be distinguished readily, by their different qualities and extent of somatic area. The $S_1$-$S_2$ interval in this procedure refers to the time between $S_1$ (single pulse) and the *beginning* of the $S_2$ train of pulses.

The conscious sensory response obtained with $S_1$ alone completely vanished when $S_1$ and $S_2$ overlapped in time. However, response to $S_1$ continued to be masked when the $S_1$-$S_2$ interval was increased up to 125-200 msec for most subjects, and up to 500 msec in one subject, *i.e.* at these intervals only a single sensory response was reported which was identical with that elicited by $S_2$ alone. When the interval was greater, or when the strength of $S_1$ was raised sufficiently, the subject experienced both of the sensations in the same temporal order as the responsible stimuli. When $S_2$ was a single pulse stimulus to the cortex it appeared to be ineffective for masking; the minimum $S_2$ train duration that is needed in order to achieve retroactive masking, and therefore the total effective interval for masking, is yet to be determined. The latency to the beginning of the primary EP produced at somatosensory cortex by $S_1$ (when applied to the hand) is about 20 msec, and this should be subtracted from the total effective interval. Fortunately, the latent periods of the early EP components do not vary significantly with the strength of peripheral somatic stimuli (Giblin, 1964; Libet *et al.*, 1967) and are not increased even for skin stimuli at threshold-c levels or below (Libet *et al.*, 1967); this is unlike the increase in latencies of the visual evoked response with decreasing intensities of light flash (*e.g.* Vaughan *et al.*, 1966).

Retroactive masking of the peripheral ($S_1$) sensation by a later stimulus ($S_2$) applied directly to somatosensory cortex, with $S_1$-$S_2$ intervals of up to 200 msec or more, could only be due to interference with some late components of the brain responses to $S_1$ that are necessary for the mediation of a conscious sensory response. The mechanism of such interference is an open question. It seems probable that the maximum possible $S_1$-$S_2$ interval for effective backward masking is even greater than that achieved, as it is unlikely that all of the potentially effective cortex was stimulated by the $S_2$ stimulus that was applied. In any case, the extent of retroactive masking that could be demonstrated provides strong support for the hypothesis that a relatively long period of suitable cerebral activations is a necessary feature of the processes mediating conscious sensory experiences (Libet *et al.*, 1964; Libet, 1965, 1966).

*Evoked potentials (EP's) and threshold-c skin stimuli*

If a relatively long period of cortical activation is required, how does one explain the production of a conscious sensory response by a single pulse stimulus at threshold-c intensity to the skin? The answer could be that such single or brief trains of pulses at the skin in fact elicit a series of components which follow the primary (surface-positive) EP and which last for some variable period of several tenths of a second and longer, as may be seen in Figure 2 (Libet *et al.*, 1967; see also Katzman, 1964; Cobb and Morocutti, 1968; Donchin

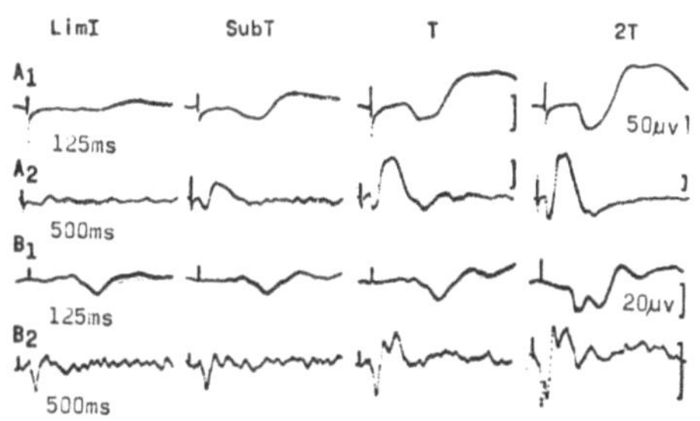


*Fig. 2* Averaged evoked potentials of somatosensory (SS-I) cortex in relation to threshold stimuli at skin. Each tracing is the average of 500 responses at 1.8 per sec. Total trace length is 125 msec in $A_1$ and $B_1$ and 500 msec in $A_2$ and $B_2$; beginning of stimulus artifact has been made visible near start of each tracing. A and B, separate subjects, both parkinsonian patients. Vertical column T = threshold stimuli, subject reporting not feeling some of the 500 stimuli. Column 2T = stimuli at twice threshold current; all stimuli felt distinctly. Column SubT = subthreshold stimuli, none felt by subject; current about 15% below T in subject A, 25% below T in B. Column Lim I = subthreshold stimuli at 'liminal intensity' (see text), about 25% below T in subject A, about 35 to 40% below T in B. Polarity, positive downward in all figures. Vertical bars in A, under T, indicate 50 μV in $A_1$ and $A_2$ respectively, but gains are different in 2T as shown; for $B_1$ and $B_2$, 20 μV bars. (Calibration obtained by summating 500 sweeps of calibrating signal.) (From Libet *et al.*, 1967, *Science;* courtesy of the editors.)

and Lindsley, 1969; Williamson *et al.*, 1970). After-waves do not appear following the direct cortical response (DCR, initial surface negative wave and after-positivity) that is elicited by a single pulse stimulus to somatosensory cortex; this is true even when the peak current is $\geq$ liminal I level that is adequate for production of conscious sensation when a train of repetitive pulses is used (see Fig. 4) (Libet *et al.*, 1967, 1972*c*). The after- or later components of the EP elicited by threshold-c *skin* stimuli may then represent the postulated neural elaboration of activities over a sufficiently long period of time. This suggestion is supported by the finding that most of the later EP components only appeared at or above threshold-c levels of skin stimuli. It is also supported more indirectly by other types of findings (Davis, 1964; Haider *et al.*, 1964; Wagman and Battersby, 1964; Satterfield, 1965; Donchin and Cohen, 1967; however, see below for further treatment of this point).

In VPLN a single localized pulse of relatively high current is ineffective for eliciting a conscious sensory experience (see above). Nevertheless, such a pulse can produce a large EP in the appropriate area of human somatosensory cortex (Libet *et al.*, 1967), as was to be expected from studies on animals (Dempsey and Morison, 1943; Andersen and Andersson, 1968). The primary surface-positive component of these EP's is presumably elicited in the same cortical elements and by the same specific projection fibers to the cortex as is the primary component elicited by a skin stimulus. The primary EP elicited by single-pulse stimuli to VPLN can attain an amplitude that is greater than the one elicited by a skin stimulus that is well above threshold-c level for skin, as seen in Figure 3 (Libet *et al.*, 1967). The absence

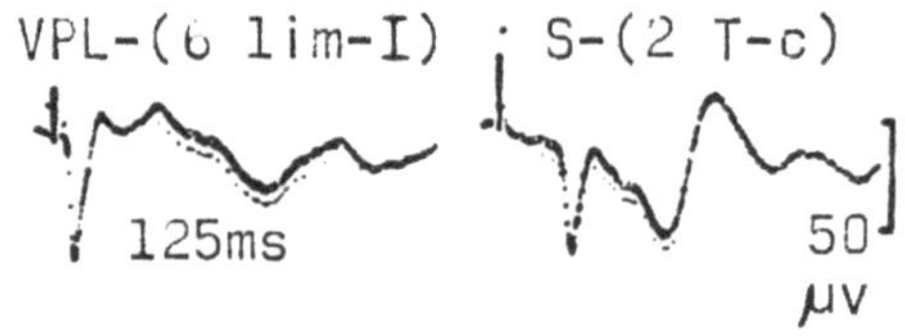


*Fig. 3* Evoked potentials of somatosensory (SS-I) cortex in response to thalamic (VPL) and skin stimuli in the same subject (patient with heredofamilial tremor). Each tracing is the average of 250 responses at 1.8 per sec; total trace length, 125 msec. VPL = stimuli in ventral posterolateral nucleus of thalamus; subject reported not feeling any of these stimuli, though current was 6 times liminal I for VPL electrode (liminal I being minimum current to elicit sensation with 60 p.p.s. train of stimuli). S = stimuli at skin; current at twice threshold, all stimuli felt. Vertical bar, 50 µV. Note the shorter latency of the primary (positive) evoked response to VPL stimulus. (From Libet *et al.*, 1967, *Science;* courtesy of the editors.)

of a conscious sensory response to a single pulse in VPLN, therefore, establishes firmly the conclusion that the primary surface positive component is not sufficient to initiate subjective experience, even in the context of alerted, attentive brain activity. This conclusion was also inferred from the ability of subthreshold-c pulses at the skin to elicit a detectable primary surface-positive EP at SS-I cortex (Libet *et al.*, 1967).

Conversely, it has been reported that both early and late EP components, as recorded extracranially with scalp electrodes, could be abolished by cyclopropane inhalation with little rise in sensory threshold for ulnar nerve stimuli (Clark *et al.*, 1969). In order to determine whether the findings of Clark *et al.* mean that responses represented by the EP's are actually not necessary for somatosensory experience, their experiments should be repeated with subdural recordings. The scalp EP recordings are only considerably attenuated versions of the subdural ones and any reduced responses could easily have been missed (*e.g.*, Libet *et al.*, 1967). In addition, we have preliminary findings which indicate that scalp EP's, particularly those recorded at or near the vertex, are elicited predominantly by the deep afferents (from muscles and joints) that are present in the peripheral nerves (Libet *et al.*, 1972*d*). If so, cutaneous afferent fibers, which probably account for the threshold conscious sensory response to peripheral nerve stimulation, could still be eliciting EP's that would have gone undetected in the extracranial recordings of Clark *et al.* (1969) both before and during inhalation of cyclopropane.

It follows from these considerations, that the presence or amplitude of an EP, particularly of the primary component, cannot be assumed to be an indicator of the occurrence or intensity of subjective sensory experience without other validation under the conditions of study. An additional example will serve to illustrate that a similar caution must be used in interpreting the functional significance of recordings of the late EP's as well. Not only are EP's recordable at supplementary motor area of human cortex in response to somatic, audio and visual stimuli, but certain late EP components can be shown to be generated there (see Libet *et al.*, 1972*d*). Yet, stimulation of this area elicits only a motor response, making it unlikely that these particular late EP's are involved in the direct mediation of conscious sensory responses. A similar doubt extends to the EP's recorded at the vertex (see Libet *et al.*, 1972*d*).

One should consider a bit further the possible relationship between the long utilization-TD's, required of threshold-c stimuli that are applied to the specific projection system (medial lemniscus, VPLN and SS-I cortex), and the role of the latter system in the production of the early and late components of the cortical EP's. The evidence on the roles of the specific (lemniscal) and non-specific (extra-lemniscal) systems in the production of the somatic EP's has been recently reviewed by Williamson *et al.* (1970). These authors conclude that the evidence favors the hypothesis that the entire contralateral EP, including the vertex-scalp recording and both the early and late EP components 'is (*a*) the result of activity in primary somatic cortex (and (*b*) is) mediated solely by the medial lemniscal system'. Point (*a*) has now been shown to be almost certainly incorrect. We have found that the supplementary motor area (SMA) in man generates its own relatively large EP's in response to somatic, audio and visual stimuli (Libet *et al.*, 1972*d*); these SMA-EP's could contribute significantly to the vertex recordings or, as seems more likely, a cortical area other than SS-I and SMA may be generating the main components of the vertex EP's.

If point (*b*) is valid, that the mediation of both early and late components is solely by the lemniscal system, it would be difficult to reconcile it with the long utilization-TD's (about 0.5 sec) that we have found to be necessary when stimulating medial lemniscus, VPLN or primary SS-I cortex, as opposed to the very short utilization-TD required when stimulating at the periphery (skin or nerve). One alternative would be to conclude that the recorded EP's (early and late) do not reflect the kinds of late activations which are necessary for conscious sensory responses and which can be elicited by single pulse peripheral stimuli but not by the cerebral ones. However, it should be noted that the evidence cited by Williamson *et al.* (1970), both from their work and that of others, only indicates that the functioning of the lemniscal system is necessary for the production of the full somatic EP; it would still be possible for the extra-lemniscal pathways normally not only to be also necessary, but actually more directly involved in the mediation of the late EP components (and of the conscious sensory response). Furthermore, since primary sensory cortex receives both specific and non-specific ascending inputs (*e.g.* Chow and Leiman, 1970) the effects on EP's of destruction of this cortical area only have a bearing on the question of cortical mediation of EP generation (Stohr and Goldring, 1969); the effects do not distinguish between the roles of lemniscal and extra-lemniscal inputs. (The long utilization-TD required by liminal I stimuli at SS-I cortex could reflect an inability of such stimuli to engage adequately the full panoply of neuronal elements normally responding to the various ascending inputs that are activated by a peripheral single pulse stimulus.) Lesions restricted to VPLN-VPMN, or to medial lemniscus, would be more relevant to the conclusion urged by Williamson *et al.* (1970), but adequately documented situations of that kind have apparently not been studied in human subjects. (The reduction in EP's noted by Domino *et al.* in 1965 after therapeutic lesions were made in the ventral posterior thalamic nuclei was restricted in their study to the earlier EP components.) Finally, while single-pulse stimuli to VPLN do elicit later EP waves at somatosensory cortex in man (in addition to the expected primary components), these later components tend to be different from those elicited by skin or nerve stimuli (Libet *et al.*, 1967, 1972*b*).

*Direct cortical responses (DCR's) with threshold-c cortical stimuli*

Since a train of pulses is required for eliciting a conscious sensory experience by stimulation of somatosensory cortex (or of VPLN or medial lemniscus), are there any unique neuronal responses in the cortex, as detected by electrical recordings, which appear with threshold-c trains but do not appear with subthreshold-c cerebral stimuli? The answer to this question would have a bearing on at least two alternative possibilities to describe the nature of the physiological 'coding' of the conscious sensory experience, in connection with the relatively long utilization time requirement of about 0.5 sec: (*a*) some special neuronal event is triggered at the end of the adequate period; or (*b*) the occurrence of a series of certain similar neuronal events for a given period itself constitutes the code (see Libet, 1966, 1972).

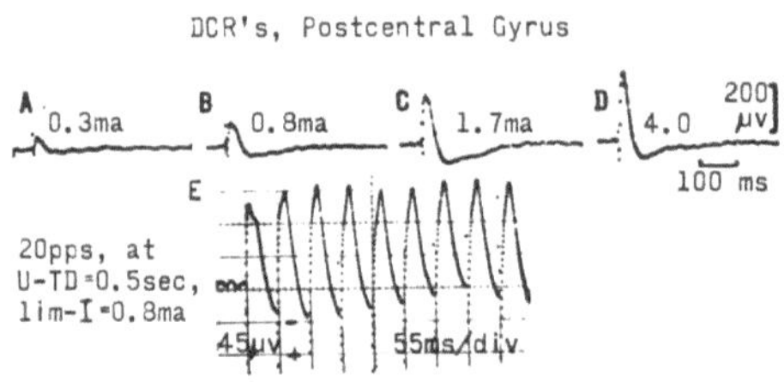


*Fig. 4* Direct cortical responses evoked in somatosensory cortex by adjacent direct stimuli (0.3 msec pulses). Subject is a parkinsonian patient, unanesthetized. Each tracing in A-D is the average of 18 responses at 0.5 per sec; horizontal bar in D indicates 100 msec. A, stimulus current 0.3 mAmp; B, 0.8 mAmp (equal to liminal I, see E); C, 1.7 mAmp; D, 4.0 mAmp. Subject reported not feeling any of these 'single pulse' stimuli, in A to D. Vertical bar, 200 μV. (A to D from Libet *et al.*, 1967, *Science;* courtesy of the editors.)

E, averaged response to 10 separate trains, 0.5 sec TD, PF of 20 p.p.s. and I of 0.8 mAmp peak currents; *i.e.* stimulus with utilization-TD at liminal I, each eliciting a conscious sensory response in the same subject used in the A-D recordings. Horizontal axis indicates time intervals of 55 msec; vertical axis 45 μV per division. Recordings were dc to 5 kHz, and positive downwards in all cases.

The threshold-I levels needed by single pulses for eliciting DCR-N and P waves (see Fig. 4) were found to be below the liminal I values that were required by cortical stimulus trains (20 or 30 p.p.s. and train-durations $\geqq$ utilization-TD) for eliciting conscious sensory responses (Libet *et al.*, 1967). However, liminal I for eliciting sensory responses was found to be below the strength required for producing maximal N and P waves, and it was below the threshold I for producing any slow negativity in single DCR responses. There were no striking changes in the DCR responses during or at the end of a utilization-TD train of liminal I pulses that was just adequate for eliciting a conscious sensory response, as in Figure 4E (Libet, 1965; Libet *et al.*, 1972 *c*). In addition, no after-waves could be detected immediately following such stimulus trains. Nor was there any correlation between threshold-c stimulus adequacy and the appearance of a negative steady potential shift during such trains. There is, therefore no evidence in the surface recorded DCR's of any unique neuronal response that appears to be correlated with the attainment, by the cortical stimulus, of adequacy for eliciting subjective sensory experience.

The absence of distinctive changes in these recorded electrical responses during and after the utilization-TD is in accord with the absence of any progressive build-up in subjective intensity of the sensory experience when cortical stimulus trains longer than the utilization-TD are delivered. Similarly, EP's elicited at sensory cortex by a utilization-TD stimulus to

VPLN did not show unique or consistently different components at the end or after the train (Libet *et al.*, 1972 *b*). Such evidence appears to favor alternative (*b*) for the 'coding mechanisms', but this can only be regarded as indicative rather than conclusive. Obviously, DCR's and EP's are incomplete representations of the neuronal events that may be occurring in the somatosensory cortex or elsewhere in the brain at such times.

Regardless of which of the 'physiological coding' requirements (*a*) or (*b*) above, should turn out to be valid, another class of questions about the nature of the requirement would remain. Does the moment of 'appearance' of the subjective sensory experience actually occur at or near the start of the appropriate cortical activation *i.e.* is the continued repetitive activation actually required for 'memory fixation' of an already ongoing subjective experience, so that one can recall the experience later and thus make possible a continual awareness of having been aware of something? Or, does the subjective experience actually not 'appear' at all until a sufficient period of suitable activation has occurred? If the latter were true it would introduce a relatively long 'latent period' of several tenths of a second between the onset of cortical activation (or sensory input) and the subjective experience elicited by it (Libet, 1965, 1966). Obtaining experimental answers to questions such as these poses formidable difficulties but offers potentially fascinating rewards.


## SUMMARY

The temporal parameters of electrical stimuli that are adequate for eliciting conscious sensory responses, when applied to cerebral and peripheral sites, and their relationship to electrophysiological responses evoked by such stimuli, are analyzed. The requirement of repetition of pulses of liminal intensities for the relatively long 'utilization' period of about 0.5 sec led to a hypothesis that prolonged cortical activation is a fundamental requirement in mediation of conscious experiences. In support of this hypothesis it was shown that when a supraliminal stimulus was delivered to somatosensory cortex up to 200 msec or more *after* a near-threshold peripheral (skin) stimulus, the conscious sensory experience of the skin stimulus could be blanked out, *i.e.* retroactively masked. In addition, the primary short-latency evoked potential elicited by a peripheral stimulus was shown to be not sufficient for the full mediation of a conscious sensory response. This and other findings also demonstrated that evoked potentials, especially when recorded extracranially and/or at the vertex, cannot be assumed without direct validation to represent processes that mediate subjective sensory experience. The evidence, from evoked surface responses of the cortex to stimulus trains at cerebral sites, suggests that the occurrence of an appropriate series of similar neuronal events may itself constitute the physiological coding for a conscious sensory experience.



## RÉSUMÉ

Les paramètres temporaux des excitations électriques suffisantes pour provoquer des réponses sensorielles conscientes lorsqu'elles sont appliquées à des localisations cérébrales et périphériques, sont analysés ainsi que la relation entre ces paramètres et les réponses électrophysiologiques déclenchées par ces stimulations. On a été conduit à formuler l'hypothèse qu'une activation corticale prolongée est une condition essentielle pour l'élaboration des expériences conscientes si l'on veut expliquer la nécessité de la répétition d'impulsions d'intensité liminaire pour la période relativement longue d'utilisation qui est d'une 0,5 seconde environ. A l'appui de cette hypothèse, on a montré que la perception sensorielle consciente pouvait être effacée, c'est-à-dire retrospectivement masquée, si une excitation supraliminaire était transmise au cortex somato-sensoriel dans un délai allant jusqu'à 200 millisecondes ou plus *après* une excitation périphérique (cutanée) voisine du seuil. De plus, le potentiel, évoqué primaire à courte latence déclenchée par une excitation périphérique, s'est révélé être insuffisant pour assurer l'élaboration complète d'une réponse sensorielle consciente. Cette constatation et d'autres observations ont, également, fait apparaître que les potentiels évoqués ne peuvent pas être considérés sans validation directe, comme représentant des processus qui assurent l'élaboration de la perception sensorielle subjective surtout quand ils sont enregistrés à l'extérieur du crâne et ou au niveau du vertex. Les indices, fournis par les réponses évoquées de surface du cortex à des trains d'excitations à certains endroits du cerveau, laissent penser que l'apparition d'une série appropriée de phénomènes neuronaux du même genre peut constituer, par elle-même, la mise en code physiologique pour une perception sensorielle consciente.

## REFERENCES


ADAMS, J. K. (1957): Laboratory studies of behavior without awareness. *Psychol. Bull.*, *54*, 383.

ANDERSEN, P. and ANDERSSON, S. A. (1968): *Physiological Basis of the Alpha Rhythm.* Appleton-Century-Crofts, New York.

BEVAN, W. (1964): Subliminal stimulation: a pervasive problem for psychology. *Psychol. Bull.*, *61*, 81.

BUCHTHAL, F. and ROSENFALCK, A. (1966-67): Evoked action potentials and conduction velocity in human sensory nerves. *Brain Res.*, *3*, 1.

CHOW, K. L. and LEIMAN, A. L. (eds.) (1970): The structural and functional organization of the neocortex. *Neurosci. Res. Progr. Bull.*, *8*, 153.

CLARK, D. L., BUTLER, R. A. and ROSNER, B. S. (1969): Dissociation of sensation and evoked responses by a general anesthetic in man. *J. comp. physiol. Psychol.*, *68/3*, 315.

COBB, W. and MOROCUTTI, C. (eds.) (1968): The evoked potentials. *Electroenceph. clin. Neurophysiol.*, *Suppl. 26.*

CRAWFORD, B. H. (1947): Visual adaptation in relation to brief conditioning stimuli. *Proc. roy. Soc. B*, *134*, 283.

DAVIS, H. (1964): Enhancement of evoked cortical potentials in humans related to a task requiring a decision. *Science*, *145*, 182.

DEMPSEY, E. W. and MORISON, R. S. (1943): The electrical activity of a thalamocortical relay system. *Amer. J. Physiol.*, *138*, 283.

DOMINO, E. F., MATSUOKA, S., WALTZ, J. and COOPER, I. S. (1965): Effects of cryogenic thalamic lesions on the somesthetic evoked response in man. *Electroenceph. clin. Neurophysiol.*, *19*, 127.

DONCHIN, E. and COHEN, L. (1967): Averaged evoked potentials and intramodality selective attention. *Electroenceph. clin. Neurophysiol.*, *22*, 537.

DONCHIN, E. and LINDSLEY, D. B. (1969): Averaged evoked potentials. Methods, results and evaluation. *National Aeronautics and Space Administration*, *SP-191.* Washington, D.C.

FOERSTER, O. (1936): The motor cortex in man in the light of Hughling Jackson's doctrines. *Brain*, *59*, 135.

GALBRAITH, G. C. and HEATH, R. G. (1970): Backward visual masking during direct brain stimulation in man. *Vision Res.*, *10*, 911.

GIBLIN, D. (1964): Somatosensory evoked potentials in healthy subjects and in patients with lesions of the nervous system. In: *Sensory Evoked Responses in Man.* Editor: R. Katzman. *Ann. N.Y. Acad. Sci.*, *112*, 93.

GUIOT, G., ALBE-FESSARD, D., ARFEL, G., HERTZOG, E., VOURC'H, G., HARD, Y., DEROME, P. and ALEONARD, P. (1962): Interpretation of the effects of thalamus stimulation in man by isolated shocks. *C. R. Acad. Sci. (Paris)*, *254*, 3581.

HAIDER, M., SPONG, P. and LINDSLEY, D. B. (1964): Attention, vigilance and cortical evoked-potentials in humans. *Science*, *145*, 180.

HENSEL, H. and BOMAN, K. K. A. (1960): Afferent impulses in cutaneous sensory nerves in human subjects. *J. Neurophysiol.*, *23*, 564.

JASPER, H. H. (1963): Studies of non-specific effects upon electrical responses in sensory systems. In: *Progress in Brain Research*, *Vol. I*, pp. 272-293. Editors: G. Moruzzi, D. Albe-Fessard and H. H. Jasper. Elsevier, Amsterdam.

KATZMAN, R. (ed.) (1964): Sensory evoked responses in man. *Ann. N.Y. Acad. Sci.*, *112*, 1.

LIBET, B. (1965): Cortical activation in conscious and unconscious experience. *Perspect. Biol. Med.*, *9*, 77.

LIBET, B. (1966): Brain stimulation and the threshold of conscious experience. In: *Brain and Conscious Experience*, p. 165. Editor: J. C. Eccles. Springer-Verlag, New York.

LIBET, B. (1972): Electrical stimulation of cortex in human subjects, and conscious sensory aspects. In: *Handbook of Sensory Physiology, Vol. II, Somatosensory System.* Editor: A. Iggo. Springer, Heidelberg, in press.

LIBET, B., ALBERTS, W. W., WRIGHT JR, E. W., DELATTRE, L., LEVIN, G. and FEINSTEIN, B. (1964): Production of threshold levels of conscious sensation by electrical stimulation of human somatosensory cortex. *J. Neurophysiol.*, *27*, 546.

LIBET, B., ALBERTS, W. W., WRIGHT JR, E. W. and FEINSTEIN, B. (1967): Responses of human somatosensory cortex to stimuli below threshold for conscious sensation. *Science*, *158*, 1597.

LIBET, B., ALBERTS, W. W., WRIGHT JR, E. W. and FEINSTEIN, B. (1972*a*): Masking of skin sensation by a cortical stimulus delivered simultaneously with or after the peripheral stimulus. In preparation.

LIBET, B., ALBERTS, W. W., WRIGHT JR, E. W. and FEINSTEIN, B. (1972*b*): Human cortical evoked potentials in relation to adequacy of the stimulus for eliciting conscious sensation: stimulation in ventroposterolateral nucleus of thalamus. In preparation.

LIBET, B., ALBERTS, W. W., WRIGHT JR, E. W. and FEINSTEIN, B. (1972*c*): Human direct cortical responses in relation to adequacy of the stimulus for eliciting conscious sensation: stimulation of somatosensory cortex. In preparation.

LIBET, B., ALBERTS, W. W., WRIGHT JR, E. W., LEWIS, M. and FEINSTEIN, B. (1972*d*): Some cortical mechanisms mediating conscious sensory responses and the somatosensory qualities in man. In: *Somatosensory System.* Editor: H. H. Kornhuber. Georg Thieme, Stuttgart, in press.

MELZACK, R. and WALL, P. D. (1963): Masking and metacontrast phenomena in the skin sensory system. *Exp. Neurol.*, *8*, 35.

RAAB, D. (1963): Backward masking. *Psychol. Bull.*, *60*, 118.

SATTERFIELD, J. H. (1965): Evoked cortical response enhancement and attention in man. A study of responses to auditory and shock stimuli. *Electroenceph. clin. Neurophysiol.*, *19*, 470.

SHAGASS, C. and SCHWARTZ, S. (1961): Evoked cortical potentials and sensation in man. *J. Neuropsychiat.*, *2*, 262.

SHEVRIN, H. and FRITZLER, D. E. (1968): Visual evoked response correlates of unconscious mental processes. *Science*, *161*, 295.

STOHR, P. E. and GOLDRING, S. (1969): Origin of somatosensory evoked scalp responses in man. *J. Neurosurg.*, *31*, 117.

VAUGHAN, H. G., COSTA, L. D. and GILDEN, L. (1966): The functional relation of visual evoked response and reaction time to stimulus intensity. *Vision Res.*, *6*, 645.

VON BEKESY, G. (1971): Auditory backward inhibition in concert halls. *Science*, *171*, 529.

WAGMAN, I. and BATTERSBY, W. S. (1964): Neural limitations of visual excitability. V. Cerebral afteractivity evoked by photic stimulation. *Vision Res.*, *4*, 193.

WILLIAMSON, P. D., GOFF, W. R. and ALLISON, T. (1970): Somatosensory evoked responses in patients with unilateral cerebral lesions. *Electroenceph. clin. Neurophysiol.*, *28*, 566.

Reprinted from:

# CEREBRAL CORRELATES OF CONSCIOUS EXPERIENCE

Proceedings of an International Symposium on Cerebral Correlates of Conscious Experience, held in Senanque Abbey, France on 2-8 August 1977

*Editors:* PIERRE A. BUSER
*and*
ARLETTE ROUGEUL-BUSER

INSERM SYMPOSIUM No. 6

INSTITUT NATIONAL DE LA SANTÉ
ET DE LA RECHERCHE MÉDICALE

*Cerebral Correlates of Conscious Experience*
*INSERM Symposium No. 6*
*Editors: Buser and Rougeul-Buser*



# NEURONAL VS. SUBJECTIVE TIMING FOR A CONSCIOUS SENSORY EXPERIENCE

BENJAMIN LIBET
Mt. Zion Neurological Institute and Hospital; and the Department of Physiology, University of California, San Francisco, California 94143 (U.S.A.)

In order to investigate a relation between conscious experience and specific kinds of neuronal activities, it is virtually necessary to study brain function intracranially in the awake, responsive human subject. Obviously, any such direct experimental investigation is subject to the severe limitations imposed by the rights of the subject and the ethical responsibility not to add unwarranted risks to the therapy. However, it has been possible to utilize gainfully the opportunities afforded by the surgical implantation of electrodes intracranially for therapeutic purposes[1,2], in informed and consenting human subjects*. When electrode contacts are located in various parts of the cerebral somatosensory system, innocuous electrical stimulation procedures can be employed in a controlled fashion to manipulate and investigate neuronal function in a causative, rather than merely correlative, relationship to conscious sensory responses. A subdural stimulus to primary somatosensory cortex (S I) initiates an input different in its entry path and pattern from that generated by a peripheral sensory stimulus[3]. Nevertheless, we have found it possible to elicit conscious sensory experiences with natural-like somatosensory qualities in most subjects by careful regulation of electrical stimulus parameters (particularly of intensity, train duration, and pulse frequency) to near-liminal values[3,4,5]. This is in contrast to the paresthesias more commonly reported in the pioneering studies of Penfield and others[3].

Two temporal features for specifically required cerebral processes have emerged from the study. One presents a substantial minimum time period of activation of cerebral neuronal systems, in order to achieve an adequate state for production of a conscious sensory experience[3,4,6]. The other invokes a mechanism for referral of the subjective timing of

*All the investigational procedures have been reviewed (in relation to any possible risk factors and to the conditions of informed consent) and approved by an independent Committee on Human Experimentation in Mt. Zion Hospital, in accordance with guidelines set out by the National Institutes of Health, U.S. Public Health Service.

the experience retroactively, to a time close to the initiation of the sensory signal[7].

Minimum train duration for effective stimuli. Discovery of these temporal features began with the finding that direct electrical stimulation of postcentral gyrus (S I cortex) required repetition of pulses for surprisingly long periods of time in order to elicit any conscious sensation. The minimum train duration (T.D.) of the stimulus varied with pulse intensity, but was generally about 0.5 sec for the liminal intensity below which nothing could be felt (Fig. 1)[4,8].

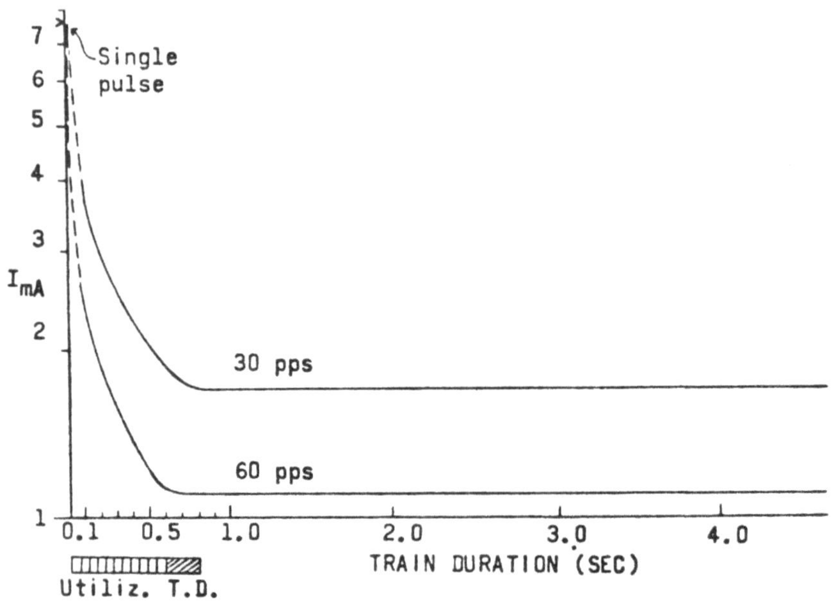


Fig. 1. Intensity/train duration combinations for stimuli (to postcentral gyrus) just adequate to elicit a threshold conscious experience of somatic sensation. Curves are presented for two different pulse repetition frequencies, employing rectangular pulses of 0.5 msec duration. (From Libet, 1966)[8].

This form of intensity-T.D. relationship, including the value of about 0.5 sec for utilization T.D. at liminal intensity, was independent of variations in other stimulus parameters, including pulse frequency. It was also found to hold for stimuli applied to subcortical sites in the specific projection system, i.e., in white matter below S I, and in brain stem targets judged to be n.VPL (or VPM) of thalamus or medial lemniscus (LM). However, this was in sharp contrast to stimuli in the periphery (skin or nerve)[4,6] and in dorsal columns[9]; at these levels a single pulse is quite effective and repetition is relatively inconsequential in its effects on threshold intensity.

Insufficiency of primary evoked response. Even though the initial stimulus pulses in the specific sensory system were ineffective for eliciting sensation, they nevertheless could elicit substantial electrophysiological responses. "Direct cortical responses" were recordable, adjacent to the stimulus at S I cortex, in response to the first as well as to each succeeding pulse of the stimulus train[3]. Also, a primary evoked potential response at S I, similar to that elicited by a peripheral sensory stimulus, is produced by a pulse in n.VPL or LM (Fig. 2). It followed from the latter that the cerebral

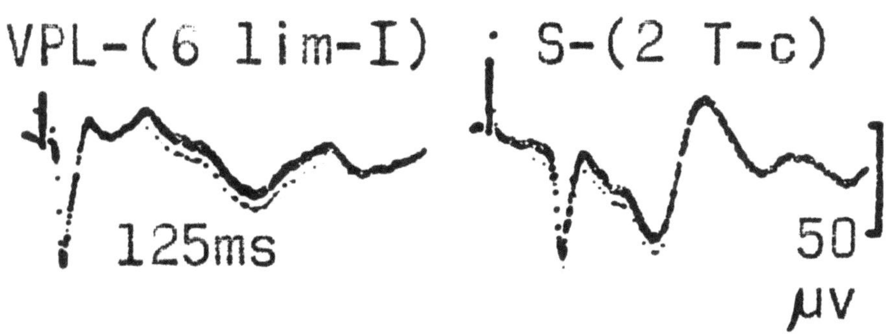


Fig. 2. Evoked potentials of somatosensory cortex, recorded subdurally, in response to thalamic (n.VPL) or skin stimuli in the same subject (patient with heredofamilial tremor). Each tracing is the average of 250 responses at 1.8 per sec; total trace length, 125 msec VPL: stimuli in ventral-posterolateral-nucleus of thalamus; subject reported not feeling any of these stimuli, although current pulses were at 6 times liminal-I for VPL electrode (liminal-I being minimum peak current per pulse to elicit sensation when delivering a > 0.5 sec train of pulses, 60 pps in this case; see Fig. 1). S: stimuli at skin; current at twice threshold, every stimulus pulse was felt. Note the shorter latency of the primary (surface positive) evoked response to VPL stimulus. (From Libet et al., 1967)[10].

processes associated with the primary evoked response of S I cortex to a specific projection volley, even in the context of awake and attentive brain function, were insufficient to induce a conscious sensory experience[5,6,10]. Also in accordance with this conclusion were the findings on evoked potentials recorded subdurally at S I cortex in response to single pulse stimuli to the skin. A pulse below threshold for producing any sensation could still elicit a primary evoked potential but little or no later components; a pulse at or above threshold elicited the later evoked components as well[10].

These findings led to the hypothesis that suitable neuronal "activations" at cerebral levels must proceed for a substantial minimum time period (of up to about 0.5 sec) in order to give rise to any conscious sensory experience. It was inferred that an adequate single volley of peripheral nerve impulses meets this requirement by its capability for eliciting a series of evoked cortical responses, early and late[5,10]. The later components have been thought to depend upon extralemniscal pathways[11] and presumably would not be elicited by a single volley generated in the specific lemniscal system well above the medullary nuclei. However, no definitive statement can presently be made about the ultimate locations and neuronal nature of any prolonged activations that may directly "represent" the conscious sensory experience[3,6].

Retroactive effects of a conditioning cortical stimulus, on a peripherally-induced sensation. How could one further test the possibility that the proposed hypothesis is in fact applicable to the case of a minimally adequate peripheral stimulus, a single threshold pulse to skin? The hypothesis would predict the existence of a period of up to 0.5 sec, following such a single pulse stimulus to skin, during which alterations of cortical activity might be able to modify the elaboration of the conscious sensory response to that pulse. It had been reported[12] that the conscious perception of a normally effective peripheral sensory stimulus could often be suppressed by simultaneous electrical stimulation of the appropriately related cortical sensory area. We were now able to show that such a suppression could also be retroactive. The delayed or conditioning stimulus consisted of a train of pulses applied to S I cortex; to be effective it had to have a minimal intensity of 1.1-1.2 times the liminal intensity needed for eliciting its own conscious sensation, and a train duration > 100 msec. The conscious experience normally elicited by a preceding test pulse to the skin could be abolished when the cortical conditioning train was begun after the skin pulse, up to 200 msec in most subjects, and up to 4-500 msec later in some[3,6]. To obtain this effect, the skin pulse itself had to be near threshold in intensity and also applied spatially in or adjacent to the peripheral field of the stimulated cortex (i.e., in or near the somatic area to which the sensation elicited by the cortical conditioning stimulus was referred). Retroactive inhibition producible by a peripheral (sensory) conditioning stimulus was of course already well-known[13], especially with visual stimuli. Our findings demonstrated that it could be initiated at the direct cortical site which received the afferent projection of the preceding sensory test stimulus and the quantitative timing relationships at the cortical level.

Retroactive enhancement was also detected in a number of subjects, perhaps in association with a somewhat different positioning of the cortical electrode. This type of conditioning effect was demonstrable when the test consisted of two separate pulses applied to the same site on the skin but delivered about 5 sec apart, and the subject asked to compare their subjective intensities. When the two delivered skin pulses were electrically equal or unequal (by about 10%), subjects reported them respectively to be subjectively equal or unequal (in the appropriate direction). However, if a conditioning cortical stimulus was begun at any time up to 200-500 msec following the second of two equal skin pulses, some subjects consistently reported that the second test pulse to the skin felt distinctly stronger than the first. (These same subjects never exhibited retroactive inhibition by the conditioning cortical stimulus, which was applied at the available electrode located on a fixed cortical site in each case.)

The finding of retroactive enhancement helps to select among some possible alternative interpretations of the retroactive actions. We must recognize that our operational definition of a conscious sensory experience is a report by the subject that he is presently aware of having had such an experience during the preceding test period, a few seconds earlier; this obviously adds the processes of short-term memory and of recallability to those that may be required to produce the subjective sensory experience. Generalized electroconvulsive shock treatment is known to produce some retroactive amnesia for recent events. In the case of retroactive inhibition by a cortical conditioning stimulus, it might be argued that there is in fact an early conscious experience of the test stimulus, but that the later cortical stimulus "disrupts" the processes of short-term memory for the experience. However, no such memory disruption could apply to retroactive enhancement. But, even for the case of retroactive enhancement, the view that there is no significant delay, for the conscious experience of a skin stimulus, may be retained by employing another sort of argument; it has been suggested that the experience of the test skin stimulus may later be reported as having been more intense, than it was in fact, because the delayed conditioning cortical stimulus may in the interim have induced some kind of "reinforcement" of the processes involved in recalling the intensity of the experience. Although this argument cannot be presently excluded, it introduces ad hoc assumptions and mechanisms that are not required by our hypothesis. Also, it does not explain why the initial primary cortical response, evoked by a stimulus to skin or to VPL/LM, elicits no conscious sensory experience at all (see above). We suggest that the simplest tenable conclusion from our findings on the retroactive effects of the cortical stimuli is that, as hypothesized, the threshold skin stimulus initiates cerebral processes which proceed for a substantial time before becoming adequate, and that during this delay period other imposed alterations in cortical activity can affect the nature of the conscious sensory experience that eventually "appears".

Is there a subjective delay for a peripheral sensation? We come now to the intriguing question of whether the subjective experience of a peripheral stimulus is actually delayed for the same substantial period of time that is postulated to be necessary for achieving the neuronal adequacy (to elicit the experience). It seemed to be impossible to establish experimentally the absolute timing of the subjective experience. Values that derive from the speed of any behavioral response to a stimulus, i.e., reaction time, may be invalid; clearly, it is possible for the subject to have made the response to a stimulus unconsciously and to have become subjectively aware of the stimulus afterwards. Instead, we adopted a procedure involving relative timing order of two sensory experiences. The validity of this procedure was based on the availability of a "reference" sensation elicited by a stimulus to S I cortex; since a minimum or utilization T.D. is known to be required in this case, in order to produce any conscious sensation, it seemed reasonable to assume that this subjective experience could not begin before the end of such a stimulus train (though it might begin after the end). The single test pulse at threshold intensity to the skin can be delivered at any desired time relative to the end of the minimum T.D. applied to S I cortex (see Fig. 3), and the subject asked to report which of the two sensations appeared first, that elicited at the skin or the one (recognizably different) elicited by the liminal cortical stimulus. The results of several such experiments with different subjects indicated clearly that there was in fact essentially "no delay" for the subjective sensory experience induced by the skin pulse. That is, the skin pulse-sensation was reported to occur "first" even when this pulse was delivered at a time as little as 100 msec ± before the end of the minimum cortical T.D.; when the pulse was delivered at or near the end of the cortical T.D., the two sensations were reported to appear synchronously.

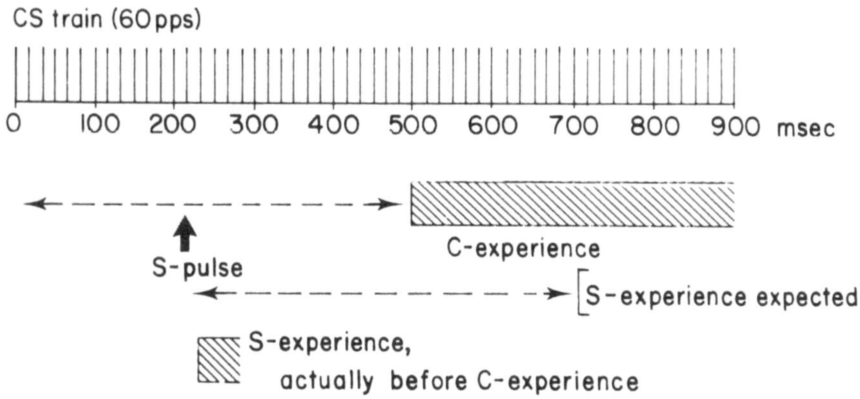


Fig. 3. Subjective timing of sensory experience for a skin pulse relative to that for stimulus train at S I cortex. Diagram shows cortical stimulus train at liminal intensity, producing a sensory (C-) experience no earlier than the utilization T.D. of 500 msec (see Fig. 1). With a near-threshold single pulse to skin (S- pulse) delivered 200 msec after onset of CS train, subjective sensory experience of it (S-experience) was expected to occur after the "C-experience" (if a similar period of cerebral activations was required). Actually, S-experience was reported to precede C-experience. (From Libet et al.7)

The absence of any substantial delay for the subjective experience of the skin pulse appeared to falsify our general hypothesis; but, strictly speaking, the hypothesis had dealt with a delay in neuronal adequacy rather than in subjective timing. Continuing tentatively with the assumption that the timing of neuronal adequacy for both the peripheral and cortical stimuli have similar delays, a modified working hypothesis was developed that might explain the discrepancy between the timings of the subjective experiences for the two stimuli.

Modified hypothesis: Subjective referral in time, retroactive to primary evoked response. In the modified hypothesis there are two new features which apply to peripheral sensory inputs: (1) The early or "primary" evoked response of S I cortex, to the afferent volley delivered by the fast, specific lemniscal projection system, is postulated to serve as "time-marker"; and (2), there is an automatic subjective referral of the conscious experience backwards in time to this time-marker, after the delayed neuronal adequacy has been achieved. With such an arrangement this sensory experience would appear subjectively to occur with no significant delay from the arrival of the fast projection volley. It should be recalled that the primary evoked response at S I begins about 10-25 msec, depending on the bodily location, after delivery of a single pulse peripheral stimulus (e.g., Fig. 2); also, that a near liminal stimulus train applied to the pial surface of S I cortex does not elicit a response resembling the primary evoked potential, presumably because it does not initially excite the ascending projection fibers (from ventrobasal thalamus) which are responsible for this response[4,5]. Fortunately, it was possible to subject these newly added postulates to experimental tests that could potentially falsify or contradict them.

Subjective timing for thalamic vs. cortical stimulus trains. One type of test is based on the unique conditions associated with a stimulus applied in a cerebral but subcortical level of the specific projection pathway. As noted above, a stimulus applied to LM or n.VPL requires the same kinds of minimum train durations as one applied to S I cortex, in order to elicit a conscious sensory experience. However, unlike S I cortex, each volley in LM or n.VPL should generate a primary evoked response in S I cortex (e.g., Fig. 2); this primary response should supply the same putative timing signal as does the single skin pulse. Consequently, the modified hypothesis would lead to a startling prediction: The subjective timing of a sensory experience elicited by a stimulus train in LM or n.VPL should be essentially similar to that for a skin pulse (i.e., as if there were no perceptible delay from the onset of the LM train); this should occur in spite of the experimental fact that the stimulus to LM or n.VPL does not become adequate until its train duration has achieved a substantial value of up to about 500 msec, depending on the intensity employed.

A diagram of the experimental paradigm for these tests with pairs of temporally-coupled stimuli is shown in Fig. 4. For the cerebral stimulus train at 60 pps, whether cortical (S I) or subcortical (LM or n.VPL), the intensity (peak current per pulse) was adjusted, for practical experimental reasons, so that a minimum train duration of about 200 msec was required in order to elicit any conscious sensory response. When a peripheral stimulus pulse (P) to the skin is delivered synchronously with the start of the stimulus train to S I cortex (C), we would expect the subject to report that the skin-elicited sensation appears before the cortically-induced one, as in previous experiments (see section I above). On the other hand, a similarly synchronous delivery for a skin stimulus relative to onset of a stimulus train in n.VPL or in LM should result in an apparent synchrony for the subjective onset of both conscious sensations, according to the modified hypothesis (Fig. 4). Related predictions can be made for cases of other timings of the skin stimulus relative to the onset of cortical and thalamic stimulus trains.

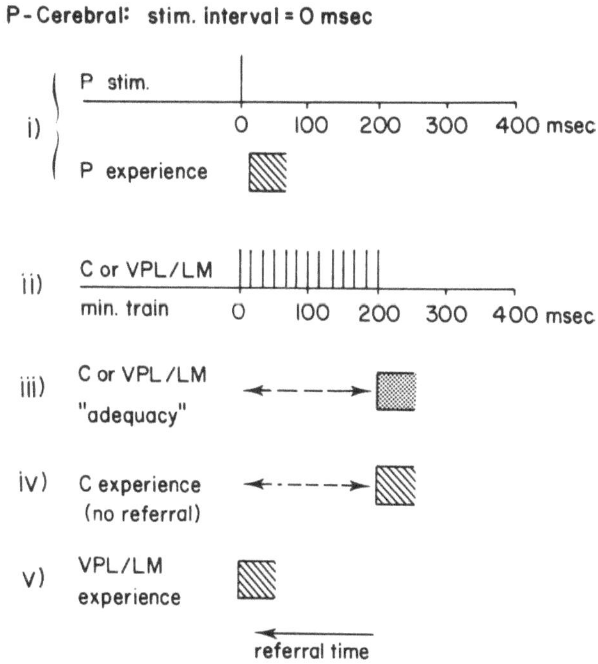


Fig. 4. Timing relationships of reported subjective experiences, for peripheral vs. cerebral stimuli, as predicted by the modified hypothesis. Each pair of stimuli consists of a peripheral pulse (P) and a cerebral train of pulses. The cerebral stimulus train may be either at S I cortex (C) or in n.VPL (or LM). Therefore, timing relationships may be compared for two types of pairs: (a) P paired with S I cortex (C); and (b) P paired with n.VPL/LM. In this set the time interval, between (i) P stimulus and (ii) the onset of a cerebral train (C or VPL/LM) = 0. P usually consisted of a single pulse to skin of the hand, on the side of body opposite to that in which a referred sensation was elicited by the cerebral stimulus. The cerebral stimulus was a train of pulses, 60 pps, with peak current I adjusted so that a minimum train duration of about 200 msec was required in order to produce a conscious sensory experience; this meant that "neuronal adequacy" for either C or for VPL/LM could not be achieved before 200 msec had elapsed (as in iii). The subjective timing of the experience of C stimulus should be delayed for a similar time (iv); but the postulated subjective referral retroactively, to a time associated with the primary evoked cortical response elicited by the afferent specific projection volley, should place the experience of both P and of VPL/LM earlier, as shown in (i) and (v). (From Libet et al.[7])

The experimental results of the studies carried out support the modified hypothesis. (The full experimental study will be reported elsewhere[7].) For each subject, the subjective timing orders were determined not only for test pairings of a peripheral and cerebral stimulus, but also for test pairs in which both stimuli were peripheral. With either type of test pairing, three or more different time intervals between the two stimuli in each pair were employed. Almost all subjects were essentially 100% "accurate" in their subjective timing orders for two peripheral stimuli delivered with intervals of $\pm$ 200 msec, but only partially accurate for $\pm$ 100 msec intervals. When a skin stimulus was paired with a stimulus train to LM/n.VPL, the relative time orders reported for the two subjective sensations tended to approximate those reported for similar intervals between two peripheral stimuli. For example, consider the tests in which the two stimuli of each pair were delivered synchronously, stimulus interval = 0. If a P pulse to right hand was delivered synchronously with another P pulse to left hand, the subject usually reported that the sensory experiences for right and left hand appeared subjectively at the same time. Similarly, as in Fig. 4, if a P pulse to right hand was delivered coincidentally with the <u>starting</u> of a train of stimulus pulses in right VPL (which elicits a sensation referred to the left hand, with electrode placement in a given subject), the subject usually reported that the right and left hand sensations began together. This occurred in spite of the fact that the VPL stimulus intensity was set so as to require a minimum train duration of 200 msec in order for the subject to report any conscious sensory experience for it at all. This result with VPL/LM contrasted sharply with relative time orders for skin vs. S I cortex. If a P pulse to right hand was delivered coincidentally with the start of a train of stimulus pulses (minimum of 200 msec required) to the right S I cortical area representing the left hand, the subject usually reported feeling the sensation in the right hand before the onset of that in the left hand (as diagrammed in Fig. 4).

Another type of test series of the hypothesis is in progress. It is based upon the consequences of unilateral but severe pathological destruction of the specific projection pathway at a cerebral level, i.e., upon a loss or deficiency in the source of the putative timing signal for one side of the body. The hypothesis would predict that subjective timing of a sensation produced by a peripheral stimulus to the affected side of the body would show a substantial delay relative to that for a peripheral stimulus to the normal side. Thus far, only one patient has been studied who fulfilled the pathological criteria. In tests for subjective sensory timing orders resulting from paired stimuli, one to each hand, this subject exhibited a "delay" of roughly 200-400 msec for the abnormal (left) side.

General discussion. The present proposal of a new functional role for the specific projection system, in the subjective timing of a sensory experience, would extend the already accepted role for this system in mediating subjective localization of a sensory stimulus in space. To the already familiar though still mysterious "automatic" subjective referral of sensory perceptions (bodily image, visual images, etc.) in the spatial dimension, we are now adding an "automatic" subjective referral in the temporal dimension. A number of further interesting possible implications and inferences stem from the temporal features postulated here for the conscious sensory process:

(a) A requirement for a substantial period of neuronal "activations" could act as a "filter" mechanism in keeping much ongoing sensory inputs from reaching conscious levels, if most of these only elicit shorter-lasting activations of the cortex[6].

(b) One may speculate that this requirement may have a more general significance for the mechanisms that differentiate all kinds of conscious from unconscious mental functions, not merely those in sensory responses. A controlling physiological factor in the transition between unconscious and conscious experiences generally might simply involve a shorter vs. an adequately longer duration respectively for the suitable neuronal activities[14].

(c) A delay in achieving neuronal adequacy would provide an opportunity for modifying or modulating a perception during the delay interval. Experimental examples of modulation by delayed inputs to the sensory system were already discussed above in connection with the retroactive effects of delayed conditioning stimuli to S I cortex. However, possible delayed modulators might also encompass endogenous cerebral actions. For example, psychological suppressor or censor actions, as postulated by Freud, would require that unconscious cerebral processes intervene to block or modify the production of a conscious sensory or other experience. The presently postulated delay in achieving neuronal adequacy for the experience provides an opportunity for the interaction of such physiological mechanisms, even after the arrival of the initial afferent projection message at the cortex.

(d) A substantial delay in the actual production of a conscious sensory experience, as distinguished from its subjective timing, would imply that quick behavioral responses, even when complex and purposeful, would arise unconsciously[8]. This could put limitations on the kinds of actions in which conscious free choice could be exercised.

(e) The findings relative to a retroactive subjective referral of an experience in time, appear to present us with a temporal dissociation between a subjective sensory experience on the one hand, and the adequate neuronal state that elicits or accompanies the experience on the other. (Professor Donald MacKay raised the possibility that the subjective antedating of the experience might be due to an illusory judgment by the subject when he reports the timings after a test, and not to an actual difference in subjective and neuronal timings when they occur. It could be argued that during the recall process, cerebral mechanisms might "read back" to the primary evoked response and then construe the timing of the experience to have occurred prior to the time of the actual occurrence of the original experience. Again, such a possibility cannot presently be excluded; it would, however, suffer from an inability to explain the absence of asynchrony or "jitter" among the immediate subjective experiences for a variety of synchronous sensory inputs of differing intensities, etc., if the postulated delays in neuronal adequacy are accepted.) Some possible philosophical implications of such a temporal dissociation for the mind-brain relationship can only be touched upon here; they will be developed more fully elsewhere. On the face of it, an apparent lack of synchrony between the "mental" and the "physical" would appear to provide an experimentally-based argument against "identity theory", as the latter has been formulated by Feigl, Pepper, etc.[15]. Of course, it could be argued that a dissociation or discrepancy in the temporal dimension is, in principle, no different from the well-known discrepancy in the spatial dimension (the discrepancy between the subjective referral of an image in space and the spatial pattern of the neuronal activities that "give rise to" the spatially patterned experience). Since there is presently no logical explanation of the relationship between the "mental" and the "physical" (cerebral) events[16], there are currently no necessary restrictions on the manner in which subjective experiences may relate either spatially or temporally to the neuronal activities that "underlie" them. Yet, a temporal dissociation between the mental and physical events would further stretch the concept of psychophysiological parallelism or, if one prefers[15], of co-occurrence of corresponding mental and neuronal states. It could thus have an impact on the philosophical interpretations of such parallelisms or co-occurrences when formulating alternative theories of the mind-brain relationship.

SUMMARY

Two temporal features for the cerebral processes that lead to a conscious sensory experience are proposed:

(a) A substantial delay (up to about 0.5 sec) before "neuronal adequacy is achieved.

(b) A subjective referral backwards in time, after neuronal adequacy is achieved, which antedates the experience to correspond to the time of early cortical responses to specific afferent projection signal.

Evidence for these postulated features, obtained with intracranial stimulation and recording techniques in awake and attentive human subjects, is presented. Some implications of these features for perceptual and conscious mental functions are suggested.

ACKNOWLEDGMENTS

The experimental work described here was carried out in the Mt. Zion Neurological Institute, Mt. Zion Hospital, San Francisco, California, in collaboration with Dr. Bertram Feinstein, E. W. Wright, Jr., and in part Drs. W.W. Alberts and Curtis Gleason. I am indebted to all the patients who cooperated so splendidly and willingly in making the studies possible. The preparation of this paper was in part carried out while I was a scholar-in-residence at the Bellagio Study and Conference Center in Italy, supported by the Rockefeller Foundation.

REFERENCES

1. Feinstein, B., Alberts, W.W. and Levin, G. (1969) in Proceedings of the Third Symposium on Parkinson's Disease, Gillingham, F.J. and Donaldson, I.M.L. eds., Livingstone, Edinburgh, pp. 232-237.
2. Feinstein, B., Alberts, W.W., Wright, E.W., Jr. and Levin, G. (1960) J. Neurosurg., 17, 708-720.
3. Libet, B. (1973) in Handbook of Sensory Physiology, vol. II, Iggo, A. ed., Springer-Verlag, Heidelberg, pp. 743-790.
4. Libet, B., Alberts, W.W., Wright, E.W., Jr., Delattre, L.D., Levin, G. and Feinstein, B. (1964) J. Neurophysiol., 27, 546-578.
5. Libet, B., Alberts, W.W., Wright, E.W., Jr., Lewis, M. and Feinstein, B. (1975) in The Somatosensory System, Kornhuber, H.H. ed., Geo. Thieme, Stuttgart, pp. 291-308.

6. Libet, B., Alberts, W.W., Wright, E.W., Jr. and Feinstein, B. (1972) in Neurophysiology Studied in Man, Somjen, G.G. ed., Excerpta Medica, Amsterdam, pp. 157-168.

7. Libet, B., Wright, E.W. and Feinstein, B. (submitted to Brain).

8. Libet, B. (1966) in Brain and Conscious Experience, Eccles, J.C. ed., Springer-Verlag, New York, pp. 165-181.

9. Nashold, B., Somjen, G. and Friedman, H. (1972) Exp. Neurol., 36, 273-287.

10. Libet, B., Alberts, W.W., Wright, E.W. and Feinstein, B. (1967) Science, 158, 1597-1600.

11. Albe-Fessard, D. and Besson, J.M. (1973) in Handbook of Sensory Physiology, vol. II, Iggo, A. ed., Springer-Verlag, Heidelberg, pp. 489-560.

12. Penfield, W.W. (1958) The Excitable Cortex in Conscious Man, Liverpool University Press, Liverpool.

13. Békésy, G. von (1971) Science, 171, 529-536.

14. Libet, B. (1965) Perspectives in Biology and Medicine, 9, 77-86.

15. Feigl, H. (1960) in Dimensions of Mind, Hook, S. ed., New York University Press (Washington Square), New York, pp. 24-34; Pepper, S.C., ibid., pp. 37-56.

16. Nagel, T. (1974) Philosophical Rev., 83, 435-450.

AN OFFPRINT FROM

# Brain

A Journal of Neurology

Volume 102 Part 1 March 1979

# SUBJECTIVE REFERRAL OF THE TIMING FOR A CONSCIOUS SENSORY EXPERIENCE

## A FUNCTIONAL ROLE FOR THE SOMATOSENSORY SPECIFIC PROJECTION SYSTEM IN MAN

*by* BENJAMIN LIBET, ELWOOD W. WRIGHT, JR.
BERTRAM FEINSTEIN[1] *and* DENNIS K. PEARL

OXFORD: AT THE CLARENDON PRESS

*Brain* (1979), **102**, 193-224

# SUBJECTIVE REFERRAL OF THE TIMING FOR A CONSCIOUS SENSORY EXPERIENCE

## A FUNCTIONAL ROLE FOR THE SOMATOSENSORY SPECIFIC PROJECTION SYSTEM IN MAN

*by* BENJAMIN LIBET, ELWOOD W. WRIGHT, JR.
BERTRAM FEINSTEIN[1] *and* DENNIS K. PEARL

(*From the Neurological Institute, Department of Neuroscience, Mount Zion Hospital, San Francisco, Ca. 94115, the Department of Physiology, School of Medicine, University of California, San Francisco, Ca. 94143, and the Department of Statistics, University of California, Berkeley, USA*)

### INTRODUCTION

PREVIOUS studies had indicated that there is a substantial delay, up to about 0·5 s, before activity at cerebral levels achieves 'neuronal adequacy' for eliciting a conscious somatosensory experience (Libet, Alberts, Wright, Delattre, Levin and Feinstein, 1964; Libet, 1966). The delay appeared necessary not only with stimulation of medial lemniscus, ventrobasal thalamus, or postcentral cortex, but even when the stimulus was a single electrical pulse at the skin (Libet, Alberts, Wright, and Feinstein, 1967, 1972; Libet, 1973). The present investigation began with an experimental test of whether there is in fact also a subjective delay in the conscious experience for a peripheral sensory stimulus. That is, is there a delay in the subjective timing of the experience that would correspond to the presumed delay in achieving the neuronal state that 'produces' the experience? The results of that test led to a modified hypothesis; this postulates (a) the existence of a subjective referral of the timing for a sensory experience, and (b) a role for the specific (lemniscal) projection system in mediating such a subjective referral of timing. Experimental tests of the new proposal were carried out and are reported here.

The timing of a subjective experience must be distinguished from that of a behavioural response (such as in reaction time), which might be made before conscious awareness develops; or even from minimum time intervals that are perceptually discriminable, since the question of *when* the subject becomes introspectively *aware* of the stimuli or of the discrimination is not answered by such measurements. There seemed to be no method by which one could determine the absolute timing of a subjective experience. Instead, we adopted a procedure in

[1] Deceased April 15, 1978.

which the subject reported the subjective timing order of two sensory experiences. The validity of this procedure depended on the availability of a 'reference' sensation, with known constraints on its subjective timing, to which the timing for the peripherally-induced sensation could be meaningfully related. The reference sensation employed was that elicited by a stimulus applied directly to postcentral, somatosensory (SI) cortex. This cortical stimulus could be set experimentally so as to require a minimum train duration of up to about 0·5 s before it could elicit any conscious sensory experience (Libet *et al.*, 1964; Libet, 1966). Consequently, it could be assumed that the cortically-induced subjective experience could not 'arise' before the end of the experimentally fixed minimum train duration. (Preliminary reports of some of these findings have been made before the Society for Neuroscience in Toronto, Canada, November 7–11, 1976—*see* summary in *Brain Information Service Conference Report* No. 45, 1977, UCLA, Los Angeles, Ca. 90024, pp. 103–121; and in a symposium on *Cerebral Correlates of Conscious Experience*, *see* Libet, 1978).

## METHODS AND PROCEDURES

### *Subjects*

Subjects were drawn from among two groups of patients in whom a stereotactic neurosurgical procedure was to be carried out, one with dyskinesias (chiefly parkinsonism or hereditary familial tremor) and another with chronic intractable pain. Only those patients were selected for the present studies whose physical and mental condition permitted them to give their fully informed consent, as well as the necessary degree of attention and responsiveness to somatosensory stimuli. Upon any indication of fatigue or loss of interest, or of a preference not to continue, the study session was terminated. All the investigational procedures have been reviewed (in relation to any possible risk factors and to the conditions of informed consent) and approved by an independent Committee on Protection of Human Subjects in Mount Zion Hospital (in accordance with guidelines set out by the National Institutes of Health, US Public Health Service). We note here our deep gratitude to the patients whose co-operation and interested participation made this type of investigation possible.

### *Stimulation and Recording Electrodes*

The procedures employed were similar to those employed previously (Libet *et al.*, 1964, 1967, 1972). The subdural electrode assembly (Delgado, 1955) contained five or seven separate wires enclosed in a flat, flexible plastic carrier. Each wire has one uninsulated region about 1 mm long, and the exposed contacts for the separate wires are spaced 3 to 10 mm apart. (These assemblies are similar to those which have been used routinely for subdural explorations of cortical epileptigenic areas in certain epileptic patients).

The carrier is inserted posteriorly via a frontal burr hole made for therapeutic purposes; the insertion is done slowly and gently so that, in the occasional case when any obstruction is encountered, it could be halted and withdrawn. Such insertions of these subdural electrode assemblies and the temporary periods of stimulation with them have, in our experience with hundreds of patients since 1957, never resulted in any detectable signs of damage. To determine whether a given contact was located on the pre- or postcentral gyrus the motor and sensory responses to subsequent stimulation were tested. The subcortical contacts, in the ventroposterolateral or ventroposteromedial nuclei of thalamus (n.VPL or n.VPM), and in the medial lemniscus (LM) at a point just a few mm below the thalamus, were part of electrode assemblies (0·5 mm diameter) that were inserted to reach therapeutic targets (Feinstein, Alberts and Levin, 1969). Each uninsulated contact had an exposure length of 0·5 mm at right angles to the long axis of the subcortical assembly. Obviously the precise structure in which a contact lay could not be identified histologically. However, physiological evidence for location in n.VPL (or VPM) or in LM was considered to be confirmatory when stimulus trains of pulses with relatively low peak currents (0·1–0·2 mA) could produce a purely sensory response (usually a 'tingling', referred to a contralateral portion of the body), and when relatively small increases in stimulus strength gave large increases in the portion of the body to which the sensory response was referred. Production of any motor response by a cerebral stimulus had to be avoided, as the peripheral sensation generated by a muscular contraction could confuse the subjective timings for the test responses; fortunately, stimulus intensities must be raised far above the liminal sensory level at SI cortex of non-epileptic patients to produce any motor effects (Libet *et al.*, 1964), while in LM even very strongly supraliminal stimuli produce no motor responses (Libet *et al.*, 1967).

In each test trial two separate stimuli were presented in a temporally coupled fashion. The pair could consist of two peripheral stimuli ($P_1$-$P_2$); or, a 'Cerebral' stimulus replaced $P_2$, producing the pair $P_1$-Cerebral. Peripheral electrical stimuli (0·2–0·5 ms pulses) to the skin or median nerve were applied via two disc electrodes (Grass EEG type) separated by 1–2 cm. For a peripheral stimulus purely to skin the two electrodes were commonly applied to the back of the hand along a mediolateral axis; this usually results in the production of a sensation that is local and superficial in the skin, in contrast to one produced when stimulating a nerve bundle (Libet *et al.*, 1967). (In some experiments a visual flash constituted the $P_1$-peripheral stimulus; for this, a brief (0·01 ms) flash was delivered from a Grass photic stimulator, set at its lowest intensity and placed a metre or more away. The subject sat in a lighted room and usually did not gaze directly at the lamp. Under these conditions any visual after-image was minimized).

The 'Cerebral' stimulus could be applied either at the cortical (C) electrode, located subdurally on postcentral or SI cortex, or via a subcortical contact in n.VPL/VPM or in LM. Cerebral stimuli consisted of brief trains of constant current pulses, each 0·2–0·5 ms and usually at 60 pps, applied via one contact (unifocally); a large metal armband (over saline-soaked gauze) served as the second electrode. Peak currents for liminal stimuli were usually in the range of 1–3 mA for SI cortex and 0·1–0·2 mA for LM. The total coulombs passed was kept to a value well below that regarded as the threshold for producing irreversible tissue damage (Pudenz, Bullara, and Talalla, 1975; Bartlett, Doty, Lee, Negrão and Overman, 1977). This becomes especially important when purely unidirectional pulses are employed, as for the cortical stimuli. Stimuli to LM were usually biphasic in nature; either the successive pulses alternated in polarity (*see* Results, Section III-C, 2) or each pulse was capacitatively coupled.

*Instructions to Subject*

The subject was asked to report, within a few seconds after the delivery of each pair of temporally-coupled $P_1$-C stimuli, whether he subjectively experienced the peripherally-induced sensation (for $P_1$) first; or whether the cerebrally-induced sensation (for C) was first; or whether both appeared to him to start 'together' (at the same time). The subject was asked to pay attention to the *onset* of the two sensations.

In most experimental series, the sensation induced by a cerebral stimulus (sensation referred to an area on the contralateral side, commonly in the hand or arm region) was subjectively timed with respect to a $P_1$ stimulus that was applied to a related area of the hand or limb on the side opposite to that for the referred cerebrally-induced sensation; this allowed the subject to report simply—'right first' or 'left first', or 'together'. With many subjects in the earlier part of the study, the $P_1$ was a brief, weak flash of light; in these cases the subject reported 'flash first' or 'hand first' (the latter being the cerebrally-induced sensation) or 'together'. For the control or 'comparison' series of trials, in which the temporally-coupled stimuli were both peripheral ones ($P_1$-$P_2$), the nature of the subject's task and reporting was a similar one; the $P_2$-peripheral stimulus that replaced the experimental cerebral stimulus was placed on the skin generally within the referral area of the cerebrally-induced sensation.

*General Procedure in an Experiment*

The following features were common to most of the experiments in Section III of the 'Results and Discussion'. (They also apply with some obvious modifications to Section I, in which studies were carried out acutely in the operating room). After the patient recovered from the acute effects of the surgical procedure for intracranial implantation of electrodes, usually in two to three days, the responses to electrical stimulation at each electrode site were checked for the actual placement with respect to the intended structures.

The initial experimental phase usually consisted of a brief 'training' period. This consisted of 15 to 25 trials with temporally-coupled pairs of two peripheral stimuli; these were similar to trials to be used for the 'control' series except that, after his report of subjective order for the $P_1$-$P_2$ stimuli in each training trial, the subject was told the 'correct' answer (that is, the order in which the two peripheral stimuli were actually delivered in that trial). In addition to providing some familiarity with the procedure, the training series often quickly improved the consistency of responses obtainable in the following control series portion of the experiment when no information was given to the subject. Trials in the 'experimental series', with the temporally-coupled peripheral vs. cerebral stimuli, were subsequently started without further training; subjects were never given any information about the actual order of stimulus deliveries for peripheral-cerebral couplings.

In most series of trials with temporally-coupled stimuli, three different time intervals were employed for delivery of one stimulus of the pair relative to the $P_1$-peripheral stimulus, usually $-200$, 0 and $+200$ ms. For a number of subjects additional intervals were used (as in Tables 1-3). The particular coupling interval used in each given trial of a series was set by the operator just before each trial on the basis of a randomized sequence of numbers. Inter-trial intervals were kept to about 20 s with tests involving n.VPL or LM, 30 s with C (SI cortex), and 10-15 s when both coupled stimuli were purely peripheral.

For each individual trial, the subject was pre-alerted by the word 'ready', given orally by the observer who was in the room with the subject. Within a second or two, the actual trial period was initiated by an operator who was located in a control booth just outside the closed room in which the subject and the observer sat. The trial was initiated by a brief alerting tone signal and was followed by the two test stimuli which were separated by the preset coupling time interval. A fixed time period of 600 ms between the alerting tone and the $P_1$ test stimulus helped the subject to focus his attention during the required time. Neither the observer nor the subject were given any indication (either before or after a trial) of the specific coupling time interval used in any of the tests. However, the subject was told in advance that the time intervals, between each pair of test stimuli, might be changed in a random manner for each succeeding trial in the series; and that he was therefore to try to ignore and not be influenced by any particular pattern of the reports that he might make. Guessing about the sequential order of the two sensations was discouraged; the subject was asked to report his actual conscious experience of the order. When a subject occasionally 'missed', that is, found himself unable to report the timing order without guessing, because of a lapse in attention or memory, the same trial was repeated. The latter procedure conforms to our objective of studying the relative subjective timings of reportable awarenesses, not the ability to detect a timing order without necessarily consciously experiencing the order.

*Establishment of Stimulus Values for Tests in an Experiment*

The procedure for determining the threshold intensity (I) and minimum train duration (TD) at a given cerebral site were similar to those previously employed (Libet *et al.*, 1964; Libet, 1973). (Liminal I is the lowest peak-current level for a train of stimulus pulses which can elicit any conscious sensory experience; for SI cortex, n.VPL or LM, a minimum train duration of 'utilization TD' [U-TD] of about 0·5 s is required with liminal I—*see* Libet *et al.*, 1964; Libet, 1973). For stimuli to SI cortex, liminal I and U-TD values remain consistent (to within about $\pm 10$ per cent or less) when individual stimulus tests are repeated, if the intervals between tests are about 30 s or more (Libet *et al.*, 1964). However, when the stimulus to one of the cerebral sites was temporally coupled with a peripheral stimulus, as required in the present study, and when such paired stimuli were presented repeatedly at regular intervals in an experimental series, the subject's conscious responses to the cerebral stimulus tended to drop out (reversibly). In order to retain positive responses in the series without intolerable interruptions, intensity was raised somewhat above liminal I, to a level which reduced the minimum TD requirement to 200 to 300 ms. (Increases in TD above the U-TD of about 500 ms, when using a liminal I train, did not appear to eliminate this difficulty).

In most experimental series, therefore, the pulse intensity 'I' for the cerebral stimulus was set at a value such that the minimum required train duration was reduced to (but not below) a value of 200 ms; this minimum peak current for a 200 ms train is termed $I_{200}$. It should be recognized that a cerebral stimulus with intensity of $I_{200}$ and with TD of 200 ms produces the same near-threshold subjective experience as does one with TD of 500 ms at the somewhat lower, liminal I current (*see* I-TD relationship; Libet *et al.*, 1964; Libet, 1966, 1973). The difference is that the 200 ms stimulus train obviously becomes adequate earlier than does the 500 ms one.

The minimum TD was judged to be at least 200 ms only when reduction of TD to 150 ms produced flatly negative (as opposed to 'uncertain') responses in every test. There were thus often instances in which the adopted minimum TD (of 200 ms) produced an inconsistent and uncertain subjective response ('maybe something was there'), and when consistent and not uncertain responses ('I felt it even though very weak') might require minimum train durations of 300 ms or more. In order to retain positive responses, suitable for subjective timings during a series of trials, it was found necessary to employ a TD of 500 or 600 ms, rather than the minimum effective one of 200 ms for the intensity of $I_{200}$. The use of such cerebral *test stimuli* (peak current intensity of $I_{200}$ but TD of 500 or 600 ms) would imply the following: (a) the test stimulus produces a conscious sensory experience that is somewhat stronger than threshold, and it lasts longer than one produced by a stimulus with $I_{200}$ and TD of 200 ms (*see* Libet *et al.*, 1964); (b) the test stimulus however *cannot become adequate before at least a 200 ms portion* of the total train duration of 500 to 600 ms has elapsed. (With some 'bobbling' of threshold effectiveness some test stimuli might require up to about 300 ms for adequacy, as noted above.)

It was not possible to match precisely the temporal and spatial features of the P-induced sensations with those of the cerebrally-induced sensations. The peak current of the 0·2 to 0·5 ms pulse to the skin could be set so that the subjective intensity of its sensation roughly matched the relatively weak subjective sensory experience produced by the test cerebral stimulus. However, the peripherally-induced sensation was sharper both in its spatial localization and onset, as well as briefer in duration and different in quality. The area of subjective spatial referral for the cerebral test stimuli (whether at SI cortex, n.VPL or LM) was not only larger but, not uncommonly, it could shift to some degree in successive tests of a given series (*see also* Libet *et al.*, 1964). It was also not uncommon (for example, subject G.S. in Table 2) for each cerebrally-induced sensation to have a 'spreading' character, i.e., it could start in a given smaller referral area and quickly extend to a wider referral area before terminating. Subjects in fact reported feeling that it was distinctly easier to perceive and report timing orders for $P_1$–$P_2$ rather than for $P_1$-cerebral couplings; in accordance with this there was a greater scatter (degree of inconsistency of reported timings for the same coupling interval) for the subjective timing orders reported with $P_1$-cerebral couplings than with $P_1$–$P_2$ couplings. In addition, a less sharp onset for a cerebrally-induced sensory experience might tend to bias the reported timing order in the direction of this experience starting relatively later than its actual time of onset might warrant. It should be noted that, in tests in which the P stimulus is coupled with one in n.VPL or LM (as in Table 2), any such bias would alter the reports in a direction contrary to that predicted by our modified hypothesis (*see* Results and Discussion); that is, it would tend to weaken the support for the hypothesis. However, in tests coupling a P with a cortical (SI) stimulus, such a bias would operate in the same direction as the hypothetically predicted one, and could thus provide some measure of false support for this particular prediction.

The longer duration of the cerebrally-induced test sensation, and to a lesser extent the less sharp onset, can be roughly matched by the peripherally-induced one if a suitable train rather than a single pulse is applied to skin. (Intensity of the pulse train to skin is also adjusted so that subjective intensities of the two different sensations also match). For such peripheral stimulus trains, the (skin TD) = (total TD of the cerebral test stimulus) minus (min TD required by cerebral test stimulus); for example, with a cerebral stimulus employing a test TD of 500 ms but requiring a minimum TD of 200 ms, the TD for the peripheral stimulus should be 500 minus 200, i.e., 300 ms. (This assumes that the peripheral stimulus requires a negligibly small minimum TD; *see* Libet *et al.*, 1964; Libet, 1973, and below). This arrangement (*see* Table 2) in fact did make the duration of the two sensations (skin vs. LM stimulus) appear subjectively similar, and it seemed to reduce the inconsistency of reported timings.

*Data Analysis*

The useful data that could be obtained with a given subject were limited in amount and scope in most cases. The first part of the study, reported in Section I of Results, was carried out in the operating room, during second stage acute procedures in which therapeutic electrodes were inserted and used for treatment of dyskinesias in the awake, responsive patients (Feinstein, Alberts, Wright and Levin, 1960). A number of the earlier cases involved in Section III of the Results were also studied in the operating room. Most of the studies in Section III, however, were carried out in sessions outside the operating room, with patients in whom therapeutic procedures required the electrodes to remain implanted for approximately a week (Feinstein *et al.*, 1969). Even in this group of ambulatory but in-hospital patients the quality and number of the rather demanding experimental test series that could be achieved with each subject was limited by their condition (recovering from the intracranial implantation procedure and undergoing therapeutic procedures during the week); in most cases it was not feasible to evolve and employ a fully adequate and standardized experimental series of tests. The latter became more readily possible in only a few of a small group of patients, most recently available, in whom stimulating electrodes were chronically implanted in LM for treatment of intractable pain (Feinstein *et al.*, in preparation). Two of this group of patients, in whom the implanted electrodes were still retained with apparent therapeutic benefit after some two to four years, were alert, younger men (H.S., G.S.) whose pain was now generally under control and of negligible significance, and who were able to return and participate adequately in suitably complete experiments (*see* Table 2). Even under optimal conditions, however, the total number of suitable trials in a given session of a few hours is limited by the nature of the subject's task in the experiment. The required attention to very weak sensory experiences itself imposes a considerable 'information load' (*see* e.g., Desmedt and Robertson, 1977); in the present experiments the weak and brief sensations induced by the cerebral test stimuli were even more elusive and required greater concentration by the subject than did the weakest effective peripheral stimuli. In addition, there was the task of remembering and reporting the temporal order for two experiences closely coupled. Even our normal subjects, who had to deal only with coupled peripheral stimuli, found each session somewhat demanding, and the total number of trials were kept to a minimum for best results (*see* Table 1). Therefore, statistical evaluation of results under even the best obtainable conditions required development of a statistical procedure that could deal successfully with the relatively small numbers both of different time intervals (for coupled stimuli) and of trials at each time interval. The statistical treatment was developed by Dennis Pearl, in consultation with Prof. Elizabeth Scott of the Department of Statistics, University of California at Berkeley; a discussion of the treatment is given below in connection with the data in Tables 1, 2, and 3. For the large number of other subjects, we had to rely on in-depth evaluation of more limited data in each case to make qualitative judgments of what such data appeared to demonstrate.

## RESULTS AND DISCUSSION

### I. *Subjective Timing Order for Couplings of a Threshold Peripheral Stimulus* (*P*) *with a Liminal Cortical* (*C*) *Stimulus Train*

When train duration of a cerebral stimulus with liminal intensity is reduced by 10 to 20 per cent below the average value of about 500 ms for the minimum or 'utilization' train duration (U-TD), the subject reports with certainty that he feels nothing (Libet *et al.*, 1964, 1972). It seemed justifiable, therefore, to assume that the conscious experience for a liminal C stimulus could not begin before the end of the U-TD, though it might begin afterwards (*see* fig. 1). The end of the U-TD is an empirically determinable value for a given subject. Consequently, the onset of the conscious sensory experience elicited by the liminal C stimulus appeared to be utilizable as a 'reference' time. The timing of a peripherally-induced sensory experience might then be meaningfully compared to the 'reference' time provided by the onset of such a cortically-induced sensation. (It should be recalled that conduction delays for the arrival of the peripherally-initiated neural message at the sensory cortex are trivial, in the context of the hundreds of milliseconds required for the 'reference' stimulus at cortex. The latency for the primary evoked potential, recorded on SI cortex in response to a stimulus on the hand, is about 15 ms or so).

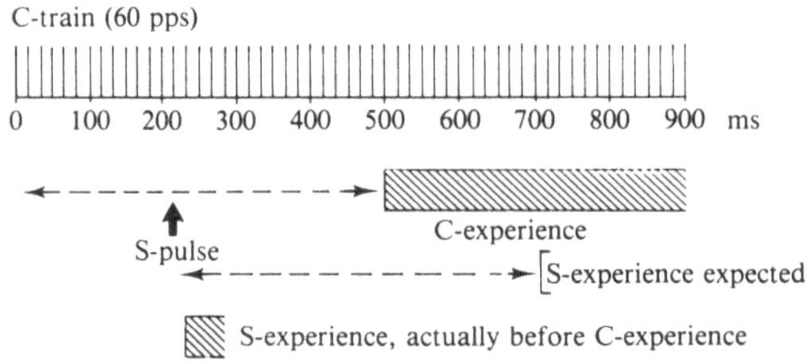


FIG. 1. Diagram of experiment on subjective time order of two sensory experiences, one elicited by a stimulus train to SI cortex (C) and one by a threshold pulse to skin (S). C consisted of repetitive pulses (at 60 pps) applied to postcentral gyrus, at the lowest (liminal) peak current sufficient to elicit any reportable conscious sensory experience. The sensory experience for C ('C-experience') would not be initiated before the end of the utilization-train duration (U-TD, average about 500 ms), but then proceeds without change in its weak *subjective* intensity for the remainder of the applied liminal C train (*see* Libet *et al.*, 1964; Libet, 1966, 1973). The S-pulse, at just above threshold strength for eliciting conscious sensory experience, is here shown delivered when the initial 200 ms of the C train have elapsed. (In other experiments, it was applied at other relative times, earlier and later.) If S were followed by a roughly similar delay of 500 ms of cortical activity before 'neuronal adequacy' is achieved, initiation of S-experience might have also been expected to be delayed until 700 ms of C had elapsed. In fact, S-experience was reported to appear subjectively before C-experience (*see* text).

On our original hypothesis (Libet *et al.*, 1964, 1972; Libet, 1966, 1973), a single pulse stimulus to the skin at just above threshold level should also be followed by a period of about 500 ms before cerebral neuronal adequacy for the conscious sensation would be achieved. If such a *skin pulse* (*S*) were to be applied some time (say 200 ms) *after* the beginning of the C stimulus train, then neuronal adequacy for the peripherally-induced sensation should be achieved after the end of the U-TD of the C train (in this example, at about 700 ms after the beginning of the C train; *see* fig. 1). If the *subjective experience* were to occur at the same time as the achievement of *neuronal adequacy* in the case of either stimulus, one would expect the subject to report that the conscious sensory experience for the C stimulus began before the appearance of that for the threshold S pulse (fig. 1).

Actual tests of this kind, with S delayed for variable times after the start of the C train, were carried out with six patients. In each of these subjects only a limited number of observations could be made. However, the pooled reports were predominantly those of sensory experience for the C (cortical) stimulus beginning *after*, not before, that for a delayed threshold S pulse; this was true even when the delivery of the S pulse was delayed from the start of the C stimulus train by almost the full value of the U-TD (that is, by up to 400 to 500 ms when U-TD was 500 ms). These findings indicated that the subjective experience of the skin stimulus occurs relatively quickly after delivery of the S pulse, rather than after the expected delay of up to about 500 ms for development of neuronal adequacy following the S input.

Study of this point in this way was carried out chiefly at a time (before 1969) when chronic implantations of therapeutic electrodes were not being made. Thus, only a relatively small number of tests could be made during the surgical procedure (with local anæsthesia) in the operating room, with each of the 5 subjccts involvcd (4 parkinsonians, 1 spasmodic torticollis). A sixth subject, an amputee being treated for intractable pain in 1970, was tested via implanted electrodes outside the operating room. However, the results were qualitatively consistent among the different cases in this group.

II. *Modified Hypothesis, to Relate Subjective Timing to the Timing of Neuronal Adequacy for an Experience*

There were now two possible alternative conclusions that might be drawn from the experimental result in Section I, above. Alternative (1): the substantial delay empirically required to achieve neuronal adequacy with the cortical stimulus might not apply to the case of a sensory experience elicited by a peripheral stimulus. However, our previous evidence strongly supported the hypothesis of such a cerebral delay for eliciting even a peripherally-induced sensation (Libet *et al.*, 1972; Libet, 1973), and it argued against adopting this alternative (*see also* General Discussion). Alternative (2): there is a discrepancy between the subjective timing (of a conscious sensory experience) and the expected time at which 'neuronal adequacy' for eliciting the experience is achieved. In considering the apparent paradox posed by alternative (2) it was necessary to recognize that the original hypothesis dealt directly only with the time to achieve the *adequate neuronal state* that elicits the experience. The two timings, for subjective experience vs. neuronal adequacy, might not necessarily be identical. But if there were a discrepancy between the two kinds of timings, why should it appear in the case of peripheral skin stimuli and not with a cortical stimulus? A possible answer to this question lay in the difference between the initial cortical responses elicited by peripheral vs. cortical stimuli. The S (skin) pulse leads to a volley in the ascending specific projection (lemniscal) system; the latter elicits a relatively localized 'primary' (initially surface-positive) evoked potential in the SI cortex, with an onset latency of about 15 ms after a stimulus to the hand (*see*, Fig. 2; *see also* Jasper, Lende and Rasmussen, 1960; Desmedt, 1971; Goff, Matsumiya, Allison and Goff, 1977). The liminal C stimulus, applied subdurally at SI cortex (postcentral gyrus), does not elicit a similar type of response (*see* Libet *et al.*, 1967, 1972; Libet, 1973). With these considerations in mind we developed the following postulates, to be added as modifiers to our original hypothesis on the cortical processing time for a conscious sensory experience:

(1) Some neuronal process associated with the early or *primary evoked response*,

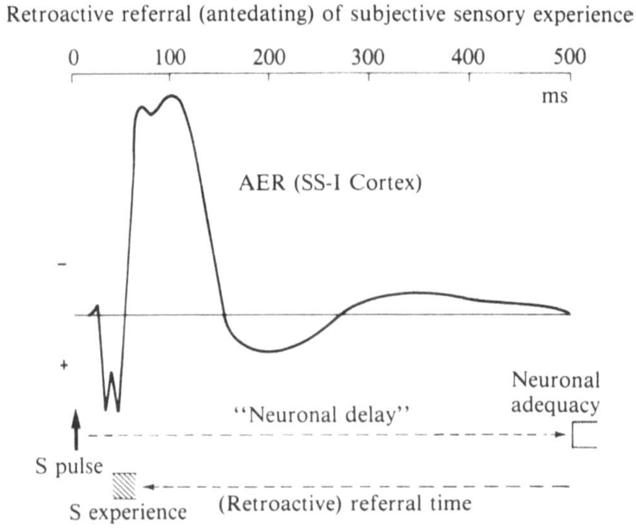


FIG. 2. Diagram representing the 'averaged evoked response' (AER) recordable on the surface of human primary somatosensory cortex (SI), in relation to the modified hypothesis on timing of the sensory experience. Below the AER, the first line shows the approximate delay in achieving the state of 'neuronal adequacy' that appears (on the basis of other evidence) to be necessary for eliciting the sensory experience. The second line shows the postulated retroactive referral of the subjective timing of the experience, from the time of 'neuronal adequacy' backwards to some time associated with the primary surface-positive component of the evoked potential. The primary component of AER is relatively highly localized to an area on the contralateral postcentral gyrus in these awake human subjects, as had been shown in anæsthetized patients (Jasper, Lende and Rasmussen, 1960). The secondary or later components, especially those following the surface negative component after the initial 100 to 150 ms of the AER, are wider in distribution over the cortex and more variable in form even when recorded subdurally (*see*, e.g., Libet *et al.*, 1975). It should be clear, therefore, that the present diagram is not meant to indicate that the state of 'neuronal adequacy' for eliciting conscious sensation is restricted to neurons in primary SI cortex of postcentral gyrus; on the other hand, the primary component or 'timing signal' for retroactive referral of the sensory experience would be a function more strictly of this SI cortical area.

The AER shown here is a composite drawing, based on actual recordings by us in several unanæsthetized patients, in whom a subdural metal contact was located on the pia-arachnoid surface of the postcentral gyrus (SI cortex) at a 'good spot'; the latter was defined by the low intensity required by a direct cortical stimulus train at this site in order to elicit a relatively localized somatic sensation (*see* Libet, 1973; Libet *et al.*, 1964), and by the fact that stimulation of another contact, located 5 to 10 mm anterior to it, could elicit the localized pyramidal-type motor response typical with primary motor cortex. In each case the evoked potentials were recorded (relative to an indifferent electrode usually on ear lobes) in response to single pulse stimuli (S). The latter were applied to a contralateral area of skin within the referral area for the sensation elicited when stimulating via the recording SI cortical electrode. Skin stimuli were just above the threshold for eliciting a sensation in 100 per cent of the trials. The AER for each subject was the average of SI evoked potentials from usually 256 such stimuli, delivered at about 1/s. (The later components of these AERs are probably relatively minimal in their amplitudes, compared to what could presumably be obtained if the rate of stimulus repetition were lower than 1/s and if the subjects had been asked to perform some discriminatory task in relation to the series of 256 stimuli, as in Desmedt (1971). The composite form of these AERs to skin stimuli may be compared with those obtainable as responses to stimulating the median nerve, both with intracranial recordings (Hirsch, Pertuiset, Calvet, Buisson-Ferey, Fischgold and Scherrer, 1961) and with scalp recordings (*see*, e.g., Desmedt, 1971; Goff, Matsumiya, Allison and Goff, 1977)).

of SI (somatosensory) cortex to a skin stimulus, *is postulated to serve as a 'time-marker'*. (2) There is an automatic *subjective referral of the conscious experience backwards in time* to this time-marker, after the delayed neuronal adequacy at cerebral levels has been achieved (*see* fig. 2). The sensory experience would be 'antedated' from the actual delayed time at which the neuronal state becomes adequate to elicit it; and the experience would appear subjectively to occur with no significant delay from the arrival of the fast projection volley. Fortunately, it was possible to put these newly added postulates to experimental tests that could potentially falsify them.

*Predictions from the hypothesis.* The chief test of the modified hypothesis was based on the unique conditions associated with a cerebral stimulus applied to LM or n.VPL, as contrasted with one to SI cortex. As already noted, a stimulus applied to LM or n.VPL requires the same kinds of minimum train durations as one applied to SI cortex, in order to elicit a conscious sensory experience. However, unlike SI cortex, each volley in LM or n.VPL should and does elicit a primary evoked response in SI cortex equivalent to the early components in the response to a skin pulse, which are seen in fig. 2 (Libet *et al.*, 1967). The primary cortical response to an LM volley should supply the same putative timing signal as does the single skin pulse. Consequently, the modified hypothesis would lead to a startling prediction: the subjective timing of a sensory experience elicited by a stimulus train in LM (or n.VPL) should be essentially similar to that for a skin pulse (i.e., as if there were no perceptible delay from the onset of the LM train); this should occur in spite of the experimental fact that the stimulus to LM or n.VPL does not become adequate until its train duration has achieved a substantial value of up to about 500 ms, depending on the intensity employed.

A diagram of the experimental paradigm for these tests with pairs of temporally-coupled stimuli is shown in fig. 3. For the cerebral stimulus train at 60 pps, whether cortical (C) or subcortical (LM or n.VPL), the intensity (peak current per pulse) is adjusted so that a minimum train duration of about 200 ms is required in order to elicit any conscious sensory response (*see* Methods).

---

FIG. 3. Diagram of timing relationships for the two subjective experiences when a peripheral stimulus (P) is temporally coupled with a Cerebral stimulus, as predicted by the modified hypothesis. The Cerebral stimulus train is located either at SI cortex (C) *or* in n.VPL (or LM). Therefore, timing relationships may be compared for two types of coupled pairs: (a) P paired with SI; and (b) P paired with n.VPL/LM. In set I, the *time interval* between (i) P stimulus and (ii) *onset* of a cerebral train (whether C or VPL/LM) = 0; in set II, this interval = −200 ms (i.e., cerebral stimulation starts 200 ms before P); in set III, the interval = +200 ms.

P usually consisted of a single pulse applied to skin of the hand (but trains used for experiments in Tables 1 and 2) on the side of body opposite to that in which a referred sensation was elicited by the cerebral stimulus. The P-experience (*see* (i)) is timed subjectively to appear within 10 to 20 ms after the P stimulus (*see also* fig. 2). Each cerebral stimulus (in (ii)) is a train of pulses, usually 60 pps, with peak current adjusted so that a minimum train duration of about 200 ms is required in order to produce any conscious sensory experience; this means that the state of 'neuronal adequacy' (*see* (iii)) with either C or VPL/LM stimuli could not be achieved before 200 ms of stimulus train duration had elapsed. The subjective timing of the experience of C stimulus (iv) should be delayed for a time similar to this minimum TD of 200 ms. But the experience of VPL/LM(v) should be timed earlier; i.e., it should be subjectively referred retroactively, to a time associated with the primary evoked cortical response that is elicited even by the first pulse of a stimulus train in the VPL/LM portion of the specific projection pathway.

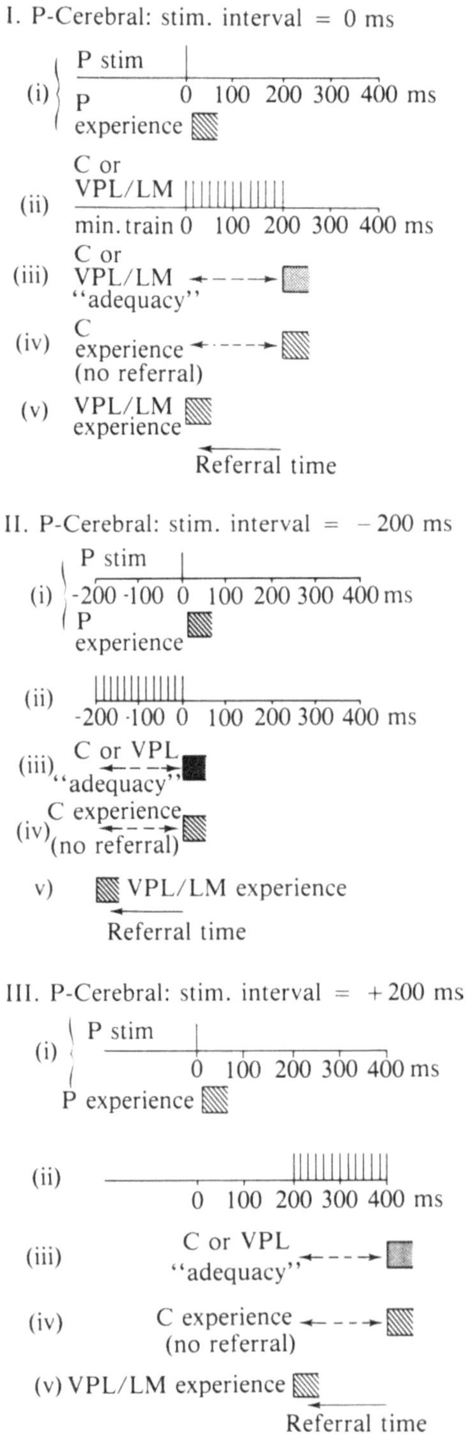
I. P-Cerebral: stim. interval = 0 ms
(i)
P stim
P experience
0 100 200 300 400 ms
(ii)
C or VPL/LM
min. train 0 100 200 300 400 ms
(iii)
C or VPL/LM "adequacy"
(iv)
C experience (no referral)
(v)
VPL/LM experience
Referral time
II. P-Cerebral: stim. interval = -200 ms
(i)
P stim
-200 -100 0 100 200 300 400 ms
P experience
(ii)
-200 -100 0 100 200 300 400 ms
(iii)
C or VPL "adequacy"
(iv)
C experience (no referral)
v)
VPL/LM experience
Referral time
III. P-Cerebral: stim. interval = +200 ms
(i)
P stim
0 100 200 300 400 ms
P experience
(ii)
0 100 200 300 400 ms
(iii)
C or VPL "adequacy"
(iv)
C experience (no referral)
(v) VPL/LM experience
Referral time

For *P-Cerebral stim. interval* = *0* (fig. 3-I). When a P stimulus pulse or train (to skin or nerve) and a stimulus train to LM (or n.VPL) are begun synchronously (i) and (ii), the patient should report that the *subjective* onsets of both conscious experiences are very close together or synchronous (i) and (v). This should occur even though it is empirically established that stimulus adequacy for the LM/n.VPL stimulus cannot become adequate until after its minimum TD of 200 ms has elapsed, while the P stimulus is adequate either after 1 pulse, or within 33 ms or less when a train is used to elicit a sensation matching both the subjective intensity and duration of the LM-induced one (*see below*). *Neuronal adequacy* for the experience (as distinguished from stimulus adequacy for inducing this neuronal response) would be achieved after a roughly similar delay for both LM and P (as well as for C), as in fig. 3-I (iii) and in fig. 2, according to the hypothesis. On the other hand, with a similar coupling between P and C stimuli, sensory experience for C should be reported to appear after that for P (i) and (iv), as in experiments of Section I above. This should occur even though it is empirically established that C stimulus becomes adequate after the same minimum TD of 200 ms as does LM stimulus.

For *P-Cerebral stim. interval* = *−200 ms* (fig. 3-II). When LM/n.VPL stimulus train is begun 200 ms before P stimulus (ii) vs. (i), the patient should report that the experience for LM/n.VPL starts before that for P (i) and (v); whereas for a similarly coupled C stimulus the onset of the experience should be reported to come either at the same time or possibly after that for P (i) and (iv).

For *P-Cerebral stim. interval* = *+200 ms* (Fig. 3-III). When cerebral stimulus train is begun 200 ms after P stimulus (ii) vs. (i), patient should report that the experience for either LM/n.VPL or C stimuli (v) and (iv) appears after that for P (i).

III. *Subjective Timing Orders, Experimentally Determined, for Couplings of* $P_1$-$P_2$, $P_1$-*LM*, *and* $P_1$-*C*

A. *Couplings of two peripheral stimuli* ($P_1$-$P_2$). In order to assess the significance of the relative subjective timings for couplings of peripheral and cerebral stimuli, it is necessary to have an appropriate 'control' series in which the cerebral stimulus is replaced by another suitable peripheral one ($P_2$). Indeed, the experimental question becomes one of directly comparing the reported timing orders for a block of trials with $P_1$-$P_2$ couplings with those for a block of trials with $P_1$-cerebral couplings (both types of pairings temporally-coupled by similar intervals). Any *difference* between the two sets of subjective timing orders (for $P_1$-$P_2$ vs. $P_1$-cerebral couplings) would indicate the presence or absence of a delay in subjective timing for the cerebral stimulus relative to a comparable peripheral one (*see* Tables 2B and 3B). In this method each subject provides his own individual control or comparison data, in the patterns of his reports of subjective timing orders for two peripheral stimuli which should be processed with no differences which are significant in the present context.

Skin stimulus trains which elicit a sensation matching the subjective intensity and duration of that induced by the test cerebral stimuli (*see* Methods) were found to be below the threshold I for a single pulse and to require a minimum TD of 17 or 33 ms, that is, 2 or 3 pulses at 60 pps. When such an S train is coupled with an LM stimulus train which has a minimum requirement of 200 ms TD, the difference between the required stimulus durations for S vs. LM could, therefore, be reduced from 200 ms down to 167 ms. The possible impact of such a 33-ms reduction in the effective coupling intervals, on the pattern of reported subjective timing orders, was tested in normal subjects (Table 1).

In these, the $P_1$-$P_2$ stimuli were applied to skin of the right and left hands respectively, becoming $S_R$ (right) and $S_L$ (left) stimuli. Stimulus intensity in each case was reduced to a level that required a minimum of 2 or 3 pulses at 60 pps (TD = 17 or 33 ms) to elicit any sensory experience; but actual TD of each test stimulus was 300 ms, to simulate both the subjective intensity and duration of cerebrally-induced sensations (*see* Methods). For the A-blocks of trials, $S_R$-$S_L$ coupling intervals ranged from −200 to +200 ms (and were similar to the 'control' series with $S_1$-$S_2$ couplings in Table 2). For the 'experimental' B-block series in Table 1, delivery times of $S_R$ stimuli were modified by delaying onset of each $S_R$ stimulus by 33 ms. This reduced each $S_R$-$S_L$ coupling by 33 ms, from what the interval had been in the first or A series of the session with that subject; for example, an $S_R$-$S_L$ interval of −200 ms (that is, $S_R$ started 200 ms after $S_L$) in series A would now become −233 ms in B, while an interval of $S_R$-$S_L$ = 0 (synchronous delivery) in A would now become −33 ms ($S_R$ started after $S_L$) in B, etc. In all Tables the value of the coupling time interval between test pair of stimuli was varied in a random manner in successive trials. However, all the subjects' reports in those trials using a given coupling time interval in a given block of trials are collected together in the Table under 'Subject's timing'. In block A of Table 1, for example, subject D.A. was presented with an $S_R$-$S_L$ interval of 0 in 10 trials which were randomly distributed among the total of 50 trials in block A. Of these 10, she reported experiencing $S_R$ (skin of right hand) 'first' in 3 trials; she reported $S_R$ and $S_L$ sensations starting 'together' (T) or at the same time in 7 trials; and there were no reports of '$S_L$ first' in any of these 10 trials.

The comparison of timing orders in block A with those in block B trials of Table 1 simulates the design for the experiments in Tables 2 and 3. But in Table 1 the known delay imposed on one of the stimuli ($S_R$ in block B) should theoretically produce a similarly defined shift in subjective timing of the sensation elicited by that stimulus; whereas in Tables 2 and 3 the possibility of a shift in subjective timing relative to onset of a stimulus train in LM or SI cortex constitutes the unknown point. Therefore, the experiment in Table 1 serves as at least a partial check on the validity of this design and of the statistical method employed in the analysis, as well as to test the specific question of what effect a 33-ms shift may have on overall pattern of reports of subjective timing.

Comparing block A with block B for each subject in Table 1A, it is seen that the extra 33-ms delay for $S_R$ stimuli produced no change in subjective timing orders for the 200- and 150-ms coupling intervals (that is, the orders remained essentially the same as the actual order of stimuli delivered). However, small differences between blocks A and B did appear for shorter $S_R$-$S_L$ intervals; these indicated qualitatively that in series B there was a slight shift of reported timing orders in the expected direction of a small delay in the experience of $S_R$ relative to that

TABLE 1A. SUBJECTIVE TIMING ORDERS OF EXPERIENCES FOR TEMPORALLY-COUPLED PAIRS OF SKIN STIMULI, RIGHT VS LEFT HANDS, IN NORMAL SUBJECTS

| Subject | | $S_R$-$S_L$ interval (ms) | No. of trials | Subject's timing: $S_R$ first | Subject's timing: T | Subject's timing: $S_L$ first | Estim. 'mean shift' (ms) | Approx. SD |
|---|---|---|---|---|---|---|---|---|
| D.A. | A | −200 | 10 | 0 | 0 | 10 | | |
| (female, aged | | −100 | ↓ | 0 | 3 | 7 | | |
| 26 y) | | 0 | ↓ | 3 | 7 | 0 | −30 | 10 |
| | | +100 | ↓ | 10 | 0 | 0 | | |
| | | +200 | ↓ | 10 | 0 | 0 | | |
| | B | −233 | 10 | 0 | 0 | 10 | | |
| | | −133 | ↓ | 0 | 1 | 9 | | |
| | | − 33 | ↓ | 0 | 10 | 0 | 18 | 16 |
| | | + 67 | ↓ | 8 | 1 | 1 | | |
| | | +167 | ↓ | 10 | 0 | 0 | | |
| R.J. | A | −200 | 10 | 0 | 0 | 10 | | |
| (male, aged | | −100 | ↓ | 0 | 0 | 10 | | |
| 29 y) | | 0 | ↓ | 0 | 3 | 7 | 100 | 21 |
| | | +100 | ↓ | 3 | 6 | 1 | | |
| | | +200 | ↓ | 8 | 1 | 1 | | |
| | B | −233 | 10 | 0 | 0 | 10 | | |
| | | −133 | ↓ | 1 | 0 | 9 | | |
| | | − 33 | ↓ | 0 | 1 | 9 | 122 | 25 |
| | | + 67 | ↓ | 2 | 7 | 1 | | |
| | | +167 | ↓ | 6 | 3 | 1 | | |
| J.Wl. | A | −200 | 6 | 0 | 1 | 5 | | |
| (male, aged | | −150 | ↓ | 0 | 0 | 6 | | |
| 31 y) | | −100 | ↓ | 0 | 0 | 6 | | |
| | | − 50 | ↓ | 1 | 0 | 5 | | |
| | | 0 | ↓ | 2 | 1 | 3 | −5 | 14 |
| | | + 50 | ↓ | 5 | 1 | 0 | | |
| | | +100 | ↓ | 6 | 0 | 0 | | |
| | | +150 | ↓ | 6 | 0 | 0 | | |
| | | +200 | ↓ | 6 | 0 | 0 | | |
| | B | −233 | 6 | 0 | 0 | 6 | | |
| | | −183 | ↓ | 0 | 0 | 6 | | |
| | | −133 | ↓ | 0 | 0 | 6 | | |
| | | − 83 | ↓ | 1 | 1 | 4 | | |
| | | − 33 | ↓ | 1 | 0 | 5 | 22 | 14 |
| | | + 17 | ↓ | 4 | 0 | 2 | | |
| | | + 67 | ↓ | 6 | 0 | 0 | | |
| | | +117 | ↓ | 6 | 0 | 0 | | |
| | | +167 | ↓ | 6 | 0 | 0 | | |

Effect of added 33 ms delay for one stimulus of each pair, *see* text.

TABLE 1B. COMPARISON OF TIMING ORDERS FOR TWO BLOCKS OF $S_R$-$S_L$ PAIRS, WHEN ONE BLOCK OF STIMULUS COUPLING INTERVALS IS SHIFTED BY 33 MS.

| *Subject* | *Comparison* | *Estim. change in shift (ms)* | *Approx. SD* | *Approx. 95% confidence interval* |
|---|---|---|---|---|
| D.A. | A vs. B | −48 | 19 | (−86; −9) |
| R.J. | A vs. B | −22 | 33 | (−87; 43) |
| J.Wl. | A vs. B | −27 | 20 | (−68; 13) |
| | $(S_R-S_L)_A-(S_R-S_L)_B$ | | | |

for $S_L$. Statistical evaluation of the data in Table 1 was carried out by the same procedure that was developed for treatment of experimental data obtained with the patients, as in Tables 2 and 3 (*see* discussion of statistical terms, etc.). The 'mean shift', in the subjective timing for $S_R$ relative to $S_L$, should be close to 0 for each A-block, since the coupling intervals are grouped symmetrically, + and −, around 0. (In fact, however, the mean shift was close to 0 only for subject J.Wl.; the −30 ms shift for D.A. indicates a slight 'bias' in the direction of reporting $S_R$ first, and the +100 ms shift for R.J. indicates a bias in the opposite direction. All three subjects were right-handed).

The *change in mean shift*, when block B ($S_R$ delayed by 33 ms) is compared with block A, is given for each subject in Table 1B. The changes in mean shift are not far from −33 ms for all three subjects; the changes all indicate a statistical tendency to experience $S_R$ as delayed, relative to $S_L$, by a time not far different from the actual delay imposed on $S_R$ in the $S_R$-$S_L$ couplings in block B.

*Statistical treatment.* In this treatment, the responses for all the stimulus-coupling intervals in a given series or block of trials are used to estimate a 'mean shift'. The mean shift is, qualitatively speaking, the best time delay to use for one of the stimuli in order to get close to the centre around which the subject is reporting the relative orders for the other stimulus in all the temporally-coupled pairs in that block. (A value of zero for the mean shift would indicate no shift, in the subjective timing 'centre' of the first stimulus relative to the second stimulus of each pair, from the actual relative positions of the coupled stimuli as delivered. A negative value for the mean shift indicates that subjective timing 'centre' for the first stimulus of the pair is shifted retroactively away from the timing for the second; that is, the timing for the second stimulus is delayed from that of the first by this value).

The *difference* between the mean shifts for two blocks of trials, carried out in the same or in a closely comparable session, gives the estimated change in shift. For each such change in shift, the standard deviation (SD = square root of the variance) and the approximate '95 per cent confidence interval' is given. The latter consists of two values (in parentheses, Tables 1B, 2B, 3B) each of which is equal to two standard deviations on either side of the estimated change in shift; that is, we are about 95 per cent certain that the true value of the change in shift lies between these two values. The estimated 'change in shift' in the appropriate direction for all three subjects (Table 1B) is surprisingly close to the relatively small imposed shift of 33 ms, and it provides a kind of confirmatory test of the validity of the statistical procedure.

In Tables 2 and 3, of course, the comparisons for change in shift are between a block of trials for $P_1$-$P_2$ couplings and a block of $P_1$-cerebral couplings ('cerebral' stimuli being those to LM in Table 2, and to SI cortex in Table 3). When $P_1$ is common to both blocks, the change or difference in mean shifts between the two blocks reflects the difference between subjective timings for the second stimuli of the couplings in each block, for example, between the timings for $S_L$ and for C (*see* Table 3B). When the mean shift for block B is subtracted from that for block A, a positive value for the resulting 'change in shift' indicates a delay for timing of the second stimulus in block B (C in Table 3B) relative to the timing for the second stimulus in block A ($S_L$ in Table 3B).

A fuller, rigorous description and analysis of the statistical treatment is beyond the scope of the present paper. An alternative statistical approach which could have been adapted to our needs has since been published (Dempster, Laird, and Rubin, 1977). Three assumptions involved in formulating the statistical model employed were as follows: (i) for a particular block of data the shift function is roughly constant from trial to trial, and (ii), each trial is independent. Both of these assumptions appear to be reasonable and warranted on the basis of the care taken to maintain constant conditions within a given session, keeping the total load on the subject at levels well below any apparent fatiguability, etc., and randomizing the presentation of the coupling time intervals in successive trials; in addition, data of the exact sequence of responses were examined for subjects H.S. and G.S. in Table 2, and they showed no significant deviation from the assumption of independence of trials; (iii) the distribution function of the responses is roughly linear within the experimental range even for coupling time intervals which were not tested, and also for the range of intervals extending where necessary on either side of the experimental range. This assumption introduces a possible 'interpolation error'. This potential error tends to be reduced by the use of more different time intervals for coupling, for example, at least five intervals for the series in Tables 1 and 2 instead of the three employed in most of our other experiments. The probable actual interpolation error, due to linearizing the estimated shift function, appeared to be no more than 10 ms in the experiments using five different time intervals (as in Table 2A) and no more than 30 ms in experiments using three. Fortunately, the amounts in such errors would not seriously affect the significance of the data in relation to the hypothesis being investigated.

*'Accuracy' of timing orders for paired peripheral stimuli.* The degree of 'accuracy' (the similarity of the reported to the actually applied order of $P_1$-$P_2$ stimuli) had to be virtually 100 per cent for $P_1$-$P_2$ coupling intervals of 200 ms, in order to be useable for the comparisons with the $P_1$-Cerebral couplings. Experimental tests of the hypothesis required that the subject be capable of subjectively distinguishing timing orders for intervals equal to or less than the minimum TD of 200 ms that was required by the test cerebral stimuli. Fortunately, subjective timing orders for $S_R$-$S_L$ coupling intervals of 200 ms did turn out to be essentially 100 per cent 'accurate' for almost all the patients tested. The 'accuracy' was often somewhat reduced with 150-ms intervals, and distinctly poorer with 100-ms intervals (for which some patients could not report consistent orders). The examples in Tables 2 and 3 illustrate this point. The use of brief trains of pulses (300 to 400 ms TD's) for skin stimuli, rather than the single pulse stimuli used in most of the subjects, did not appear to affect the reported timing orders.

The foregoing was also applicable to 'normal' subjects (non-patients); some of these exhibited considerable 'accuracy' even with a 50-ms interval, but almost none with a 25-ms interval. Four normal subjects with ages in the 40's and 50's were tested with single pulse $S_R$ and $S_L$ stimuli. In 4 subjects the S stimuli were trains of weak pulses to better simulate the cerebrally-induced sensations (data for 3 of these are given in Table 1A, with similar results for the fourth, a female aged 40 years). The weaker S-induced sensations in this second group appeared to be more demanding of effort and alert attention by the subjects; indeed, in two additional subjects who were studied at the end of their regular eight-hour working period as nurses, the reported timing orders for given coupling intervals exhibited considerable variability and the 'accuracy' was poor. This factor probably helps explain some of the inconsistency of results obtained with $P_1$-Cerebral couplings in the patients; most of the patients were not in an optimal physical and mental state when they were studied with $P_1$-Cerebral trials in which the cerebrally-induced sensation was always, because of the experimental requirements, similarly weak in subjective intensity.

*Negative 'mean shifts' for $S_1$-$S_2$ in Table 2A.* For subject H.S., blocks A and C, estimated 'mean shifts' are −41 and −91 ms respectively; for subject G.S., block A, it is −35 ms even though the left-right order is reverse of that for H.S. The mean shifts for comparable blocks of trials were −21 and −9 ms for patients J.W. and C.J. in Table 3A; and −30, +100, and −5 ms for the normal subjects in Table 1A. To the extent that the negative shifts in Table 2A may be significant, they would imply that, for these patients, a stimulus to skin on the normal side ($S_L$ for H.S., and $S_R$ for G.S.) tended to be subjectively timed somewhat earlier than a comparable stimulus on the abnormal side (that was treated for pain). Most of this estimated negative mean shift appears to be associated with an asymmetry between the responses for the smaller time intervals, that is, for −100 ms (or −150) compared to those for +100 ms (or +150 ms). For both subjects, the reported timing orders were predominantly $S_1$ and $S_2$ 'together' for the −100 (−150) ms intervals; whereas, they were predominantly '$S_1$ first' for the +100 (+150) ms intervals, in accord with the actual order of delivery. This asymmetry is in fact associated with a modest but definite deficit in epicritic sensibilities on the abnormal side (right side for H.S., and left for G.S.). The deficits are ascribable to losses in the specific projection pathway probably at thalamic levels, due either to their own pathological disorders, or to small heat lesions in n.VPL or just subthalamic sites (which were made to treat their pain some years prior to implanting the present stimulating electrodes in LM), or to both. The results obtained for $S_1$-$S_2$ couplings when both stimuli were located on the normal side in subject G.S. (Table 2A, block D) are in accord with this interpretation; the smaller estimated mean shift of −19 ms has a standard deviation of ±14 and the designation of + vs. − direction was an arbitrary one in block D, in which either of the S stimuli could have been called $S_1$. For the $S_{normal}$−$S_{abnormal}$ couplings in these patients, negative mean shifts are in fact explainable by our modified hypothesis; if a small deficit in specific projection pathway were present on one side, the reduction in putative early 'timing signal' delivered by this system to the cerebral cortex should tend to reduce the degree of retroactive subjective referral of the timing for that abnormal side.

B. *Couplings of a peripheral stimulus ($P_1$) with a medial lemniscus stimulus (LM).* The technically most satisfactory experimental series of this type are presented in Table 2 (*see further details* in small print section below). They were carried out with two subjects (H.S. and G.S.) who were able to return for study a few years after the permanent implantation of electrodes in LM. (The implantation was made for the therapeutic relief of intractable pain of central origin by self-stimulation (Feinstein *et al.*, in preparation). The subjects were now outpatients with their pain controlled and presenting no interference to studies. They were in relatively good physical and psychological condition, and they were able to tolerate well a more concentrated period of successive morning and afternoon study sessions for two days each. Peripheral stimuli were matched in the best obtainable manner to the LM stimuli (*see* Methods). The experiments could be designed and completed in a manner making them as amenable to statistical evaluation as were the experiments with the normal subjects (in Table 1).

The distribution of reported subjective timing orders is seen in Table 2A to be roughly similar for both the $S_1$-$S_2$ and the $S_1$-LM couplings. For example, with a coupling interval of −200 or −250 ms for either $S_1$-$S_2$ or for $S_2$-LM pairs, most or all the reports were either '$S_2$ first' or 'LM first' respectively, as seen in the appropriate blocks. If the subjective timing for LM had actually been delayed an extra 200 ms, in accordance with the minimum TD requirement of 200 or more ms for the LM stimulus employed, one would have expected that the $S_1$-LM couplings

TABLE 2A. SUBJECTIVE TIMING ORDERS OF EXPERIENCES FOR TEMPORALLY-COUPLED STIMULI: COMPARE SKIN ($S_1$) AND MEDIAL LEMNISCAL (LM) PAIRINGS WITH $S_1$-$S_2$ PAIRINGS

| Subject | | *LM stim.* *Min. TD (ms)* | *LM stim.* *Test TD (ms)* | $S_1$-$S_2$ / $S_1$-*LM* | *Interval (ms)* | *No. of trials* | *Subject's timing* $S_1$ *first* | *Subject's timing* *T* | *Subject's timing* $S_2$/*LM first* | *Estim. 'mean shift' (ms)* | *Approx. SD* |
|---|---|---|---|---|---|---|---|---|---|---|---|
| H.S. | A | | | | −200 | 9 | 0 | 1 | 8 | | |
| (male, | | | | | −100 | 9 | 0 | 8 | 1 | | |
| aged 62 y) | | | | $S_L$-$S_R$ | 0 | 10 | 0 | 10 | 0 | −41 | 9 |
| | | | | | +100 | 11 | 9 | 2 | 0 | | |
| | | | | | +200 | 10 | 10 | 0 | 0 | | |
| | B | 200-300 | 600 | | −200 | 10 | 0 | 0 | 10 | | |
| | | | | | −100 | ↓ | 1 | 6 | 3 | | |
| | | | | $S_L$-LM | 0 | ↓ | 1 | 9 | 0 | 1 | 18 |
| | | | | | +100 | ↓ | 1 | 8 | 1 | | |
| | | | | | +200 | ↓ | 10 | 0 | 0 | | |
| | C | | | | −250 | 5 | 0 | 0 | 5 | | |
| | | | | | −150 | 6 | 1 | 3 | 2 | | |
| | | | | $S_L$-$S_R$ | 0 | 5 | 0 | 4 | 1 | −91 | 23 |
| | | | | | +150 | 5 | 5 | 0 | 0 | | |
| | | | | | +250 | 5 | 5 | 0 | 0 | | |
| | D | 300 | 600 | | −250 | 6 | 0 | 0 | 6 | | |
| | | | | | −150 | 4 | 1 | 0 | 3 | | |
| | | | | $S_L$-LM | 0 | 4 | 0 | 4 | 0 | −109 | 39 |
| | | | | | +150 | 5 | 3 | 2 | 0 | | |
| | | | | | +250 | 5 | 5 | 0 | 0 | | |
| G.S. | A | | | | −200 | 10 | 0 | 0 | 10 | | |
| (male, | | | | | −100 | 11 | 0 | 10 | 1 | | |
| aged 57 y) | | | | $S_R$-$S_L$ | 0 | 10 | 0 | 10 | 0 | −35 | 8 |
| | | | | | +100 | 10 | 8 | 2 | 0 | | |
| | | | | | +200 | 10 | 10 | 0 | 0 | | |
| | B | 200 | 500 | | −200 | 8 | 0 | 2 | 6 | | |
| | | | | | −100 | 8 | 0 | 8 | 0 | | |
| | | | | $S_R$-LM | 0 | 9 | 4 | 2 | 3 | −25 | 25 |
| | | | | | +100 | 7 | 4 | 2 | 1 | | |
| | | | | | +200 | 10 | 8 | 1 | 1 | | |
| | C | 200-300 | 500 | | −200 | 10 | 0 | 4 | 6 | | |
| | | | | | −100 | ↓ | 0 | 8 | 2 | | |
| | | | | $S_R$-LM | 0 | ↓ | 2 | 6 | 2 | −12 | 28 |
| | | | | | +100 | ↓ | 6 | 3 | 1 | | |
| | | | | | +200 | ↓ | 8 | 1 | 1 | | |
| | D | | | $S_1$-$S_2$ | −200 | 9 | 0 | 1 | 8 | | |
| | | | | (both | −100 | 11 | 0 | 4 | 7 | | |
| | | | | on R. | 0 | 10 | 2 | 7 | 1 | −19 | 24 |
| | | | | side) | +100 | 10 | 8 | 2 | 0 | | |
| | | | | | +200 | 10 | 10 | 0 | 0 | | |

TABLE 2B. COMPARISONS OF TIMING ORDERS FOR ($S_1$-$S_2$) VS. ($S_1$-LM)

| Subject | Min. TD for LM (ms) | Comparison | Estim. 'change in shift' (ms) | Approx. SD | Approx. 95% confidence interval |
|---|---|---|---|---|---|
| H.S. | 200-300 | A vs. B ($S_L$-$S_R$)—($S_L$-LM) | −42 | 20 | (−82; −2) |
| | 300 | C vs. D | 18 | 46 | (−74; 110) |
| G.S. | 200 | A vs. B ($S_R$-$S_L$)—($S_R$-LM) | −10 | 26 | (−63; 43) |
| | 200 | A vs. (B+C) | −30 | 24 | (−78; 18) |

at −200 ms should have produced more reports of 'together' (both sensations experienced about the same time) or even of '$S_1$ first'.

This qualitative impression of overall similarity between the subjective timing orders for $S_1$-$S_2$ vs. $S_1$-LM couplings is substantiated by the results of the statistical evaluation. The following should be noted in Table 2B: (a) each 'estimated change in mean shift' is relatively small, far less than the minimum TD of 200 ms or more required by the LM stimulus to elicit any sensory experience. Furthermore, most of the changes in shift are negative, although probably not significantly so. If valid, a negative change would indicate that subjective timing for LM is slightly earlier than, rather than delayed after, that for $S_2$ (each relative to $S_1$). (Such an earlier timing might even be additionally explainable by the shorter latency time for the primary cortical response to an LM volley than to an S volley, although there are other possible small modifiers in both directions); (b) the 95 per cent confidence intervals do not contain the value of the minimum TD of 200 ms or more that is required by the LM stimulus; consequently it is very unlikely that the data can be explained by a shift or delay in subjective timing equal to or determined by this minimum TD of the LM stimulus. Rather, the data are most reasonably explained on the basis of the prediction from the hypothesis, that subjective timing for an LM stimulus was roughly similar to that for a peripheral S stimulus, in spite of the empirically determined extra 200 ms or more that was required by the LM stimulus to be effective at all.

A much larger number of less adequate experimental series was carried out prior to those in Table 2 (*see* Methods). These included preliminary studies in four subjects, in which controls and experimental procedures were being developed. There were also previous extensive studies with H.S. and G.S. that were conducted under less favourable conditions than those in Table 2; these included 12 sessions with H.S., some five years earlier (during a prolonged stay of two and a half months in the hospital) plus a few sessions two years after that during a revisit, as well as 5 sessions with G.S. some two years earlier. Additionally, there were sessions of variable numbers and durations with two other patients with chronically implanted electrodes in LM. Our own in-depth analyses of each case convinced us that, when experimental conditions were at least partially adequate, the results obtained were qualitatively in support of the hypothesis, that is, they tended to show patterns of subjective timing qualitatively resembling those in Table 2.

In Table 2, the stimulus to LM (medial lemniscus) consisted of a train of 0·2 ms-pulses at 60 pps. Peak current intensity was set so that a minimum train duration (TD) of 200 or 300 ms was required (*see* 'min TD' column) in order for the stimulus to elicit any reportable sensory experience. The actual test stimulus applied to LM, in each trial with an $S_1$-LM coupling, was at this same intensity ($I_{200}$) but had a TD longer than the minimum required one (*see* values under 'test TD' column), and explanation in Methods). The peripheral stimuli, $P_1$ and $P_2$, were applied to the skin and are called $S_1$-$S_2$. Since the LM-induced sensation (that replaces $S_2$) was referred to the right side for subject H.S. and to left side for subject G.S., $S_1$ and $S_2$ were actually $S_L$ and $S_R$, respectively, for subject H.S., but $S_R$ and $S_L$, respectively, for subject G.S. Each test stimulus to skin (whether $S_1$ or $S_2$) consisted of a train of 0·2 ms-pulses, 60 pps, with peak current set so that subjective intensity of the sensation approximated that elicited by the test LM stimulus; and TD was set at 400 ms in block A for H.S. and at 300 ms in all the other $S_1$-$S_2$ blocks, so as to approximate the subjective duration of the sensation elicited by the LM test stimuli.

Reports of 'subject's timing' for either $S_2$ or LM 'first' refer, respectively, to whether a block of $S_1$-$S_2$ couplings or $S_1$-LM couplings is involved. In block D for G.S., $S_1$ and $S_2$ were both located on the normal right hand ($S_1$ on back of hand near digits 4-5, and $S_2$ on ventral aspect of wrist). *See* small print section above on 'Negative mean shifts . . .' for comparison of these results in block D with those in block A, $S_R$-normal side vs. $S_L$-abnormal. Because of this, it was not appropriate to compare blocks C and D for subject G.S. in the same way as the other comparisons in Table 2B. Instead, the two blocks with $S_R$-LM trials (blocks B and C) were combined and then compared to block A ($S_R$-$S_L$ trials) for subject G.S. In any case, however, a comparison of the D vs. C blocks in subject G.S. in fact produces a result qualitatively similar to that for A vs. B.

C. *Couplings of a peripheral stimulus (P) with an SI cortical stimulus (C).* These experiments were similar in principle to those in section III-B above, but with the cerebral stimulus applied subdurally to somatosensory, SI cortex (C), rather than to LM. Blocks of trials with paired $P_1$-C stimuli were compared in each subject with blocks of $P_1$-$P_2$ stimuli, with coupling intervals for each trial that overlapped for the two blocks. $P_1$ was either a single pulse to the skin ($S_1$) of the hand opposite to that in which C-induced sensation was referred, or in many cases it was a brief but weak flash of light (F). $P_2$ was a single pulse to skin ($S_2$), usually placed within the referral area of the C-induced sensation. As with LM, the C stimuli (pulse trains, 60 pps) were set at intensities somewhat above liminal levels so that a *minimum* TD of at least 200 ms or more was required to elicit any sensory experience; and, with these same intensities, the *test* TD's used in the actual trials were longer than the required minimum (*see* actual values in columns under 'min TD' and 'test TD'). The results obtained with C trains of unidirectional cathodal pulses ('cath') were distinctly different from those with C trains of pulses that successively alternated or reversed in polarity ('PR'). The two kinds of $P_1$-C couplings are therefore considered separately.

In Table 3 subjects J.W., C.J., W.M. and A.E. were parkinsonians; subject O.K. had 'basal ganglion disease' and M.T. had spasmodic torticollis. The $S_1$ and $S_2$ stimuli were single pulses, whose intensities were matched subjectively but were at levels distinctly above the threshold. Therefore, unlike those in Table 2, the sensations elicited by these S stimuli were not optimally matched, for subjective intensity and duration, with the cortically-induced sensations. Subjects in Table 3 were inpatients who were less able, than those in Table 2, to maintain consistency of reported sensory experiences in a long series of trials when S stimuli were set very close to threshold intensities for a conscious sensory experience. The use of a suprathreshold single pulse for S stimuli could have to some unknown degree biased the reported subjective timing orders for $S_1$-C couplings in the direction of $S_1$ first. However, any such bias would appear not to have determined the qualitative overall pattern of timing orders, as illustrated by the following: in subject C.J., the same S stimuli produced quite different results for C-cathodal (block B) when compared to C-polarity reversed pulses (block C). On the other hand, the use of a weak flash of light in place of $S_1$ (in $P_1$-C couplings with C-cathodal) produced a pattern of subjective orders (subject M.T.) qualitatively similar to those using $S_1$ and C cathodal (as in J.W., C.J. and O.K.).

For subject O.K., $S_R$ was applied to back of right hand, while the C-induced sensation (stimulus to left SI cortex) was referred to the same right side but to the vicinity of the ear. The coupling of this $S_R$ (instead of $S_L$) with C would result in the processing of the initial responses to $S_R$ and C at separate sites in the same, left postcentral gyrus. However, this did not prevent the subject from maintaining a clear distinction between the timings for the two sensations, as he reported a preponderance of '$S_R$ first' (in block B) compared to the reports for $S_L$-$S_R$ couplings (in block A). It may be added that trials (in some other subjects) with $S_1$ and $S_2$ on the same side have generally produced timing orders with an 'accuracy' comparable to series using $S_R$ and $S_L$ analysis (for example, Section III, block D for G.S. in Table 2A). However, there tended to be less confusion and less effort required on the part of the subject with comparisons of one side vs. the other.

For Table 3B, 'changes in shift' were estimated only for those experiments in Table 3A in which stimulus C consisted of cathodal pulses.

1. *$P_1$-C couplings using a cathodal pulse train for C.* Results from those experiments in which suitable comparative blocks of trials were achieved are given for subjects J.W., C.J., O.K., and M.T. in Table 3A. Patterns of subjects' timing orders (blocks A vs. B in these subjects) show a qualitative difference from those in Table 2A (with LM stimuli). For example, J.W. reported in most of the trials with $S_R$-C interval at $-300$ ms (i.e., C stimulus train begun before $S_R$ by 300 ms) that the $S_R$- and C-induced sensations subjectively started 'together'. But when two peripheral stimuli, $S_R$-$S_L$, were coupled by intervals of $-200$ and $-150$ ms he reported mostly '$S_L$ first'. Only when C stimulus was advanced to 400 ms before $S_R$ ($S_R$-C interval $= -400$) did J.W. report mostly 'C firsts'. Similarly, with coupled stimuli initiated simultaneously (coupling interval $= 0$), J.W. reported mostly '$S_R$ first' for the $S_R$-C couplings, but mostly 'together' for the $S_R$-$S_L$ couplings.

The statistical analysis of all the suitable data is in agreement with the impression of a qualitative difference. The estimated 'change in mean shift', when C stimulus (cathodal pulses) was substituted for the $S_2$ stimulus ('mean shift' for block A minus that for block B) is given in Table 3B for subjects J.W., C.J. and M.T. They all show values for a substantial delay in the subjective timing of the C-induced sensation relative to the S-induced sensation. The amount of this estimated change in shift, that is, the delay for C-induced sensation, is close to the actual required

TABLE 3A. SUBJECTIVE TIMING ORDERS OF EXPERIENCES FOR TEMPORALLY-COUPLED STIMULI: (i) PERIPHERAL ($S_1$ OR FLASH F) AND SOMATOSENSORY CORTEX (C) PAIRINGS, AND (ii) $S_1$ (OR F)-$S_2$ PAIRINGS

| Subject | | C stim.: cath/PR | Min. TD (ms) | Test TD (ms) | $S_1(F)$-$S_2$ / $S_1(F)$-C | Interval (ms) | No. of trials | Subject's timing: $S_1(F)$ first | T | $S_2$/C first | Estim. 'mean shift' (ms) | Approx. SD |
|---|---|---|---|---|---|---|---|---|---|---|---|---|
| J.W. | A | | | | | −200 | 5 | 0 | 0 | 5 | | |
| (male, | | | | | | −150 | 12 | 0 | 3 | 9 | | |
| aged | | | | | $S_R$-$S_L$ | 0 | 12 | 1 | 11 | 0 | −21 | 10 |
| 54 y) | | | | | | +150 | 9 | 9 | 0 | 0 | | |
| | | | | | | +200 | 4 | 4 | 0 | 0 | | |
| | B | cath | 200 | 500 | | −400 | 6 | 0 | 1 | 5 | | |
| | | | | | | −300 | 11 | 0 | 8 | 3 | | |
| | | | | | $S_R$-C | 0 | 12 | 11 | 1 | 0 | −241 | 24 |
| | | | | | | +300 | 5 | 5 | 0 | 0 | | |
| | | | | | | +400 | 4 | 4 | 0 | 0 | | |
| C.J. | A | | | | | −200 | 3 | 0 | 0 | 3 | | |
| (male, | | | | | | −150 | 8 | 0 | 0 | 8 | | |
| aged | | | | | $S_R$-$S_L$ | 0 | 8 | 1 | 7 | 0 | −9 | 9 |
| 60 y) | | | | | | +150 | 6 | 6 | 0 | 0 | | |
| | | | | | | +200 | 4 | 4 | 0 | 0 | | |
| | B | cath | ∼300 | 700 | | −200 | 9 | 5 | 4 | 0 | | |
| | | | | | $S_L$-C | 0 | 7 | 5 | 2 | 0 | −463 | 156 |
| | | | | | | +200 | 9 | 9 | 0 | 0 | | |
| | C | PR | ∼200 | 700 | | −200 | 6 | 0 | 6 | 0 | | |
| | | | | | $S_L$-C | 0 | 4 | 1 | 3 | 0 | | |
| | | | | | | +200 | 6 | 1 | 5 | 0 | | |
| O.K. | A | | | | | −200 | 11 | 0 | 1 | 10 | | |
| (male, | | | | | $S_L$-$S_R$ | 0 | 7 | 1 | 5 | 1 | | |
| aged | | | | | | +200 | 10 | 9 | 1 | 0 | | |
| 53 y) | | | | | | | | | | | | |
| | B | cath | ∼400 | 500 | | −500 | 10 | 9 | 1 | 0 | | |
| | | | | | | −400 | 7 | 7 | 0 | 0 | | |
| | | | | | $S_R$-C | −200 | 5 | 5 | 0 | 0 | | |
| | | | | | | 0 | 14 | 11 | 3 | 0 | | |
| | | | | | | +200 | 4 | 4 | 0 | 0 | | |
| M.T. | A | | | | | −100 | 10 | 0 | 0 | 10 | | |
| (male, | | | | | F-$S_L$ | 0 | 13 | 1 | 9 | 3 | 15 | 10 |
| aged | | | | | | +100 | 7 | 6 | 1 | 0 | | |
| 42 y) | | | | | | | | | | | | |
| | B | cath | 200-300 | 400 | | −200 | 11 | 0 | 7 | 4 | | |
| | | | | | | −150 | 11 | 0 | 7 | 4 | | |
| | | | | | F-C | 0 | 15 | 6 | 8 | 1 | −80 | 23 |
| | | | | | | +150 | 10 | 9 | 1 | 0 | | |
| | | | | | | +200 | 9 | 8 | 0 | 1 | | |

| Subject | C stim. cath / PR | C stim. Min. TD (ms) | C stim. Test TD (ms) | $S_1(F)$-$S_2$ / $S_1(F)$-C | Interval (ms) | No. of trials | Subject's timing $S_1(F)$ first | Subject's timing T | Subject's timing $S_2$/C first |
|---|---|---|---|---|---|---|---|---|---|
| W.M. | A | | | | −300 | 15 | 0 | 1 | 14 |
| (male, | | | | | −200 | 33 | 0 | 4 | 29 |
| aged | | | | F-$S_R$ | 0 | 31 | 1 | 26 | 4 |
| 47 y) | | | | | +200 | 24 | 12 | 10 | 2 |
| | | | | | +300 | 12 | 11 | 1 | 0 |
| | B PR | 200 | 400 | | −400 | 7 | 1 | 1 | 5 |
| | | 300 | 500 | | −300 | 11 | 0 | 6 | 5 |
| | | | | | −200 | 17 | 1 | 6 | 10 |
| | | | | F-C | 0 | 23 | 2 | 17 | 4 |
| | | | | | +200 | 8 | 2 | 6 | 0 |
| | | | | | +300 | 8 | 4 | 4 | 0 |
| A.E. | A | | | | −200 | 25 | 1 | 2 | 22 |
| (female, | | | | | −150 | 23 | 3 | 2 | 18 |
| aged | | | | | −100 | 13 | 3 | 10 | 0 |
| 57 y) | | | | F-$S_L$ | 0 | 41 | 2 | 38 | 1 |
| | | | | | +100 | 10 | 2 | 7 | 1 |
| | | | | | +150 | 19 | 11 | 6 | 2 |
| | | | | | +200 | 18 | 16 | 1 | 1 |
| | B PR | ~300 | 700 | | −400 | 22 | 2 | 12 | 8 |
| | | | | F-C | −200 | 16 | 3 | 12 | 1 |
| | | | | | 0 | 18 | 5 | 13 | 0 |

TABLE 3B. COMPARISONS OF TIMING ORDERS, FOR PAIRS OF PERIPHERAL STIMULI VS. PERIPHERAL-CORTICAL PAIRS (C = CATHODAL PULSES)

| Subject | Min. TD for C (ms) | Comparison | Estim. 'change in shift' (ms) | Approx. SD | Approx. 95% confidence interval |
|---|---|---|---|---|---|
| J.W. | 200 | A vs. B ($S_R$-$S_L$)—($S_R$-C) | 220 | 25 | (169; 271) |
| C.J. | ~300 | A vs. B ($S_R$-$S_L$)—($S_L$-C) | 454 | 156 | (143; 765) |
| M.T. | 200-300 | A vs. B (F-$S_L$)—(F-C) | 95 | 25 | (45; 145) |

minimum train duration (TD) of 200 ms for C stimulus in J.W.; it is greater than the minimum TD of 300 ms in C.J.; and it is less than the minimum TD of 200 to 300 ms in M.T. (on the latter difference, *see below*). Note also that the 95 per cent confidence intervals all indicate that the change in shift in each case is in the positive range only, that is, in the direction of a delay for timing of C-induced sensation. (Results for O.K. were not treated statistically because the coupling was left-right [$S_L$-$S_R$] in block A, but right-right [$S_R$- and C-induced sensations on the same side] in block B (*see* details in small print section, above). However, the results for subject O.K. in Table 3A are in qualitative agreement with those for J.W., C.J., and M.T.)

These findings are thus in general accord with the prediction from the hypothesis (*see* fig. 3) and also with the results described in Section I of the Results. The substantial relative delay in the subjective timing of the C-induced sensation is of course sharply different from the absence of such a delay for a sensation induced by a comparable LM stimulus train (*see* Section III-B of Results, *above*); for example, compare Table 2B with Table 3B. For subject M.T. the estimated change in shift of 95 ms indicated a relative delay that is less than the 200 to 300 ms minimum TD required by the C stimulus. However, subject M.T. (who was a very alert, introspective observer) reported having a 'hunch', or preconscious type of feeling, that 'something' was building up before the instant at which he himself felt that he was actually aware of a somatic sensation, when elicited with the C stimulus; this would tend to make his reported timings for onset of C earlier than warranted by the criterion for subjective experience. (This interesting phenomenon may in fact be related to an additional hypothesis, that neuronal activities which are too brief for eliciting a conscious experience may mediate unconscious mental functions, *see* Libet, 1965, 1966).

2. *$P_1$-C couplings using 'polarity reversals' (PR) in pulse train for C.* In 'PR' stimulus trains applied to the unifocal subdural electrode the polarity of each successive pulse was reversed; that is, the polarity alternated between being cathodal and anodal. PR trains should potentially reduce any tendency for the stimulus to produce electrolytic damage to tissues, and so they had been previously employed by us for cerebral stimuli in which a consistent unidirectional polarity of pulses is not essential to the study. It should be noted that PR trains generally required greater peak currents than did cathodal pulse trains, in order to elicit any conscious sensory experience (*see also* Libet *et al.*, 1964; Libet, 1973).

When PR trains were used for the cortical test stimuli in the present study, the reported subjective timings for C relative to the $P_1$ stimulus did not show the same patterns that were seen for the unidirectional cathodal pulse trains (*see* Table 3A; block C for subject C.J., and block B for W.M. and A.E.). Instead, there tended to be a preponderance of subjective timing reports of 'together', for the $P_1$- and C-induced sensations, regardless of the coupling intervals between the two stimuli. This is especially convincing in the subject C.J. for whom trials with both cathodal and PR pulses can be directly compared (blocks B vs. C). It would appear that use of PR pulses for C stimuli (a) tends to confuse or blur the subjective experience of timing orders, so that distinctions are less possible; and (b) does not produce the kind of clear evidence of a substantial delay in subjective timing for C that was seen with cathodal pulses. The curious difference between the responses with PR as opposed to cathodal cortical pulse trains may be explainable in terms of the proposed modified hypothesis (*see below*).

Each cathodal pulse would tend initially to excite neuronal elements, probably conducting fibres, that lie in the surface layers of the cortex (Libet *et al.*, 1964; Libet, 1973). The electrophysiological response to each cathodal pulse begins with a large surface negative component; this 'direct cortical response' is different from the primary evoked potential elicited by a peripheral or a lemniscal stimulus pulse (Libet *et al.*, 1967; Libet, 1973). On the other hand, each surface anodal pulse should tend to excite deeper lying nerve fibres (Hern, Landgren, Phillips and Porter, 1962; Phillips, 1969; Libet, 1973), and these might include some of the afferent specific projection fibres from the thalamus which terminate chiefly in layer IV (Colonnier, 1966). This possibility is further promoted by the relatively larger peak currents that were required by PR as opposed to cathodal pulse trains, to elicit the same minimal sensory experience. Excitation of some of these ascending fibres would, according to our hypothesis, provide at least a weak timing signal for retroactive subjective referral. A train of pulses with successively alternating polarities might then provide both the surface cortical and the ascending specific kinds of input alternately, in addition to possible other kinds. In such circumstances, it would hardly be surprising for there to be a subjective confusion about the timing of the sensory experience elicited. Indeed, the nature of these results provides some indirect support for the modified hypothesis. (This analysis suggested the possibility that one or a few surface anodal pulses of sufficient intensity, delivered at a time just before a cathodal pulse train, might provide the signal for shifting the subjective timing of the experience, from the usual position at the end of a minimum TD to a position at the onset of the cathodal C stimulus train. It has, however, been possible to carry out only a few preliminary and inconclusive tests of this kind).

## GENERAL DISCUSSION

### *Overall Evidence for the Present Hypothesis*

The results obtained in these experiments provide specific support for our present proposal, that is, for the existence of a subjective temporal referral of a sensory experience by which the subjective timing is retroactively antedated to the time of the primary cortical response (elicited by the lemniscal input). Subjective timing for onset of an LM-induced sensation did in fact appear to occur with no more delay than that for a peripheral (skin)-induced sensation, even though the minimum delay for the stimulus to achieve neuronal adequacy for the LM-induced sensory experience was experimentally set to be at least 200 ms. On the other hand, with cathodal cortical stimuli subjective timings appeared to exhibit relative delays similar to those required to achieve neuronal adequacy. Stimuli to LM (or n.VPL) of course excite specific projection afferents of the lemniscal system, while C-cathodal stimuli (at near liminal intensities) do not. The apparent confusion or blurring of subjective timings found when the C stimulus pulses were 'polarity-reversed' (train of cathodal pulses alternating with anodal ones) is compatible with this interpretation.

An even more startling experimental prediction remains yet to be tested: if the putative timing signal alone were to be delivered at the *onset* of a cortical-cathodal stimulus train, one might expect that the subjective timing of the *cortically*-induced sensation could be shifted or referred, *from* its usual position (at or after the end of the minimum required train duration) backwards *to* the onset of the cortical stimulus train. An 'isolated' timing signal could be generated by a single pulse stimulus in LM, which can elicit a large primary evoked response at the SI cortex with no conscious sensory experience (Libet *et al.*, 1967). However, this experiment would require the placement, in a given patient, of one electrode in LM (or n.VPL) and another over the precise area of the SI cortex that receives the projection of impulses electrically initiated by the LM (or n.VPL) electrode; these conditions are obviously difficult to achieve under the limitations of approaches that are clinically warranted.

Further testing of the proposal can also be sought in the effects of pathological destruction of the specific projection system at cerebral levels. Elimination of the putative signal required for retroactive subjective referral of timing should introduce a substantial delay for the subjective experience even of a peripherally-induced sensation. If the destruction were purely unilateral, one could test for such a delay by comparing the subjective timings for peripheral stimuli applied to homologous sites on the normal and abnormal sides of the body. For this purpose, the same experimental paradigm which was employed in the 'control' series of the present study could be employed; in this, the pattern of reports of relative timing orders for a skin stimulus on the normal side temporally coupled with one on the abnormal side is obtained. Some indications of inadvertent partial tests of this kind may be already apparent in the present study, with subjects who had sustained some partial unilateral sensory losses apparently due to damage in the specific projection system (*see* discussion of 'Negative mean shifts for $S_1$–$S_2$ in Table 2A' at the end of Section III-A of Results). A more thorough study of this issue, employing purely peripheral testing in responsive patients who have incurred appropriately located cerebrovascular accidents which produced a severe unilateral 'epicritic' sensory deficit, has been initiated and will be reported separately. For the one suitable patient studied thus far, there did indeed appear to be a delay of 200 to 400 ms in the subjective timing for a peripherally-induced sensation on the abnormal side, relative to one on the normal side.

### *Alternative Explanations of the Evidence*

Some possible alternative explanations of our findings should here be considered. One type of argument would hold that the delay in achieving 'neuronal adequacy' when stimulating LM/n.VPL or somatosensory cortex, as seen in the relatively long minimum train durations required, is simply due to the 'abnormal' route and/or pattern of these inputs. These cerebral stimuli might require a longer time to develop some special neuronal response, for example because of cortical inhibitory as well as excitatory patterns that they might produce. In this view, 'normal' inputs that do not require long minimum train durations (stimuli to skin, peripheral nerve, dorsal columns) would generate neuronal adequacy with no substantial delay; there would thus be no need to introduce the postulate of a subjective referral backwards in time for the experience of the 'normal' inputs, in order to account for their earlier, more immediate subjective timing. This kind of alternative view would seem to be untenable for the following reasons: (a) this view does not account for our experimental observation that the subjective timing for LM stimuli, requiring minimum TDs of 200 ms or more, appeared to show no delay relative to that for skin stimuli. It would still have to be conceded, therefore, that some retroactive subjective referral process can be engaged selectively by the LM stimulus, even if it does deliver an 'abnormal' input; (b) this view would ignore other evidence already strongly indicating that input via a normal peripheral route (skin stimulus) does require a substantial period of cerebral activities before neuronal adequacy for the conscious sensory experience is achieved. The previous evidence included demonstrations (i) of retroactive effects, on the conscious sensory experience for a near threshold skin stimulus, which could be produced by a conditioning cortical stimulus that follows the skin pulse by 200 to 500 ms (Libet *et al.*, 1972; Libet, 1973); and (ii) of the insufficiency of the early components of the cortical evoked responses to a sensory stimulus for eliciting any conscious sensory experience (Libet *et al.*, 1967); (c) other previous findings indicate that the subjective sensory experiences elicited by a stimulus to somatosensory cortex can have 'natural-like' qualities which can resemble those elicited via normal peripheral inputs (Libet, 1973; Libet *et al.*, 1975), although the temporal and spatial features are different (*see* Methods). Also, if the train duration of the cortical stimulus is extended to one longer than the minimum required 'utilization TD' of about 500 ms, the conscious sensory experience is found to continue but with no progressive increase in subjective intensity above that at its onset; this is different from primary motor cortex where any effective stimulus, no matter how weak, elicits a progressively increasing motor response if the stimulus train duration is extended (Libet *et al.*, 1964; Libet, 1966, 1973).

Another alternative explanation accepts our proposal that there is substantial delay in achieving neuronal adequacy with all inputs, peripheral or central; but it would argue that, in those cases where there is apparent antedating of the subjective timings of the sensory experience, the subjective referral backwards in time may be due to an illusory judgement made by the subject when he *reports* the timings. This possibility was raised by Professor Donald M. MacKay in a discussion with the senior author, B.L. On such a basis, the timings of the subjective experience and of the achievement of neuronal adequacy could be actually identical at the time each sensation is elicited, that is, they would both be delayed. However, in those cases in which the neuronal response includes a component due to the fast specific projection, the subject's later report of how he perceives the timing of the sensory experience is assumed to be affected by the previous presence of the primary cortical response. For example, it could be argued that during the recall process, cerebral mechanisms might 'read back' via some memory device to the primary evoked response and now construe the timing of the experience to have occurred earlier than it in fact did occur. Such a possibility cannot be excluded at present, but it requires added assumptions and appears to be less satisfactory than our own hypothesis: (a) for example, if any 'read back' to the primary timing signal does occur, it would seem simpler to assume that this takes place at the time when neuronal adequacy for the experience is first achieved, when the 'memory' of the timing signal would be fresher; such a process would then produce the retroactive subjective referral we have proposed. Whether the later report of antedated timing of the experience is due to an immediate referral (as postulated by us) or to a later 'illusory judgment', the processes involved would be unconscious and 'automatic' in nature and would not be distinguishable by the subject; (b) the alternative explanation based upon later, illusory judgment of timing has a serious deficiency with respect to an important feature of subjective sensory experiences. By retaining delays for the immediate subjective sensory experiences, when they initially and actually occur, this alternative explanation becomes unable to explain the absence of subjective 'jitter' or asynchrony in our experience, when a variety of peripheral sensory stimuli are applied synchronously. At least one factor that should produce differences in the delays for achieving neuronal adequacy with different stimuli, is the strength of the stimulus. (This is based upon the intensity/train-duration relationship for stimuli to LM (or n.VPL), as well as to SI cortex—*see* Libet *et al.*, 1964, 1972; Libet, 1966, 1973; and on the tendency for the subjective timing of a cortically-induced sensation to approximate the end of the minimum train duration, whether the latter is set at 500 or at 200 to 300 ms—*see* Sections I and III-C in Results). One attractive feature of our modified hypothesis is in fact its ability to deal with this difficulty. Subjective referrals, that are retroactive to the early primary evoked response to each sensory input, would make irrelevant any differences among the timings for neuronal adequacy in a group of synchronously initiated inputs; delays for the primary evoked potential are short (10 to 20 ms), and the differences produced by differing intensities of peripheral somatic stimuli are known to be so small as to be negligible for the purpose of subjective timing (*see* Desmedt, 1971).

*Roles of Specific Projection System*

The specific projection system is already regarded as the provider of localized cerebral signals that function in fine spatial discrimination, including the subjective referral of sensory experiences in space. Our present hypothesis expands the role for this system to include a function in the temporal dimension. The same cortical responses to specific fast projection inputs would also provide timing signals. They would subserve subjective referral in such a way as to help 'correct' the subjective timing (relative to the sensory stimulus), in spite of actual substantial delays in the time to achieve neuronal adequacy for the 'production' of the conscious sensory experience. The temporal functions of the specific projection system need not be restricted to the one postulated here for subjective referral; for example, the fineness of temporal information it provides is probably utilized in behavioural responses to stimuli that involve spatiotemporal sequences (*see*, e.g., Azulay and Schwartz, 1975).

The role at present postulated for the specific projection system would presumably be significant in the subjective awareness of the timing *order* of two inputs (that is, which one appeared first), and not merely of the existence and duration of some time interval between the inputs. In our studies, even young alert normal subjects were not subjectively aware of the order of two somatosensory inputs when the time interval between them was less than 25 to 50 ms; i.e., with these short intervals most of the timing reports tended to be 'together' or 'same time'. This suggests that even if latencies differed by as much as 25 ms or so for primary evoked cortical potentials elicited by two different peripheral stimuli, the two stimuli would be consciously experienced as being either synchronous or having an ambiguous order. Conscious experience of temporal order should be distinguished from forced-choice judgments of order; for example, in the forced choice paradigm the subject is not given the options of reporting that two sensations appear to be either simultaneous or subjectively not definable as to their order. However, even with forced-choice judgments of order the 'difference threshold' for two temporally-coupled stimuli of the same modality has been found to be about 18 ms—*see* Sternberg and Knoll, 1973.

*Some Implications for the Mind-brain Relationship*

That the time factor in neural coding and decoding of experience could raise fundamental questions for the mind-brain relationship had already been recognized (*see* Lord Brain, 1963). The presently modified hypothesis deals with the problem of a substantial neuronal time delay, apparently required for the 'encoding' of a conscious sensory experience, by introducing the concept of a subjective referral of sensory experience in the temporal dimension. This would introduce an asynchrony or discrepancy between the timing of a subjective experience and the time when the state of 'neuronal adequacy' associated with the experience is achieved. However, the concept of subjective referral in the spatial dimension, and the discrepancy between subjective and neuronal spatial configurations, has long been recognized and accepted; that is, the spatial form of a subjective sensory experience need not be identical with the spatial pattern of the activated cerebral neuronal system that gives rise to this experience. Indeed, both temporal and spatial referrals are here postulated to depend in part upon the ability to generate the same physiological signal, the primary cortical response to the specific projection input. Philosophically, a discrepancy between the 'mental' and the 'physical' in the temporal dimension can be regarded, in a manner analogous to that for the discrepancy in the spatial dimension, as not contradicting the theory of psycho-physical parallelism or correspondence. But a dissociation between the timings of the corresponding 'mental' and 'physical' events would seem to raise serious though not insurmountable difficulties for the more special theory of psychoneural identity (Popper and Eccles, 1977; Libet, 1978).

## SUMMARY

Subjective experience of a peripherally-induced sensation is found to appear without the substantial delay found for the experience of a cortically-induced sensation. To explain this finding, in relation to the putative delay of up to about 500 ms for achieving the 'neuronal adequacy' required to elicit the peripherally-induced experience, a modified hypothesis is proposed: for a peripheral sensory input, (a) the primary evoked response of sensory cortex to the specific projection (lemniscal) input is associated with a process that can serve as a 'time-marker'; and (b), after delayed neuronal adequacy is achieved, there is a subjective referral of the sensory experience backwards in time so as to coincide with this initial 'time-marker'.

A crucial prediction of the hypothesis was experimentally tested in human subjects using suitably implanted electrodes, and the results provide specific support for the proposal. In this, the test stimuli to medial lemniscus (LM) and to surface of somatosensory cortex (C) were arranged so that a minimum train duration of 200 ms or more was required to produce any conscious sensory experience in each case. Each such cerebral stimulus could be temporally coupled with a peripheral one (usually skin, S) that required a relatively negligible stimulus duration to produce a sensation. The sensory experiences induced by LM stimuli were found to be subjectively timed as if there were no delay relative to those for S, that is, as if the subjective experience for LM was referred to the *onset* rather than to the end of the required stimulus duration of 200 ms or more. On the other hand, sensory experiences induced by the C stimuli, which did not excite specific projection afferents, appeared to be subjectively timed with a substantial delay relative to those for S, that is, as if the time of the subjective experience coincided roughly with the *end* of the minimum duration required by the C stimuli.

The newly proposed functional role for the specific projection system in temporal referral would be additional to its known role in spatial referral and discrimination.

A temporal discrepancy between corresponding mental and physical events, i.e., between the timing of a subjective sensory experience and the time at which the state of 'neuronal adequacy' for giving rise to this experience is achieved, would introduce a novel experimentally-based feature into the concept of psycho-physiological parallelism in the mind-brain relationship.

## ACKNOWLEDGEMENTS

We are indebted to all the patients who co-operated so splendidly and willingly in these studies, to Dr. Curtis Gleason and Professor Elizabeth Scott (Statistics) for assistance in some aspects of the work, and to the National Science Foundation for a Research Grant No. GB-30552X that provided some support during the initial phase of this investigation. Preparation of the paper was in part carried out while the senior author (B.L.) was a scholar-in-residence at the Bellagio Study and Conference Center in Italy, supported by the Rockefeller Foundation. Reprint requests to Dr. B. Libet, Department of Physiology, University of California, San Francisco, California 94143, USA.

## REFERENCES

AZULAY, A. and SCHARTZ, A. S. (1975) The role of the dorsal funiculus of the primate in tactile discrimination. *Experimental Neurology*, **46**, 315-332.

BARTLETT, J. R., DOTY, R. W., LEE, B. B., NEGRÃO, N. and OVERMAN, W. H. (1977) Deleterious effects of prolonged electrical excitation of striate cortex in Macaques. *Brain, Behavior and Evolution*, **14**, 46-66.

BRAIN, LORD. (1963) Some reflections on brain and mind. *Brain*, **86**, 381-402.

COLONNIER, M. L. (1966) The structural design of the neocortex. In: *Brain and Conscious Experience*. Edited by J. C. Eccles. New York: Springer-Verlag, pp. 1-23.

DELGADO, J. M. R. (1955) Evaluation of permanent implantation of electrodes within the brain. *Electroencephalography and Clinical Neurophysiology*, **7**, 637-644.

DEMPSTER, A. P., LAIRD, N. M. and RUBIN, D. B. (1977) Maximum likelihood from incomplete data via the EM algorithm. *Royal Statistical Society Journal, B*, **39**, 1-38.

DESMEDT, J. E. (1971) Somatosensory cerebral evoked potentials in man. In: *Handbook of Electroencephalography and Clinical Neurophysiology*. Edited by A. Rémond. Amsterdam: Elsevier, Vol. 9, pp. 8-82.

—— and ROBERTSON, D. (1977) Differential enhancement of early and late components of the cerebral somatosensory evoked potentials during forced-paced cognitive tasks in man. *Journal of Physiology, London*, **271**, 761-782.

FEINSTEIN, B. F., ALBERTS, W. W. and LEVIN, G. (1969) The use of implanted thermistor electrodes in the therapy of parkinsonism. In: *Proceedings of the Third Symposium on Parkinson's Disease*. Edited by F. J. Gillingham and I. M. L. Donaldson. Edinburgh: Livingstone, pp. 232-237.

—— —— WRIGHT, E. W., JR. and LEVIN, G. (1960) A stereotaxic technique in man allowing multiple spatial and temporal approaches to intracranial targets. *Journal of Neurosurgery*, **17**, 708-720.

GOFF, G. D., MATSUMIYA, Y., ALLISON, T. and GOFF, W. R. (1977) The scalp topography of human somatosensory and auditory evoked potentials. *Electroencephalography and Clinical Neurophysiology*, **42**, 57-76.

HERN, J. E. C., LANDGREN, S., PHILLIPS, C. G. and PORTER, R. (1962) Selective excitation of corticofugal neurons by surface-anodal stimulation of the baboon's motor cortex. *Journal of Physiology, London*, **161**, 73-90.

HIRSCH, J. F., PERTUISET, B., CALVET, J., BUISSON-FEREY, J., FISCHGOLD, H. and SCHERRER, J. (1961) Etude des réponses électrocorticales obtenues chez l'homme par des stimulations somesthésiques et visuelles. *Electroencephalography and Clinical Neurophysiology*, **13**, 411-424.

JASPER, H., LENDE, R. and RASMUSSEN, T. (1960) Evoked potentials from the exposed somatosensory cortex in man. *Journal of Nervous and Mental Diseases*, **130**, 526-537.

LIBET, B. (1965) Cortical activation in conscious and unconscious experience. *Perspectives in Biology and Medicine*, **9**, 77-86.

—— (1966) Brain stimulation and the threshold of conscious experience. In: *Brain and Conscious Experience*. Edited by J. Eccles. New York: Springer-Verlag, pp. 165-181.

—— (1973) Electrical stimulation of cortex in human subjects and conscious sensory aspects. In: *Handbook of Sensory Physiology*. Edited by A. Iggo. Heidelberg: Springer-Verlag, Vol. 2, pp. 743-790.

—— (1978) Neuronal vs. subjective timing, for a conscious sensory experience. In: *Cerebral Correlates of Conscious Experience*. Edited by P. Buser and A. Rougeul-Buser. Amsterdam: Elsevier, pp. 69-82.

—— ALBERTS, W. W., WRIGHT, E. W., JR., DELATTRE, L. D., LEVIN, G. and FEINSTEIN, B. (1964) Production of threshold levels of conscious sensation by electrical stimulation of human somatosensory cortex. *Journal of Neurophysiology*, **27**, 546-578.

—— —— —— and FEINSTEIN, B. (1967) Responses of human somatosensory cortex to stimuli below threshold for conscious sensation. *Science*, **158**, 1597-1600.

—— —— —— —— (1972) Cortical and thalamic activation in conscious sensory experience. In: *Neurophysiology Studied in Man*. Edited by G. G. Somjen. Amsterdam: Excerpta Medica, pp. 157-168.

—— —— —— LEWIS, M. and FEINSTEIN, B. (1975) Cortical representation of evoked potentials relative to conscious sensory responses, and of somatosensory qualities in man. In: *The Somatosensory System*. Edited by H. H. Kornhuber. Stuttgart: Georg Thieme, pp. 291-308.

NASHOLD, B., SOMJEN, G. and FRIEDMAN, H. (1972) Parasthesias and EEG potentials evoked by stimulation of the dorsal funiculi in man. *Experimental Neurology*, **36**, 273-287.

PHILLIPS, C. G. (1969) Motor apparatus of the baboon's hand. *Proceedings of the Royal Society, B*, **173**, 141-174.

POPPER, K. R. and ECCLES, J. C. (1977) *The Self and Its Brain*. Berlin: Springer, pp. 256-259, 362-365.

PUDENZ, R. H., BULLARA, L. A. and TALALLA, A. (1975) Electrical stimulation of the brain. *Surgical Neurology*, **4**, 265-270, 389-400.

STERNBERG, S. and KNOLL, R. L. (1973) The perception of temporal order: fundamental issues and a general model. In: *Attention and Performance*. Edited by S. Kornblum. New York: Academic Press, No. 4, pp. 629-685.

(*Received June 1, 1978*)

CONSCIOUSNESS AND COGNITION 1, 367–375 (1992)

# Retroactive Enhancement of a Skin Sensation by a Delayed Cortical Stimulus in Man: Evidence for Delay of a Conscious Sensory Experience

B. LIBET[1]

*Department of Physiology, University of California, San Francisco, California 94143-0444*

E. W. WRIGHT AND B. FEINSTEIN†

*Department of Neurosurgery, Mount Zion Hospital and Medical Center, San Francisco, California 94115*

AND

D. K. PEARL

*Department of Statistics, Ohio State University, Columbus, Ohio 43210*

Sensation elicited by a skin stimulus (S) was subjectively reported to feel stronger when followed by a stimulus to somatosensory cerebral cortex (C), even when C was delayed by up to 400 ms or more. This expands the potentiality for retroactive effects beyond that previously known as backward masking. It also demonstrates that the content of a sensory experience can be altered by another cerebral input introduced after the sensory signal arrives at the cortex. The long effective S–C intervals support the thesis that a duration of cortical activity of up to 0.5 s is required before awareness of a sensory stimulus is developed. 

Retroactive (or backward) masking of the awareness of a "target" sensory signal by another delayed sensory signal has been well known (e.g., Raab, 1963; Kahneman, 1968). We have reported retroactive masking at the cerebral level, when a near-threshold stimulus to the skin was followed by a conditioning stimulus train of pulses applied subdurally to somatosensory cerebral cortex (postcentral gyrus) (Libet, Alberts, Wright, & Feinstein, 1972). Retroactive masking was obtained even when onset of the conditioning cortical stimulus was delayed as much as 200 to 500 ms after the skin stimulus. Retroactive *enhancement* of awareness of the target stimulus in a comparable manner has apparently not been reported, except for some minimal indications (Pieron & Segal, 1939; Dember & Stefl, 1972). We now report that definite retroactive enhancement of the subjective intensity of a skin-induced sensation can be produced by a following stimulus to somatosensory cortex in human subjects, even when the onset of the cortical

[1] To whom correspondence should be addressed.

† Deceased.

stimulus is delayed by at least 400 ms (preliminary report in Libet, 1978). Such a finding (a) eliminates the argument that retroactive effects are simply explainable as amnesic actions that interfere with the memory of a target stimulus, and (b) it provides further support for the view that a substantial duration of cortical activity is required before conscious awareness of a sensory input is developed (Libet, 1965; Libet et al., 1967, 1972; Libet, Pearl, Morledge, Gleason, Hosobuchi, & Barbaro, 1991).

## METHODS

*Subjects.* Four patients with dyskinesias in whom intracranial electrodes were temporarily implanted for therapeutic purposes served as subjects (Feinstein, Alberts, & Levin, 1969). At the time of study the patients were ambulatory but in residence in the hospital. The subjects selected were in good postoperative condition, alert, responsive, of at least normal intelligence, and with no somatosensory abnormalities. They were taking little or no significant medications. (In any case, each subject acted as his/her own control in the experiments; see below.) Electrodes for stimulation and RF-heat lesions were inserted into VL-thalamus and/or one or two basal ganglia to make graduated lesions over several days for control of tremor and/or rigidity. The cortical electrodes were introduced subdurally via the same opening that admitted the subcortical ones; they were part of a separate clinical study of the relation between the somatosensory cortex and the production of tremor. Subjects gave their informed consent after the procedures and their possible consequences were explained to them. They were also told that they could terminate any given study session or the entire study at any time, without any prejudice to their course of therapy. Fully cooperative and nonstressed subjects were essential to the study. The risks of applying brief trains of near-threshold electrical pulses to the cortex were regarded as negligible by us and by the Medical Board of Mount Zion Hospital, which approved the study. (The experiments were performed during 1972–1974. Full publication was delayed in the hope of obtaining more data, but due to the illness and death, in 1978, of the neurosurgeon, Bertram Feinstein, this was not possible. However, a fully detailed analysis of the available data has now shown that statistically adequate statements of the results were possible.)

*Stimuli.* Methods for cortical stimulation and procedures have been described (Libet, Wright, Feinstein, & Pearl, 1979). The cortical stimulating electrode in all the tabulated trials was a 1-mm uninsulated portion of Pt wire, part of an electrode assembly previously introduced subdurally. (In some separate series, mentioned below, a 10-mm stainless steel disk replaced the 1-mm Pt electrode.) Cortical (C) monopolar stimuli consisted of 0.2- to 0.5-s trains of pulses (each a 0.2-ms cathodal pulse followed by a polarity-reversed phase) at 60 pps; intensities were 1.2 to 1.5 times the liminal intensity. (Liminal intensity is the level of peak current required by repetitive pulses in a 1-s train to elicit a threshold sensory report.) For skin stimulation two identical single-pulse stimuli ($S_1$ and $S_2$), at an intensity near threshold for each to elicit a local sensation, were applied sequentially to the same biopolar electrodes but separated in time by a 5-s interval. (Subjects

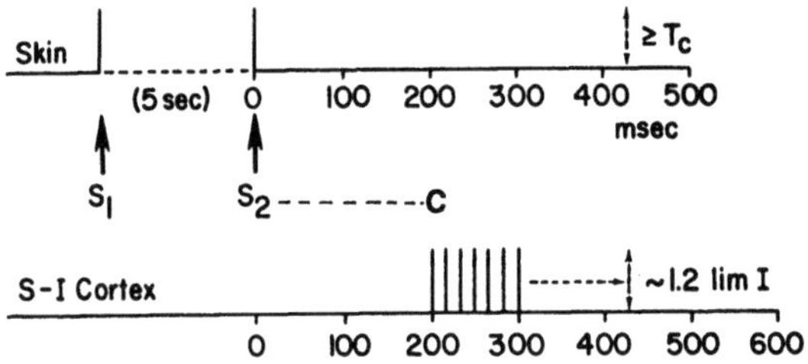


FIG. 1. Diagram of test for retroactive effects by a cortical stimulus on a sensation elicited by a preceding skin stimulus. Stimuli to skin ($S_1$ and $S_2$), each a single pulse near threshold for conscious sensation ($T_c$). Stimulus to somatosensory cortex (S–I), a train of pulses at intensity about 1.2 to 1.5 times the liminal intensity required to elicit a sensation. Cortical stimulus begins at some interval after $S_2$ ($S_2$–C interval shown as 200 ms in this example). Subject reports, within a few seconds after the cortical stimulus, whether $S_2$ felt stronger than (*st*), the same as (*s*), or weaker than (*w*) $S_1$.

had been tested with skin stimuli of varying intensities in trial runs, and they were not told that experimental $S_1$ and $S_2$ were identical.) The S electrode was placed on the skin within the area to which the cortically induced sensation was referred, unless otherwise indicated. The C stimulus was delivered *after* $S_2$. The $S_2$–C interval refers to the time between $S_2$ and *onset* of the C train of stimulus pulses (see Fig. 1). Actual $S_2$–C intervals tested were 25, 100 (a relatively small number at 50 and 150 ms), 200, 300, 400, 500, 600, 800, and 1000 ms. For statistical tabulations, all trials with $S_2$–C intervals of 25 through 150 ms were combined, as the intervals longer than those were of greater interest here; the number of trials with 400-ms intervals was roughly one-fourth those with 300-ms intervals and so the two were combined. Because numbers for each interval between 500 and 1000 ms were small, these were combined into one group; of the 86 trials in this group, 30 were at 500 ms and 42 at 1000 ms.

*Subject responses.* Subjects were asked after each trial whether $S_2$ felt stronger than (*st*) the same as (*s*), or weaker than (*w*) the preceding $S_1$ stimulus. Responses were made leisurely; this was *not* a reaction-time design. The *deviation* from the ideal of 100% *s* responses (since $S_1$ and $S_2$ were identical) was tested (a) in the absence of a C stimulus and (b) with C delivered after some $S_2$–C interval. Differences between the results for the control condition (a) and those for (b) would indicate the kind and amount of retroactive effects of C on the subjective reports of $S_2$. Subjects were studied in five to six separate sessions on different days, with about 50 to 90 trials in each session. (Subject 4 was studied in two sessions.)

*Statistical treatment.* To provide a quantitative statistical analysis, we formed an index of the chance of feeling $S_2$ more strongly: index = percentage of trials in which $S_2$ felt stronger (*st*) + half the percentage of trials in which $S_1$ and $S_2$ were felt to be the same (*s*). This addresses the question of whether applying the C stimulus after $S_2$ leads to a *shift* in the subjects' reports, *from* the idealized case of $S_2$ same as $S_1$ (which, in fact, it was at the stimulus level) *to* either $S_2$ stronger or weaker than $S_1$. If the responses are regarded as a forced choice between $S_2 = st$ and $S_2 = w$, then responses of $S_2 = s$ would be half-way between those two choices and may be divided equally between them. Actually, any alter-

TABLE 1
Summary of Responses by Subject and $S_2$–C Interval[a]

| | Controls (no C) | | | $S_2$–C intervals in ms: 25–150 | | | 200 | | | 300–400 | | | ≥500 | | |
|---|---|---|---|---|---|---|---|---|---|---|---|---|---|---|---|
| Subject | *st* | *s* | *w* | *st* | *s* | *w* | *st* | *s* | *w* | *st* | *s* | *w* | *st* | *s* | *w* |
| 1 | | | | | | | | | | | | | | | |
| No. trials | 3 | 37 | 14 | 43 | 14 | 0 | 71 | 41 | 1 | 34 | 21 | 0 | 6 | 10 | 0 |
| Percentage | 6 | 69 | 26 | 75 | 25 | 0 | 63 | 36 | 1 | 62 | 38 | 0 | 38 | 62 | 0 |
| 2 | | | | | | | | | | | | | | | |
| No. trials | 11 | 25 | 8 | 12 | 3 | 0 | 20 | 16 | 1 | 46 | 17 | 4 | 19 | 25 | 7 |
| Percentage | 25 | 57 | 18 | 80 | 20 | 0 | 54 | 43 | 3 | 69 | 25 | 6 | 37 | 49 | 14 |
| 3 | | | | | | | | | | | | | | | |
| No. trials | 4 | 35 | 10 | 65 | 30 | 5 | 10 | 6 | 1 | 12 | 6 | 0 | 2 | 1 | 0 |
| Percentage | 8 | 71 | 20 | 65 | 30 | 5 | 59 | 35 | 6 | 67 | 33 | 0 | 67 | 33 | 0 |
| 4 | | | | | | | | | | | | | | | |
| No. trials | 8 | 15 | 3 | 20 | 0 | 1 | — | — | — | 11 | 0 | 0 | 12 | 3 | 1 |
| Percentage | 31 | 58 | 12 | 95 | 0 | 5 | — | — | — | 100 | 0 | 0 | 75 | 19 | 6 |
| Total trials | 26 | 112 | 35 | 140 | 47 | 6 | 101 | 63 | 3 | 103 | 44 | 4 | 39 | 39 | 8 |
| Total percentage | 15 | 65 | 20 | 73 | 24 | 3 | 60 | 38 | 2 | 68 | 29 | 3 | 45 | 45 | 9 |

[a] Responses are to the two identical skin stimuli, $S_1$ followed after 5 s by $S_2$. $S_2$ may be followed by the cortical (C) stimulus train after the designated $S_2$–C interval; no C follows $S_2$ in the control group. Reported responses are for $S_2$ felt stronger than $S_1$ (*st*); $S_2$ felt the same as $S_1$ (*s*); $S_2$ felt weaker than $S_1$ (*w*).

native approach of giving equally spaced numerical values to each response, e.g., 1, 2, and 3, without making assumptions about the meaning of the subjects' responses can be shown to produce statistical results identical to those formed by the method adopted here. Statistical theory tells us that our index will have an approximately normal sampling distribution with the variance of the index given by

$$\text{Var[index]} = \frac{p_1(1\text{-}p_1)}{n} + \frac{p_2(1\text{-}p_2)}{4n} - \frac{p_1 p_2}{n},$$

where $p_1$ is the true chance of feeling $S_2$ stronger, $p_2$ is the true chance that $S_1$ and $S_2$ are of equal intensity, and $n$ is the number of trials under the given conditions. Thus, the index could be analyzed as the dependent variable using a weighted randomized block design as the statistical model. This model assumes no difference within each subject between trials of different days (which was supported by our data). The weights used in the analysis were inversely proportional to a continuity corrected estimate of the variance of the index. Inferential statements regarding the difference between the various $S_2$–C intervals and the control situation with no C after $S_2$ were made using the "multiple comparison versus control" technique described by Hsu (1992). Finally, an examination of

TABLE 2
Estimates of "Forced-Choice" Chance That $S_2$ Is Felt Stronger (*st*) Than $S_1$ (Calculated from "Index" Definition as Described in Text)

| $S_2$–C Interval | Estimated chance for *st* value (from "index") | Simultaneous test of difference with controls | Estimated *increases* over controls, for chance of *st* ("index"-ed) |
|---|---|---|---|
| Controls (no C) | 48.9% | | |
| 25–150 ms | 90.9% | $p < .0001$ | 42 ± 5% |
| 200 ms | 83.1% | $p < .0001$ | 34 ± 5% |
| 300–400 ms | 87.4% | $p < .0001$ | 38 ± 5% |
| ≥500 ms | 67.0% | $p < .01$ | 18 ± 6% |

residual plots supported the validity of the distributional assumptions underlying our model.

A value of the index for each day of trials at each of the tested $S_2$–C intervals was obtained for each of the subjects. Altogether, this produced 64 observations of the index (21 on subject no. 1, 20 on no. 2, 15 on no. 3, and 8 on no. 4).

## RESULTS

A summary of the responses for each of the four subjects, using the 1-mm C electrode only, is given in Table 1. In the control trials, with *no* C stimulus after $S_2$, *s* responses ($S_2$ feeling the same as $S_1$) predominated. When C followed $S_2$, the responses were radically different, with *st* ($S_2$ feeling stronger than $S_1$) predominating at all $S_2$–C intervals up to 300–400 ms. Even with intervals in the range 500–1000 ms, there was a significantly higher percentage of *st* reports than in control trials.

The index values for all subjects combined are given in Table 2. Some marginally significant differences among the percentages of the four subjects were found ($p = .04$), but we adjusted for this intersubject effect by using the randomized block statistical model. Note that the estimated percentage for the control trials (no C after $S_2$) is very near 50%, as should be expected.

The estimated *increases* (over the controls) in the "forced-choice" chance that $S_2$ is felt stronger (*st*) than $S_1$ are also given for each category of $S_2$–C interval (Table 2). The increases are roughly similar for all $S_2$–C intervals between 25–150 and 300–400 ms, being close to 40%. There appears to be some increase in the chance of *st* reports even at ≥500 ms, although the total number of trials with $S_2$–C intervals was relatively small. Combining all interval groups (including the ≥500 ms) gives a chance of reporting $S_2$ stronger than $S_1$ that is increased by 33.6 ± 4.1% ($p < .0001$), when C followed $S_2$ compared to controls (no C after $S_2$). This estimated value is in fact close to that observable upon direct inspection of the raw data in Table 1: For the control group about two-thirds of the reports were $S_2$ same as $S_1$; but in the experimental groups, with $S_2$–C intervals up through 300–400 ms, the "$S_2$ same" reports dropped to roughly one-third of the total while "$S_2$ stronger" reports constituted about two-thirds of the total responses.

The subjects' reports of $S_2$ stronger than $S_1$ were not due to a combination or confusion of the $S_2$ sensation with the C-induced sensation. Subjects regularly reported feeling both the $S_1$ and the $S_2$ sensations at the same site and that $S_2$ clearly preceded the C-induced sensation. For most series of trials the S electrode was placed on the skin within the area of the hand or arm to which the C-induced sensation was referred. Subjects easily distinguished between the two sensations, however, by the difference in qualities of the two sensations; S often gave a localized "tap" sensation, while C induced a parasthesia (tingling, etc.), a wave-like sensation, or other non-parasthesia (Libet, 1989). In a few series in two of the subjects, the C-induced sensation disappeared when C followed $S_2$. The meaning of this observation is currently obscure, but the increased reports, even in those series, of "$S_2$ stronger" (over the incidence of *st* in control trials) indicate further that retroactive enhancement did not involve a confusion of the $S_2$ and C sensations. In one session for each of two subjects (nos. 1 and 4) the S electrode was placed on the skin of the same limb as but outside the referral area for the C-induced sensations. In these series there was no chance for any "mixing" of $S_2$ and C sensations and there was never any abolition of the C sensation, but the incidence of retroactive enhancement appeared to be similar to that in the usual series with overlapping areas for S and C sensations. If this finding with nonoverlapping areas holds up under further study, it would add the important point that retroactive interaction can occur between different cortical areas. (In one subject, no. 1, the effect of C at the left cortex on the response to S was compared with S electrode either on the right hand or on the left hand, in one series of trials. With an $S_2$–C interval of 200 ms, no retroenhancement appeared with S on the left hand but the usual predominance of "$S_2$ stronger" responses appeared with S on the right hand.)

## DISCUSSION

There is little doubt, then, that a cortical stimulus can retroactively enhance the sensation for a preceding skin stimulus, even though the C train does not begin until 300–400 ms or more after that skin stimulus. Since the initial cortical response begins within 10–20 ms after the skin stimulus pulse (Libet, Alberts, Wright, & Feinstein, 1967) it is clear that the delayed cortical stimulus begins its action well after the cortex has begun to process the sensory signal. Previous evidence indicated that the duration of the cortical processing needed to develop *awareness* of the signal is relatively long, up to 500 ms or more (Libet, 1965, 1973), whereas durations for *detection* without awareness can be much shorter (Libet et al., 1991). The observation of retroactive enhancement even with delays of 400 ms or more between target and onset of conditioning cortical stimuli provides further support for the postulated delay in sensory awareness. That is, if the appearance of a sensory experience is delayed by 400 ms or more, a delayed cortical stimulus could modify the content of the experience before the experience finally appears.

As an alternative possibility, it has been argued that a conscious sensory experience may develop without such delays and that the retroactive effect of a delayed conditioning stimulus is explained as an alteration in the memory of the experience (e.g., Dennett & Kinsbourne, 1992). Retroactive *masking* or inhibition may be thought of as due to a partial or complete extinction of a memory, although the fact that electroconvulsive shock stimulation does this is hardly applicable to the case of a brief near-threshold localized stimulus that produces no convulsive/seizure activity locally or elsewhere in the brain. Another study (Dember & Purcell, 1967) had already made such an explanation dubious even for retroactive masking; in that study, a target visual stimulus was masked by a delayed visual stimulus, but it could be restored to awareness by a second even more delayed stimulus that masked the first delayed masking stimulus. In any case, retroactive *enhancement* clearly cannot involve a loss of memory.

However, the proponents of a memory change may still maintain that a *falsification* of the memory of the sensory intensity could account for this effect (Dennett & Kinsbourne, 1992). That would shift the proposal from a retroactive effect on memory *retention* to one on the *content* of the memory retained. There is no experimental evidence for such a proposed process, so that it becomes a gratuitous ad hoc construction; the experimental finding is simply that the content of the experience is retroactively altered. Second, the other strong evidence for an actual substantial delay in the development of a sensory experience would simply make such a proposal inapplicable or moot (see Libet et al., 1979; Libet, 1989). That evidence supported a *neural* delay even for a skin-induced experience but provided for a *subjective* referral backward in time to the arrival time of the initial fast signal at the cortex.

It is well known that the content of a reportable sensory experience may be modified considerably (or even fully repressed) in relation to the actual signal presented. For example, the context of an emotionally laden picture may be (honestly) reported by the subject in an altered form or it may be even fully repressed and reported as nonexistent. In order for this to happen it seems essential that there be cerebral intervention after the initiating event arrives at the cortex but before the appearance of the conscious experience. Our findings of retroactive enhancement by a cortical input delayed by up to 0.5 s provide an experimental example of such a modulatory action by cerebral cortex, introduced well after the initial cortical receipt of the signal (Libet, 1978, 1989). The putative delay for developing a sensory experience thus provides the physiological opportunity for endogenous cerebral modulation of the content or nature of the experience before it appears.

It remains to be considered why retroactive *masking* was not observed in the present experiments, whereas we had observed and reported it in an earlier series (Libet et al., 1972). Our earlier suggestion was that enhancement may be associated with a fortuitous placement of the cortical electrode somewhat removed from the sharply localized somatosensory cortex (Libet, 1978). A more complete tabulation of responses now does not support that suggestion. There were, however, two technical differences between the experiments that produced retromasking and those that gave retroenhancement: (1) In all experiments that gave retromasking the cortical stimulating electrode was the large 10-mm disk, but in the enhancement series it was the more common small wire contact. This suggested that the large difference in the area of cortex affected by the stimulus current could account for the opposing effects, masking vs enhancement. However, the results of a relatively small number of trials with the large cortical disk electrode in two of the present subjects indicated that retroenhancement could occur even with the disk electrode, at least with a 100-ms $S_2$–C interval; with an $S_2$–C interval of 100 ms these added up to 36/56 (64%) reports of $S_2$ stronger than $S_1$; 16/56 (29%) $S_2$ same as $S_1$; 4/56 (7%) $S_2$ weaker than $S_1$. (It should be added that stimulus intensities in these trials with the disk electrode were close to liminal intensity in one subject and 1.2× liminal intensity in the other. Retromasking had previously been found with disk electrode intensities at 1.3 to 1.5× liminal intensity.) (2) The earlier series that produced retromasking utilized only one S stimulus and the subject was asked whether the S-induced sensation was present or absent when S was followed by a C stimulus. That is, the question was not one of a graded difference in subjective intensity, as was the case in the present retroenhancement experiments (when the subject compared $S_2$ with $S_1$ sensations). It is possible, therefore, that retroenhancement could have occurred in some of the earlier trials using a single S pulse without being reported. The present tentative suggestion, then, is that a small (1-mm wire) cortical electrode may produce only retroactive enhancement, even with stronger intensities, but that a very large electrode (10-mm disk) may produce either retromasking or enhancement, depending perhaps on the intensity of the C stimulus. This issue merits further investigation employing the $S_1$–$S_2$–C design for testing.

## REFERENCES


Dember, W. N., & Purcell, D. G. (1967). Recovery of masked visual targets by inhibition of the masking stimulus. *Science,* **157,** 1335–1336.

Dember, W. N., & Stefl, M. (1972). Backward enhancement? *Science,* **175,** 93–95.

Dennett, D., & Kinsbourne, M. (1992). Time and the observer: The where and when of consciousness in the brain. *Behavioral & Brain Science,* **15,** 183–248.

Feinstein, B., Alberts, W. W., & Levin, G. (1969). The use of implanted thermistor electrodes in the therapy of Parkinsonism. In F. J. Gillingham & I. M. L. Donaldson (Eds.), *Proceedings of the Third Symposium on Parkinson's Disease,* (pp. 232–237). Edinburgh: Livingstone.

Hsu, J. C. (1992). The factor analytic approach to simultaneous inference in the general linear model. *Journal of Computational and Graphical Statistics,* **1**(2) in press.

Kahneman, D. (1968). Method, findings, and theory in studies of visual masking. *Psychological Bulletin,* **70,** 404–425.

Libet, B. (1965). Cortical activation in conscious and unconscious experience. *Perspectives in Biology & Medicine,* **9,** 77–86.

Libet, B. (1973). Electrical stimulation of cortex in human subjects and conscious sensory aspects. In A. Iggo (Ed.), *Handbook of sensory physiology: Vol. II, Chap. 19. Somatosensory system* (pp.743–790). New York: Springer-Verlag.

Libet, B. (1978). Neuronal vs. subjective timing, for a conscious sensory experience. In P. A. Buser & A. Rougeul-Buser (Eds.), *Cerebral correlates of conscious experience* (pp. 69–82). Amsterdam: Elsevier/North-Holland Biomedical Press.

Libet, B. (1989). Conscious subjective experience vs. unconscious mental functions: A theory of cerebral processes involved. In R. M. J. Cotterill (Ed.), *Models of brain function* (pp. 35–49). London/New York: Cambridge Univ. Press.

Libet, B. (1992). Models of conscious timing and the experimental evidence. *Behavioral & Brain Sciences*, **15**, 213–215.

Libet, B., Alberts, W. W., Wright, E. W., & Feinstein, B. (1967). Responses of human somatosensory cortex to stimuli below threshold for conscious sensation. *Science*, **158**, 1597–1600.

Libet, B., Alberts, W. W., Wright, E. W. Jr., & Feinstein, B. (1972). Cortical and thalamic activation in conscious sensory experience. In G. G. Somjen (Ed.), *Neurophysiology studied in man* (pp. 157–168). Amsterdam: Excerpta Medica.

Libet, B., Pearl, D. K. Morledge, D. E., Gleason, C. A., Hosobuchi, Y., & Barbaro, N. M. (1991). Control of the transition from sensory detection to sensory awareness in man by the duration of a thalamic stimulus: The cerebral 'Time-on' factor. *Brain*, **114**, 1731–1757.

Libet, B., Wright, E. W., Feinstein, B., & Pearl, D. K. (1979). Subjective referral of the timing for a conscious sensory experience. *Brain*, **102**, 193–224.

Pieron, H., & Segal, J. (1939). Sur un phenomene de facilitation retroactive dans l'excitation electrique de branches nerveuses cutanees (sensibilite tactile). *Journal of Neurophysiology*, **2**, 179–191.

Raab, D. (1963). Backward masking. *Psychological Bulletin*, **60**, 118–129.

Received May 6, 1992

# THE EXPERIMENTAL EVIDENCE FOR SUBJECTIVE REFERRAL OF A SENSORY EXPERIENCE BACKWARDS IN TIME: REPLY TO P. S. CHURCHLAND

BENJAMIN LIBET

*Department of Physiology*
*University of California, San Francisco*

Evidence that led to the hypothesis of a backwards referral of conscious sensory experiences in time, and the experimental tests of its predictions, is summarized. Criticisms of the data and the conclusion by Churchland that this hypothesis is untenable are analysed and found to be based upon misconceptions and faulty evaluations of facts and theory. Subjective referral in time violates no neurophysiological principles or data and is compatible with the theory of "mental" and "physical" correspondence.

**I. Introduction.** Our experimental investigations of cerebral neuronal activities that might uniquely be part of the processes that elicit a conscious sensory experience, produced evidence for the hypothesis that two remarkable temporal factors govern this kind of mind-brain relationship (see Libet 1973, 1978a; Libet *et al.* 1979). (1) *There is a substantial delay before cerebral activities,* initiated by a sensory stimulus, *achieve "neuronal adequacy"* for eliciting any resulting conscious sensory experience. For stimuli close to the threshold level for sensation the delay would average approximately 500 msec; for stronger stimuli this delay could be reduced, possibly to as little as 100 msec ±. The response in question is the introspective awareness of a localized somatic sensation elicited by the stimulus, as reported by the subject. It is in the category of a subjective "raw feel" (e.g., Feigl 1960), as distinct from behavioral responses to a stimulus that could reflect a form of detection that may be unaccompanied by a conscious, subjective experience (Libet 1965). (2) After neuronal adequacy is achieved, the *subjective timing* of the experience *is (automatically) referred backwards in time,* utilizing a "timing signal" in the form of the initial response of cerebral cortex to the sensory stimulus. This initial response is represented by the primary evoked potential, an electrophysiological response recordable at the primary sensory cortical area that receives the earliest (as well as the most localized) neural message, within 10 to 20 msec after the peripheral sensory nerve fibers are excited by the stimulus. The experience would thus be "antedated", and its timing would appear to the subject to occur with-

*Philosophy of Science*, 48 (1981) pp. 182–197.

out the actual substantial delay required before neuronal adequacy for eliciting the experience is achieved.

Subjective referral backwards in time is a strange concept and perhaps is not readily palatable on initial exposure to it. But there is a major precedent for it in the long recognized and accepted concept of subjective referral in the spatial dimension. For example, the visual image experienced in response to a visual stimulus has a subjective spatial configuration and location that are greatly different from the spatial configuration and location of the neuronal activities that give rise to the ("subjectively referred") image. The characteristics of the relationship between the "subjective" (introspectively experienced) and the "objective" (externally observable) can only be discovered, if at all, by direct study of the two phenomena conjointly, in the same human subject who, we believe, can report some information about his introspective experience to us (e.g., Libet 1966; Thorpe 1966).

A discrepancy between the subjective timing of an experience and the time of its neural "production" does introduce a novel experimentally-based feature into our views of psycho-physiological correspondence, with some interesting philosophical implications that merit analysis. However, Churchland's chief argument (pp. 169–76) is that our neurophysiological data are themselves insufficient to support the hypothesis of subjective-referral-in-time in the first place. Since the validity of our hypothesis is basic to any philosophical questions it may raise, and since Churchland's presentation and treatment of our neurophysiological studies contain serious misconceptions and distortions, I shall first briefly re-summarize the experimental evidence for the hypothesis. Following that I shall directly address those criticisms by Churchland that attack the general features of our methods, data, and conclusions. (I do not feel it is worthwhile to deal here with Churchland's many subsidiary criticisms of our experimental conditions, types and numbers of subjects employed, etc. I would merely note that the experimental paper in question (Libet *et al.* 1979) passed the stern tests of scientific quality that are normally applied by the authoritative editors of the journal *Brain*.)

**II. Experimental Evidence Summarized.** It was known from earlier work by the neurosurgeons Cushing, Foerster and Penfield, that electrical stimulation of the postcentral gyrus in an awake human subject could elicit a somatic sensory experience. The site of this sensation is subjectively referred to a localized region in the contralateral side of the body. The area of somatosensory cortex so stimulated receives a localized fast projection of ascending neural input; the latter is normally initiated by sensory stimulation in the same region of the body in which the subject feels the sensation elicited by the cortical stimulus. This "specific" sensory projection pathway begins with the larger sensory fibers, which carry messages for fine touch, pressure and position sensibilities that originate in skin, joints, etc.; these sensory fibers proceed up the spinal cord (dorsal columns) and terminate on neurones in the lower end of the brain stem; the latter neurones send out fibers that cross to the other side and ascend in a bundle (called the medial lemniscus) that terminates in the thalamus, at the base of the forebrain; the neurones in this thalamic structure project fibers directly to the primary somatosensory cortex (in postcentral gyrus). The topographical segregation of the originating fibers is maintained throughout. The earliest nerve impulses reach postcentral gyrus within 10–25 msec after a sharp sensory stimulus; they initiate a characteristic response in neurones in the sensory cortex which is recordable electrically as the "primary evoked potential". Electrical stimulation at any point in this specific ascending pathway can elicit both a somatically localized sensation and the characteristic primary evoked potential in sensory cortex. On the other hand, appropriate electrical stimulation at the outer surface of the sensory cortex (postcentral gyrus) can elicit the localized sensation and certain electrophysiological changes but *not* the "primary evoked potential" (nor the specific neuronal responses represented by this potential) (e.g., Libet 1973). The foregoing brief background is essential for an adequate understanding of the experimental tests of the hypothesis.

1. *Delayed Neuronal Adequacy with Cerebral Stimuli.* When stimulating the surface of sensory cortex with brief electrical pulses at their liminal (minimum threshold) intensity, we found that pulses had to be repetitive for a minimum period of about 500 msec to elicit any sensory experience. A shorter train (say 400 msec) of the same pulses produced no sensory experience whatsoever. Stimulation with pulse trains in the cerebral parts of the specific ascending pathway, in medial lemniscus or above, was found to show the same long time requirements; i.e., the surprisingly long period of required repetitive activations was not altered by the ability of these subcortical stimuli to elicit the "primary evoked potentials". (The requirement of long minimum stimulus durations was similar whether tested in awake patients under operating room conditions with local anesthetics, as is therapeutically customary, or tested outside the operating room in fully ambulatory patients, in whom appropriate electrodes had been implanted intracranially for therapeutically desirable reasons. It was obviously easier to obtain more fully adequate and controlled data on the ambulatory patients; all the tabular data reported in Libet *et al.* 1979, were obtained with such patients—a point that appears to have escaped Churchland.)

2. *Is There Delayed Neuronal Adequacy with Skin Stimuli?* In contrast

to cerebral stimuli, a single pulse electrical stimulus applied to skin is quite effective for eliciting a conscious sensory experience; repetition of pulses has a relatively unimportant effect on threshold intensity. Nevertheless, neuronal activities at the cerebral level continue to develop for more than 500 msec following even the single pulse stimulus to skin, and additional evidence strongly supports our hypothesis that these continued activities are similarly necessary in order to elicit the sensory experience.

a. *Primary vs. Late Evoked Potentials in Cerebral Cortex.* The cortical processes represented by the primary evoked potential are not sufficient for eliciting conscious sensation; only when the appropriate later components are evoked does the skin stimulus become effective (Libet *et al.* 1967; Libet 1973). Although the late evoked potentials, after a single effective skin pulse, do have a total duration of more than 500 msec, this experiment by itself only specifies that the duration of cortical activities must exceed the initial 100–150 msec.

b. *Retroactive Masking and Enhancement of Skin-Induced Sensations.* When a threshold single pulse stimulus to skin (S) is followed by a train of relatively strong stimulus pulses applied to somatosensory cortex (C), the conscious sensation otherwise induced by the skin stimulus can be either suppressed ("retroactive masking") or made to feel stronger ("retroactive enhancement"), probably depending on the location of the cortical stimulus (Libet 1973, 1978a). The conditioning cortical stimulus could be started more than 500 msec following the skin pulse and still modify the skin sensation, although in most cases retroactive effects were not observed with S-C intervals greater than 200 msec. However, since the cortical stimulus train was not effective at all unless its duration was greater than 100 msec, one must add at least 100 msec to all of the S-C intervals measured between S pulse and the *start* of the C stimulus pulses, i.e., the average effective S-C intervals were at least 300 msec. This means that the sensory experience induced by a brief stimulus to skin is fully developed only after a substantial delay, since neural inputs delayed for 300-600 msec after the skin stimulus can still mask or enhance the experience.

c. *Reaction Times.* In the usual measurements of minimum reaction times, for a subject to respond to a sensory stimulus by a simple button press, etc., the emphasis is of course on speed of response. When Jensen (1979) asked his subjects to try deliberately to lengthen their reaction time in a graded manner, he was startled to find that they could not do this. Instead, the reaction times jumped, *discontinuously*, from usual minimum values of about 250 msec to *minimum* values of 500–1000 msec. If it is assumed that deliberate control of reaction time would first require that the subject become consciously aware of the stimulus, Jensen's data provide further evidence that the production of the conscious experience involved a substantial cerebral delay; it also implies, in accordance with other kinds of studies, that the usual, quick (non-deliberated) reaction is made before conscious experience of the stimulus has developed, and that the cognitive processing and decision to act can be developed unconsciously (e.g., Libet 1965).

3. *Subjective Timing vs. Time for Neuronal Adequacy.* If there is a substantial cerebral delay to achieve neuronal adequacy for any sensory experience, whether induced by a brief pulse to the skin (S) or by a train of pulses to somatosensory cortex (C), one might have expected that the subjective timings for both the S and C experiences would be similarly delayed. In a direct test of this, the subject was asked to report his experience of the time order of the two sensations, when an S pulse and a C stimulus were coupled in a given trial (see Fig. 1 in Churchland's paper, as modified from Libet *et al.* 1979). It was found that S was experienced before C even when the S pulse was delivered well after the onset of the C stimulus, in fact at any time before the end of the C train of pulses required at sensory cortex. This was true whether S and C stimuli were both at the liminal threshold strengths (when minimum duration of C would be about 500 msec), or S and C were both set for a somewhat greater (matched) intensity at which minimum duration of C was 200 msec.

One possible inference from these experiments would obviously be that experience of a skin-induced sensation is elicited at cerebral levels after a much shorter delay than that for the cortically-induced sensation. This alternative was explicitly considered by us (Libet *et al.* 1979), contrary to the impression given by Churchland that we failed to do so. However, in view of the other strong evidence supporting the proposal for long cortical delays after a skin pulse (see above), as well as of our reluctance to assign fundamentally different cerebral processes to producing similar sensory experiences induced by different stimuli, we generated a second alternative hypothesis. This is temporal factor #2 (in the Introduction, above), i.e., a subjective referral backwards in time, after the delayed development of "neuronal adequacy". Such a referral would not occur in the case of the cortical stimulus because no specific "timing signal" (the primary evoked potential) is generated by that stimulus.

The crucial experimental test of the second hypothesis (subjective referral) resided in the coupling of S (skin) with LM (medial lemniscus) stimuli (Libet *et al.* 1979)[1]. Each pulse of the LM stimulus, including the

[1]Churchland mistakenly believes that the S-C (skin-cortical couplings) constituted our experimental test of this hypothesis (see also point #3 in section III of the text). She argues about this at some length, irrelevantly, but ignores the crucial S-LM test completely!

very first one, should and does elicit the same primary evoked potential that the skin pulse elicits; thus the LM stimulus should resemble the skin stimulus in providing the putative timing signal (for referral) at the very start of a train of pulses. On the other hand, the LM stimulus resembles the cortical stimulus in requiring similarly long durations of up to about 500 msec; unlike the skin stimulus, a single LM pulse is completely ineffective in producing a conscious sensory experience (Libet *et al.* 1967, 1972). The hypothesis for subjective referral thus makes a startling experimental prediction: If a skin pulse (S) is delivered so as to be synchronous with the *beginning* of a stimulus in medial lemniscus (LM, in brain stem), the subject should report that both of the resulting sensory experiences began at the same time; the subjective timings should be the same in spite of the empirically established fact that the LM stimulus train of pulses, like the cortical stimulus, could not become adequate for eliciting any sensory experience unless it was allowed to continue for up to about 500 msec. (For technical reasons, intensities were set in these experiments so that an absolute minimum of 200 msec or more was required by the LM stimuli.) Related predictions were made for different S-LM coupling intervals. The experimental results of such tests, even when subjected to a rigorous statistical analysis, clearly confirmed these predictions (Libet *et al.* 1979). Furthermore, additional independent lines of evidence in support of the hypothesis were cited (Libet *et al.* 1979).

**III. Reply to Churchland's Chief Criticisms.** 1. *Experimental Definition of Subjective Timing.* Churchland argues that our "exclusive reliance on after-the-trial reporting [by the subject, about his experiences during the trial] is an error in the methods". She claims that a more valid criterion consists in measuring the time for "the subject to say 'go' as soon as he is aware of the skin sensation". Indeed, Churchland and Martin (cited by her as unpublished) made their own study using this immediate response criterion, and she reports a "mean response time across nine subjects [of] 358.22 msec" (sic). (Incidentally, this value is given without any standard deviation, and with no control data on "false positive" answers when stimuli are omitted, etc.) But Churchland and Martin's approach is fundamentally equivalent to that of the many "reaction time" studies in the literature. There is nothing magical or uniquely informative when the motor response is a vocalization of the word "go" instead of the more usual one of a finger tapping a button; with the latter, simple reaction times can be down in the 100 to 200 msec range. Except for the longer time probably required to neurally organize the motor outflow generating the verbal response, the same cerebral processes in the response, cognitive and decision-making, are required in either type of reaction-time paradigm. The ability to detect a stimulus and to react to it purposefully, or be psychologically influenced by it, without any reportable conscious awareness of the stimulus, is widely accepted (e.g., Shevrin and Fritzler 1968; Shevrin 1978). When the emphasis is on speed of response, there is clearly the possibility that the subject may react when he is still only preconscious or unconscious of the stimulus, i.e., before he is actually consciously aware of the stimulus. Such a possibility would have to be excluded by other independent evidence, if an experimenter wants to use the immediate reaction time as a valid, primary indicator of *when* awareness (not simply detection, etc.) appears. Churchland makes the extraordinary argument that until someone else proves that the possibility (of response before awareness) is actually operative in her experiments, she can regard her criterion as a valid measure of time of awareness. (There is in fact experimental evidence other than our own for this possible order of reaction before awareness. For example, reaction time responses could be made even when the awareness of the stimulus is masked by a conditioning stimulus that was applied *after* the reaction (Fehrer and Raab 1962)! Churchland chooses to dismiss such evidence.)

Because "immediate" or reaction-time type of response has, at the least, an uncertain validity as a primary criterion of the timing of a subjective experience, we adopted a procedure in which the subject reported the relative timing order that he subjectively experienced when two different sensory stimuli were temporally coupled. The report is made unhurriedly within a few seconds after each trial, allowing the subject to introspectively examine his experience. Reporting after the trial of course requires that processes of short-term memory and recallability be operative, but this presents no difficulty for subjects with no significant defects in these abilities. Although the usual mode of reporting was verbal (e.g., "right hand first"; or, "left hand first"; or "same time"; etc.) for convenience sake, the verbal feature is not an essential one. The same results would be obtainable if the subject were asked to raise an appropriate finger to indicate the relative time order. The essential point in any such criterion is that the subject should understand the question (what did he actually experience or feel?) and that the nature of his report permits him to give a valid answer to the question. (As indicated, the latter may be compromised by insistence on maximum speed of report. The whole issue of the validity of operational indicators of subjective experiences generally, as opposed to other forms of detection, has already been discussed at length—e.g., Libet 1965, 1973, 1978b.)

Admittedly, a report of relative timing order cannot, in itself, provide an indicator of the "absolute" time (clock-time) of the experience; as suggested, there is no known method to achieve such an indicator. However, the cerebral stimuli (to sensory cortex, LM, etc.) possess built-in and experimentally controllable constraints on the minimum times required for neuronally eliciting an experience, and these provide meaningful reference times for the analysis (see point 3, below).

2. *Cerebral Delay After a Skin Stimulus.* Churchland argues that we have not produced any acceptable evidence for a substantial cerebral delay in neuronal adequacy (referred to as "latency" by Churchland) after a single pulse stimulus to skin (see #2 under Section II on "Experimental evidence", above). Churchland erroneously states we have "just one method to fix the figure for latency of skin stimuli", the one for retroactive masking (see section 2-b, under "Experimental evidence", above). She believes she can demolish the significance of this method, but her analysis is partly irrelevant and partly misleading.

Churchland argues at some length that the conditioning cortical stimuli, which we showed could retroactively mask the sensory experience otherwise elicited by a threshold skin stimulus, are "abnormal" inputs. Of course they are "abnormal" inputs! We employed them deliberately to attempt to interfere with the normal cerebral activities that are initiated by the peripheral sensory input from the skin. The significance of our experiment lay in the finding that this "abnormal" cortical input could alter (mask or enhance) the skin-induced experience even when the cortical train of pulses did not begin for up to 500 msec after the skin pulse. This indicated that normal, cerebral processes as late as 500 msec or more after the skin pulse were involved in producing the experience induced by the skin pulse. (Incidentally, it is not correct to state that masking by a cortical stimulus is specific for order, i.e., that there is no masking of S when C *precedes* S. If the same supraliminal C stimulus required for backward (S-C) masking is employed for the C-S order, then the latter does produce "forward masking" (Libet 1973, pp. 747–8). The absence of such masking in the C-S coupling trials used to test the referral hypothesis is due to the use of near-threshold intensities for C. These experimental facts were called to Churchland's attention, in response to her inquiry about it!)

Churchland believes there is something wrong with the fact that for skin-skin couplings, the maximum interstimulus intervals ($S_1$-$S_2$) producing retroactive masking have been reported to be only 50–100 msec (when $S_1$ and $S_2$ were on separate hands). These values are in contrast to the 300–600 msec intervals we were able to obtain for skin-cortex couplings. The length of the maximum effective interval is clearly a function of the location of the second or conditioning stimulus. The effective interval even for $S_1$-$S_2$ couplings is reportedly reduced to only 5–10 msec when both stimuli are on the same hand. The much larger effective S-C interval, for retroactive masking, is no doubt due to our placement of the "abnormal" conditioning C stimulus directly on that area of sensory cortex that is involved with the initial processing of the skin input as well as with firing out secondary messages to other areas; i.e., our conditioning C stimulus acts along the same pathways in the brain that have responded to the skin input which is being masked. Even at this C-stimulus site, there can be no assurance that the conditioning stimulus, limited in its area and intensity by the potential risks to the patient, will be able to exert the maximal retroactive effects in all cases; it was in fact surprisingly gratifying that we were able to establish that S-C intervals could be as much as several hundreds of msec and still show retroactive interactions. Churchland's "final and telling point" about our (allegedly) having only one subject with a backward masking S-C interval of 500 msec is a distasteful distortion. She was informed (i) that at least 100 msec must be added to all those numbers, in order to arrive at the effective intervals (see section 2-b in "Experimental evidence" above), and (ii) that there were *several* subjects who exhibited retroactive effective masking with S-C intervals in the range of 300 to 500 msec, which actually becomes at least a 400–600 msec range when appropriately corrected. There were also the additional cases in which retroactive *enhancement* was obtained with S-C intervals of 200 to 500 msec or more (Libet 1978a), becoming at least 300–600 msec when corrected, as above.

Churchland also suggests that retroactive (backward) masking may simply reflect an interference with the short-term memory of the skin-induced sensation, when S is followed by the C stimulus. This is a reasonable argument but we had ourselves already specifically raised it and we had countered it with our observations on retroactive enhancement (Libet 1978a). In cases of retroactive (backward) *enhancement* the subjective intensity of the skin-induced experience was markedly greater, when S was followed by C at intervals similar to those producing retroactive masking. With retroactive enhancement there is obviously no loss of memory or disruption of function by our "abnormal" C stimulus; the subject is aware of and recalls the skin-induced sensory experience, except that the latter is stronger than otherwise. Churchland alludes in her footnote number 11 to our counter-argument with retroactive enhancement, but she dismisses it simply on the basis that she has not seen the full data on retroactive enhancement.[2]

[2]The full research papers on the retroactive effects by cortically conditioning stimuli have, unfortunately, not yet been prepared and published. In my correspondence with Churchland I stated that I was writing my off-hand, informal recollections without checking back into the original data. When Churchland wrote to inquire about these details of data, she gave no indication that she expected to use my reply in a paper by herself. My permission for such publication use was never requested by her or offered by me. Nor was I given an opportunity to check on the accuracy of the quotations or on the potentially

3. *Cerebral Delay After the Minimum Duration of a Cortical Stimulus.* Churchland believes that we have adopted the 500 msec figure as a rigid value for what she terms the "cortical latency", the time before appearance of the cortically-induced sensation, as well as for the "skin latency". She argues that this "arithmetic . . . is crucial, for if the cortical latency should be *longer* than postulated, or if the skin latency should be *shorter,* then the temporal displacement phenomenon becomes a figment of miscalculation, . . ". Churchland appears to be partially confused on both the facts and their significance.

Firstly, the empirical determination of the minimum duration required for an applied stimulus should not be confused with the time sequences of the neuronal or subjective responses elicited by this stimulus. The 500-msec figure is the average "utilization train duration" of stimulus pulses when the weakest effective or *liminal intensity* is used. Our published data-curves of the minimum train durations that are required at different stimulus intensities, showed that minimum duration decreases progressively as intensity of stimulus is raised above liminal levels (e.g., Libet 1966, 1973, 1978a). Churchland implies that we arbitrarily change what we regard as the minimum cortical latency to suit our purposes. For example, in her footnote number 4 Churchland states: "Libet sometimes (Libet *et al.* 1979) uses 200 msec as the latency for both skin and cortical sensations. My arguments apply *mutatis mutandis.*" But, as clearly pointed out in Libet *et al.* (1979), the 200 msec minimum stimulus duration employed in that study was one *empirically* established, by raising stimulus intensity above the liminal level; it was not any arbitrary theoretical construct. (Stimulus intensity was set empirically to require a minimum stimulus duration of 200 msec, rather than 500, because, as noted there, that made a longer series of consistent trials more feasible.) All experimental predictions and interpretations in that study were of course based on the minimum stimulus durations experimentally required in any given set of trials.

Secondly, we were indeed aware of the hypothetical possibility that the responses to the stimulus, i.e., achievement of an adequate state of neuronal activity and the sensation this elicits, could appear at some unspecified time after the end of a required minimum duration of stimulus pulses (e.g., Libet 1978; Libet *et al.* 1979). But the experimental results of the "ordering tests" reported by Libet *et al.* (1979) indicated that there

---

misleading implications of omitting certain qualifying statements that I had included in my replies. (It should also be noted that Churchland has no hesitation about citing her own unpublished work, done together with Martin, and she offers this unpublished mean response time value as part of her major argument with our estimates of subjective timing of a skin sensation.)

is a definite significance to the minimum stimulus time in this regard. It was found that the difference between the subjective timings for the skin-induced and the cortically-induced sensations, was approximately equal to the actual minimum duration of the cortical stimulus train (whether the latter was set at 200 msec or at 500 msec, etc.)[3]! Such findings by themselves admittedly do not establish the absolute time of the cortically-induced sensation, since the time of the skin-induced sensation is not specifiable from them (see Churchland's "postponement hypothesis", below). However, selection of different minimum durations for the cortical stimulus cannot be expected to influence the latency of the sensation elicited by the *skin*-stimulus, especially not in a linear fashion. Nor is there any physiological reason to expect that any hypothetical additional delay that may extend beyond the end of the stimulus train before the cortically-induced sensation appears, should be strictly related to the amount of the minimum duration of the stimulus employed. Consequently, the simplest and best interpretation of the consistent experimental relationship between the latency differential (skin-cortical sensations) and the minimum duration of cortical stimulus employed, is the one implicit in our hypothesis; that is, subjective timing of the skin-induced sensation would involve no substantial delay after a skin stimulus pulse, whereas the cortically-induced sensation would appear after a delay equal to the minimum required duration of the stimulus that elicits it.

Thirdly, what is crucial to our hypothesis for backward referral in time is the test with couplings of skin (S) and medial lemniscus (LM) stimuli (as discussed in section II-3, above), which Churchland ignores in favor of the less discriminative S-Cortical couplings. LM stimuli require the same kinds of minimum train durations as the cortex, but Churchland's argument about precise values for the latency of the sensation, relative to the stimulus duration, becomes meaningless in the case of LM stimuli. The experimental prediction, borne out by the results of the study, was that LM-induced sensations would be subjectively timed as if the *beginning* (not the end) of the LM stimulus train matched the timing of a skin-induced sensation.

[3]This was true for surface cathodal cortical stimulus pulses but not when alternate pulses were anodal. Our explanation of this difference was based on a well known and thoroughly accepted physiological analysis by C. G. Phillips (as referenced in Libet *et al.* 1979). The curious results when anodal stimulus pulses were included were not "rejected" by us (contrary to Churchland's statement), but were shown actually to provide additional support for our hypothesis. Churchland's lack of caution about neurophysiological issues leads her into stating that our interpretation of these results "is unacceptable" and "is sheer speculation". It happens that Professor Phillips is the editor-in-chief of the journal *Brain* and had himself participated in the thorough review of our 1979 experimental paper before accepting it for publication! I think we can safely match Phillips' approval of our interpretation on this issue against Churchland's outcries.

4. *Alternative Explanations*. Putting aside her "reservations" about the issues which are addressed above, Churchland suggests hypotheses other than subjective referral backwards in time that are also compatible with our experimental results, etc. One can of course devise any number of possible alternative explanations for a given set of observations. One selects the "best" or most attractive hypothesis on the basis of minimum requirements of ad hoc assumptions, compatibility with and interdigitating support by other physiological evidence and concepts, experimental verifiability and predictive value, etc. On these grounds it will be argued below that our combined hypothesis (items #1 and 2 in Introduction) stands far above the available alternative hypotheses; this is especially so in its successful prediction of the results found with couplings of skin and medial lemniscus stimuli (and also of the findings, so far preliminary, of a discrepancy in subjective timings, for right and left sides, in patients with a massive unilateral pathological loss of the specific projection at cerebral levels (Libet *et al.* 1979).

a. *"Postponement Hypothesis"*. Churchland offers this hypothesis as an equally acceptable alternative to explain the data showing a delay for C-induced relative to S-induced sensations. In this hypothesis both S and C experiences would be delayed, but the cortically-induced one (C) would be "put on hold"—until it can be "admitted to consciousness"; i.e., there would be an *extra* delay for the C experience (see Churchland's Fig. 3). Let me improve on her hypothesis by eliminating the mysterious "holding" mechanism; it is quite feasible to make the neurophysiological assumption that neuronal activities set into motion by a 500 msec C stimulus may have to continue for some additional time after the stimulus, before the cerebral neuronal state achieves adequacy to elicit the experience.

Since our experiments only measured relative timing orders for the two experiences, not absolute (or clock-) times, such a postponement hypothesis would be a viable alternative if no other evidence were available. But Churchland has missed or ignored the other evidence that argues against this alternative hypothesis: (i) The *amount* of the delay in subjective timing of C relative to S experience was roughly equal to the minimum required duration for the C (cortical) stimulus. The significance of this experimental finding in relation to delay of cortical sensation is discussed above (see section III-3). (ii) Churchland's postponement hypothesis cannot explain the findings with the couplings of S-LM stimuli (see section II-3 and section III-3, above). The LM (medial lemniscus) stimuli required the same long minimum durations as C stimuli; but LM-experiences were subjectively timed as if there was *no delay* relative to S-experiences (Table 2, Libet *et al.* 1979), as predicted by our hypothesis. The postponement hypothesis would have additionally to assume that there are no extra delays for LM, in contrast to C stimuli. (iii) Churchland's postponement hypothesis is based on an acceptance of our thesis that there is a substantial cerebral delay even for a skin-induced (S) experience. However, the postponement hypothesis does not accept our added hypothesis for subjective referral of the S-experience backwards in time. Thus, in order to account for the absence of subjective "jitter" when multiple skin stimuli with different intensities are applied synchronously, her hypothesis would have to make the additional assumption that the periods of cerebral delay are the same for all such stimuli. This is a very unlikely situation, given the facts (i) that minimum durations of LM and C stimuli do vary with intensity, and (ii) that S-LM couplings show no difference in timings for the two stimuli even when intensity and minimum duration of LM stimulus are deliberately changed. On the other hand, if Churchland wants to abandon our hypothesis of a substantial cerebral delay for S-experience (she argues strongly against such a delay elsewhere in her paper), the S-LM experimental results would require that, at the very least, there would have to be subjective referral backwards in time for the experience of the medial lemniscus stimuli (see Libet *et al.* 1979).

Indeed, the absence of subjective jitter probably presents a problem for all hypotheses, unless one adds the hypothesis that subjective timing is referred to some rigidly timed signal. It is neurophysiologically very unlikely that any required neuronal processes, other than the primary evoked response, would have the rigidly same timing after all forms and intensities of stimuli; this must be coupled with the fact that the fairly rigidly timed primary evoked response was experimentally shown to be insufficient, by itself, for eliciting a conscious sensory experience. A chief attractive feature of our subjective referral hypothesis is its ability to deal with this difficulty. If sensory experiences are referred backwards to the time of the early primary evoked response that develops after each respective sensory input, any differences among the times to achieve neuronal adequacy in a group of synchronously initiated inputs become irrelevant; delays for primary evoked potentials are short (10–20 msec) and the differences in these delays with differing intensities of peripheral somatic stimuli are known to be too small (Desmedt 1971) to be experienced separately in time (see study of normal subjects in Libet *et al.* 1979).

b. *"Illusions"*. Churchland begins and ends her critique of our work by calling attention to the well-known existence of perceptual illusions. (i) Firstly, it should be clear that simply referring to the discrepancy between subjective timing and the time of the adequate neuronal action as an "illusion" constitutes at best only a semantic but not any scientific challenge to our hypothesis. The only meaningful challenge of this type (made by D. M. MacKay) suggested that the apparent subjective referral backwards in time may be due to an illusory judgment, made by the subject when he reports the timing order after the trial; i.e., there may not be an actual discrepancy at the time the experience is elicited. This particular suggestion has been analysed and was rejected as an adequate alternative to our hypothesis (Libet *et al.* 1979). It should be recognized that there is a fundamental distinction between what are usually termed "sensory illusions" and the phenomenon of "subjective referral" (whether referral is in the spatial or temporal dimension). Sensory illusions occur when one stimulus input is juxtaposed to another stimulus input in such a way that the *perception* of at least one of the inputs *becomes a distortion of the real stimulus* configuration that we know actually exists. At least for some sensory illusions it seems clear that the juxtaposed inputs have produced a modification of the neural responses which thus leads to a correspondingly modified perception. On the other hand, the phenomena of subjective referral describe fundamental relationships between the spatio-temporal configuration of any conscious sensory experience or perception and that of the neural actions that elicit or are associated with the experience. In subjective referrals there are no distorted perceptions of the real stimulus configuration, in contrast to sensory illusions; referral backwards in time, for example, results *not* in an illusion of a "prior entry" before the objective time of the stimulus, but rather in a more accurate perception of the actual time of the stimulus!

(ii) Secondly, Churchland's arguments against invoking any non-physical mental functions, in relation to spatial and temporal "illusions", appear to betray a fundamental philosophical misconstruction of the mind-brain problem and of the "identity theory" to explain it. She implies that if one only had sufficient information about the observable neural events one could then satisfactorily explain and presumably predict all the paradoxical mental phenomena. But to take a more readily understandable example—no amount of knowledge of the neural processes observable in the brain of a bat can ever tell us what it (the experience) is like to be a bat (Nagel 1974). The general principle is that there are no *a priori* rules governing the relationship between mental events and brain events; the rules must be discovered (Libet 1979). Subjective referrals for spatial and temporal features of sensory experiences provide striking examples of this principle; they could not have been predicted simply from the array of neural events. What we are discussing is not any denial of correspondence between mental and physical events, but rather the way in which the correspondence is actually manifested.

Churchland is entitled to hold to a belief that there is no meaningful independent category of mental as opposed to physical phenomena, but it should not be represented as neuroscientific. On the contrary, it is precisely the discovery of the relationships between mental (subjective) and physical (externally observable) phenomena that constitutes a major challenge to neuroscience. For example, contrary to Churchland's belief, there is experimental evidence for the view that the subjective or mental "sphere" could indeed "fill in" subjective spatial and temporal gaps. How else, for example, could one view the already mentioned enormous discrepancy *that is known to exist* between a subjective visual image and the configuration of neuronal activities that gives rise to the experience of the image? (See also Sherrington 1940.)

The distinction between mental and physical phenomena is of course quite compatible with the theory of a correspondence between "mental" and "physical" (brain) events. The view that "conscious experience is a primary datum of existence and as such cannot be fully defined" (Thorpe 1974) by reference to other phenomena does not necessarily distinguish "monist" from "dualist" viewpoints, contrary to what appears to be Churchland's argument. Indeed, the whole thrust of monist "identity theory" is an attempt to find a satisfactory way of reconciling the two independent categories of "the mental" and "the physical" (e.g., Feigl 1960; Pepper 1960; Sperry 1980). In identity theory the "mental" and the "physical" are reconciled by theorizing that they are both different aspects ("inner quality" vs. "externally observable") of a single "substrate". The question of how our hypothesis for subjective referral in time may affect alternative monist and dualist theories of the mind-brain relationship merits serious considerations (e.g., Popper and Eccles 1977). My own view (presented only in a summary form, Libet 1978; Libet *et al.* 1979) has been that the temporal discrepancy creates relative difficulties for identity theory, but that these are not insurmountable. Unfortunately, Churchland is hung up on her perception of our alleged inadequacies on the experimental issues, and she has not adequately addressed the philosophical issues. The latter will have to await appropriate treatment on another occasion.

### REFERENCES


Churchland, P. S. (1981), "On the alleged backwards referral of experiences and its relevance to the mind-body problem", *Philosophy of Science 48:* 165–181.

Desmedt, J. E. (1971), "Somatosensory cerebral evoked potentials in man" in *Handbook of Electroencephalography and Clinical Neurophysiology 9*, A. Rémond, (ed.), Amsterdam: Elsevier, pp. 8–82.

Fehrer, E. and Raab, D. (1962), "A comparison of reaction time and verbal report in the detection of masked stimuli", *Journal of Experimental Psychology 64*: 126–130.

Feigl, H. (1960), "Mind-body, *not* a pseudoproblem" in *Dimensions of Mind*, S. Hook, (ed.), Washington Square: New York University Press, pp. 24–36.

Jensen, A. R. (1979), "*g*: Outmoded theory or unconquered frontier?", *Creative Science & Technology II:* 16–29.

Libet, B. (1965), "Cortical activation in conscious and unconscious experience", *Perspectives in Biology and Medicine 9:* pp. 77–86.

Libet, B. (1966), "Brain stimulation and the threshold of conscious experience" in *Brain and Conscious Experience*, J. C. Eccles, (ed.), New York: Springer-Verlag, pp. 165–181.

Libet, B. (1973), "Electrical stimulation of cortex in human subjects and conscious sensory aspects" in *Handbook of Sensory Physiology*, vol. II, A. Iggo, (ed.), Berlin: Springer-Verlag, pp. 743–790.

Libet, B. (1978a), "Neuronal vs. subjective timing, for a conscious sensory experience" in *Cerebral Correlates of Conscious Experience*, P. A. Buser and A. Rougeul-Buser, (eds.), Amsterdam: Elsevier/North Holland Biomedical Press, pp. 69–82.

Libet, B. (1978b), "What is conscious sensory experience, operationally?", Commentary, p. 156, to P. E. Roland, "Sensory feedback to the cerebral cortex during voluntary movement in man", *Behavioral and Brain Sciences 1:* 129–171.

Libet, B. (1979), "Can a theory based on some cell properties define the timing of mental activities?", Commentary, pp. 270–271, to G. S. Wasserman and K.-L. Kong, "Absolute timing of mental activities", *Behavioral and Brain Sciences 2*: 243–304.

Libet, B., Alberts, W. W., Wright, E. W., and Feinstein, B. (1967), "Responses of human somatosensory cortex to stimuli below threshold for conscious sensation", *Science 158:* 1597–1600.

Libet, B., Alberts, W. W., Wright, E. W., and Feinstein, B. (1972), "Cortical and thalamic activation in conscious sensory experience" in *Neurophysiology Studied in Man*, G. G. Somjen, (ed.), Amsterdam: Excerpta Medical, pp. 157–168.

Libet, B., Wright, E. W., Jr., Feinstein, B., and Pearl, D. K. (1979), "Subjective referral of the timing for a conscious sensory experience: a functional role for the somatosensory specific projection system in man", *Brain 102:* 191–222.

Nagel, T. (1974), "What is it like to be a bat?", *Philosophical Review 83:* 435–450.

Pepper, S. C. (1960), "A neural-identity theory of mind" in *Dimensions of Mind*, S. Hook, (ed.), Washington Square: New York University Press, pp. 37–56.

Popper, K. R. and Eccles, J. C. (1977), *The Self and Its Brain*. Berlin: Springer.

Sherrington, C. S. (1940), *Man on his Nature*. London: Cambridge University Press.

Shevrin, H. (1978), "Evoked potential evidence for unconscious mental processes: a review of the literature" in *The Unconscious: Nature, Function, Methods of Study*, A. S. Prangishvili, A. E. Sherozia and F. V. Bassin, (eds.), Tbilisi: Metsniereba Publishing House, pp. 610–625.

Shevrin, H. and Fritzler, D. E. (1968), "Visual evoked response correlates of unconscious mental processes", *Science 161:* 295–298.

Sperry, R. W. (1980), "Mind-brain interaction: Mentalism, yes; dualism, no", *Neuroscience 5:* 195–206.

Thorpe, H. W. (1966), "Ethology and consciousness" in *Brain and Conscious Experience*, J. C. Eccles, (ed.), New York: Springer-Verlag, pp. 470–505.

Thorpe, H. E. (1974), *Animal Nature and Human Nature*. London: Methuen.

Human Neurobiol (1982) 1:235–242

**Human Neurobiology**


# Brain Stimulation in the Study of Neuronal Functions for Conscious Sensory Experiences

B. Libet

Neurological Institute-Department of Neuroscience, Mt. Zion Hospital, and Department of Physiology, University of California, San Francisco, California 94143

**Summary.** Some features of cerebral neuronal functions that are uniquely related to generation of conscious sensory responses, have been discovered. Included are two temporal factors: (1) Substantial delays, of up to about 0.5 sec, before achieving cerebral "neuronal adequacy" appear to be required for eliciting a sensory experience. This includes the demonstration that a cortical stimulus (C) can retroactively modify a skin (S)-induced sensation even when C stimulus begins up to 500 msec after S stimulus. (2) However, there appears to be a subjective referral of the experience back to the time of the cortical primary evoked response to S; subjectively the skin sensation would thus appear to have no delay. Stimuli that are inadequate for eliciting conscious sensory experience can nevertheless evoke considerable neuronal activity, including that represented in direct cortical responses or in primary (early) components of evoked potentials. Virtually all the qualities of somatic sensation (except pain) can be elicited by stimuli at post-central gyrus, when intensities of suitable trains of pulses are kept down to liminal levels. Some additional implications, for the manner in which mental and neural events are related, are discussed.

**Key words:** Electrical stimulation – Human brain – Somatosensory cortex – Somatosensory qualities – Conscious sensory experience – Subjective referral of sensations

## Introduction

Much of the waking brain's activities and responses can proceed at unconscious levels, without subjective experiences directly associated with them (e.g., Libet 1965, 1973, 1981a; Shevrin and Dickmann 1980). This suggests that specifically unique neural actions are required, within the context of the normally functioning brain, to elicit conscious sensory experiences and presumably subjective experiences generally (Libet, 1965). To study the causal relationship between specific neural actions and subjective experience requires an ability to manipulate neural function in a controlled manner; direct intracranial electrical stimulation in the conscious human subject is one of the very few approaches available for such purposes. The utilization of intracranial stimulation is of course limited to opportunities presented by invasive neurosurgical procedures for therapeutic purposes and by the risk factors it may introduce; and, in any case, it should only be employed with the properly informed consent and cooperation of the subject. These factors, and the inherent modes by which electrical stimuli can elicit or control neural actions, impose severe limits on the possible scope of meaningful experiments. The acceptably brief electrical pulses presumably excite some axonal elements almost exclusively; the functional effectiveness of the stimulus is then clearly dependent on whether axonal impulses initiated by the stimuli can effectively activate or lead to the activation of an appropriately large and organized spatio-temporal configuration of neural elements needed for eliciting an experience (see Libet 1973 and below).

## How do Electrical Stimuli Initiate or Alter Cortical Cerebral Functions?

In general, electrical stimuli have been found (1) to elicit some organized overt functional response and (2) to interfere with, and/or (3) to modulate an ongoing functional activity.

The first type of action defines the "excitable" cortex, most of the cortex being "silent" in this regard (e.g., Penfield 1958). Functional motor and sensory responses are most easily elicitable at the primary motor and sensory areas, respectively; the former is located in precentral gyrus ("area 4") and the latter, in the case of somatic sensibilities, in postcentral gyrus ("areas 3–1–2", also termed SI). These areas are connected to subcortical lower motoneurone (motor) or primary sensory neurones (sensory) by the most direct and fastest "specific" pathways, which also maintain a high degree of topographical, spatial segregation relative to the peripheral effector (muscle) and sensory structures. After the sensory stimulus, the earliest neural messages reach the appropriate primary sensory cortex first, within 10–25 msec; they initiate a characteristic response confined to primary cortex, recordable electrically as the "primary evoked potential"; this is followed by slower event-related potentials exhibiting wider cortical distributions and more related to cognitive aspects of the sensory response. The so-called association cortex, surrounding primary areas and occupying the vast intervening areas, is functionally involved with the more complex aspects of motor and sensory integrations and of higher functions

0721-9075/82/0001/0235/$ 01.60

generally. These areas are generally "silent" in response to electrical stimuli, particularly in non-epileptic patients (Libet 1973); some of the important exceptions to this are the ability of stimuli to elicit generalized bodily movements at supplementary motor cortex (located on mesial surface, inside the midline of the brain) and hallucinatory psychic responses termed "experiential" by Penfield (1959) at temporal lobe cortex. However, silent cortex can respond to electrical stimuli with observable responses other than direct motor acts or reports of subjective experiences. These responses include electrophysiological ones, the so called direct cortical responses (DCR); the establishment of conditioned behavioral responses to stimuli applied to virtually any cortical area in cats and monkeys (e.g., Doty 1969); and interference with or modulation of various ongoing functions (see below). It is perhaps more appropriate to view the absence of overt functional responses to stimulation of silent cortex as a reflection of (a) inadequacy of stimuli employed and/or (b) inadequacy of the observations employed to detect changes in behavioral or unconscious psychical processes (Libet 1973).

The second or "interference" effect of electrical stimuli can appear when stimulating excitable cortex, for example by producing an anesthesia for or masking of normal peripheral sensory inputs when applied to primary sensory cortex (see Penfield 1968; Libet 1973). These interference actions appear to require stronger electrical stimuli than those needed for eliciting a near threshold sensory experience (e.g., Libet 1973; Libet et al. 1975; see further below). Stimuli can also interfere with or disrupt normal functional responses even when applied to "silent" cortex; this provides a potentially powerful tool for studying all cortical areas (e.g., Penfield 1959, and more recently Ojemann, see this volume). The third or modulatory action is represented by some simpler examples, e.g., stimulation in the vicinity of postcentral gyrus producing enhancement of a somatic sensation (Libet 1978, see further below); and by more complex psychical changes, e.g., illusionary changes in present or ongoing experience induced by temporal lobe stimulations (termed "interpretive" responses by Penfield 1959).

Even with these limited potentialities of brain stimuli, it has been possible to pry out some of the significant parameters of neuronal action involved in eliciting a conscious sensory experience (see Libet 1973), and to obtain evidence for the existence of two remarkable temporal factors governing the relation of neural activity to subjective perception of a sensory stimulus (see Libet 1978, 1981b; Libet et al. 1979). The various electrical parameters (e.g., intensity, polarity, electrode size, pulse duration, pulse frequency, train duration), that help determine the effectiveness of a stimulus for this purpose (see Libet 1973; Libet et al. 1964), will not be reviewed in detail here, except for the specific relationship between intensity (peak current) and train duration of a stimulus consisting of repetitive pulses, each of brief duration (0.5 msec or less). This relationship has turned out to be the most interesting and productive one among the significant stimulus parameters.

### The Intensity (I)-Train Duration (TD) Relationship

When the intensity (I) of stimulus pulses is adjusted for each train duration (TD), so that the same, just barely threshold, conscious sensory experience is produced by each I-TD combination (Fig. 1), one finds there is a minimum (liminal) intensity below which no sensation can be elicited no matter how long the TD. Conversely, the liminal intensity stimulus elicits no reportable sensory experience at all unless its repetitive pulses are continued for at least an average of 0.5 sec (Libet 1966, 1973; Libet et al. 1964). This remarkably long minimum stimulus train duration required at liminal I, termed the *utilization TD*, was relatively independent of changes in pulse frequency or in other stimulus variables. With TD's shorter than the utilization TD, the required I begins to rise steeply (s. Fig. 1). The resulting I-TD curve and the approximately 0.5 sec value for utilization TD appeared to be a property of the cerebral sensory system generally, and not simply a function of stimulating at the pial surface of cortex. Similar I-TD curves were found with stimuli in the subcortical pathways rostral to the medullary nuclei, e.g., in ventroposterolateral nucleus (n.VPL) of thalamus and in medial lemniscus (LM); they were not found at peripheral nerve or skin (Libet 1973; Libet et al. 1964; 1967; 1972), or in the dorsal columns of spinal cord (Nashold et al. 1972) which of course contain ascending collateral axons of primary afferent fibers.

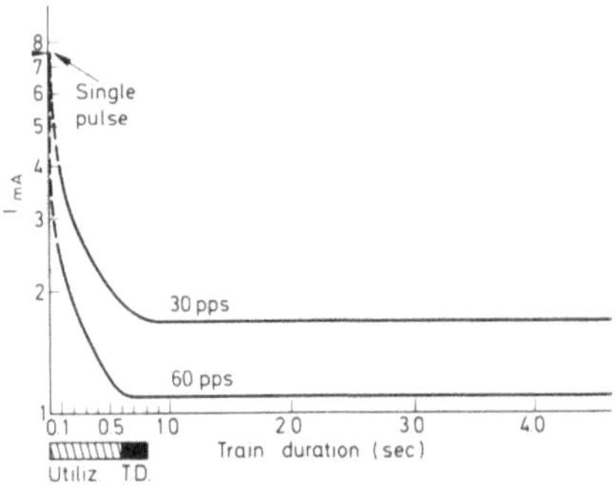


**Fig. 1.** Intensity/train duration combinations for stimuli just adequate to elicit a threshold conscious experience of somatic sensation, when applied to primary somatosensory cortex SI (post-central gyrus). Curves are presented for two different pulse repetition frequencies, employing rectangular pulses of 0.5 msec duration. Bar for "utilization TD" indicates the minimum train duration required (or "utilized") in order to elicit any conscious sensation when intensity is at the minimum effective level ("liminal I"). Note that liminal I remains constant even if TD is increased above the utilization TD. (From Libet, 1966.)

### Qualities of Somatic Sensations elicited by Cortical Stimuli

When stimulus values are kept to liminal levels, virtually all kinds of the naturally experienced somatosensory qualities, except for pain, can be elicited by stimulating postcentral gyrus (Libet 1973; Libet et al. 1975); s. Table 1. However, when intensity of a given liminal stimulus was raised by 50–100% or more, a naturallike quality of the liminal response could regularly be changed to a paresthesia (tingling, electric shock, pins and needles, numbness, etc.); this could help to explain the preponderance of paresthesia-like

**Table 1.** Somatosensory Qualities of Sensations elicited by Stimulation of Postcentral Gyrus

| | | |
|---|---|---|
| A) *Incidence of qualities, among 124 non-epileptic subjects*[a] | | |
| I. *"Paresthesia-like"* (total) | 88 | |
| II. *"Natural-like"* (total) | 115 | |
| a) something "moving inside" | | 32 |
| b) feeling of movement of part | | 32 |
| c) deep pressures | | 20 |
| d) surface mechano-type | | 12 |
| e) vibration | | 6 |
| f) warmth | | 11 |
| g) coldness | | 2 |

B) *Subject's descriptions,* within each category of qualities listed in (A).

I. *Paresthesia-like* sensory responses: tingling; electric shocks; pins and needles; prickling; numbness.

II. *"Natural-like"* sensory responses:

a) something "moving inside": wave moving along inside through the affected part; or wavy-like feeling inside; or wavy "like a snake back's" motion; rolling or flowing motion inside; moving back and forth inside; circular motion inside; crawling under the skin, or more deeply.

b) feeling of movement of the part (but with no actual motion observable to outside): quiver; trembling; shaking; flutter; twitching; jumping; rotating; jerking; pushing; pulling; straightening; floating; or sensation of hand raising up, or lifting.

c) deep pressures: throbbing; pulsing; swelling; squeezing; tightening.

d) surface mechano-type: touch; tapping; hairs moving; rolling (a ball, etc.) over surface; water running over surface; talcum powder sprinkling on; light brushing of skin; holding a ball of cotton; rubbing something between thumb and index finger.

e) vibration: vibration, buzzing (distinct from tingling, etc.).

f) warmth, or warming.

g) coldness.

[a]Each type of quality is counted only once for a given subject, even if it was elicited repeatedly by multiple tests in that subject. However, if a given subject experienced more than one type of quality, he was listed in each appropriate category. Referral sites for almost all responses were in the upper extremity, mostly in the regions of the hand and fingers. Stimuli were at liminal I in 60 subjects, and at or somewhat > liminal I in 64 subjects. Stimulus pulse frequencies were between 15 and 120 pps, train durations > 0.5 sec.

Modified from Libet et al. (1975) In: Kornhuber HH (ed) The somatosensory system. Thieme, Stuttgart, pp 291–308

sensations generally elicited by others when stimulating postcentral gyrus (e.g., Penfield 1958). Such findings suggest that not merely spatial localizability but also specific qualities are individually represented in somatosensory cortex, presumably in the form of columnar modules (Mountcastle 1967). The production of the rather nonspecific paresthesialike qualities by supraliminal stimuli could be due to activation of mixed types of columns, as in the case of supraliminal stimulation of a peripheral nerve containing a mixture of sensory fibers (s. Fig. 2) (e.g., Libet 1973; Libet et al. 1975).

## Neural Responses that do Not Lead to Sensory Experience

With a surface electrode on primary somatosensory (SI) cortex (postcentral gyrus) stimulus pulses at even below liminal I elicit substantial "direct cortical responses"

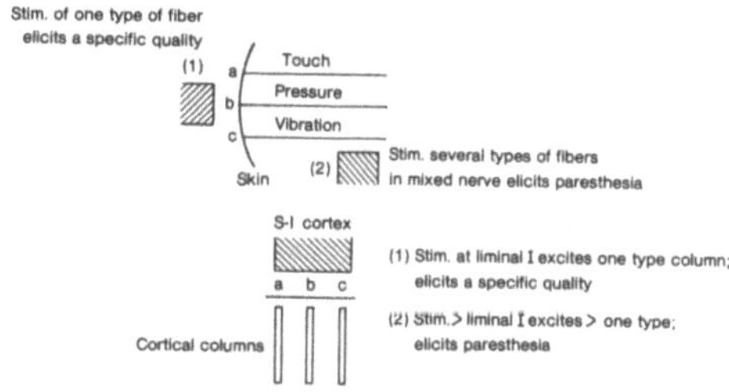


**Fig. 2.** Schema of hypothesis to explain conversion of specific quality of sensation to paresthesia, when strength of stimulus to SI cortex is raised above liminal intensity (s. Table 1, and text). The hypothesis states that excitation of one type of sensory unit, whether in periphery or primary SI cortex, elicits a specific "natural" quality; whereas excitation of a mixture of types (not simply numbers) of units elicits a paresthesia. The upper schema shows the similar explanation for the known difference between threshold stimuli at skin (1) vs. surface stimulation of the peripheral mixed nerve (2)

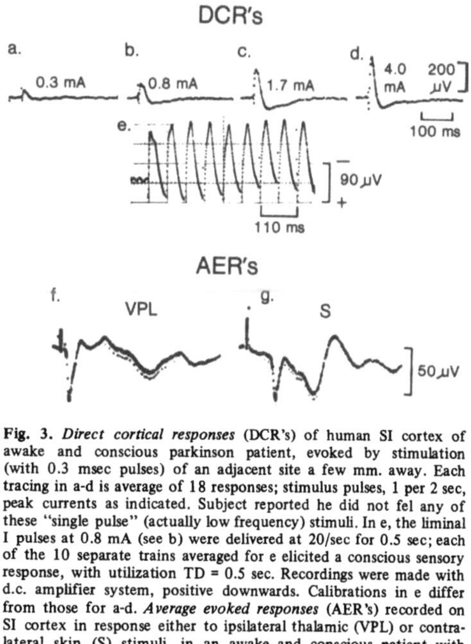


**Fig. 3.** *Direct cortical responses* (DCR's) of human SI cortex of awake and conscious parkinson patient, evoked by stimulation (with 0.3 msec pulses) of an adjacent site a few mm. away. Each tracing in a-d is average of 18 responses; stimulus pulses, 1 per 2 sec, peak currents as indicated. Subject reported he did not fel any of these "single pulse" (actually low frequency) stimuli. In e, the liminal I pulses at 0.8 mA (see b) were delivered at 20/sec for 0.5 sec; each of the 10 separate trains averaged for e elicited a conscious sensory response, with utilization TD = 0.5 sec. Recordings were made with d.c. amplifier system, positive downwards. Calibrations in e differ from those for a-d. *Average evoked responses* (AER's) recorded on SI cortex in response either to ipsilateral thalamic (VPL) or contralateral skin (S) stimuli, in an awake and conscious patient with heredofamilial tremor. Each tracing is average of 250 responses at 1.8 per sec; total length of trace is 125 msec. In f, stimuli applied in VPL (ventro-postero-lateral nucleus); subject reported not feeling any of these stimuli, even though peak currents were 6 times the liminal I that was adequate for sensory experience when a train of 60/sec and TD > 0.5 sec was applied. In g, stimuli S (skin of back of hand); peak currents were at 2 times threshold for subjectively feeling a single pulse, and all stimuli were felt. (a-d and f-g, from Libet et al., 1967, reprinted by courtesy of *Science*; e, from Libet et al., 1972, with permission of *Excerpta Medica*.)

(DCR), recordable at within a few mm of adjacent cortex. At liminal I or above, each pulse may elicit even larger DCR's and other electrophysiological responses, but no conscious sensory experience at all unless TD's are sufficiently long (Fig. 3). Similar DCR's can be elicited at "silent cortex", with no reportable experiences. Initial negative components of DCR's could even be abolished at SI cortex, by surface application of GABA (gamma amino butyric acid), with no effects on the subjective sensory experiences elicited either by direct stimulation of postcentral gyrus or by natural peripheral sensory input (Libet 1973).

Cortical neuronal activities represented by the primary evoked response, to afferent input via the specific projection pathway, also appear *not* to be sufficient for sensory experience. A single pulse stimulus in n.VPL of thalamus or in LM (medial lemniscus) can elicit at SI cortex a large primary evoked potential, apparently identical in its form and neuronal basis with the primary evoked potential elicited by a peripheral stimulus (Fig. 3). But this single pulse in LM is completely ineffective for eliciting conscious sensation, regardless of how high the peak current (I) is raised and of how large the evoked potential elicited by it (Libet et al. 1967). (A sufficiently strong single pulse at SI cortex can elicit a muscular twitch response, and this may then indirectly generate a sensation by exciting sensory structures.) Even a skin stimulus below threshold level for any conscious sensation can still elicit a small primary evoked potential but no later components (Libet et al. 1967). The association of sensory experience with appearance of later components of evoked potential and the wider cortical distribution of the latter, suggest that the regions as well as the kinds of neuronal activities required to elicit sensation are much broader than those available in primary sensory cortex alone (e.g., Libet et al. 1975).

Clearly, there can be substantial neuronal responses to stimuli in the sensory pathways that are not sufficient, and at least in some cases also not necessary, for eliciting conscious sensory experience. On the other hand, it seems probable that some such responses could be involved in behavioral and psychological detection at unconscious levels (e.g., Libet 1965; Libet et al. 1967, 1972; Shevrin et al. 1971). In this connection, it is essential that behavioral detection not be confused with subjective experience of a stimulus; the former may be manifested with or without the latter (see Libet 1973; 1981a). Indeed, the present evidence (see further below) supports our contention that to elicit subjective experience requires specific and unique kinds and durations of neural activities not essential to "unconscious" forms of detection.

### Cerebral Delay in "Neuronal Adequacy" for a Sensory Experience

As indicated above, stimuli at any of the cerebral levels in the somatosensory specific projection system require substantial train durations, varying with intensity, in order to become effective. For example, a subject who requires the usual utilization-TD of about 500 msec at liminal I reports that he feels absolutely nothing with the same stimulus train shortened to 400 msec. Since there is no reportable experience after the stimulus unless stimulus TD is raised to 500 msec, in this example, it seems clear that the state of "neuronal adequacy" for eliciting the conscious experience is not achieved until at least the end of the required TD of stimulus pulses. ("Neuronal adequacy" is used in the broad sense of integrated patterns of neural activity needed to mediate the function in question, rather than of stimple uncomplicated excitation – following the example of Jasper, 1963.) The possibility that such neuronal adequacy is delayed even beyond the minimum stimulus TD appears to be an unlikely one, in view of further evidence (see below). The question of what is neuronally unique about the activity elicited by the minimum stimulus train remains an open one. One viable hypothesis suggests that it is sufficient duration *per se*, of appropriate neuronal activities, that gives rise to the emergent phenomenon of subjective experience (Libet 1965; Libet et al. 1972).

For peripheral sensory fibers (in skin, nerve, or ascending collaterals in dorsal columns) a single threshold stimulus pulse is sufficient to elicit a sensory experience. It might, therefore, be argued that the long TD's required of cerebral stimuli may be due to an abnormal processing of such inputs, for example because they may induce cortical inhibitory patterns. (Actually, it would seem very improbable that the very different kinds and patterns of inhibitory and excitatory responses elicited by a surface cortical as opposed to a medial lemniscal, LM, stimulus train would result in the same utilization-TD's, about 500 msec, needed by either one to achieve neuronal adequacy; see Libet 1973). Although more indirect, the evidence obtainable for the case of a peripheral stimulus also indicates that similarly long delays are required to achieve cerebral neuronal adequacy. This evidence has been discussed elsewhere (Libet 1973; 1978; 1981b; Libet et al. 1972; 1979). Only the portion involving retroactive effects of a cortical stimulus on a peripherally-induced sensation will be briefly summarized in the following.

#### *Retroactive Modulatory Effects of a Cortical Stimulus*

It had already been shown that a sufficiently strong stimulus to primary sensory cortex could interfere with sensations induced by simultaneous peripheral stimuli (e.g., Penfield 1958; 1959). The hypothesis that suitable cortical activities must persist for up to about 500 msec following a threshold skin pulse, before becoming adequate for conscious sensation, would predict that even when a sufficient cortical stimulus (C) *follows* a skin pulse (S) by some hundreds of msec, C should still be able to interfere with or otherwise modify the peripherally-induced sensation. This was indeed found to be the case (Libet 1978; Libet et al. 1972). Retroactive effects were produced by a delayed "conditioning" C stimulus (applied to postcentral gyrus) even when C did not begin until 200 to 500 msec after an S pulse. The effect was one of retroactive masking or inhibition in some subjects, or of retroactive enhancement of the S-induced sensation in others. (Whether one obtained masking or enhancement was thought to be a function of a difference in location of the C stimulus with respect to the most "excitable" sites for a C-induced sensation.) The delayed conditioning "C" stimulus was applied to a relatively large area of postcentral gyrus, via a 10 mm disc electrode. In order to exert retroactive effects, the C stimulus train required a supraliminal intensity (I), at least 1.2–2 times the liminal I needed for C itself to elicit a sensation; this brings the C stimulus into a range that can interfere with responses to peripheral inputs. The C stimulus also required a minimum TD of at least 100 msec for this purpose; this would raise the effective conditioning interval, by which C can follow S, to at least 300–600 msec. The fact that neural inputs (from C) can either interfere with or modulate the S-induced sensory experience, even when delayed by 300–600 msec after an S stimulus, indicates that the S-induced experience is not finally developed neuronally until after such delays.

### Subjective Timing vs. Time for Neuronal Adequacy

If a brief peripheral sensory stimulus leads to a state of cerebral neuronal adequacy for a sensory experience only after substantial delays of up to about 500 msec, as postulated on the evidence, then one may ask whether the *subjective timing* of the experience is similarly delayed. In a direct test of this question, the subject was asked to report the subjective timing order of skin (S)- and cortically (C)-induced sensations, when an S pulse and a C stimulus were coupled with different time intervals (Libet et al. 1979). It was found that S was experienced before C even when the S pulse was delivered well after the onset of the C stimulus, in fact at any time before the end of the C train of pulses that was required at sensory cortex (s. schema in Fig. 4). This was true whether S and C stimuli were both at the liminal threshold strengths (when minimum duration of C would be about 500 msec), or S and C were both set for a somewhat greater (matched) intensity at which minimum duration of C was 200 msec. These results indicated (a) that there is

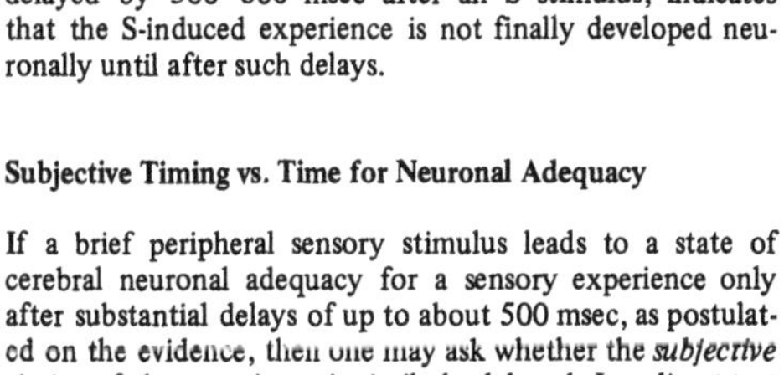


**Fig. 4.** Diagram of experiment on subjective time order of two sensory experiences, one elicited by a stimulus train to SI cortex (C) and one by a threshold pulse to skin (S). C consisted of repetitive pulses (at 60/sec) applied to postcentral gyrus, at the lowest (liminal) peak current sufficient to elicit any reportable conscious sensory experience. The sensory experience for C ("C-experience") would not be initiated before the end of the utilization-train duration (average about 500 msec), but then proceeds without change in its weak *subjective* intensity for the remainder of the applied liminal C train (see Libet et al., 1964; Libet, 1966; 1973). The S-pulse, at just above threshold strength for eliciting conscious sensory experience, is here shown delivered when the initial 200 msec of the C train have elapsed. (In other experiments, it was applied at other relative times, earlier and later.) If S were followed by a roughly similar delay of 500 msec of cortical activity before "neuronal adequacy" is achieved, initiation of S-experience might have also been expected to be delayed until 700 msec of C had elapsed. In fact, S-experience was reported to appear subjectively before C-experience (see text). (From Libet et al., 1979, by permission of *Brain*.)

essentially no delay in the subjective timing of an S-induced experience (unless one wishes to add the unnecessary assumption that both S- and C-experiences are "postponed" by some similar time that is additional to the train-duration requirement in the case of C); and (b) that neuronal adequacy for the C-experience is achieved at or near the end of the required stimulus TD, whether this be 500 or 200 msec or other tested values.

The foregoing results on subjective timing order for S and C stimuli appeared to be in conflict with the other evidence that strongly supports the proposal for long cortical delays even after a skin pulse (see above). In view of this, as well as of our reluctance to assign fundamentally different cerebral processes to producing similar sensory experiences induced by different stimuli, we generated a second alternative hypothesis to account for the paradoxical difference between delayed neuronal adequacy and non-delayed subjective timing for the experience (Libet et al. 1979), as is described subsequently.

### Subjective Referral of Sensory Experience Backwards in Time

There were two components in the revised hypothesis: 1. After the delayed achievement of neuronal adequacy, there occurs an *automatic referral of the experience backwards in time*, to a time that approximates that for delivery of the stimulus. 2. The initial cortical response to the fast specific (lemniscal) projection message generated by the sensory input, as represented by the *primary evoked potential* at SI cortex, *serves as timing signal* for this backward referral (s. Fig. 5). Latencies for the primary evoked potential are brief enough, 10–20 msec, so that differences among these latent periods cannot be differentiated subjectively (see Libet et al., 1979). The experience would thus be "antedated", and its timing would appear to the subject to occur without the actual substantial delay required for achieving the neuronal adequacy for the experience.

A crucial experimental test of this hypothesis was made possible by the special features of the responses to stimulation of medial lemniscus (LM). Each pulse of the LM stimulus, including the very first one, should and does elicit the same primary evoked potential that the skin pulse elicits; thus the LM stimulus should resemble the skin stimulus in providing the putative timing signal (for referral) at the very start of a train of pulses. On the other hand, the LM stimulus resembles the cortical stimulus in requiring similarly long durations of up to about 500 msec; unlike the skin stimulus, a single LM pulse is completely ineffective in producing a conscious sensory experience (Libet et al. 1967; 1972). The hypothesis for subjective referral thus makes a startling experimental prediction: If a skin pulse (S) is delivered so as to be synchronous with the *beginning* of a stimulus train in medial lemniscus (LM, in brain stem), the subject should report that both of the resulting sensory experiences began at the same time; the subjective timings should be the same in spite of the empirically established fact that the LM stimulus train of pulses, like the cortical stimulus, could not become adequate for eliciting any sensory experience unless allowed to continue for up to about 500 msec (s. Fig. 6). (For technical reasons, intensities were set in these

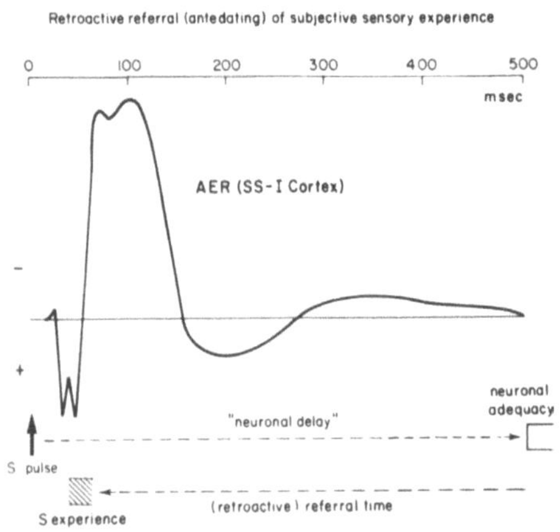


**Fig. 5.** Diagram of hypothesis for subjective referral of a sensory experience backwards in time. The average evoked response (AER) recorded at SI cortex was evoked by pulses just suprathreshold for sensation (at about 1 per sec, 256 averaged responses) delivered to skin of contralateral hand. Below the AER, the first line shows the approximate delay in achieving the state of "neuronal adequacy" that appears (on the basis of other evidence) to be necessary for eliciting the sensory experience. The second line shows the postulated retroactive referral of the subjective timing of the experience, from the time of "neuronal adequacy" backwards to some time associated with the primary surface-positive component of the evoked potential. The primary component of AER is relatively highly localized to an area on the contralateral postcentral gyrus in these awake human subjects. The secondary or later components, especially those following the surface negative component after the initial 100 to 150 msec of the AER, are wider in distribution over the cortex and more variable in form even when recorded subdurally (see, e.g., Libet et al., 1975). It should be clear, therefore, that the present diagram is not meant to indicate that the state of "neuronal adequacy" for eliciting conscious sensation is restricted to neurons in primary SI cortex of postcentral gyrus; on the other hand, the primary component or "timing signal" for retroactive referral of the sensory experience would be a function more strictly of this SI cortical area. (The later components of the AER shown here are small compared to what could be obtained if the stimulus repetition rate were lower than 1/sec and if the subjects had been asked to perform some discriminatory task related to the stimuli, as seen, for example, in Desmedt and Robertson 1977.) (From Libet et al., 1979, by permission of *Brain*.)

experiments so that an absolute minimum of 200 msec or more was required by the LM stimuli.) Related predictions were made for different S-LM coupling intervals. The experimental results of such tests, even when subjected to a rigorous statistical analysis, clearly confirmed these predictions (Libet et al. 1979). Furthermore, additional independent lines of evidence in support of the hypothesis are already in evidence (Libet et al. 1979).

## Some Further Implications in Relating Mind and Brain

The presently generated concept of subjective referral-in-time is analogous to the long recognized subjective referral-in-space. The spatial form of a subjective sensory experience is known to be markedly different from the spatial configuration of the activated cortical neuronal system that gives rise to the experience. A simple direct demonstration of this is routinely obtainable upon stimulating primary sensory cortex, e.g., S-I (postcentral) cortex. Stimulation there gives rise to conscious sensations experienced in the contralateral body part, with an inverted vertical orientation and with proportions of the sensory field grossly distorted from the neural representation (homunculus) at the cortex (e.g., Penfield 1958); there is of course no experience located subjectively in the head, where the neural activation is actually occurring. The role of the specific projection system, in providing signals that discriminate functions in the spatial referral process, is now expanded to one that functions also

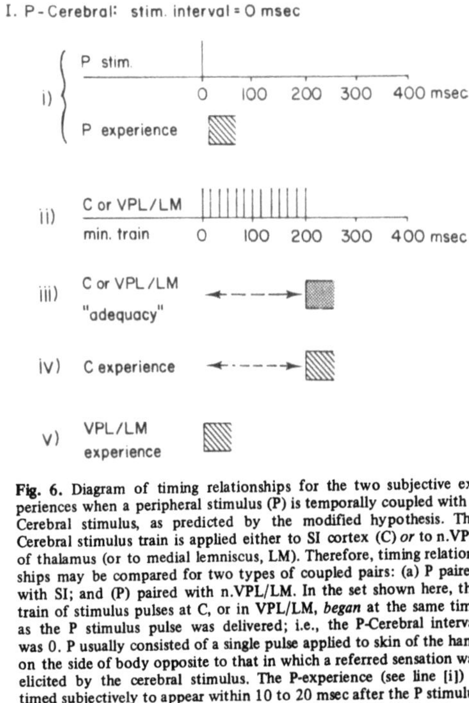


**Fig. 6.** Diagram of timing relationships for the two subjective experiences when a peripheral stimulus (P) is temporally coupled with a Cerebral stimulus, as predicted by the modified hypothesis. The Cerebral stimulus train is applied either to SI cortex (C) *or* to n.VPL of thalamus (or to medial lemniscus, LM). Therefore, timing relationships may be compared for two types of coupled pairs: (a) P paired with SI; and (P) paired with n.VPL/LM. In the set shown here, the train of stimulus pulses at C, or in VPL/LM, *began* at the same time as the P stimulus pulse was delivered; i.e., the P-Cerebral interval was 0. P usually consisted of a single pulse applied to skin of the hand on the side of body opposite to that in which a referred sensation was elicited by the cerebral stimulus. The P-experience (see line [i]) is timed subjectively to appear within 10 to 20 msec after the P stimulus (see also Fig. 4). Each cerebral stimulus (in line [ii]) is a train of pulses, usually 60/sec, with peak current adjusted so that a minimum train duration of about 200 msec is required in order to produce any conscious sensory experience; this means that the state of "neuronal adequacy" (see line [iii] with either C or VPL/LM stimuli could not be achieved before 200 msec of stimulus train duration had elapsed. The subjective timing of the beginning of a sensory experience elicited by the C stimulus (line iv) is delayed until the end of the minimum TD (200 msec) of this stimulus. But the experience of VPL/LM (line iv) is timed earlier; i.e., the sensation elicited here is subjectively referred, retroactively, to a time associated with the primary evoked cortical response. The latter is elicited even by the first stimulus pulse in VPL/LM and resembles the primary evoked response to a peripheral stimulus (s. Fig. 3, f-g), but it is not elicited by stimuli to SI cortex (see Fig. 3, a–e). (From Libet et al. 1979, by permission of *Brain*.)

in temporal referral. Both spatial and temporal referrals actually help to "correct", at the subjective level, the distortion that is inherent in the way in which activities of cerebral neurons represent the real spatial and temporal sensory configurations impinging at the peripheral levels (see Libet 1981b). The "corrections" also extend to the experience of subjective synchrony for a number of sensory stimuli, differing in intensity and even modality, when these are delivered synchronously at different bodily sites. The subjective synchrony, or absence of "subjective jitter", is experienced in spite of the probability that, for the different stimuli, the individual delays for neuronal adequacy may differ by hundreds of msec (Libet 1981b; Libet et al. 1979). If sensory experiences are each referred backwards to the time of the early primary evoked potentials elicited by each sensory input, any differences among the times to achieve neuronal adequacy become subjectively irrelevant. It is also noteworthy that subjective referral in time would not have been discovered without the experimental opportunity to directly manipulate neural actions by means of intracranial stimulations.

It is important to realize that these subjective referrals and corrections are apparently taking place at the level of the *mental* "sphere"; they are not apparent, as such, in the activities at neural levels (Libet 1981b; Libet et al. 1979; Sherrington 1951; Eccles 1980). The distinction is based on the observed phenomena; its validity is therefore independent of the theory one may adopt for the mind-brain relationship, whether this be monist-emergent (e.g., Sperry 1980) or dualist (e.g., Popper and Eccles 1977; Eccles 1980). The distinction does not deny that there are orderly relationships between the two spheres, i.e., between the inner subjective experience (the "mental") and the externally observable (the "physical") side; but it does imply that a complete knowledge of the "physical" (neural events) could not, in itself, have described or predicted the "mental" (subjective experiences) (Libet 1981b).

The postulated requirement for substantial delays, in achieving cerebral neuronal adequacy for sensory experiences, raises additional potential functional inferences that are separate from the issue of referral in time. These psychologically interesting inferences have been discussed elsewhere (Libet 1965; 1966; 1978; 1981b; Libet et al. 1972). They include the following: 1. Quick responses to a stimulus, as in a reaction time test, are performed initially unconsciously, with the conscious awareness of the stimulus, of its significance and of the response appearing after the cognitive and behavioral motor events. 2. Signal detection may occur without awareness. This is already implicit in item #1, and experimental evidence for it has appeared (e.g., Fehrer and Raab 1962; Shevrin et al. 1971). 3. The provision of substantial time, between the stimulus and the eventual cerebral adequacy for awareness, provides an opportunity for alterations in the eventual nature or appearance of the conscious experience; modulating processes could be exogenous (as in the experiment of retroactive effects of a C stimulus), or endogenous (based upon previous experiences, personality bias, repressive factors, etc.) 4. Finally, a possible neural basis to account for the distinction between conscious experience and unconscious mental operations may be postulated, in terms of durations of neuronal activities (Libet 1965; Libet et al. 1972); in this, the suitable neural activities, wherever they may be occurring in the brain, would have to endure for or appear at sufficiently delayed times, in order to achieve the transition from an unconscious mental operation to a conscious experience.

## References


Desmedt JE, Robertson D (1977) Differential enhancement of early and late components of the cerebral somatosensory evoked potentials during forced-paced cognitive tasks in man. J Physiol (Lond) 271:761–782

Doty RW (1969) Electrical stimulation of the brain in behavioral cortex. Ann Rev Physiol 20:289–320

Eccles JC (1980) The Human Psyche. Springer, Berlin Heidelberg New York, 279 pages

Fehrer E, Raab O (1962) A comparison of reaction time and verbal report in the detection of masked stimuli. J Exp Psychol 64:126–130

Jasper HH (1963) Studies of non-specific effects upon electrical responses in sensory system. In: Moruzzi G, Albe-Fessard D, Jasper HH (eds) Progress in brain research, Vol. 1 Elsevier, Amsterdam, pp 272–293

Libet B (1965) Cortical activation in conscious and unconscious experience. Perspect Biol Med 9:77–86

Libet B (1966) Brain stimulation and the threshold of conscious experience. In: Eccles JC (ed) Brain and conscious experience. Springer-Verlag, Berlin Heidelberg New York, pp 165–181

Libet B (1973) Electrical stimulation of cortex in human subjects and conscious sensory aspects. In: Iggo A (ed) Somatosensory System. Springer, Berlin Heidelberg New York (Handbook of sensory physiology, vol. II, pp 743–790)

Libet B (1978) Neuronal vs. subjective timing, for a conscious sensory experience. In: Buser PA, Rougeul-Buser A (eds) Cerebral Correlates of Conscious Experience. Elsevier/North Holland Biomedial Press, Amsterdam, pp 69–82

Libet B (1981a) ERPs and conscious awareness. In: Galambos R, Hillyard SA (eds) Electrophysiological Approaches to Human Cognitive Processing. (Neurosci Res Program Bull 20:171–175). The MIT Press Journals, Cambridge

Libet B (1981b) The experimental evidence for subjective referral of sensory experience backwards in time. Philos of Sci 48:182–197

Libet B, Alberts WW, Wright EW, Delattre LD, Levin G, Feinstein B (1964) Production of threshold levels of conscious sensation by electrical stimulation of human somatosensory cortex. J Neurophysiol 27:546–578

Libet B, Alberts WW, Wright EW, Feinstein B (1967) Responses of human somatosensory cortex to stimuli below threshold for conscious sensation. Science 158:1597–1600

Libet B, Alberts WW, Wright EW, Feinstein B (1972) Cortical and thalamic activation in conscious sensory experience. In: Somjen GG (ed) Neurophysiology studied in man. Excerpta Medica, Amsterdam, pp 157–168

Libet B, Alberts WW, Wright EW, Lewis M, Feinstein B (1975) Cortical representation of evoked potentials relative to conscious sensory responses and of somatosensory qualities in man. In: Kornhuber HH (ed) The somatosensory system. Thieme, Stuttgart, pp 291–308

Libet B, Wright EW Jr, Feinstein B, Pearl DK (1979) Subjective referral of the timing for a conscious sensory experience: a functional role for the somatosensory specific projection system in man. Brain 102:191–222

Mountcastle VB (1967) The problem of sensing and the neural coding of sensory events. In: Quarton GC, Melnechuk T, Schmitt FO (eds) The neurosciences. Rockefeller University Press, New York, pp 393–408

Nashold B, Somjen G, Friedman H (1972) Paresthesias and EEG potentials evoked by stimulation of the dorsal funiculi in man. Exp Neurol 36:273–287

Penfield W (1958) The Excitable Cortex in Conscious Man. Liverpool University Press, Liverpool, 42 pp

Penfield W (1959) The interpretive cortex. Science 129:1719–1725

Popper KR, Eccles JC (1977) The self and its brain. Springer, Berlin Heidelberg New York, 597 pp

Sherrington CS (1951) Man on his Nature. Cambridge University Press, London, 300 pages (2nd edition)
Shevrin H, Dickman S (1980) The psychological unconscious: a necessary assumption for all psychological theory? Am Psychol 35: 421–434
Shevrin H, Smith WH, Fritzler D (1971) Average evoked responses and verbal correlates of unconscious mental processes. Psychophysiology 8:149–162
Sperry RW (1980) Mind-brain interaction: mentalism, yes; dualism, no. Neuroscience 5:195–206

*Electroencephalography and clinical Neurophysiology*, 1982, 54: 322–335
Elsevier Scientific Publishers Ireland, Ltd.

# READINESS-POTENTIALS PRECEDING UNRESTRICTED 'SPONTANEOUS' VS. PRE-PLANNED VOLUNTARY ACTS [1]

B. LIBET, E.W. WRIGHT, Jr. and C.A. GLEASON

*Neurological Institute – Department of Neuroscience, Mount Zion Hospital and Medical Center, and Department of Physiology, University of California, San Francisco, Calif. 94143 (U.S.A.)*

(Accepted for publication: May 19, 1982)

Discovery of the 'Bereitschafts'- (BP) or 'readiness'-potential (RP), a scalp-recorded potential change that starts up to a second or more before a self-paced motor act (Kornhuber and Deecke 1965; Gilden et al. 1966), appeared to provide an electrophysiological indicator of neuronal activity that specifically precedes and may initiate a freely voluntary movement. There have been at least two kinds of uncertainties in this proposition.

(1) The conditions imposed on the subject by the practical requirements of the experiment could often have compromised the 'freely voluntary,' i.e., the fully endogenous nature of the acts. For recording RPs with self-paced acts, a minimum period of a few seconds is required before each act; and the repetition of the act often hundreds of times for averaging purposes tends to impose a limit on the time in which to perform each act, as well as boredom in the subject. These timing factors, as well as the physical rigors (especially 'no blinking') required during each trial, may act as external controlling influences on the subject's initiation of the act, even though the precise instant for each self-initiated act may remain somewhat irregular and self-determined within these limitations. An example of an approach to these difficulties was that of Vaughan et al. (1968) who 'adopted as a standard procedure contractions at 3–4 sec intervals determined by the sweep second hand of a clock mounted several feet in front of the subject.'

(2) The slow negative RP (termed $N_1$ by Vaughan et al. 1968), that precedes the final −90 msec before a self-paced movement, itself apparently contains two different components. A recent analysis by Shibasaki et al. (1980) has definitively distinguished between an initial portion beginning at variable times of 1–1.5 sec before EMG-0 time, and a second component (NS′) that is signaled by a rapid increase in the negative gradient starting at about −500 msec. The earlier ramp-like component is maximal at or near the vertex but shows a wide and symmetrical distribution from frontal to parietal region, while the NS′ component is asymmetrical, the increase being maximal over the contralateral precentral region. Previous reports had already indicated that such a second, later component exists (Deecke et al. 1969, 1976; Kutas and Donchin 1974, 1980). The existence of two RP components, having different timing features and evidently different neuronal sources, raises the question of which component, if any, is directly related to the initiation of a freely, unrestricted voluntary act.

The present study investigated RPs under conditions designed to minimize or eliminate all external factors that might affect the immediate initiation of a freely voluntary motor act. 'Self-initiated' RPs were in fact exhibited, and they tended to fall into two separable types. Since they appeared separately, the two types of self-initiated RPs could be examined in relation to possible subjective factors. These appeared to involve mental factors of 'pre-planning' vs. 'spontaneity' for the volitional acts. The analysis included comparative experiments involving pre-planning but without volun-

[1] This work was supported in part by the Research Support Program of the Mount Zion Hospital and Medical Center.

0013-4649/82/0000–0000/$02.75 

tary choice of when to act. Evidence bearing on the process underlying these RPs, i.e., whether a 'preparation to move' vs. a change in attentiveness, etc., was studied by substituting passive reception of task-related skin stimuli for the motor acts, under otherwise similar conditions. These findings lead to a proposal that voluntary motor acts can involve two types of neuronal processes preparatory to movement, only one of which is associated more uniquely with the spontaneously voluntary phase of preparing to act.

## Methods

### *Subjects*

Six right-handed college students were studied as two separate groups of three each. Group 1 comprised 3 females (SS, CM and MB), but the quality of the EEGs and the minimal amplitude of the RPs of one precluded using much of her data. Group 2 consisted of two males and one female (SB, BD, and GL). Study of this group began a few months after completing the study of group 1. A few important procedural changes, whose significance was only discovered during the study of group 1, were systematically employed in all sessions with group 2, as described below. One of the authors (BL) served as a pilot subject throughout and some of his data obtained in the later regularized sessions were deemed satisfactory for inclusion.

### *Recording*

EEGs were recorded with a DC system, using pre-amplifiers set for high frequency cut-off at 35 Hz. Electrodes were sintered Ag-AgCl pellets recessed in cylinders filled with electrode paste. After cleaning the skin with acetone, electrode paste was applied in a mound and the cylinder inserted into it and covered with a wad of absorbent cotton and paste to prevent drying during each ±2 h of recording time. Electrode locations were at vertex ($C_z$), left parietal ($P_3$) and prefrontal ($F_{p1\ \mathrm{and}\ 2}$) of the 10-20 system. A left lateral central electrode ($C_c$), contralateral to the right hand used in movements, was located 6 cm from the midline and 1–2 cm anterior to the frontal plane of the auditory meati and vertex. This position is over or close to the prerolandic motor area for the right hand (equivalent to 'LHM' of Shibasaki et al. 1980). The ipsilateral ($C_i$) was over an homologous position on the right side. Linked mastoid electrodes served as a reference lead. A ground electrode was placed on the left ear lobe. A large ground electrode was placed on the arm when stimulating the skin of the hand.

The DC electro-oculogram (EOG) was recorded from supraorbital and canthal electrodes routinely during the first 4 regular sessions of group 1 during RP recordings with the $C_c$ electrode. Since EOG potentials rarely occurred, the EOG was discontinued and an observer's monitoring of eye fixation during each trial was substituted so that the EOG DC channel could be used for EEG recording. An additional check on extraneous potentials was provided by the ERPs to skin stimuli, which were recorded always in all subjects from the same EEG electrodes used for RPs. Visual tasks and conditions of head and neck musculature were the same for skin stimulation as for voluntary movement, but no significant potential changes occurred during the 1.4 sec before the stimulus pulse to the skin (see below).

Bipolar EMGs were recorded with EEG disc electrodes fixed longitudinally over the activated muscle of the right forearm. The EMG amplifier had a time constant of 0.01 sec and high frequency cut-off at 3 kHz. When EMG amplitude reached a pre-set value, it triggered the PDP-12 computer which was programmed to activate a computer of average transients (CAT) and to record the time of the EMG trigger. (With series involving skin stimulation, a synch pulse from the stimulator served as trigger.) Subjects were asked to make the flexion of the fingers for the voluntary response brisk enough so that the trigger value was achieved within no more than the 10–20 msec after the beginning of any EMG potentials. This, as well as arm muscle relaxation, was practiced in training sessions with the loud speaker on to hear muscle potentials. The time delay of EMG activation for the trigger was monitored throughout and kept relatively constant by reminding subjects to make movements briskly.

Skin stimulation was applied through a pair of

electrodes on the back of the hand (Libet et al. 1967). After establishing the threshold level for a single, 0.5 msec constant-current pulse, the peak current was raised by about 15–20%. Subjective intensity was thus sufficiently weak to make recognition somewhat difficult, but sufficiently above threshold to eliminate equivocation about stimulus delivery (as in related paradigms of Desmedt and Debecker 1979).

Averaging of the 1.4 sec portion of EEG preceding an event was achieved simply by having the computer begin its 2 sec period of storage with that leading portion of recording tape between the recording and playback heads of the recorder running at 3.75 in./sec (see McAdam and Seales 1969). For each regular series 40 trials were performed and averaged. Analysis intervals were 10 msec in the 2000 msec of stored sweep on the CAT. The DC baseline level was readjusted for any progressive DC drift by a compensatory voltage after each trial. From the total voltage compensation over a 40 trial series, the baseline drift for the 2 sec period of EEG recorded with each event was estimated.

*Procedure*

The subject sat in a partially reclining position on a lounge chair with an observer present in the room. Each trial was started only when the subject considered himself comfortably ready. The trial began with a brief 'get-ready' tone. This signaled that during the next 1–3 sec the subject should relax his muscles, especially those of the head, neck and forearm, blink his eyelids if he wished and fix his gaze on the center of the 5 in. circular screen of a cathode ray oscilloscope (CRO) that was positioned at about 1.95 m away in his direct line of vision. At the end of these irregular get-ready periods the operator activated the PDP-12 computer to initiate a circular revolution of the beam of the CRO. The CRO spot of light revolved in a clockwise circle near the circumference of the screen starting from the '12-o'clock' position; this motion simulated a sweep second hand of a clock but each revolution was completed in 2.56 rather than 60 sec. A circular scale, with numbers at each '5 sec' position, was mounted at the external edge of the CRO screen, and a plastic grill on the peripheral portion of the screen displayed illuminated radial lines spaced at '2.5 sec' intervals (each equal to 106 msec of actual time). Subjects were asked to maintain their gaze fixed on the center of the CRO screen and not to follow the CRO spot around, even though they were to report information relating the 'clock position' of the spot to the events (see below). The visual angle subtended between the center and the peripheral position of the moving spot was small enough (1.8°) to present no difficulty from loss of visual acuity. The 'clock-time' of the CRO spot at each event, i.e., EMG with motor act or synch pulse with stimulus, was recorded by the computer.

The subject was asked not to blink from the time the CRO spot started revolving until after the event. To minimize the possibility that the need to blink might become a controlling 'external' factor that compels or impels him to act, the subject was told that he may blink during the trial if the need arose; but that, if he did blink (or made some other extraneous motion), he should then simply wait for the CRO spot to make at least another full revolution before performing the quick voluntary movement, as at the start of the trial.

Three different kinds of series were studied:

*(1) 'Self-initiated' voluntary acts.* The subject was asked to wait for one complete revolution of the CRO spot and then, at any time thereafter when he felt like doing so, to perform the quick, abrupt flexion of the fingers and/or the wrist of his right hand. Each subject was allowed to choose the precise extent of this flexion movement that felt most convenient to him, but was asked to maintain the same form throughout. The movement was a phasic one; subject was not required to maintain the flexed position of the fingers. An additional instruction to encourage 'spontaneity' of the act was given routinely to subjects in group 2 but only in the latter half to two-thirds of sessions with group 1. For this, the subject was instructed 'to let the urge to act appear on its own at any time without any pre-planning or concentration on when to act,' i.e., to try to be 'spontaneous' in deciding when to perform each act; this instruction was designed to elicit voluntary acts that were freely capricious in origin.

Within a few seconds after he had flexed the

fingers and/or wrist, the subject was asked to report the 'clock position' in 'seconds' of the CRO spot, in relation to performing the self-initiated act. (This condition of attentiveness and reporting paralleled the requirements in series type 3, with 'skin stimuli at unknown times;' see below. A further analysis of the significance of these timing reports will be published elsewhere.) On the other hand, it was emphasized that only an after-the-event recall of the experience was required, and that the subject should not worry about the task in advance of each event. The requirement to recall and report the clock-time of his voluntary movement did not appear to have any obvious effect on the RPs. This could be seen from comparisons to the RPs recorded in the initial training session, when no such reports were requested; see also direct comparisons in a trained subject, in Fig. $1_{S.S._2}$.

In addition, after almost every regular series in group 2, but only after a small number in group 1, subjects were asked whether they were aware of any pre-planning of acts as opposed to complete 'spontaneity' (all acts appearing without forethought, 'out of nowhere'), and also about the incidence of 'surprises' for the voluntary acts. ('Surprises' mean those trials after which the subject was surprised by having moved; these would thus be self-initiated acts presumably accomplished without a prior subjective awareness of the urge to act.)

The voluntary acts in these series are referred to as 'self-initiated' rather than as self-paced, because: (1) there was no limitation on the time in which to perform the act; (2) the available option to blink, if necessary, minimized any impulsion from this source; (3) each trial was initiated as an independent event, after a flexible delay suitable to the readiness of the subject.

*(2) 'Pre-set' motor acts.* In this condition subject was asked to perform the same brisk flexion of the fingers and/or wrist when the revolving CRO spot reached a pre-set 'clock time,' the value of which was given to him before the trial began. Within each such series of 40 trials, the pre-set clock times for acting were at '10 sec,' '20 sec,' '40 sec' and '50 sec' for the successive 4 blocks of 10 trials respectively. One complete revolution of the spot was to precede each of these pre-set clock times, so that the 4 pre-set time intervals were in actual time approximately 3.0, 3.4, 4.2 and 4.7 sec. This distribution of time intervals and 'clock positions' for action was intended to straddle the intervals and spot positions mostly adopted by the subjects when making the self-initiated voluntary acts. The subject was encouraged to try to make his movement coincide as closely as possible with the arrival of the spot at the pre-set time; the general accuracy of his 'shots,' in terms of how closely his EMG times actually coincided with the pre-set clock times, was reported to the subject after the end of each such 40 trial series.

*(3) Skin stimuli 'at unknown times.'* For such a series the subject expected to receive a near-threshold stimulus pulse on the hand instead of performing a voluntary motor act. Delivery of the pulse was to be made by the operator at any time after the first revolution of the CRO spot; they were actually delivered randomly during the 2nd or 3rd revolution of the spot (i.e., between about 2.6 and 7.6 sec from the start). The subject's task was to report, after each trial, the 'clock time' of the CRO spot at which he experienced the weak stimulus. These conditions closely paralleled the attentive and other requirements associated with performing and recalling the CRO clock time for 'spontaneous' self-initiated voluntary acts. The subject was further encouraged to 'do well' in this task by giving him, after each such 40 trial series, a rough indication of how close he was to the actual delivery times for the stimuli.

*Progression of the experiments*

The first (and, in some cases, the second) half-day sessions were purely for training purposes. Subsequently, each subject was studied in 6–8 regular half-day sessions, usually 1/week. In each of the first 4 regular sessions there were two 40 trial series of self-initiated acts and one 40 trial series with skin stimuli delivered at unknown times; each 40 trial series was preceded by a briefer 10 trial series for re-training purposes, including re-familiarization with the instructions. Pre-set series and recording sites other than $C_z$ and $C_c$ were studied in the additional sessions.

## Results

*Self-initiated motor acts and their types of RPs*

Under our conditions RPs varied considerably in form and duration, even for a given subject in the same session. The RPs appeared to fall into 3 distinct types, distinguished chiefly by the form of the RP and the time at which the main rise in negativity begins.

In *type I RPs* a gradually or steadily rising, ramp-like form begins distinctly prior to −700 msec (Table I). In the more extreme examples of type I (e.g., Fig. $1A_{S.S._1}$ and BL), which appeared in earlier sessions with group 1 and BL before introducing the instruction for 'spontaneity,' the RP appeared to have begun rising well before the −1400 msec of the available pre-event recording time. RPs classifiable as type I with more modest features continued to appear irregularly after introducing this instruction, more in some subjects than in others (see Table I). Examples are seen in Figs. 2B, $3A_{G.L._2}$ (borderline I vs. II), $4A_{G.L.}$ and $5A_{S.B.,G.L.}$. The mean onset time for main negative shift was −1055 ± 173 msec for the type I RPs in group 2. In some type I RPs, an inflection at about −500 msec indicating two components appears to be discernible (e.g., Figs. $1A_{B.L.}$ and $5A_{G.L.}$); such an inflection has been more routinely observed by others in self-paced RPs (Shibasaki et al. 1980).

In *type II RPs*, the main rise of negativity starts in the range of about −400 to −700 msec, with a mean value of −577 ± 151 msec for group 2. (For subject S.S. in group 1, mean onset for 4 such RPs was −391 msec.) The main portion of this RP is often somewhat dome-shaped rather than ramp-like in form.

A striking example of a switch from type I to type II RP in the same session, after introducing the instruction for spontaneity, is seen in Fig. 1A vs. B for $S.S._1$. Examples of type II RPs are seen in Figs. $1B_{S.S._1}$; 1A, $C_{S.S._2}$, $2A_{G.L.,S.B.,B.D.}$, $3A_{S.B.,G.L._1}$, $4A_{S.B.}$, $5A_{B.L._1}$ ($4A_{S.B.}$ grades into type I). Some definitely distinguishable negativity does appear before the −400 to −700 msec range in many but not all of the type II RPs, but this tends to have a relatively irregular, low amplitude and not the steadily rising ramp-like form of type I (e.g., Figs. $1B_{S.S._1}$, $2A_{G.L.,S.B.}$ and $4A_{S.B.,S.S.}$). Such earlier low level negativities preceding the main rise in type II RPs may represent weaker forms of the early ramp in type I RPs.

In *type III RPs*, the main rise of negativity does not appear until about −250 to −200 msec (see examples in Fig. $4A_{B.D.}$). Total durations of any detectable negativity and especially total areas of RP are also low. Type III RPs virtually never appeared in the initial session for any subject.

*Randomness of 'clock-times' for self-initiated acts.* The distribution of the actual EMG-0 times for these acts was plotted from the clock positions of the revolving CRO spot at which the acts occurred. This was done for 38 of these series (40 trials

TABLE I

Onset (msec) of main negative SP shift for each type of self-initiated RP (recorded at vertex) *. $\overline{X}$ = mean value; M.D. = mean deviation; S.D. = standard deviation.

| Subjects in group 2 | Type I RP | | | Type II RP | | | Type III RP | | |
|---|---|---|---|---|---|---|---|---|---|
| | n | $\overline{X}$ | M.D. | n | $\overline{X}$ | M.D. | n | $\overline{X}$ | M.D. |
| S.B. | 7 | −1032 | 106 | 5 | −655 | 156 | | | |
| G.L. | 7 | −1114 | 184 | 5 | −560 | 112 | | | |
| B.D. | 1 | −800 | | 4 | −500 | 50 | 5 | −240 | 38 |
| | | $\overline{X}$ | S.D. | | $\overline{X}$ | S.D. | | $\overline{X}$ | S.D. |
| Grand averages | 15 | −1055 | 173 | 14 | −577 | 151 | 5 | −240 | 47 |

* Onset times, relative to 0 for the EMG trigger, were judged by the beginning of the rise for the main negative shift visible in each RP; small shifts that did not build progressively were ignored for this purpose.

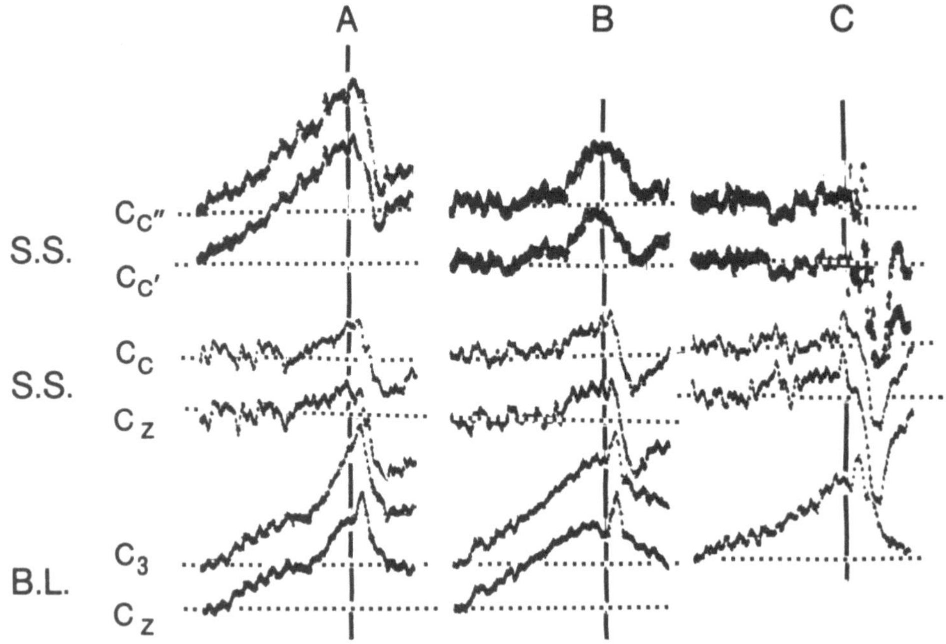


Fig. 1. Self-initiated RPs with different instructions. In this and other figures, each horizontal row gives averaged potentials, 40 trials, for different series of a given session with the subject whose initials, and the sites of the 2 monopolar leads, are shown in left margin. ($C_{c''}$ and $C_{c'}$, in top row refer to contralateral precentral leads, the usual 6 cm lateral and 2 cm anterior for $C_{c''}$ and 6 cm lateral only for $C_{c'}$; $C_z$ at vertex.) The solid vertical line through each column represents 0 time, given by EMG or skin stimulus trigger (see Methods). The dashed horizontal line in each tracing represents the estimated DC baseline. Time and voltage calibrations in Fig. 2 apply to all figures. For S.S., row 1 was obtained in an early session: RP in A was followed by subject's report of some 'pre-planning' (see text); before B, subject was instructed to let urge to act come on its own, 'spontaneously;' for C, skin stimulus was applied at unknown time in each trial (see Methods), with subject reporting the actual delivery time after each trial (no motor acts). For S.S., row 2, A–C, self-initiated RPs obtained in a later session, with instruction for 'spontaneity' now routinely given in all cases; for C, subject reported the clock-time of her movement after each self-initiated act, as was usual in this study; for the preceding 2 series, A and B, subject was told to ignore this usual requirement and not to report anything. This difference in reporting requirements appeared to make no obvious difference; all RPs in this row are of type II. For B.L., row 3, A is another example of self-initiated RP before the instructions for either 'spontaneity' or for reporting the time of each movement were in use; B and C are 'pre-set' RPs (see text).

each), performed during the initial 4 regular sessions by each subject. The $\chi^2$ test with 5 degrees of freedom ($\chi_5^2$), for disparity between the actual and a theoretically uniform distribution of EMG times, was applied to each series. The significance probability values indicated that timing of the acts showed an acceptable likelihood of being random, with respect to position of the revolving spot, in over 75% of these series.

*Subjective reports associated with self-initiated RPs.* The mode of questioning subjects about the state of pre-planning, spontaneity or 'surprises' for the voluntary acts in a given preceding series could only provide limited indications about the relation of these states to types of RPs, chiefly because it was impractical to question the subject after each individual trial. Nevertheless, the reports obtained did contain suggestive information.

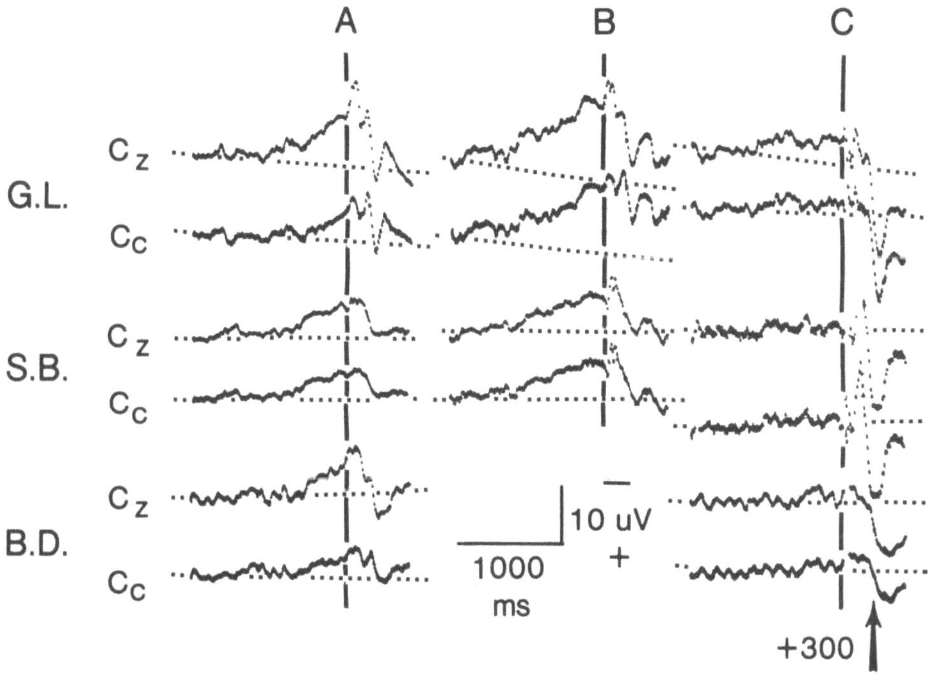


Fig. 2. Self-initiated RPs, types I and II; ERPs for skin stimuli. Vertical column A shows type II RPs, and column B type I RPs. (Subject B.D. exhibited only one type I RP, Table I.) For S.B., tracing in A was in a different session from that for B and C.) Column C shows ERPs with skin stimuli delivered at unknown times; arrow indicates position of +300 msec time, after the stimulus.

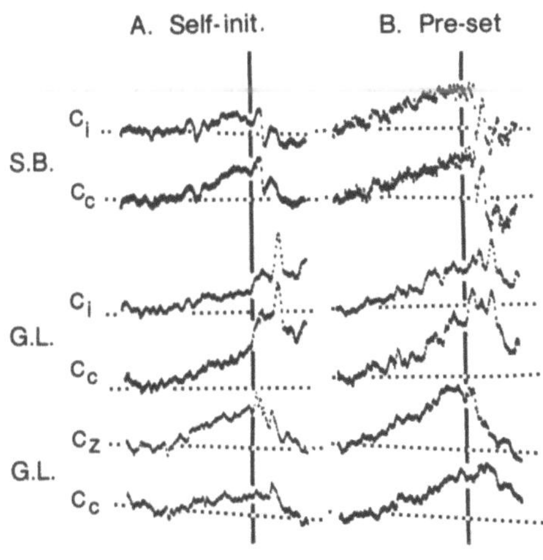


Fig. 3. Self-initiated RPs (A) and pre-set RPs (B), with bilateral precentral comparisons. Each row gives series A and B done in same session for that subject. Contralateral ($C_c$) and ipsilateral ($C_i$) precentral, simultaneous recordings for S.B. and G.L.$_1$ (middle row); vertex ($C_z$) and $C_c$ for G.L.$_2$ (bottom row; different session from G.L.$_1$).

'Pre-planning' or 'pre-intention' in some form was reported by the 3 subjects in group 2 and subjects S.S. and C.M. in group 1 in a total of 9 series. In 8 of these 9 series the RP was a type I (the exception was a type II, in subject B.D.). The subjective contents of these recalled awarenesses contained some important features. Subjects reported being aware of some 'pre-planning' in only a minority of the 40 self-initiated acts that occurred in the series for that averaged RP. The subjective

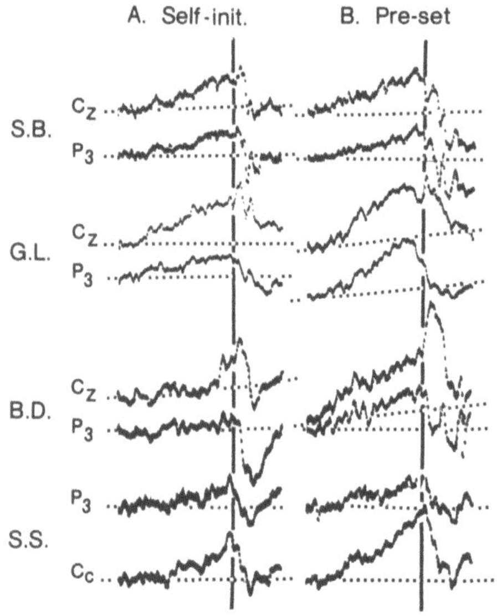


Fig. 4. Self-initiated RPs (A) and pre-set RPs (B), with parietal-central comparisons. Upper dashed line in $B_{B.D.}$ is baseline for the upper, $C_z$, tracing.

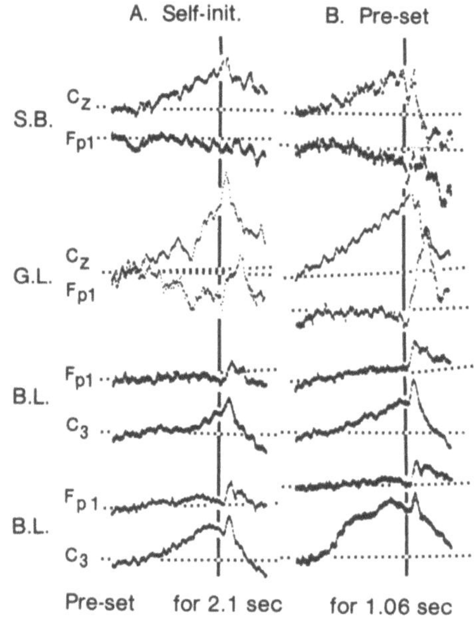


Fig. 5. Self-initiated RPs (A) and pre-set RPs (B), with prefrontal-central comparisons. For $BL_2$ (bottom row), both A and B are pre-set RPs, for which the time interval, between starting the revolving circular motion of the CRO spot and the pre-set clock-time for the act, was reduced to 2.1 sec and 1.06 sec in A and B, respectively; the usual, longer time intervals (see Methods) were used for $B_{B.L._1}$ (third row). (With the 1.06 sec interval, the big negative shift begins at about the start of this interval, while with the longer intervals the shift becomes more ramp-like.) The 3 pre-set RP series for B.L. (B.L.$_1$, B.L.$_2$-A,B) were performed in that sequence, followed by the self-initiated RP series in B.L.$_1$-A.

recollection was most often one of having a general intention or anticipation of performing the act during a forthcoming period of time, when the moving CRO spot would have entered a specific portion of its revolving circle (e.g., between '12 and 6 o'clock,' or between '6 and 12 o'clock'). Every subject clearly and definitively distinguished such advance feelings, of intention to move sometime during the immediate 1–2 sec in the future, from the specific urge or desire to actually move that appeared more immediately before the movement. That is, even when some pre-plannings were recalled and reported, subjects insisted that the more specific urge or intention to actually move did not arise in that pre-planning stage. Subject S.B. described his advance feelings as 'pre-tensions,' rather than as pre-plannings to act.

Reports of 'spontaneity' and/or 'no planning' meant that subjects recalled that each urge to act arose on its own, as if appearing capriciously out of nowhere, with no recollections of advanced plannings for any of the 40 acts in the series. For the 14 series with such reports, essentially none of the RPs were type I (one of these RPs was regarded as borderline I–II), while 9 were type II and 4 type III. It should be added that some subjects volunteered reports that, in some of the individual trials, they felt an urge or intention to move which did not consummate in an actual movement, accompanied in some cases by an experience of voluntarily suppressing the act; in such cases, the subject waited until another urge to act arose suddenly, without deliberate planning (see Discussion of 'Volitional processes').

Recollections of feeling 'surprised' by at least some of the acts were reported in only 3 series by 3 subjects with only a few 'surprises' out of 40 events in each of those series. The 5 RPs of type III in B.D.'s case were mostly produced without reports of 'surprise' experiences; indeed B.D. reported advance awareness of the urge to perform all 40 acts in 4 of these type III RPs. These results do not show a specific correlation between automaticity of action (being 'surprised') and type III RPs, but a more adequate analysis of this possibility would be desirable.

The subjects' reports of the time of their movement were generally within < 100 msec of the actual time of the EMG for the event. Mean difference between reported and EMG time, in 32 series of 40 trials each obtained from all 6 subjects, was $-64$ msec $\pm 18$; in each individual series the standard deviation of the mean difference was also small, all in a range between $\pm 9$ and $\pm 25$ msec.

*'Pre-set' motor acts and their pre-potentials*

'Pre-set' motor series were carried out in the same session as 'self-initiated' motor series, with the same electrode placements, etc. Examples of such coupled series are shown in columns A vs. B, in Figs. 3, 4 and 5, and in Fig. $1_{B.L.}$, A vs. B, C.

'Pre-set' RPs were almost always ramp-like in form. Although durations were usually close to or greater than the available analysis time of $-1400$ msec and amplitudes usually considerably larger than for most self-initiated RPs, pre-set RPs were generally similar to type I RPs (see Figs. $1_{B.L.}$, $3_{G.L._2}$, $4_{G.L.}$ and $5_{S.B.,G.L.}$). On the other hand, the incidence of type I vs. type II RPs was not related to the inclusion of a pre-set series in that session. The relative incidences of type I and type II RPs in the first 4 regular sessions, when even the concept of performing pre-set motor acts had not yet been introduced to the subjects, were similar to those in all other sessions. In those sessions in which pre-set series were included, the self-initiated motor series was carried out first in many cases (as they were in Figs. $1_{B.L.}$, $3_{G.L._2}$ and $5_{S.B.,B.D.}$), with no advance indication to the subject that a pre-set motor series was going to be employed and, when a pre-set motor series was performed first in a session, the succeeding self-initiated motor series produced both type I (as in Figs. $4_{G.L.}$ and $5_{G.L.}$) and type II (or III) RPs (as in Figs. $3_{S.B.,G.L._1}$, $4_{S.B.,B.D.,S.S.}$ and $5_{B.L.}$).

The pre-set RPs bear some resemblance to the ramp-like component that preceded $S_2$ in a CNV-type paradigm employing an $S_1$–$S_2$ interval of 4 sec (Sanquist et al. 1981). When the present time intervals, from onset of motion of CRO spot to clock-time for action, were reduced in successive 40 trial series within the same session so as to approach values closer to 1 sec (Fig. $5_{B.L.}$), the pre-set RP changed to a briefer, more sharply rising dome-lime potential that resembled the usual CNV or forewarned response with 1 sec interstimulus intervals (e.g., Walter et al. 1964; Kutas and Donchin 1980).

*Skin stimuli 'at unknown times'*

A 40 trial series of this type was carried out at the end of each of the first 4 regular study sessions in which all 6 subjects each participated, and additionally in some subsequent sessions for each subject. During the 1400 msec of pre-stimulus time, the tracings for these 40 trial averages were usually flat, except for some irregular bumpiness; examples are shown in Figs. $1C_{S.S.}$ and 2C. (The small negative shift apparently exhibited by G.L. (Fig. 2C) was not seen in 3 similar series in other sessions with this subject.) The absence of any significant pre-potential is in striking contrast with the self-initiated RPs exhibited by the same subject in the same session.

*Subjects' reports and P300 ERPs.* The subjects' reports of the time of stimulus were, with some exceptions, within < 100 msec of the actual time of delivery. The average of the mean differences, between reported and actual times in 4 series of 40 trials for each subject, were: S.B., $-40 \pm 14$ msec; G.L., $-167 \pm 28$ msec; B.D., $+83 \pm 17$ msec; S.S., $-111 \pm 20$ msec; C.M., $-20 \pm 19$ msec; M.B., $+46 \pm 21$ msec. (Recall that the finest clock marker, '2.5 sec' interval, corresponded to an actual time interval of 106 msec.) The accuracy of the reports indicate good attentiveness, and cognitive decision-making. This is in accord with the consistent appearance of a large

P300 wave in the post-stimulus ERP (Figs. $1C_{S.S._1}$ and 2C). Peak amplitudes of P300 for 15 series in group 2 averaged 15.7 $\mu V \pm 4.3$ for vertex ($C_z$) and 13.0 $\mu V \pm 4.1$ for $C_c$ recordings. For 10 series in group 1 they averaged 15.2 $\mu V \pm 6.2$ in $C_c$ recordings (with 4 of these in subject S.S. averaging 26.2 $\mu V$). The ERP recording was extended to 1.8 sec of post-stimulus time in one series for each of 3 subjects. In each of these, P300 was followed by somewhat rhythmic components, initiated by a negative wave (peak at +450 to +600 msec in the different subjects, G.L., S.B., B.D.) and then a relatively small positive wave (peak at +540 to +760 msec, respectively).

*Topographic distribution of RPs*

This issue was studied in only a limited manner, so that the relationships are only qualitatively indicative. Comparisons of different sites were made only when recorded simultaneously, in view of the variations in RPs at a given electrode site even in the same session (see above). (The DC recording system required continual re-setting of each monopolarly recorded channel before each trial, limiting our practicable number of simultaneously recorded channels to two.) Simultaneous recordings for $C_z$ (vertex) and $C_c$ (contralateral precentral) electrodes were routinely made in all series in the first 4 sessions with each subject in group 2, for a total of 8 such series with each subject; in group 1, mostly $C_c$ alone was recorded in these sessions. For each of the other simultaneously recorded comparisons of electrode sites one or two 40 trial series were conducted for each of the subjects; these comparisons were for $C_z$-$P_3$ (or $C_c$-$P_3$); $C_z$-$F_p$ (or $C_c$-$F_p$); and $C_c$-$C_i$.

*$C_z$-$C_c$ comparisons.* For self-initiated RPs, such comparisons are seen in Figs. $1_{S.S._2}$, 2A, 2B and $3A_{G.L.}$. For peak amplitudes in the −200 to 0 msec interval, ratios of $C_z/C_c$ *for type I RPs* averaged approximately 1.1 for 4 series with subject S.B., 1.7 for 4 with G.L., 1.0 for 2 with S.S.; grand *average for these 10 $C_z/C_c$ ratios is 1.3.* ($C_z/C_3$ was 0.9 for 1 series with BL, $C_3$ referring to the 10-20 system.) *For type II RPs,* mean ratios of $C_z/C_c$ were 1.5 for 2 series with S.B., 2.3 for 4 with G.L., 1.3 for 2 with S.S., and 2.0 for 3 with B.D.; grand *average for these 11 $C_z/C_c$ ratios is 1.9.* (Data for type III RPs is inadequate.) Thus, type II RPs exhibited a consistent and marked predominance at the vertex, while type I showed little if any. For the 4 available *pre-set RPs* (e.g., Figs. $1B_{B.L.}$ and $3B_{G.L.}$) ratios of $C_z/C_c$ were close to 1.0 with subjects S.B. and B.D., 1.5 with G.L., and (for $C_z/C_3$), 0.8 with B.L. Thus, pre-set RPs appear to resemble type I self-initiated RPs in the $C_z/C_c$ relationship.

*$C_z$-$P_3$ (or $C_c$-$P_3$) comparisons.* RPs of all types were distinctly smaller at the $P_3$ site (Fig. 4). For the 5 *self-initiated RPs*, ratios for 3 $C_z/P_3$ and 2 $C_c/P_3$ series were all not far from 2; these included 1 RP of type I, 3 of type II and 1 of type III. For the 5 *pre-set RPs*, 3 $C_z/P_3$ ratios were approximately 1.5 (for 2 RPs) and 1.0 (Fig. $4B_{G.L.}$); while 2 $C_c/P_3$ ratios were $>2$. As in the other electrode comparisons ($C_z$-$C_c$; $C_c$-$C_i$), the areas under the RPs often exhibited larger differences than did the peak amplitudes (e.g., Fig. $4A_{G.L.,B.D.,S.S.}$ and $4B_{S.B.,G.L.,S.S.}$).

*$C_c$-$C_i$ comparisons.* In contra- and ipsilateral precentral comparisons, the two available type II RPs exhibited a distinct asymmetry (Fig. $3A_{S.B.,G.L.}$). The pre-set RPs recorded during the same session exhibited either no asymmetry (Fig. $3B_{S.B.}$) or distinctly less asymmetry than the type II RPs (Fig. 3A vs. B, for G.L.). (No $C_c$-$C_i$ comparisons were available for type I RPs.)

**Discussion**

There would seem to be no doubt that an RP can precede a 'freely voluntary' (endogenous) act, one that is essentially free of the external constraints in the usual 'self-paced' studies, and is independently 'self-initiated' and even spontaneously capricious in origin. The additional question of specificity of RP component in the volitional process will be considered below.

*Types of self-initiated RPs in relation to components of self-paced RPs*

In contrast to the consistency reported for self-paced RPs (e.g., Deecke et al. 1969, 1976), self-initiated RPs could appear with different forms and durations in the same subject. Type I self-initiated RPs resemble, in their form, timing and distribution, the self-paced RPs as commonly described, especially the ramp-like component that precedes the −500 msec time (e.g., Shibasaki et al. 1980). Both of these RPs are maximal at the vertex but are almost equally large at lateral precentral sites (Deecke et al. 1969, 1976; Kutas and Donchin 1974). Type II RPs resemble in form and timing that component of self-paced RPs occupying the −500 to −90 msec interval, termed NS′ by Shibasaki et al. (1980) (see also Vaughan et al. 1968; Deecke et al. 1969; Kutas and Donchin 1980), although mean onset time is earlier for the main negative rise in type II. Both type II and NS′ exhibit asymmetry, for contra- and ipsi-lateral precentral sites but differ in relative amplitudes at the vertex. $C_z/C_c$ ratios averaged 1.9 for type II RPs, but, for NS′ in self-paced RPs reported ratios are 1.2 (Deecke et al. 1969) or close to 1.0 (Vaughan et al. 1968; Shibasaki et al. 1980). Perhaps a similar dominance at the vertex for the 'specific' process in NS′ is masked by an overlap with the initial ramp-like component that is only weakly present, if at all, in type II RPs. (The less commonly observed type III RP, in which little or no negative rise appeared until about −250 to −200 msec, appears to have no clear counterpart in self-paced RPs.)

While the bilateral asymmetry of type II RP suggests there is a contribution from contralateral precentral motor cortex, as had been suggested for NS′ (e.g., Kutas and Donchin 1980), the substantial dominance of type II RPs at the vertex suggests at least a major contribution from areas other than precentral cortex. Whether supplementary motor area is a source of these non-precentral contributions to vertex RPs, as suggested by Deecke and Kornhuber (1978), remains to be established by direct recordings in that area. Increases in local blood flow have now been observed in supplementary motor area in association with voluntary self-paced movements and with stimulation of median nerve (Foit et al. 1980). That some large later components of post-stimulus ERP responses can be generated by supplementary motor area had already been shown with direct subdural and subcortical leads in a human subject (Libet et al. 1975).

*RPs as indicators of preparation to move*

In the series with 'skin stimuli at unknown times' preparation to move was eliminated, but all other conditions, including those of changes in attentiveness, were as similar as possible to those in the self-initiated RP series. That the subjects performed attentively was further indicated by the comparative accuracy of their reports of the actual timings of stimuli and by the large P300 ERP in all of such series. Yet, the pre-stimulus periods exhibited essentially no slow potential shifts. In a partially related paradigm, in which interstimulus intervals (and sequence of the target stimulus modalities) were varied randomly so as to result in uncertainty of stimulus time, Desmedt and Debecker (1979) also found no evidence of any pre-stimulus slow potential shift. (The much larger P300s in our series might be due to the heightened attentiveness associated with our subjects having continuously to monitor the pre-event progression of time, and the relaxation of this attention level after making a cognitive decision about the stimulus; see Desmedt 1981.)

Clearly, therefore, processes associated with attention and cognition could not themselves account for the actual slow potentials in the self-initiated RPs. In the paradigm for pre-set RPs there is an expectancy as the monitored time indicator approaches the time to act; but further experiments with pre-set timing of skin stimuli, in which strong expectancy is also present, appear to rule out this factor in RPs (Libet, Wright and Gleason, unpublished; to be reported separately). The evidence thus indicates, by exclusion, that all RPs, whether associated with pre-planned or spontaneously voluntary acts, are generated by neuronal processes specifically involved in the preparation to perform a motor act. This would apply not merely to RP components apparently contributed by precentral motor cortex (as argued by Kutas and Donchin 1980), but also to the more diffusely distributed RP components which appear to be generated elsewhere.

*Volitional processes*

For an act to be described as voluntary, the conditions must allow the subject to have some independence in choosing what to do and/or when to act. The condition of intention and preparation to act without such 'free volition' is represented in this study by the 'pre-set motor' series. In these, the voluntary choice of when to act was eliminated, but preparation and intention to act had to remain strong. Yet, pre-set RPs resembled type I self-initiated RPs and the early ramp of self-paced RPs. On the other hand, self-initiated acts arising 'spontaneously,' with no experience of preparatory pre-planning or pre-intention to act, were associated with type II RPs, in which the early ramp potential characteristic of type I and pre-set RPs is lacking.

On the present evidence, therefore, we propose that voluntary acts may involve more than one general process. Process I is associated with either endogenous or externally cued development of a general planning or intention to act at some loosely defined time approaching in the near future; it can begin one or more seconds before the act, as indicated by the ramp-like type I RP and the related component in self-paced RPs. Process II is associated with the more specific urge or intention to act; it immediately precedes the act, with onset at about −575 msec or less as indicated in type II RPs or by the NS′ component of self-paced RPs. Process II may be regarded as the one more uniquely associated with a fully independent volitional act, as opposed to a pre-intentionality that is not necessarily endogenous. Perhaps the akinesia of some parkinsonian patients, who are unable to initiate a voluntary act unless cued by an external stimulus (Kornhuber 1978), represents a pathological deficiency in process II, while process I is retained; this would predict that the NS′ component would be deficient or absent in self-paced RPs of such patients.

Although processes I and/or II presumably lead to the development of the final 'motor potential' (Deecke et al. 1969, 1976; Shibasaki et al. 1980) and to the actual motor outflow to the motoneurons, this last progression may not be necessary or mandatory. It was not uncommon for subjects to feel an urge to move that was not consummated in an actual movement, as if that urge was 'vetoed,' and then to wait for a new urge that did lead to movement. One may propose that each such covert or unfulfilled urge to move should also be associated with an appropriate RP, devoid of the final motor potential, but testing for such RPs would require a novel experimental technique.

Self-paced RPs, as studied by others in man, are probably dominated by process I, with any contribution from process II being partially obscured but capable of producing the net NS′ component (as designated by Shibasaki et al. 1980). Studies of self-paced movements in monkeys have involved training the animal to wait some minimum interval of time (2–4 sec) before executing the learned movement that leads to a reward of fruit juice (Gemba et al. 1979; Hashimoto et al. 1980; Pieper et al. 1980). Although the animals were apparently allowed to perform the act at some irregular time after the minimum waiting interval, it seems probable they would act relatively promptly in order to speed the delivery of their reward; indeed, Pieper et al. (1980) noted that a well-trained animal paused only a fraction of a second longer than the imposed 3 or 4 sec waiting period before squeezing. Such conditions become in effect similar to those in a straightforward reaction-time experiment that includes a forewarned period of waiting, in which slow prepotentials have also been recorded (e.g., Rosen and Stamm 1969; Donchin et al. 1971; Rebert 1972); the volitional feature of endogenous timing, if present at all, would be more restricted even than that in human self-pacing studies and certainly remote from the unrestricted, spontaneous voluntary actions in the present study.

**Summary**

The nature of readiness-potentials (RPs) that may be associated with fully endogenous, 'freely' voluntary acts was investigated. Restriction on when to act were eliminated and instructions fostered 'spontaneity.' The 'self-initiated' RPs exhibited in these conditions were categorizable into two (possibly three) types, all of which could be exhibited by the same subject.

Type I had an early onset at about −1050 ± 175 msec and a long ramp-like form, resembling self-paced RPs. In type II the main negative shift began at about −575 ± 150 msec, and at about

−240 ± 50 msec in type III. Type II partially resembled the similarly timed NS′ component in self-paced RPs. For acts produced at known, pre-set times, in which freedom of choice was eliminated but planning to act was required, RPs resembled self-initiated type I RPs and self-paced RPs. All RPs were maximal at the vertex, especially type II even though it was also bilaterally asymmetrical. These distributions suggest that cortical areas other than area 4 and 6 contribute importantly, especially to type II.

All RPs, whether in self-initiated or pre-planned acts, appear related specifically to preparation for a motor action. When task-related skin stimuli replaced self-initiated movements, under similar conditions of attentiveness (and expectancy), there were either no or relatively small event-preceding-slow potential shifts. All post-stimulus P300 waves were very large.

Two volitional processes are postulated: process I is associated with development of pre-planning or preparation to act in the near future (seconds), whether voluntary choice is present (type I RPs) or absent (pre-set RPs); process II, with an onset at roughly 0.5 sec before the act, is associated more uniquely with voluntary choice and with the more specific as well as endogenous urge or intention to act; it can be present in the comparative absence of or in sequence and overlapping with process I.

**Résumé**

*Comparaison des potentiels de préparation motrice précédant un acte moteur volontaire soit spontané libre, soit préprogrammé*

La nature des potentiels de préparation motrice (PPM) pouvant être associés à des actes moteurs volontaires 'libres', entièrement 'endogènes' a été étudiée. Toutes restrictions concernant l'instant de l'action étaient supprimées et les instructions poussaient à la 'spontanéité'. Les PPM 'auto-initiés', enregistrés dans ces conditions ont pu être classés en 2 (ou éventuellement 3) catégories, un même sujet pouvant présenter tous les types.

Le type I débutait précocément, à −1050 ± 175 msec et avait une forme de longue rampe, ressemblant ainsi à un PPM autocontrôlé. La principale composante négative du type II débutait à environ −575 ± 150 msec, et celle du type III à −240 ± 50 msec. Le type II ressemblait en partie à la composante négative du PPM autocontrôlé à décours temporel similaire. Pour des mouvements effectués à des instants prédéterminés et connus, sans liberté de choix, et exigeant une programmation du mouvement, le PPM ressemblait aux PPM auto-initiés de type I et aux PPM autocontrôlés. Tous les PPM étaient maximaux au vertex, particulièrement ceux de type II bien qu'ils fussent bilatéralement asymétriques. Ces distributions suggèrent que des aires corticales autres que 4 et 6 contribuent de manière importante aux PPM surtout celui de type II.

Tous les PPM, qu'ils soient auto-initiés ou préprogrammés apparaissent spécifiquement liés à la préparation de l'acte moteur. Lorsque des stimulus cutanés liés à la tâche remplaçaient le mouvement auto-initié dans des conditions similaires d'attention (et d'attente), les potentiels lents précédant l'événement étaient absents ou relativement petits. Toutes les ondes P300 post stimulus étaient très amples.

Deux processus volitionnels sont postulés: le processus I est associé au développement de la préprogrammation ou de la préparation de l'action dans un proche avenir (secondes), que le choix volontaire existe (PPM de type I) ou non (PPM préprogrammés); le processus II avec un début à environ 0,5 sec avant le mouvement, est associé plus électivement à un choix volontaire et avec une intention plus spécifique et plus endogène de l'acte. Ce processus II peut soit exister en absence du processus I soit lui faire suite et interférer avec lui.

We thank Dennis K. Pearl (University of California, Berkeley) for statistical assistance, and Drs. Enoch Callaway and Charles Yingling (Department of Psychiatry, University of California, San Francisco) for their helpful reading of the manuscript.

## References

Deecke, L. and Kornhuber, H.H. An electrical sign of participation of the mesial 'supplementary' motor cortex in human voluntary finger movement. Brain Res., 1978, 159: 473–476.

Deecke, L., Scheid, P. and Kornhuber, H.H. Distribution of readiness potential, pre-motion positivity, and motor potential of the human cerebral cortex preceding voluntary finger movements. Exp. Brain Res., 1969, 7: 158–168.

Deecke, L., Grözinger, B. and Kornhuber, H.H. Voluntary finger movement in man: cerebral potentials and theory. Biol. Cybernet., 1976, 23: 99–119.

Desmedt, J.E. Scalp-recorded cerebral event-related potentials in man as point of entry into the analysis of cognitive processing. In: F.O. Schmitt, F.G. Worden, G. Adelman and S.D. Dennis (Eds.), The Organization of the Cerebral Cortex. M.I.T. Press, Cambridge, Mass. 1981: 441–473.

Desmedt, J.E. and Debecker, J. Wave form and neural mechanism of the decision P350 elicited without pre-stimulus CNV or readiness potential in random sequences of near-threshold auditory clicks and finger stimuli. Electroenceph. clin. Neurophysiol., 1979, 47: 648–670.

Donchin, E., Otto, D., Gerbrandt, L.K. and Pribram, K.H. While a monkey waits; electrocortical events recorded during the foreperiod of a reaction time study. Electroenceph. clin. Neurophysiol., 1971, 31: 115–127.

Foit, A., Larsen, B., Hattori, S., Skinhøj, E. and Lassen, N.A. Cortical activation during somatosensory stimulation and voluntary movement in man: a regional cerebral blood flow study. Electroenceph. clin. Neurophysiol., 1980, 50: 426–436.

Gemba, H., Hashimoto, S. and Sasaki, K. Slow potentials preceding self-paced hand movements in the parietal cortex of monkeys. Neurosci. Lett., 1979, 15: 87–92.

Gilden, L., Vaughan, Jr., H.G. and Costa, L.D. Summated human EEG potentials with voluntary movement. Electroenceph. clin. Neurophysiol., 1966, 20: 433–438.

Hashimoto, S., Gemba, H. and Sasaki, K. Premovement slow cortical potentials and required muscle force in self-paced hand movements in the monkey. Brain Res., 1980, 197: 415–423.

Kornhuber, H.H. Cortex, basal ganglia and cerebellum in motor control. In: W.A. Cobb and H. Van Duijn (Eds.), Contemporary Clinical Neurophysiology, Electroenceph. clin. Neurophysiol., Suppl. No. 34. Elsevier, Amsterdam, 1978: 449–455.

Kornhuber, H.H. und Deecke, L. Hirnpotentialänderungen bei Willkürbewegungen und passiven Bewegungen des Menschen: Bereitschaftspotential und reafferente Potentiale. Pflügers Arch. ges. Physiol., 1965, 284: 1–17.

Kutas, M. and Donchin, E. Studies of squeezing: handedness, responding hand, response force, and asymmetry of readiness potential. Science, 1974, 186: 545–548.

Kutas, M. and Donchin, E. Preparation to respond as manifested by movement-related brain potentials. Brain Res., 1980, 202: 95–115.

Libet, B., Alberts, W.W., Wright, E.W. and Feinstein, B. Responses of human somatosensory cortex to stimuli below threshold for conscious sensation. Science, 1967, 158: 1597–1600.

Libet, B., Alberts, W.W., Wright, E.W., Lewis, M. and Feinstein, B. Cortical representation of evoked potentials relative to conscious sensory responses and of somatosensory qualities — in man. In: H.H. Kornhuber (Ed.), The Somatosensory System. George Thieme, Stuttgart, 1975: 291–308.

McAdam, D.W. and Seales, D.M. Bereitschaftspotential enhancement with increased level of motivation. Electroenceph. clin. Neurophysiol., 1969, 27: 73–75.

Pieper, C.F., Goldring, S., Jenny, A.B. and McMahon, J.P. Comparative study of cerebral cortical potentials associated with voluntary movements in monkey and man. Electroenceph. clin. Neurophysiol., 1980, 48: 266–292.

Rebert, C.S. Cortical and subcortical slow potentials in the monkey's brain during a preparatory interval. Electroenceph. clin. Neurophysiol., 1972, 33: 389–402.

Rosen, S.C. and Stamm, J.S. Cortical steady potential shifts during delayed response performance by monkeys. Electroenceph. clin. Neurophysiol., 1969, 27: 684–685.

Sanquist, T.F., Beatty, J.T. and Lindsley, D.B. Slow potential shifts of human brain during forewarned reaction. Electroenceph. clin. Neurophysiol., 1981, 51: 639–649.

Shibasaki, H., Barrett, G., Halliday, E. and Halliday, A.M. Components of the movement-related cortical potential and their scalp topography. Electroenceph. clin. Neurophysiol., 1980, 49: 213–226.

Vaughan, Jr., H.G., Costa, L.D. and Ritter, W. Topography of the human motor potential. Electroenceph. clin. Neurophysiol., 1968, 25: 1–10.

Walter, W.G., Cooper, R., Aldridge, V.J., McCallum, W.C. and Winter, A.L. Contingent negative variation: an electrical sign of sensorimotor association and expectancy in the human brain. Nature (Lond.), 1964, 203: 380–384.

*Electroencephalography and clinical Neurophysiology*, 1983, 56: 367–372
Elsevier Scientific Publishers Ireland, Ltd.


# PREPARATION- OR INTENTION-TO-ACT, IN RELATION TO PRE-EVENT POTENTIALS RECORDED AT THE VERTEX [1]

BENJAMIN LIBET *, ELWOOD W. WRIGHT, Jr. ** and CURTIS A. GLEASON **

* *Department of Physiology, University of California, San Francisco, Calif. 94143, and* ** *Neurological Institute, Department of Neuroscience, Mount Zion Hospital and Medical Center, San Francisco, Calif. 94115 (U.S.A.)*

(Accepted for publication: April 28, 1983)

A slow negative shift in potential, recordable extracranially, appears during the second or so preceding a signal to which a subject is supposed to respond. The 'contingent negative variation' (CNV) was the first discovered example of this (Walter et al. 1964). Similar pre-event potentials were found in a variety of delayed reaction paradigms, in both man (Low et al. 1966a; Donchin et al. 1971, 1972; Kutas and Donchin 1980; Rohrbaugh et al. 1980; Sanquist et al. 1981; Libet et al. 1982) and monkeys (Low et al. 1966b; Rebert 1972; Rosen and Stamm 1972). Several classes of processes may be reflected in these pre-event potentials: (1) a non-motor process; suggestions for this have included expectancy or anticipation (Walter et al., 1964), or an orienting or possibly more 'general response to salient or novel stimuli' (see Rohrbaugh and Gaillard, 1983) which may involve a shift in attentive or arousal state (Desmedt 1981) and leads to some behavioral/mental response (Donchin et al. 1971, 1972); (2) a general cerebral motor process, as represented in the readinesspotential (RP), that leads to the response (e.g., Rohrbaugh et al. 1980; Sanquist et al. 1981); (3) a combination of (1) and (2) (see Kutas and Donchin 1980; Rohrbaugh and Gaillard 1983); (4) a conative process of preparing or intending to act, independent of any actual movement (e.g., Low et al. 1966a; Libet et al. 1982). The present work helps to distinguish more definitively among these alternatives for the case of vertex-recorded potentials. The same human subject was tested with closely related procedures all of which developed mental sets of strong expectancy, orientation and attentiveness but differed in whether motor or non-motor responses were required. The study utilizes the subjects' introspective reports, in conjunction with the behavioral tasks, to define the mental sets in question.

## Methods

Recording and other procedures are fully described in Libet et al. (1982); only a reduced description of features essential to understanding the present text is presented. The EEG was recorded monopolarly at the vertex, referred to linked mastoid electrodes, utilizing a DC system. Storage of each 2 sec period, for averaging of 40 trials, began 1.4 sec before a trigger signal for 'zero-time.' This trigger was either the electromyogram (EMG), recorded over the appropriate forearm muscle in the case of motor trials, or the stimulator synch pulse in sensory non-motor trials. In each 2–3 h session at least one motor and one sensory series of 40 trials were studied, after some training runs. Each trial began with a warning tone at which the subject was to relax, fix his gaze on a point on the 5 inch screen of a cathode ray oscilloscope (CRO) about 2 m away and not blink until after the trial. (Absence of eye movements and blinking was monitored visually. The identical procedure had been found to obviate any significant contribution from ocular potentials with EOG recordings; see Libet et al. 1982. In any case, ocular functions were the same for all series and

[1] This work was supported in part by the Research Support Program of the Mount Zion Hospital and Medical Center, San Francisco.

could not account for systematic differences among recorded vertex potentials.) Within 1–2 sec the operator initiated continuous circular clockwise revolutions of the CRO spot of light, each starting at '12 o'clock' and completed in 2.56 sec, rather than in the conventional 60 sec.

For 'pre-set motor series' (M), the subject was to observe the spot reaching a pre-arranged time at which he would suddenly and sharply flex the fingers of the right hand (see Libet et al. 1982), similar to the acts used in studying readiness potentials (RP). For each successive block of 10 trials, the pre-set 'clock-time' to act was '70 sec,' '80 sec,' '100 sec' and '50 sec.' Subjects were encouraged 'to hit the moving spot as it crossed the pre-set time position.' Their EMG trigger times were recorded and, in most trials for all subjects, these were within 1-2 'sec' (actually about 50–100 msec) of the pre-set time. For both 'M-veto' and 'pre-set sensory' (S) series, procedures and pre-set times were the same as in M series, except that no motor act was produced by the subject. In M-veto series subject vetoed his intention to act just prior to the pre-set time for action (see fuller description in Results), while in S series delivery of a task-related stimulus to the back of the right hand at the pre-set time replaced the requirement to move. The stimulus was a single pulse at about 10–20% above threshold; it was weak enough to be missed unless the subject was properly attentive to it at the pre-set time of delivery, but strong enough to be perceived without equivocation when attending to it. In a small percentage of trials (5–8 of the 40) selected at random, the stimulus was omitted at the pre-set time. Subjects' task was to recognize any such omissions and, when later asked, to report their totals. All subjects were close to 100% accurate in the stimulus task; a relatively large P300 potential (Figs. 1 and 2) further attested to the suitability of attentive state and response.

The 5 subjects (2 males, 3 females) were students in their 20's. Four subjects (BD, SS, SB, GL) were familiar with the general procedures from related previous studies (see Libet et al. 1982, 1983), but none were aware of the experimental design or hypotheses in the earlier or present studies. BD and SS were studied in 3 sessions each, SB in 2, GL and AH in 1 each.

## Results

*Pre-set motor (M) series*

All subjects exhibited a ramp-like slow negative shift preceding EMG-0 time by > 1.4 sec (M in Figs. 1 and 2). Such pre-event potentials resemble those reported when forewarned time to act is signaled by a discrete stimulus (e.g., Low et al. 1966a,b; Donchin et al. 1971, 1972; Kutas and Donchin 1980; Rohrbaugh et al. 1980; Sanquist et al. 1981), instead of by the monitoring of 'clock-time' by our subjects. They also resemble the RPs that precede either self-paced movements (e.g., Kornhuber and Deecke 1965; Vaughan et al. 1968; Deecke et al. 1976) or those freely self-initiated movements that involve mental pre-planning (Libet et al. 1982).

*M-veto series*

For each of the 40 events in these series, the subject was instructed (a) to adopt the same mental set as in M series, of preparing to move at the

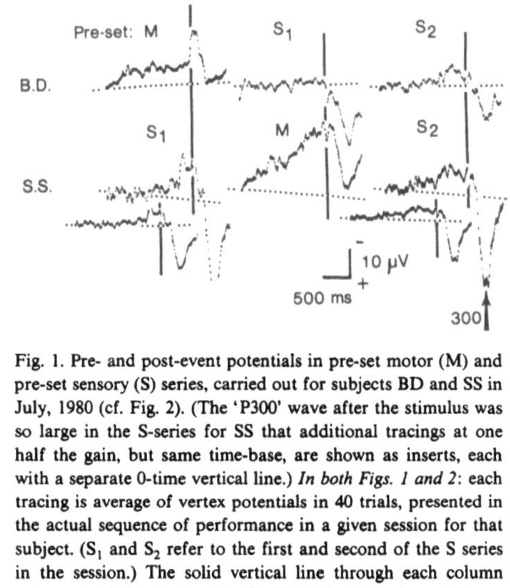


Fig. 1. Pre- and post-event potentials in pre-set motor (M) and pre-set sensory (S) series, carried out for subjects BD and SS in July, 1980 (cf. Fig. 2). (The 'P300' wave after the stimulus was so large in the S-series for SS that additional tracings at one half the gain, but same time-base, are shown as inserts, each with a separate 0-time vertical line.) *In both Figs. 1 and 2*: each tracing is average of vertex potentials in 40 trials, presented in the actual sequence of performance in a given session for that subject. ($S_1$ and $S_2$ refer to the first and second of the S series in the session.) The solid vertical line through each column represents '0-time,' given by the EMG for M series and by an externally supplied trigger pulse in M-veto and S series; '0-time' coincided with pre-set time for each event (see text). The dashed horizontal line in each tracing is estimated DC baseline.

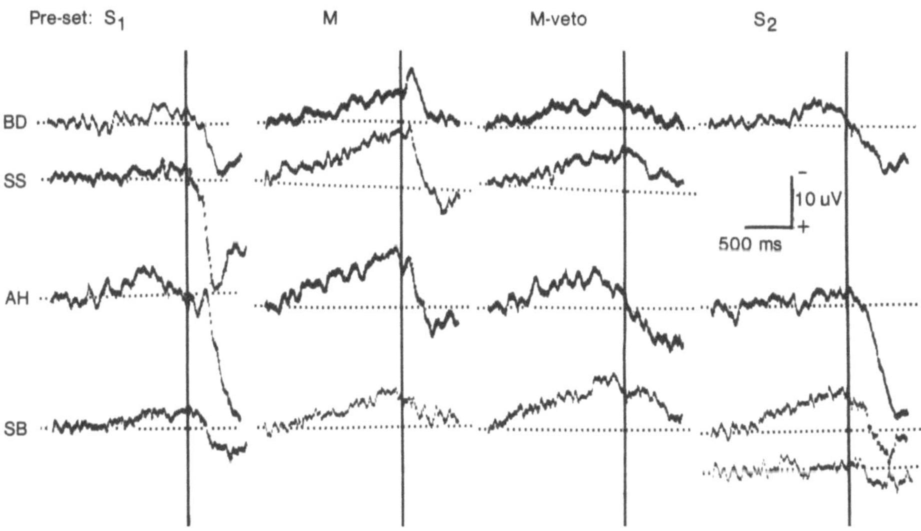


Fig. 2. Pre-event vertex potentials in S, M and M-veto series (see text), in 1 session for each of 4 subjects (in spring of 1982). Subjects BD and SS were same as in Fig. 1. In M-veto series there were no actual motor acts (as in S series), but there was preparation/intention to act before each pre-set time (as in M series). For subject SB, the tracing below $S_2$ represents a final ($S_3$) series of pre-set skin stimuli for which the subject was told to ignore the previous instruction to attend to any omitted stimuli, as there would be no omissions of stimuli. (The relatively flat tracing indicates that potential shift in his previous ($S_2$) series was related to something in his psychological set.)

designated pre-set time, but (b) 'to veto' this intention when the revolving CRO spot arrived within about '2.5–5 sec' (actually about 100–200 msec) before pre-set time. The absence of any observable motor activation was confirmed by monitoring the EMG at sufficiently high gain. The computer trigger for pre-set 0-times, in the absence of an EMG, was supplied by an operator in another room. In spite of the strangeness of this request, every subject reported experiencing varying degrees of preparation and intention before vetoing the action. A few subjects reported that the urge to move extended to some additional parts, like the left hand or a leg, but no actual movements of these were observed or experienced. Each subject in fact did produce a ramp-like pre-potential in one or more such series (Fig. 2, M-veto). Amplitude was usually though not always (SB, Fig. 2) smaller than for M in the same subject. Also, while 'pre-set RPs' in M series typically rise until close to 0-time (see also Libet et al. 1982), in M-veto series they tend to reverse direction within some 150–250 msec before 0-time; this suggests that the negative rise terminates at about the presumed time for reversing the intention to act.

*Pre-set stimulus (S) series*

Each S series could be directly compared with an M series performed in succession by the same subject in each session. For making task-related reports, the subject was instructed simply to remember the occurrences of stimulus omissions, but not to think about making a motor report about them until requested by the observer; this was to minimize the possibility that any preparation to move might begin before pre-set 0-time and thus induce its own pre-event potential, as in the M and M-veto series. The subjects reported an experience of moderate to strong expectancy or anticipation as the revolving CRO spot approached the pre-set time for the stimulus. In each session they were asked to compare the subjective intensities of their expectancies for events in S vs. M series after these were performed in sequence. They reported that their expectancies in S events were roughly comparable to those in their M series. That is, the S expectancies were reported to be equal to, somewhat greater or somewhat less than M expectancies, in roughly similar proportions, and even by the same subject in different sessions.

Pre-event potentials in S series ranged from being almost 0 (i.e., flat DC, Fig. 1, BD-$S_1$; Fig. 2, SS, AH-$S_2$) to exhibiting some negative shifts that were much smaller and briefer than and lacking the ramp-like form of those in M series (Fig. 1, BD-$S_2$, SS-$S_1$ and $S_2$; Fig. 2, BD-$S_2$, AH-$S_1$, SB-$S_1$; the one exception was exhibited by subject SB for $S_2$). Such variations between 0 and some small negative shift could be exhibited by each subject, in different S series even within the same session. The presence or absence of small pre-event potentials in different S series did not correlate with the reported intensities of expectancy. In some S series with essentially 0 pre-event potential, experiences of expectancy were reported to be as great or greater than in M series.

Comparing all 3 types of series, a prominent and consistent pre-event slow negative shift appeared only in M or M-veto series, but was absent or relatively small and brief in S series. This relationship appears definitive, even though statistically quantitative values cannot be derived. In addition to the striking qualitative differences between S vs. M or M-veto series in the present total of 10 sessions on 5 subjects, the nature and range of pre-event potentials under related conditions, in many previous sessions with 4 of these subjects (see Libet et al. 1982), helped establish the reliability of the presently reported ones.

## Discussion

The results lead to the conclusion that those non-motor processes common to all present series (M, M-veto, and S) appeared to be not sufficient to produce the vertex-recorded pre-event potential. The cerebral processes additionally present in M and M-veto series, those for preparation or intention to perform a motor act even if not consummated in a movement, appeared to provide the factor necessary for the appearance of most if not all of the pre-event potential recorded under our conditions.

The present findings are in accord with those in studies of CNV utilizing intervals of 3–4 sec rather than 1–1.5 sec, between the warning ($S_1$) and the imperative ($S_2$) stimuli. The late or terminal phase of such CNVs resembles RPs, in topographical distribution and form (see Rohrbaugh and Gaillard 1983). The early 'O' wave in such CNVs is largest at the midline frontal ($F_z$) electrode, a site not studied here. Also, the early phase appears to terminate within a second or two after the warning stimulus. This could account for its apparent absence during the averaged 1.4 sec intervals preceding our task-related S event; any general type of orienting potential in our paradigm could have appeared some seconds earlier, when the revolution of the CRO spot was begun. Our evidence does not exclude the possibility that a non-motor 'O' type component may contribute to a vertex-recorded CNV when the interval between $S_1$ and $S_2$ is only 1–1.5 sec, but it argues against even such CNVs consisting solely of such a non-motor process. Recent evidence (Libet et al. 1982) supports a related view that the more completely endogenous RPs in self-initiated 'spontaneous' voluntary movements represent preparation/intention for moving, not some non-motor process.

An absence of pre-event potential when task-related stimuli were delivered, at regular intervals known to the subject, had already been reported by Desmedt and Debecker (1979); but their subjects did not continuously monitor approaching time for the stimulus and the issue of expectancy/anticipation was not raised by that study. The pre-set S event in our study produced either no pre-event SP or a small brief one. Perhaps even the latter would be absent if it were possible to eliminate any tendency, conscious or unconscious, for the subject to prepare or intend to make a motor response even though the latter was not requested. This is in part supported by the findings that subjects produced both an essentially zero pre-event potential and a small, brief pre-event

SP in different pre-set S series performed in the same experimental session, without correlative changes in expectancy or attention. On such a view, any small SPs in S series might represent 'covert RPs' related to those in M-veto series. It also suggests that in evaluating reports that some CNV can appear even when an immediate motor response is not requested (see Donchin et al. 1971, 1972; but compare Desmedt and Debecker 1979), it is clearly important to consider possible RP contributions from covert preparation or intention to produce an unrealized motor response.

Of especial interest is our finding that a developing preparation or intention to move can clearly be accompanied by a substantial 'RP' even when the subject knows he will eventually veto the intention to act, and when in fact he does not activate the muscle. Such a 'covert RP' might be a general feature of non-consummated urges or intentions to act (Libet et al. 1982). Mental performance of imaginary movements has been found to be associated with an increase of blood flow in mesial supplementary motor cortex (Lassen et al. 1978). Such evidence could not establish whether the increased neural actions coincided with the preparational phase just before each imagined act, as could be done for the present covert RPs preceding vetoed actions. But it is in accord with other evidence that supplementary motor area may be a generator of vertex-recorded RPs (Deecke and Kornhuber 1978; Eccles 1982; Libet et al. 1982).

## Summary

Pre-event potentials were compared in the same subject, for 3 types of forewarned events, in which the foreperiod for orienting or attention began several seconds before the event. All of these trials involved similar non-motor components (expectancy, attentiveness, general orienting to a salient stimulus) but differed in whether motor or non-motor responses were required. A prominent and consistent slow negative shift preceded the pre-set time for a motor response, even when the subject 'vetoed' his intention to act shortly before the pre-set time. Pre-event potentials were absent, or small and brief, when the event was a task-related skin stimulus not involving preparation to move. The findings selectively support the view that mental preparation/intention to act is a necessary and perhaps dominant process associated with the vertex-recorded pre-event, slow negative potential. They also show that such a pre-event potential can appear even when the subject knows he is going to veto his developing intention to act and does not actually move.

## Résumé

*Mise en relation de la préparation ou de l'intention d'effectuer un acte avec les potentiels précédant l'événement recueillis au vertex*

Les potentiels précédant l'événement ont été comparés chez le même sujet pour 3 types d'événements avec avertissement préalable, dans lesquels cette période préalable d'orientation ou d'attention commençait plusieurs secondes avant l'événement. Tous ces essais comportaient les mêmes composantes non motrices (attente, attention, orientation générale à un stimulus brusque) mais différaient en ce qu'une réponse motrice était ou non, demandée. Une variation négative lente, importante et stable, apparaissait avant le moment fixé pour la réponse motrice, même lorsque le sujet décidait de ne pas agir peu de temps avant ce moment. Les potentiels précédant l'événement étaient absents, ou petits et brefs lorsque l'événement était un stimulus tactile lié à la tâche, n'impliquant pas de préparation au mouvement. Ces données confirment l'idée selon laquelle la préparation mentale d'un acte ou l'intention d'agir est un processus nécessaire et probablement dominant associé au potentiel négatif lent précédant l'événement qui est enregistré au vertex. Elles montrent également qu'un potentiel négatif pré-événement peut apparaître même quand le sujet sait qu'il s'interdira l'intention d'agir et qu'il ne développera pas de mouvement.

We thank Drs. Enoch Callaway and Alan Gevins for their helpful criticism of the manuscript.

## References


Deecke, L. and Kornhuber, H.H. An electrical sign of participation of the mesial "supplementary" motor cortex in human voluntary finger movement. Brain Res., 1978, 159: 473–476.

Deecke, L., Grözinger, B. and Kornhuber, H.H. Voluntary finger movement in man: cerebral potentials and theory. Biol. Cybernet., 1976, 23: 99–119.

Desmedt, J.E. Scalp-recorded cerebral event-related potentials in man as point of entry into the analysis of cognitive processing. In: F.O. Schmitt, F.G. Worden, G. Adelman and S.D. Dennis (Eds.), The Organization of the Cerebral Cortex. M.I.T. Press, Cambridge, Mass., 1981: 441–473.

Desmedt, J.E. and Debecker, J. Slow potential shifts and decision P350 interactions in tasks with random sequences of near threshold clicks and finger stimuli delivered at regular intervals. Electroenceph. clin. Neurophysiol., 1979, 47: 671–679.

Donchin, E., Otto, D., Gerbrandt, L.K. and Pribram, K.H. While a monkey waits; electrocortical events recorded during the foreperiod of a reaction time study. Electroenceph. clin. Neurophysiol., 1971, 31: 115–127.

Donchin, E., Gerbrandt, L.K., Leifer, L. and Tucker, L. Is the contingent negative variation contingent on a motor response? Psychophysiology, 1972, 9: 178–188.

Eccles, J.C. The initiation of voluntary movements by the supplementary motor area. Arch. Psychiat. Nervenkr., 1982, 231: 423–441.

Kornhuber, H.H. and Deecke, L. Hirnpotentialänderungen bei Willkürbewegungen und passiven Bewegungen des Menschen: Bereitschaftspotential und reafferente Potentiale. Pflügers Arch. ges. Physiol., 1965, 284: 1–17.

Kutas, M. and Donchin, E. Preparation to respond as manifested by movement-related brain potentials. Brain Res., 1980, 202: 95–115.

Lassen, N.A., Ingvar, D.H. and Skonhøj, E. Brain function and blood flow. Scient. Amer., 1978, 239: 62–71.

Libet, B., Wright, Jr., E.W. and Gleason, C.A. Readiness potentials preceding unrestricted "spontaneous" vs. preplanned voluntary acts. Electroenceph. clin. Neurophysiol., 1982, 54: 322–335.

Libet, B., Gleason, C.A., Wright, E.W. and Pearl, D.K. Time of conscious intention to act in relation to onset of cerebral activities (readiness-potential); the unconscious initiation of a freely voluntary act. Brain, 1983, 106: 623–642.

Low, M.D., Borda, R.P. and Kellaway, P. "Contingent negative variation" in rhesus monkeys: an EEG sign of a specific mental process. Percept. Motor Skills, 1966a, 22: 443–446.

Low, M.D., Borda, R.P., Frost, Jr. J.D. and Kellaway, P. Surface-negative, slow-potential shift associated with conditioning in man. Neurology (Minneap.), 1966b, 16: 771–781.

Rebert, C.S. Cortical and subcortical slow potentials in the monkey's brain during a preparatory interval. Electroenceph. clin. Neurophysiol., 1972, 33: 389–402.

Rohrbaugh, J.W. and Gaillard, A.W.K. Sensory and motor aspects of the contingent negative variation. In: A.W.K. Gaillard and W. Ritter (Eds.), Tutorials in ERP Research: Endogenous Components. North-Holland Publishing Company, Amsterdam, 1983: 269–310.

Rohrbaugh, J.W., Syndulko, K., Sanquist, T.F. and Lindsley, D.B. Synthesis of the contingent negative variation brain potential from noncontingent stimulus and motor elements. Science, 1980, 208: 1165–1168.

Rosen, S.C. and Stamm, J.S. Cortical steady potential shifts during delayed response performance by monkeys. Electroenceph. clin. Neurophysiol., 1972, 33: 389–402.

Sanquist, T.F., Beatty, J.T. and Lindsley, D.B. Slow potential shifts of human brain during forewarned reaction. Electroenceph. clin. Neurophysiol., 1981, 51: 639–649.

Vaughan, Jr., H.G., Costa, L.D. and Ritter, W. Topography of the human motor potential. Electroenceph. clin. Neurophysiol., 1968, 25: 1–10.

Walter, W.G., Cooper, R., Aldridge, V.J., McCallum, W.C. and Winter, A.L. Contingent negative variation: an electrical sign of sensorimotor association and expectancy in the human brain. Nature (Lond.), 1964, 203: 380–384.

*Brain* (1983), **106**, 623–642

# TIME OF CONSCIOUS INTENTION TO ACT IN RELATION TO ONSET OF CEREBRAL ACTIVITY (READINESS-POTENTIAL)

## THE UNCONSCIOUS INITIATION OF A FREELY VOLUNTARY ACT

*by* BENJAMIN LIBET, CURTIS A. GLEASON, ELWOOD W. WRIGHT *and* DENNIS K. PEARL[1]

*(From the Neurological Institute, Department of Neuroscience, Mount Zion Hospital and Medical Center, the Department of Physiology, School of Medicine, University of California, San Francisco, CA 94143 and the Department of Statistics, University of California, Berkeley, CA)*

SUMMARY

The recordable cerebral activity (readiness-potential, RP) that precedes a freely voluntary, fully endogenous motor act was directly compared with the reportable time (W) for appearance of the subjective experience of 'wanting' or intending to act. The onset of cerebral activity clearly preceded by at least several hundred milliseconds the reported time of conscious intention to act. This relationship held even for those series (with 'type II' RPs) in which subjects reported that all of the 40 self-initiated movements in the series appeared 'spontaneously' and capriciously.

Data were obtained in at least 6 different experimental sessions with each of 5 subjects. In series with type II RPs, onset of the main negative shift in each RP preceded the corresponding mean W value by an average of about 350 ms, and by a minimum of about 150 ms. In series with type I RPs, in which an experience of preplanning occurred in some of the 40 self-initiated acts, onset of RP preceded W by an average of about 800 ms (or by 500 ms, taking onset of RP at 90 per cent of its area).

Reports of W time depended upon the subject's recall of the spatial 'clock-position' of a revolving spot at the time of his initial awareness of wanting or intending to move. Two different modes of recall produced similar values. Subjects distinguished awareness of wanting to move (W) from awareness of actually moving (M). W times were consistently and substantially negative to, in advance of, mean times reported for M and also those for S, the sensation elicited by a task-related skin stimulus delivered at irregular times that were unknown to the subject.

It is concluded that cerebral initiation of a spontaneous, freely voluntary act can begin unconsciously, that is, before there is any (at least recallable) subjective awareness that a 'decision' to act has already been initiated cerebrally. This introduces certain constraints on the potentiality for conscious initiation and control of voluntary acts.

[1] Present address: Department of Statistics, Ohio State University, Columbus, Ohio.
Reprint requests to Dr B. Libet, Department of Physiology, University of California, San Francisco, CA 94143, USA.

## INTRODUCTION

The 'readiness-potential' (RP), a scalp-recorded slow negative potential shift that begins up to a second or more before a self-paced act (Kornhuber and Deecke, 1965; Gilden *et al.*, 1966), can also precede self-initiated 'freely' voluntary acts which are not only fully endogenous but even spontaneously capricious in origin (Libet *et al.*, 1982). The appearance of preparatory cerebral processes at such surprisingly long times before a freely voluntary act raises the question of whether conscious awareness of the voluntary urge or intention to act also appears with such similar advance timings. The present study attempts to answer this question experimentally.

In the present study, the experience of the time of the first awareness of the urge to move was related by the subject to his observed 'clock-position' of a spot of light revolving in a circle; the subject subsequently recalled and reported this position of the spot. Thus the experience of timing of the awareness was converted to a reportable, visually related spatial image, analogous to reading and recalling the clock-time for any experience. (The reliability and validity of this operational criterion are further considered below.) This indicator of the time of the conscious experience could then be related (1) to the actual time of the voluntary motor act, as indicated by the electromyogram (EMG) recorded from the appropriate muscle, and (2) to the time of appearance of the simultaneously recorded RP that is generated by the brain in advance of each act. The voluntary motor acts under study were those produced with minimal or no restrictions on the subject's independent choice of when to act, and under instructions that encouraged spontaneity of each volitional urge to act (Libet *et al.*, 1982).

The present findings thus provide experimental evidence on the timing of the conscious intention to act relative to the onset of cerebral activity preparatory to the act, and on the roles of conscious processes in the initiation of a freely voluntary motor act.

## METHODS AND PROCEDURES

### *Subjects*

Six right-handed college students were studied as two separate groups of three each. Group 1 comprised 3 females (S.S., C.M. and M.B.), but the quality of the EEGs and the minimal amplitude of the RPs of one precluded using much of her data. Group 2 consisted of 2 males and 1 female (S.B., B.D. and G.L.). Study of this group began a few months after completing the study of Group 1.

### *Recording*

The d.c. recording and averaging of the EEG has been described (Libet *et al.*, 1982). For present purposes, analysis of RPs is made for those recorded at the vertex, where they were all maximal. (For the first 4 experimental sessions with Group 1, only the contralateral precentral recording site is available.) Linked mastoid electrodes served as the reference lead, with a ground electrode on the left ear lobe. Controls excluded the electro-oculogram as a source of the slow potentials. In each experimental series, 40 trials were performed and averaged by a computer of average transients (CAT 400B). The 2 s period of EEG stored by the CAT with each trial included a 1.4 s period already on the recording tape before 'O-time'. The latter was signified by the EMG, recorded with bipolar electrodes on the skin over the activated muscle of the right forearm.

*Procedure*

The subject sat in a partially reclining position on a lounge chair with an observer present in the room. Each trial was started only when the subject considered himself comfortably ready. The trial began with a brief 'get-ready' tone. This signalled that during the next 1–3 s the subject should relax his muscles, especially those of the head, neck and forearm, blink his eyelids if he wished, and fix his gaze on the centre of the 5 inch circular screen of a cathode ray oscilloscope (CRO) that was positioned at about 1.95 m away in his direct line of vision. At the end of these irregular get-ready periods the operator activated the PDP-12 computer to initiate circular revolution of the beam of the CRO. The CRO spot of light revolved in a clockwise circle near the circumference of the screen starting from the '12-o'clock' position; this motion simulated a sweep secondhand of a clock but each revolution was completed in 2.56 rather than 60 s. A circular scale, with numbers at each '5 s' position, was mounted at the external edge of the CRO screen, and a plastic grille on the peripheral portion of the screen displayed illuminated radial lines spaced at '2.5s.' intervals (each equal to 107 ms of actual time). Subjects were asked to maintain their gaze fixed on the centre of the CRO screen and not to follow the CRO spot around, even though they were to report information relating the 'clock-position' of the spot to the events (*see below*). The visual angle subtended between the centre and the peripheral position of the moving spot was small enough (1.8 deg) to present no difficulty from loss of visual acuity. The 'clock-time' of the CRO spot at each event, namely EMG with motor act or stimulator synch pulse with stimulus to skin, was recorded by the PDP-12 computer. Subjects were trained to make their self-initiated movement sufficiently brisk so that within no more than 10–20 ms from the start of any EMG potentials they achieved the amplitude pre-set to trigger the computer.

The subject was asked not to blink from the time the CRO spot started revolving until after the event. To minimize the possibility that the need to blink might become a controlling 'external' factor that compels or impels him to act, the subject was told that he may blink during the trial if the need arose; but that, if he did blink (or made some other extranaeous motion), he should then simply wait for the CRO spot to make at least another full revolution before performing the quick voluntary movement, as at the start of the trial.

Two different kinds of series were studied.

(1) *Self-initiated voluntary acts*. The subject was asked to wait for one complete revolution of the CRO spot and then, at any time thereafter when he felt like doing so, to perform the quick, abrupt flexion of the fingers and/or the wrist of his right hand (*see* Libet *et al.*, 1982). An additional instruction to encourage 'spontaneity' of the act was given routinely to subjects in Group 2 and only in the latter half to two-thirds of sessions with Group 1. For this, the subject was instructed 'to let the urge to act appear on its own at any time without any preplanning or concentration on when to act', that is, to try to be 'spontaneous' in deciding when to perform each act; this instruction was designed to elicit voluntary acts that were freely capricious in origin.

(2) *Skin-stimuli 'at unknown times'*. For such a series the subject expected to receive a near-threshold stimulus pulse on the back of the right hand. Delivery of the pulse was made by the operator at irregular times that were unknown to the subject, but only after the CRO spot completed its first revolution. They were actually delivered randomly during the second or third revolution of the spot (that is, between about 2.6 and 7.6 s after the spot began to revolve); this range overlapped with that for the times of the self-initiated movements. These conditions closely parelleled the attentive and other requirements associated with performing and recalling the CRO clock time for 'spontaneous' self-initiated voluntary acts (*see also* Libet *et al.*, 1982).

*Subjects' reports of the time of an event*. The 'clock position' of the revolving CRO spot at the time of the subject's awareness of an event was observed by the subject for later recall. Within a few seconds after the event, the subject was asked for his report of that timing, as in recalling a spatial image of ordinary clock time in conjunction with another event. It was emphasized that only an after-the-event recall of the experience was required, and that the subject should not worry about the task in advance of each event. Subjects became rapidly accustomed to this task during the training runs and did not find it to be taxing or stressful; nor did this task have any detectable effect on RPs (Libet *et al.*, 1982).

*Modes of recall.* Although each report depended on the subject having continously monitored the revolving CRO spot and visually noting, to himself, the position of the spot at the actual time of his awareness (of the event under study—*see below*), two different modes were employed for his after-the-event recall of that spot position. With the (A) or 'absolute' mode, the subject was asked to look back on the circular time scale mounted on the CRO and report the 'clock-time' of the spot position in 'seconds'. (Each 'second' on this scale corresponded to an actual time of 2560/60 or about 42.7 ms.) With the (O) or 'order' mode, the subject was asked to report the order of the final stopped position of the CRO spot, at the end of the trial, relative to his recalled position of the moving spot at the time of his awareness. For this, the subjects simply reported 'CRO spot (stop-position) first', at an earlier clock-position than the event-awareness; or 'awareness first', or 'together' (same position for both, insofar as the subject could discriminate). The (O) mode of recall was found by most subjects to be somewhat less demanding than the (A) mode.

The final stop position for the CRO spot following each event was arrived at in a complex manner that differed for (A) or (O) mode of recall. When either of the modes of recall was to be used, the computer continued the clockwise motion of the spot for a period beyond the time triggered by the event; this was called the 'contination interval'. The continuation interval could have one of 20 different values, all in the range between +500 and +800 ms (approximately 12 to 19 'seconds' of clock-dial). One of these continuation values was selected by the computer, from a randomized series of the 20, for use in a given trial. No violations of independence of answer were found in relation to the randomized continuation intervals.

For the (O)-order-mode only, however, the CRO spot did not stop after reaching its continuation interval; instead the spot jumped discontinuously, to stop at clock positions that were both before and after the subject's recalled positions for his awareness. (1) 'Stopping range'. The clock-times within which all of these final stop positions were included (the 'stopping range') ordinarily spanned 600 ms of real time. The positive and negative end points of the stopping range, relative to zero trigger time for each event (EMG or S-synch pulse), were chosen so as to span the entire range of times (relative to O) in the reports for a given awareness (W, M or S, *defined below*). We usually succeeded in setting the positive end of the stopping range well beyond the stop-times of the CRO spot for which all reports were 'W-first' (or M- or S-first), that is, earlier than the stopped spot position; and the negative end of the stopping range well beyond the stop-times for which all reports were 'spot first' (as in fig. 1B). Actual distributions varied with the subject. The beginning and end points of the 'stopping range' were therefore set individually before each series of trials, depending on previous results with the kind of awareness to be reported and with that subject (*see* examples in fig. 1); the training series of 10 trials that routinely preceded each regular series of 40 trials was useful for this purpose. (2) 'Stop-times'. Within each selected stopping range that spanned 600 ms, one of 40 different actual stopping times for the CRO spot, at intervals of 15 ms, was randomly selected by the computer for use after each of the 40 events. A different sequence of these randomized stop times could be preselected for the successive 40 trials in given series, so that a given sequence of stop-times was not repeated in a given session. The length of the 'continuation-interval', that precedes the final jump of the spot to its stop-time (*see above*), was randomly varied in a fashion that was independent of the randomized sequences selected for the final stop-times. The objective of all this was of course to avoid providing any clues that might relate the stop-time position of the CRO spot to the clock-time of the event itself.

The O procedure would not seem to be subject to the kind of artefactual difficulty described by Garner (1954). In the latter's case, subjects were asked to judge 'half-loudness' referred to a standard acoustic stimulus; each subject gave reliably consistent judgements, but these turned out to be appropriate only with reference to each different range of stimulus intensities presented rather than to the standard stimulus. In our case, the stopping range of reference times was determined for each subject from his own range of reported W times, as indicated by initial trials, rather than *vice versa*.

There were some series for which the adopted stopping range did not appropriately span the full range of potential reports by the subject; that is, the numbers of 'W first' and 'spot first' reports were far from equal, with few or no reports for one of these possible responses. This indicates that the subject was not shaping his reports to correspond to the adopted range of stop-times for the CRO spot. In such instances, the series had to be repeated, in the same or a later session, with a more suitable stopping range that could result in a statistically usable analysis for the mean W time in the series. Additionally, the subject had no prior knowledge of or consistent experience with the actual stopping ranges that were used. In a usual given experimental session, each separate series for W, M and S awareness times required a different stopping range. This meant that the subject did not have any consistent stopping range with which he was repeatedly presented in successive series.

*Type of 'awareness' to be timed by the subject.* Three different subjective experiences were to be timed. Each series of 40 trials involved reports of only one of these for all 40 events.

(W). In W series, the subject was asked to note and later report the time of appearance of his conscious *awareness of 'wanting' to perform* a given self-initiated movement. The experience was also described as an 'urge' or 'intention' or 'decision' to move, though subjects usually settled for the words 'wanting' or 'urge'. Subjects were told there were no right or wrong or inaccurate answers possible in the W and M series; that we simply wanted their best recollection and report of their own experience. Subjects were also asked, after completing a series of 40 self-initiated acts, whether any of the acts occurred without their experiencing any prior awareness of an urge or intention to act, that is, whether they were 'surprised' to realize a voluntary act had occurred without a conscious prior intention. In three series, subjects (3 of the 5 studied) did report that some, usually a small number, of the 40 acts 'surprised' them (Libet *et al.*, 1982), although the mean W time for such series of 40 was negative to EMG zero time, as usual. Asking subjects about 'surprise' acts should have indicated to them that it was acceptable even to have and to report the absence of a conscious urge or intention to act prior to a self-initiated act. The fact that instances of 'surprises' were reported increases confidence that the reports of timing prior to the act represented endogenous experiences not defined or induced by the instructions.

It was emphasized that the reported time should refer to the earliest appearance of the W awareness that might precede an actual movement; the subjects easily distinguished this specific urge to act from any awareness of a more general preplanning nature that might precede W, as in association with type I RPs (*see* Libet *et al.*, 1982), and also from an occasional urge to move that was not followed by an actual movement.

(M). In M series, the time of subject's *awareness* that he/she '*actually moved*', in the self-initiated act, was to be noted and reported after each trial. This was intended to provide an instruction for timing an experience related to but different from that of W, and thus to act as a partial test of the validity of the W timings. Some subjects stated, on their own, that their mental set differed somewhat in W *vs* M series. In a W series there was a feeling of active attention to or 'watching for' the awareness of wanting to move, so as to be able to note the time of its appearance, although the urge to move arose spontaneously with no preplanning; but M series proceeded without such advance 'watching' for the event to occur. It should be emphasized that any such difference in mental set did not appear to affect the associated RPs, which had similar forms and onset times for W and M series (*see* Results, *below*).

(S). In S series, time of *awareness of the sensation* elicited by the near-threshold stimulus pulse to the back of the hand, delivered at randomly irregular times unknown to the subject, was to be noted and reported after each trial. The attentive and other conditions of this task closely paralleled those for the W and M series, except that the event was an externally-induced sensory instead of a self-initiated motor one. After each S series, whether for training or experimental purposes, the subject was given a rough indication of how close he was to the actual times for the stimuli, as an aid in improving accuracy in all the experiments. If the awareness times reported in an S series were to differ significantly from the actual times of the stimulus pulse in each respective trial, the mean difference between the reported and actual stimulus times may be regarded as a measure of that subject's 'bias', when observing and reporting awareness times under the experimental conditions employed (*see* Libet *et al.*, 1979). This would include errors in making simultaneous judgements of the times for a mental event (sensation here) and a visual event (position of the CRO spot) (*see* Discussion). There were in fact mean differences or 'shifts' for S that were characteristic for each subject.

*Progression of the experiments.* The first (and, in some cases, the second) half-day session was purely for training purposes. Subsequently, each subject was studied in 6 to 8 regular half-day sessions, usually 1 per week. Each of the first 4 regular sessions began with a training series of 25 trials with skin stimuli intended as a retraining of reporting the recalled times of a subjective experience. This series differed from the experimental S series (of 40 trials) in that, after each 5 trials of this series, the subject was told what the actual 'clock-times' were for the skin stimuli, whose randomly irregular delivery times were not known before the trials. Following this there were two 40-trial series of self-initiated acts, one W and one M, each preceded by a briefer 10-trial series for retraining purposes, and then one 40-trial series of skin stimuli delivered at unknown times (S). The order of W and M series within a session was alternated for each successive session. A given mode for subjects to recall clock-positions of the revolving spot, (A) *vs* (O), was used for all series in a given session, but the modes were alternated for successive sessions. In sessions after the first 4, few M or S series were studied.

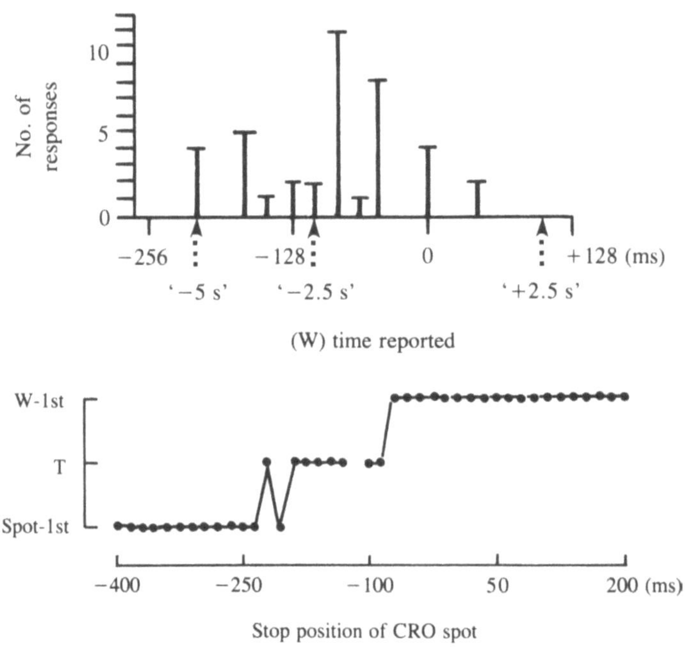


FIG. 1. Examples of plots of a subject's reports of (W) time after each trial in series of 40 self-initiated, voluntary movements. A, series in session 3 for subject S.B. mode of recall is 'absolute' (A); i.e. subject reported 'clock-time' of CRO spot at time of first awareness of 'wanting' to move. From this value, the 'clock-time' of the EMG-trigger was subtracted; the resultant net (W) time, relative to EMG zero time, was plotted. Abscissa indicates both the net 'clock-time' in seconds on the dial (arrows) and the corresponding net real times for (W). B, series in session 2 for subject S.B. mode of recall is by 'order' (O). The subject reported one of three alternatives for his recalled position of the revolving spot at the initial awareness of W. As indicated on the ordinate these were 'W-first' (W time earlier than the final stop-position of the CRO spot); or 'Spot-first'; or 'together', T (W time indistinguishable from or same as for the final stop-position of the spot). The 'stopping range', within which lay all the 40 different stop-positions of the spot, randomly sequenced in the series of 40 trials, was set between 200 ms positive to (i.e., later than) and 400 ms negative to the EMG zero time in each trial. (Since both ends of the 600 ms stopping range were included, the computer actually designated 41 stop-positions at 15 ms intervals, but the subject did not report W time for the 41st trial of the series. This accounts for an absence of a report at the stop-time of − 115 ms in the figure, which happened to be the randomly sequenced stop-time of the CRO-spot in the 41st trial of this series.)

*Statistical handling of the response times.* This is described in relation to the data in two actual series, one in (A) and one in (O) mode of recall (plotted in fig. 1). For each event in a series of 40 with (A) mode of recall, the 'clock-time' in 'seconds' for EMG-trigger time is subtracted from the 'clock-time' of the CRO spot at the time of awareness, as reported by the subject; this gives a net reported clock-time (relative to 'EMG zero time'). Each net reported clock-time is then converted to real time, and these W times (reported real time of each awareness of wanting to move, relative to EMG zero time) are plotted for that series. (Each 60 s of 'clock-time' = 2560 ms actual time.) For example, in fig. 1A we see that subject S.B. reported W times of 43 ms twice, O ms 4 times, −43 ms 8 times, etc. (as converted from net reported clock-times of 1, 0, −1 'seconds', respectively). Averaging these values for the whole series gives a mean shift for W, relative to EMG-O time, of −2.1 'seconds' of clock-time or −90 ms of real time.

For the (O) mode of recall we have extended to trinomial data the idea presented by Church and Cobb (1971). With this technique, the mean W shift was calculated as:

(upper, positive end of 'stopping range') − (time interval between 'stop times') × (number of points − 1/2)

'Stopping range' and 'stop times' are defined above. 'Number of points' is calculated by giving 1 point for each time the subject says 'W first' (i.e. spot position at time of awareness has an earlier 'clock-time' than the final 'stop-time' of the CRO spot), and 1/2 point for each response of 'together' (i.e. W time and stop-time of spot appear the same to the subject). (In some series a trial was 'aborted' or a subject's report was not available, for some technical reason. In such a case that trial was considered to contribute a number of points equal to our estimate of the probability that the subject would say 'W first' for that particular 'stop-time' of the CRO spot.) In the example shown in fig. 1B, the responses for each of the different 'stop-times' are plotted. In that series, the upper end of the 'stopping range' was 200 ms after EMG-trigger time (lower end was −400 ms), with the usual minimum 15 ms time interval between stop-times within the 'stopping range' of 600 ms. There were 19 'W first' responses and 8 'Together' responses; the remaining responses were of course 'Spot first'. There was one trial with a missing report in this series, at the CRO stop-time of −115 ms; we estimate the probability of saying 'W first' as 1/2 in this case. Putting this together, our estimate of the mean shift for W is

$$200 - 15\,(19 + 8/2 + 1/2 - 1/2) = -145 \text{ ms}$$

The mean shifts for the awareness in an M or S series were computed in an analogous manner for A or O mode of recall, respectively, using the stimulator-synch trigger for zero time in the S series.

## RESULTS

### *I. Subjective Timings*

The mean values of the 40 reported times of awareness (whether for W, M or S), for each series in a given study session, are presented in Table 1, A. Each value is for net time relative to 'zero-time' for each event, that is, W or M relative to EMG zero time for activation of muscle in a self-initiated movement, or S relative to stimulus-pulse time in the case of skin stimuli delivered at irregular times. There were no obvious or consistent differences between sessions in which mode of recall of time was 'absolute' (A) or by order (O) relative to final 'clock-position' of the spot (*see* Methods).

The mean value of W in each series, that is of the recalled times for being aware of 'wanting' to move, was invariably in advance of or negative to the EMG zero time. The average of all such mean Ws was about −200 ms (Table 2D). Except for the nature of the event, the basic procedures for attentive monitoring of the revolving spot and of noting visually and later recalling the clock-position of the spot, in connection with appearance of an awareness, were the same for M ('actually moved') or for S (skin sensation produced by irregularly timed, stimulus pulse) as they were for W. Reported times for S might be expected on the average to be close

TABLE 1

Column A. Awareness times (W) and, column B, RP-onset times, for each 'W series' of 40 self-initiated movements. Awareness times also given for (M) and (S) series in same session. RP onsets are given for both the 'main negative shift' (MN) and for time at which 90 per cent of total area under the RP begins (*see* text for definitions of W, M and S). Column C. *Differences* (ms) between RP onset-times and W times ('uncorrected', and 'corrected' for S), taking RP onset either for the main negative (MN) component or for 90 per cent of the RP area, in each W series of 40 self-initiated acts. (Instances in each series of 40 trials when the W time preceded [was negative to] onset of RP are given after $RP_{MN}$ onset for the respective series.)

| | | | A. *Awareness times (ms)* | | | | | | B. *Onset of RP (W series)* | | | C. *(Onset RP) minus (W), i.e. (B)-(A), using* | | | *(Onset RP) minus (W-S), using* | |
|---|---|---|---|---|---|---|---|---|---|---|---|---|---|---|---|---|
| | | *Mode* | *(W)* | | *(M)* | | *(S)* | | | | | | | | | |
| *Subject* | *Session* | *re-call* | *x̄* | *SE*† | *x̄* | *SE* | *x̄* | *SE* | *Type RP** | *Onset 'MN'* | *Onset 90 % area* | $RP_{MN}$ | *Ws neg. to MN* | $RP_{90\%}$ | $RP_{MN}$ | $RP_{90\%}$ |
| S.B. | 1 | A | − 54 | 11 | − 21 | 15 | − 12 | 11 | II | − 550 | −1076 | − 496 | 0 | −1022 | − 508 | −1034 |
| | 2 | O | −145 | 19 | − 48 | 20 | − 53 | 15 | II | − 900 | − 729 | − 755 | 0 | − 584 | − 808 | − 637 |
| | 3 | A | − 90 | 11 | − 72 | 12 | − 42 | 12 | I | −1100 | − 863 | −1010 | 0 | − 773 | −1052 | − 815 |
| | 4 | O | −188 | 34 | − 95 | 20 | − 53 | 18 | I | −1150 | − 757 | − 977 | 0 | − 584 | −1030 | − 637 |
| | 5a | A | −123 | 16 | | | | | II | − 800 | − 876 | − 677 | 0 | − 753 | | |
| | b | A | −119 | 10 | | | | | I | − 950 | − 694 | − 831 | 0 | − 575 | | |
| | 6a | A | −118 | 13 | | | +157 | | I | − 900 | − 685 | − 782 | 0 | − 567 | − 625 | − 410 |
| | b | A | −161 | 15 | | | | | II | − 600 | − 484 | − 439 | 0 | − 323 | − 282 | − 166 |
| G.L. | 1 | O | −208 | 28 | −213 | 28 | −147 | 28 | II | − 500 | − 380 | − 292 | 0 | − 172 | − 439 | − 319 |
| | 2 | A | −422 | 24 | −172 | 22 | −184 | 30 | I | −1200 | − 755 | − 778 | 0 | − 333 | − 962 | − 517 |
| | 3 | O | −377 | 27 | −220 | 26 | −217 | 30 | I | − 900 | − 635 | − 523 | 0 | − 258 | − 740 | − 475 |
| | 4 | A | −258 | 21 | −201 | 25 | −120 | 25 | II | − 800 | − 593 | − 542 | 0 | − 335 | − 662 | − 455 |
| | 5a | O | −213 | 42 | | | | | I | − 900 | − 599 | − 687 | 0 | − 386 | | |
| | b | O | −283 | 34 | | | | | I | −1200 | − 866 | − 917 | 0 | − 603 | | |
| | 6a | O | −221 | 34 | | | −164 | 20 | II | − 600 | − 563 | − 379 | 0 | − 342 | − 543 | − 506 |
| | b | O | −271 | 40 | | | | | I | −1400 | − 765 | −1129 | 0 | − 494 | −1293 | − 658 |
| B.D. | 1 | A | −225 | 19 | + 92 | 10 | +135 | 13 | II | − 400 | − 295 | − 175 | 3 | − 70 | − 40 | + 65 |
| | 2 | O | −145 | 24 | − 3 | 25 | + 45 | 26 | III | − 225 | − 157 | − 80 | 1 | − 12 | − 35 | + 33 |
| | 3 | A | −152 | 14 | + 76 | 12 | + 61 | 9 | II | − 500 | − 401 | − 348 | 0 | − 249 | − 287 | − 188 |
| | 4 | O | −142 | 18 | + 40 | 21 | + 90 | 20 | II | − 425 | − 469 | − 283 | 0 | − 327 | − 195 | − 239 |
| | 5a | O | −145 | 29 | | | | | III | − 250 | − 716 | − 105 | 0 | − 571 | | |
| | b | O | −108 | 46 | | | | | III | − 325 | − 210 | − 217 | 1 | − 102 | | |
| | 6a | O | −146 | 30 | | | | | II | − 650 | − 468 | − 504 | 0 | − 322 | | |
| S.S. | 1 | O | −235 | 31 | −168 | 25 | −130 | 27 | III$_c$ | − 250 | − 806 | − 15 | 3 | − 571 | − 145 | − 691 |
| | 2 | O | −253 | 28 | − 33 | 19 | − 83 | 17 | II$_c$ | − 400 | − 282 | − 147 | 0 | − 31 | − 230 | − 112 |
| | 3 | O | −255 | 26 | −153 | 24 | − 75 | 24 | II$_c$ | − 400 | − 281 | − 145 | 2 | − 26 | − 220 | − 102 |
| | 4 | A | −283 | 19 | −113 | 9 | −157 | 13 | III$_c$ | − 300 | − 695 | − 17 | 17 | − 412 | − 174 | − 569 |
| | 7a | A | −248 | 20 | | | | | I$_v$ | − 900 | − 915 | − 652 | 0 | − 667 | | |
| | b | A | −236 | 17 | | | | | II$_c$ | − 400 | − 604 | − 164 | 4 | − 368 | | |
| | c | A | −209 | 16 | | | | | I$_v$ | −1100 | − 805 | − 891 | 0 | − 596 | | |
| C.M. | 1 (30 trials) | O | −287 | | −138 | | − 63 | | II$_c$ | − 500 | − 408 | − 213 | | − 121 | − 274 | − 184 |
| | 2 | O | −223 | 25 | −123 | 21 | − 23 | 20 | II$_c$ | − 400 | − 489 | − 177 | 0 | − 266 | − 200 | − 289 |
| | 3 | O | −245 | 25 | − 83 | 20 | − 23 | 27 | (no RP available) | | | | | | | |
| | 4 | A | −132 | 17 | − 69 | 10 | + 29 | 10 | II$_c$ | − 600 | − 520 | − 468 | 0 | − 388 | − 439 | − 359 |
| | 6 | A | −211 | 13 | | | | | II$_v$ | − 400 | − 781 | − 181 | 1 | − 570 | | |
| | | | | | | | | | II$_c$ | − 400 | − 227 | − 181 | 1 | − 16 | | |
| | 7a | A | −260 | 17 | | | | | I$_v$ | −1000 | − 703 | − 740 | 0 | − 443 | | |
| | | | | | | | | | I$_c$ | −1050 | − 694 | − 790 | 0 | − 434 | | |
| | b | A | −251 | 17 | | | | | II$_c$ | − 450 | − 368 | − 199 | 2 | − 117 | | |
| | 8 | A | −204 | 16 | | | | | II$_c$ | − 475 | − 479 | − 271 | 1 | − 275 | | |

* All values for subjects S.B., G.L. and B.D. are for RPs recorded at the vertex. For subjects S.S. and C.M. the relevant RPs were recorded only at the contralateral precentral area for the hand, as designated by subscript c, except for some vertex recordings noted by subscript v. Simultaneous values for v and c are given for sessions 6 and 7a of subject C.M. † SE = standard error for our estimate of the mean value.

to zero (actual stimulus time) or possibly delayed slightly. But the actual mean values for S were usually negative rather than positive or delayed, except for subject B.D., and they differed for each subject and with each session. The value obtained for S in a given session could be regarded as at least a partial measure of the way the subject is handling those reporting factors that S and W series do have in common. As an approximation, one may 'correct' W for the subject's 'bias' in reporting awareness time by our methods, by subtracting S from W for that given session. The average of all W values (about −200 ms) would be changed to about −150 ms by subtracting the average of about −50 ms for all S values (*see* Table 2D).

TABLE 2. *GRAND AVERAGES* (MS) FOR ALL SERIES IN EACH COLUMN OF TABLE 1, ACCORDING TO TYPE OF ASSOCIATED RP AND TO ORDER OF W AND M SERIES IN A SESSION

| *Type of RP, for W series* | | A. *Awareness times* | B. *Onset of RP (in W series)* | | C. *B-A; i.e., (Onset RP) minus (W), using* | | *(Onset RP) minus (W-S), using* | | |
|---|---|---|---|---|---|---|---|---|---|
| | *n* | *W* | $RP_{MN}$ | $RP_{90\%}$ | $RP_{MN}$ | $RP_{90\%}$ | *n* | $RP_{MN}$ | $RP_{90\%}$ |
| I | 12 | −233 | −1025 | −784 | −825 | −522 | 6 | −950 | −585 |
| II | 20 | −192 | −535 | −527 | −343 | −333 | 14 | −366 | −323 |
| III | 5 | −183 | −270 | −517 | −87 | −334 | 3 | −118 | −409 |

D. *Awareness times*

| | *W* | | *M* | | *S* | |
|---|---|---|---|---|---|---|
| | *n* | $\bar{X}$ | *n* | $\bar{X}$ | *n* | $\bar{X}$ |
| For all series | 37 | −204 | 20 | −86 | 22 | −47 |
| In sessions when W series done before M series | 10 | −191 | 10 | −92 | 10 | −41 |
| In sessions when M series done before W series | 10 | −240 | 10 | −80 | 10 | −53 |

Mean values for M series were also mostly negative (except for subject B.D.), averaging about −85 ms for all mean Ms (Table 2D). M was also slightly negative to S in almost every individual study session (*see* Table 1, column A); so that even if the average of M values (−86 ms) are 'corrected' by subtracting the average of S values (−47 ms), a small average net M of about −40 ms still remains. (Even for subject B.D., subtracting the average of his mean Ss, +83 ms, from the average of mean Ms of +51 ms in the same 4 sessions, produces a net 'corrected' average M of about −30 ms.) This produces the unexpected result that reported time of awareness of 'actually moving' generally preceded the activation of the muscle at EMG zero time! (*See* Discussion.)

It is important to note that mean W values were consistently negative to mean M values in the respective session for each subject (Table 1A), in spite of the frequently negative values for M. The average of all mean Ws (about −200 ms) indicated that awareness of wanting to move preceded average awareness of actually moving (about −85 ms) by more than 100 ms. When only those W values obtained in the same 20 sessions with M values are included, the average of mean Ws was −216 instead of −204 ms (Table 2D). Mean Ws obtained in sessions when an M series was carried out before the W series appeared to be significantly more negative than Ws obtained when a W series was carried out before an M series (*see* separate averages in Table 2D). A Wilcoxon test for this ranking order gave a one-sided $P = 0.038$, and a two-sided $P = 0.076$. (This ranking was not related to the use of (A) *vs* (O) mode of recall in the session.) However, the actual differences of about 50 ms between the two sets of Ws are relatively unimportant when comparing W times to onset times of the corresponding readiness potentials (*see* Section III, *below*).

*II. Onset Times of Readiness-potentials (RP)*

RPs associated with the freely voluntary, self-initiated movements employed in this study have been described (Libet *et al.*, 1982). They can be categorized into two or three types, based on their form and the time of the main negative (MN) shift (*see* Tables 1B and 2B). Type II (and III) RPs are obtained when all 40 self-initiated movements in the averaged series are reported by the subject to have originated 'spontaneously' and 'capriciously', with no recollections of preplanning experiences for any of the 40 events in the series. Additional experiences of a 'preplanning' phase are associated with type I RPs (Libet *et al.*, 1982). No significant association could be detected between mode of recall for W (that is (A) or (O)) and type of RP obtained.

Onset times of RPs listed in Table 1B are for RPs recorded in the same series of 40 self-initiated movements for which the reports of W times are given, in each respective session for each subject. This simultaneity, for RP and W observations, is important because there can be considerable variations of RP onsets in different series even in the same session (Libet *et al.*, 1982). The actual RPs for each W series listed in Table 1 for the Group 2 subjects (S.B., G.L. and B.D.) are presented in fig. 2. RPs were also obtained with each M series in the session, but onset times for these are not listed in Table 1. Onset times for RPs in M series were actually, on average, similar to those for RPs in the W series (*see also* Libet *et al.*, 1982).

Two values for onset time are given for each RP (W series) in Table 1B. (1) Onset time of the main negative (MN) shift was determined by 'eye-ball inspection', checked independently by a second investigator. (2) Onset time was also computed for the point at which 90 per cent of the area under the RP tracing preceded EMG zero time.

Onset time based on RP area was determined as follows. On an enlarged projected image, the area under the RP was measured for each interval of 50 ms, starting from EMG zero time and progressing to successive intervals in the negative (pre-EMG) direction until −600 ms; between −600 and −1400 ms, areas for 100 ms intervals were measured. Within each time interval, any areas below the baseline were subtracted from those above. In estimating total area, however, it was considered advisable to exclude any early brief shifts of potential that did not continue progressively into the main RP, as some of these

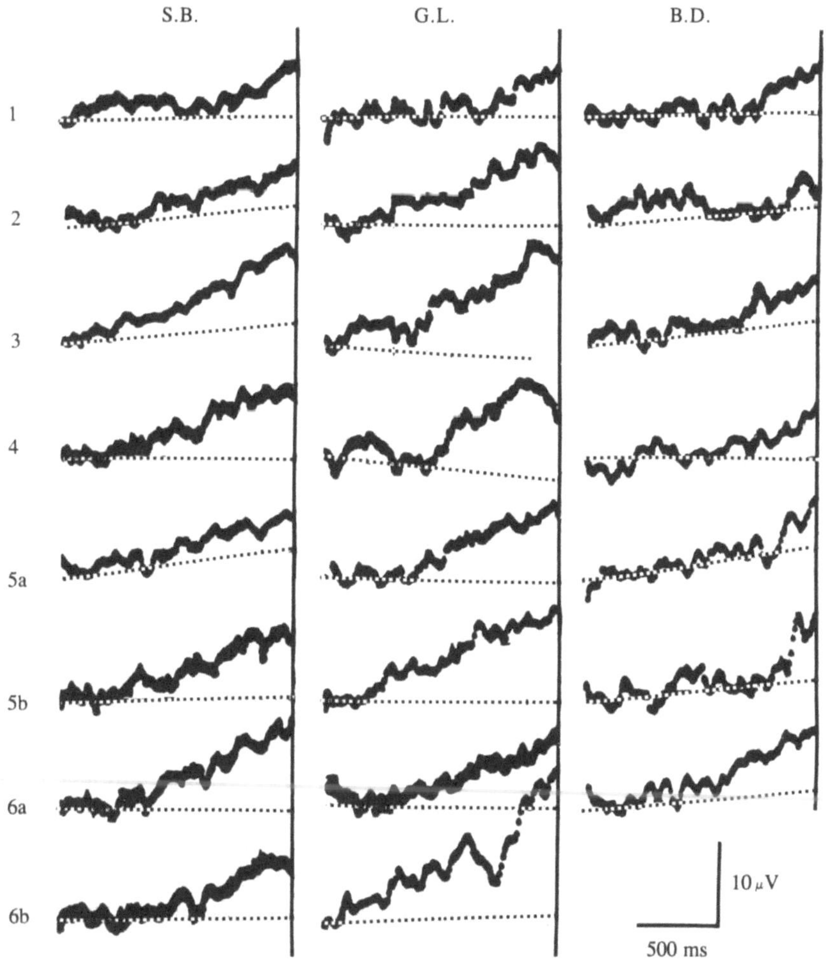


FIG. 2. Readiness-potentials (RPs) recorded at the vertex and averaged for each (W) series of 40 trials for subjects S.B., G.L. and B.D. Each RP corresponds to the respective (W) series as listed by session number in Table 1. The solid vertical line indicates the EMG zero time, marking the end of the RP. Dashed horizontal lines represent the d.c. baseline drift for the 2 s of that tracing, as estimated from the total voltage compensation for shift in d.c. level during the total time between beginning and end of that series of 40 trials.

could have been artefactual in nature. Therefore, the rule was adopted that any 200 ms segment having a total area $\leqslant$ 4 mm$^2$ (actually equivalent to 50 $\mu$V-ms) was to be regarded as zero, and that any and all areas preceding that segment were also regarded as zero. (Making the rule even more stringent, by reducing the excluding low level segment to 100 ms, rarely changed the results significantly and, when it did, the final estimates of time of onset were very little different.) It was also recognized that measurements of the beginning of the negative potential shift are subject to some possible error in judging the d.c. baseline, especially in a somewhat noisy/bumpy tracing. Therefore, after arriving at a total area for a given RP under the rule above, the time interval that included only 90 per cent of this area was computed. The 90 per cent values also fit the range of differences between the independent measurements of the areas by two different investigators.

Averages of the onset times for $RP_{MN}$ and $RP_{90\%}$ area, respectively, for the Ws series in Table 1B are given in Table 2. The initially slower but progressive ramp-like rise of type I RPs accounts for these onset times being more negative for $RP_{MN}$ than for $RP_{90\%}$ area. On the other hand, in types II and III RPs some definitely distinguishable negativity is often present even before the main (MN) shift. Such negativities tend to have a relatively irregular, low, amplitude but there was no reason to regard them as other than actual RP components in these self-initiated acts (*see* Libet *et al.*, 1982). Their inclusion in the measurements of total area makes it possible for onset of $RP_{90\%}$ area to precede onset of $RP_{MN}$ in some cases. Averages of the onsets for $RP_{MN}$ and $RP_{90\%}$ were in fact not very dissimilar for type II RPs, although individual values for the difference ($RP_{MN} - RP_{90\%}$) were in a range between −207 and +526 ms; but for type III RPs average $RP_{90\%}$ preceded $RP_{MN}$ by −247 ms (range of $RP_{MN} - RP_{90\%}$ was between −115 and +556 ms).

*III. Differences between Onset time of RP and Time of Awareness of Wanting to Move*

The data comparisons given in Table 1C are central to the objective of this study; they relate the time of appearance of the *conscious* intention to act, on the one hand, to the time of onset of the *cerebral processes* before the act (as evidenced in the RP), on the other. The *difference* between RP onset time and each W awareness-time is given for each respective series of self-initiated voluntary acts. Differences are presented when utilizing the W times as actually reported (W 'uncorrected'), giving (onset RP) minus (W); or the W times 'corrected' by subtracting the reported mean time for the S obtained in the same session, giving (onset RP) minus (W-S). 'Correcting' the W value by subtracting the S value of each subject's 'bias' in reporting, did not qualitatively change the relation of RP onset-time to W; rather it generally increased the difference by which onset of RP precedes W (as 'corrected'). For subject B.D., his *positive* values for S have the oposite effect; but even for him, the only qualitatively important change in the difference is introduced in session 1, which had a large positive S (+135 ms).

It may be seen (Table 1C) that, with few exceptions, onset of RP occurred before reported awareness time by substantial amounts of time. This was true irrespective of which measure of RP-onset or of W is employed to obtain the difference. The sizes and consistency of these differences, between onset of RP and W, indicate they are highly significant. However, it is difficult to produce a rigorous quantitative value for significance of the large differences between onset of RP and W. An SD (standard deviation) for variability among individual RPs within each series of 40 is not available, as only the average RP for the whole series could be meaningfully recorded. Consequently, only the mean W value and the averaged RP obtained in a given series can be compared for statistical purposes. Confidence in the significance of the differences (in Table 1C) is further raised by the fact that they were almost invariably very large when compared to the SEs (standard errors) for the mean values of W (Table 1A). In addition, each W series of 40 trials was examined for the incidence of individual W values that may have deviated sufficiently from the mean

W so as to be negative to (precede), rather than positive to (follow) the onset of average RP in that series (*see* Table 1c). For 36 W series of 40 trials each, instances in which individual W time preceded onset time of averaged RP numbered zero in 26 series and 1 to 3 in 8 series! (Of the remaining 2 series, in both of which RPs were recorded at contralateral precentral sites where RP onsets are often less steep than at the vertex, 1 had 4 and 1 had 17 instances. The latter large value holds only in relation to onset of MN in a type III RP but not if onset $RP_{90\%}$ is used.)

The SE of each mean value for W, as given in Table 1A, is more meaningful in relation to difference of (onset of averaged RP)-(mean W), than is the SD for the distribution of the individual W values in each series of 40 trials. (The SD for each series of 40 W values may be calculated using the respective SE, given in Table 1A; SD = $\sqrt{40} \times$ [SE].) If the individual RPs were available, the difference between each RP onset and the respective individual W value could be determined for each event, and a meaningful evaluation of variability for such individual differences could be made. However, in the absence of individual RP values, it is reasonable to assume that the W and RP-onset for each individual act are probably related in a dependent and positively correlated manner, in which onset of each RP tends to be negative to (before) each W. This condition seems likely because (1) W and RP are features of the same underlying process, (2) observed differences between averaged RP and mean W were consistently large, and (3) there was a nearly complete absence of individual W values that were negative to the averaged RP (*see above*).

For reporting each W time, subjects were asked to note the earliest awareness of the specific urge or intention to act which might occur prior to the act. All subjects reported that they could distinguish readily between this awareness and any experience of 'pre-planning' that sometimes occurred in acts associated with type I RPs (Libet *et al.*, 1982). Awarenesses of 'preplanning' were completely absent in series associated with type II (or III) RPs, in which all 40 self-initiated movements were 'spontaneous' in origin. Therefore, it is useful to consider the values for type II (and III) RP series separately from those for type I RP series, as summarized in Table 2A–C. As might be expected, series with type I RPs generally exhibit an earlier onset of RP, relative to W, than do those with type II (or III), especially for $RP_{MN}$ onsets. However, even for the series of 'spontaneous' acts with types II and III RPs, onsets of RPs generally preceded W by substantial amounts. The average of the differences [(onset $RP_{MN}$) minus (W)] for the 20 series with type II RPs was −343 ms (Table 2c). The relatively few sessions in which these differences were possibly not significant occurred mainly in association with type III RPs, for which onsets of $RP_{MN}$ averaged only −270 ms. For the 5 sessions with type III RPs, the average difference (onset $RP_{MN}$ − W) was −87 ms, although when $RP_{90\%}$ is used, the difference increased to −334 ms.

## DISCUSSION

It is clear that neuronal processes that precede a self-initiated voluntary action, as reflected in the readiness-potential, generally begin substantially *before* the reported appearance of conscious intention to perform that specific act. This temporal difference of several hundreds of milliseconds appeared fairly consistently regardless of which of the available criteria for onset of RP or for the time of awareness are adopted. Series with type II RPs are of especial interest as all of the 40 self-initiated acts arise spontaneously; on this and other evidence, the main negative (MN) shift with average onset about −550 ms was postulated to reflect the cerebral volitional process uniquely involved in initiating a freely voluntary, fully endogenous act (Libet *et al.*, 1982). Even for such series, with type II RPs, onset of RP preceded W by about 350 ms on the average. In series with type I RPs the earlier MN shift (average onset about −1025 ms) appears to reflect a more general preparation or intention to act that can be either endogenous or cued externally; it is not necessarily associated with freedom of choosing when to act (Libet *et al.*, 1982). However, actual experiences of 'preplanning' were reported for only a minority of self-initiated acts in series with type I RPs. Consequently, the much larger differences between onset of type I RP and W, on the average as much as −800 ms, may also reflect advance cerebral preparation that is generally accomplished before conscious intentionality arises. Only in the case of a small number of series with type III RPs was the difference between RP-onset and W less negative than −100 ms, when onset of MN shift (average −270 ms) is adopted. But if start of 90 per cent of RP area (average −517 ms) rather than MN shift is taken as the criterion for onset of RP, even these type III RP's precede W by an average of more than 300 ms (Table 2C). In series with type III RPs, all self-initiated acts were also spontaneous, as in type II (Libet *et al.*, 1982).

The validity of the RP as an indicator of cerebral activity had already been established (Kornhuber and Deecke, 1965; Deecke *et al.*, 1976; Pieper *et al.*, 1980). Actually, the onset of RP provides only a minimum timing for initiation of cerebral activity, as the recorded RP probably represents neuronal activity in a limited portion of the brain, possibly that of the supplementary motor area in the mesial neocortex (Deecke and Kornhuber, 1978; Eccles, 1982; Libet *et al.*, 1982). It is possible that cerebral activity is initiated at times earlier than the onset of the recorded RP in some other regions (*see* Groll-Knapp *et al.*, 1977). For the present issue, the requirement of averaging 40 individual RPs for each recordable RP might raise the question whether this obscures an actually randomized group of widely different onset times. Even in such a case, however, there are several considerations that work mathematically against an interpretation that W times are consistently equal to or precede onset times of individual RPs. Some of these have been discussed above (Section III, Results). An example might assume that the averaged RPs are contaminated by a small number of unusual individual RPs having very early onsets, compared to onsets of most RPs in each series of 40. However, if this were true, we would expect that such unusual RPs should not appear in an appreciable number of the 40-trial series, causing the order of times for W and onset of averaged RP to be reversed; but, in fact, the onset of averaged RP preceded mean W in all of the 37 W series (Table 1C, and *see above*). Furthermore, in type II RPs the main negative shift (MN) has a relatively abrupt onset and rises rapidly; any small number of individual RPs with unusually early onsets could not appreciably affect the onset time of $RP_{MN}$, as measured in type II RPs. Yet, onset of $RP_{MN}$ preceded mean W, usually by large time intervals, in all the 20 W series that exhibited type II RPs. To circumvent all these points of evidence one would have to introduce unsupported assumptions that there exist specially biased distributions of RPs.

As another possibility, it might be proposed that neural activities, represented by individual RPs with randomly variable amplitudes and onset times, must achieve some threshold, whether integrative or other, before the brain 'decides' to act voluntarily; for this one might apply a kind of 'random-walk' model for sequential decision making (*see* Audley, 1973). If applied to our case, a delayed appearance of a 'neural threshold' might coincide in time with the W time, and thus nullify any apparent discrepancy between time of cerebral decision and time of conscious intention to act. But a consistent bias or change in the random-walk distribution of the variable neural functions would have to be initiated at the onset of averaged RP, preceding the achievement of threshold for the decision. This would be analogous to the required initiation and presence of a sensory signal to produce the distribution of sensory random variables that may lead to a yes-no decision in that random-walk model (Green and Luce, 1973). Therefore, such a proposal could, at best, only separate the cerebral initiating process into two stages. An earlier stage, which would precede a final decision stage, would start with some initiating endogenous trigger. Such a model would not fundamentally affect our conclusion, that cerebral activity initiates the voluntary act before reportable conscious intention appears.

### *The Criteria for Time of Conscious Intention to Act*

The reliability and validity of these operational criteria are of course crucial to the issue of the temporal order of cerebral processes *vs* conscious intention. The reliability of the subjects' reports of 'clock-position' for the revolving CRO spot at the time of awareness appears to be fully adequate (*see* discussion of SEs of mean W values and incidence of individual W values, relative to onset of averaged RP, in Section III of Results).

Consideration of validity of our criteria begins with the premise that the subjective event in question is only introspectively accessible to the subject himself, and that this requires a report by the subject (*see* Libet, 1965, 1966; Nagel, 1974; Creutzfeldt and Rager, 1978). Any behavioural response that is not a direct function of such a report could not be used as a *primary* indicator of the subjective event (Libet, 1981*a*, *b*), although it might be found to be associated with the subjective event as studied by suitably valid reports. Acceptance of this premise, and of our specific operational precedure for the required introspective report, introduces several issues that may affect the validity of the reported W time. In particular, factors that may affect the transmission between the subject's introspective experience and his verbal report must be considered.

(1) *Simultaneity of judgements.* Our method requires the subject to observe simultaneously, for later report, the appearance of a mental event (conscious urge to move) and the visual clock-position of the revolving spot of light at that time. Reports of simultaneous events have long been known to be subject to potential errors, depending on the circumstances (differential attentiveness, in the 'prior entry' phenomenon) and on the individual subject (*see* Boring, 1957; Efron, 1973; Sternberg and Knoll, 1973). Our S experimental series, in which subjects reported awareness times for skin stimuli, were designed to serve as controls for potential

errors in such 'simultaneity' as well as for other individual biases and errors in the entire reporting procedure. Procedures and requirements for subject's attentiveness, observations and later recall, of clock-positions of the revolving CRO spot at the time of awareness of a randomly appearing skin sensation, were the same as in the W and M series. But in the S series the actual time of the stimulus was later known to the investigator, and the error in the subject's reports could be determined objectively. The bias or error found in the S series did not qualitatively alter the difference between onset of RP and W, as determined in a given session for the same subject; in fact, they generally enlarged the differences (Table 2C).

(2) *Timing of an endogenous mental event.* The subject's reported time for spontaneously arising awareness of the urge/intention to move cannot be directly or objectively validated in the manner possible for skin stimuli in the S series. The subject's report constitutes the primary evidence of his introspective experience. No other independent measure of such subjective timing is available, although some of the other available evidence can affect confidence in the validity of the observed timings; this is summarized below in point (3).

It is of course possible to conceive or postulate conditions that might introduce discrepancies between the actual and the reported initial times of such an awareness. For example, what if it were possible to judge accurately only the end of the mental event, the conscious urge to move; the actual time of its onset, in relation to a perceived clock-time, would then be unknown and in doubt. In relation to such a suggestion we note, first, that each subject was instructed to 'watch for' and report the earliest appearance of the awareness in question, and subjects did not raise any difficulties about doing this. Secondly, perceptual timings of onset and offset found for a peripherally induced sensation, at least, do not support the suggestion that only the end is accurately judged. Using a method of cross-modality simultaneity judgements, Efron (1973) found that (a) there was no difficulty in distinguishing onsets from offsets; (b) the perceptual onset latency was constant regardless of large changes in duration of the stimulus; whereas (c) perceptual offset latency could in fact vary when stimulus durations were shortened to $< 150$ ms, with the change in offset latency probably originating, in part, in the peripheral sensory structures. Somewhat related findings on perceptual onset latencies are available for a cerebrally induced mental event, elicited by a stimulus in the somatosensory system in man (Libet *et al.*, 1979). The mean differences between perceptual timings, for a brief skin stimulus relative to onset of a medial lemniscus stimulus lasting 200 ms, were only within a few tens of milliseconds. One might suggest that timing of an endogenous mental event, the spontaneously arising conscious intention to act, may be more difficult subjectively to pinpoint with accuracy than the timing even for the sensation elicited by an intracerebral stimulus in the medial lemniscus. The individual W values reported by our subjects in a given series of 40 events did show variability; but the difference between (onset of RP) and (mean W), in Table 1C, was consistently and considerably greater than the SE of mean W for the respective series. Actually, the converse possibility, of W times reported earlier than the actual time of awareness of the urge to move, presented a more real difficulty. A 'preawareness' that one is preparing to perform the voluntary act, sometime within the next second or so, does in fact accompany at least some of the events in those series that produce a type I RP, as noted above (Libet *et al.*, 1982). If such a preawareness were to have affected the report, it would mean that the reported W times were more negative, earlier, than they should have been; the difference between onset of RP and 'real' W should then be even greater than indicated by our results. However, in series giving type II RPs, all of the self-initiated acts were described as 'spontaneous'; the subjects reported that each urge or wish to act appeared suddenly 'out of nowhere', with no specific preplanning or preawareness that it was about to happen.

Finally, the possibility could be raised that an earlier nonrecallable phase of the conscious urge exists, one that is not storable as a short-term memory. If it is further assumed that the subject's report of W time requires short-term memory of the mental event, then the reported time would apply only to a later, recallable phase of awareness, given such assumptions. First, one should note that, to report W time, the subject needed to recall only the clock-position of the revolving spot at the time he became aware of the urge/intention to move, and not necessarily the conscious mental event itself. (Actually, the latter was at least often recallable, as the subjects were able to describe it, even in relation to experiences just preceding it during the trial.) Secondly, the proposal of a nonrecallable initial phase of the conscious urge to move is a hypothetical construct which, like some other potential uncertainties in timing an endogenous mental event, is at present not directly testable. Thus, although it cannot be definitively excluded, it also lacks experimental support.

(3) *Additional evidence bearing on the validity of the reported timings*. One way to test and improve confidence in the validity of the reported timings lies in using different but converging modes of observing and reporting, with each mode having independent validity without further assumptions (*see* Garner, 1954, for discussion of this approach in connection with a related issue). Two quite different modes were employed for reporting the 'clock-positions' of the CRO spot at the time of awareness, that is, the absolute (A) reading *vs* the order (O) relative to final spot positions. Yet both modes produced values for W in the same range and were essentially indistinguishable. (This also held for reported timings in the M and S series.)

Subjects definitively distinguished the experience and time of awareness of wanting to move (W) not only from those of a skin sensation (S) but also from awareness of actually moving (M). Mean values for W times were consistently negative (by > 100 ms on average) to those for M times (Tables 1A, 2D). This was true in spite of the unexpected finding that mean M values were themselves generally negative to EMG zero time, although they were only slightly negative to S values in which no movement was involved. M has some features of an endogenous mental event, rather than simply of a sensation elicited by input from peripheral sensory sources (*see below*, for discussion of M preceding the EMG). On the other hand, onset-times for type I or II RPs in series asking for W reports were generally similar to those asking for M reports, or even to those in series when no reports of awareness were requested (*see also* Libet *et al.*, 1982). This indicates that the somewhat different mental sets associated with each kind of reporting (or absence of reporting) did not affect the RP side of the time differential with respect to W.

It might be argued that subjects' reports of W times could be distorted by their awareness of the time of the actual movement; this might induce them to report W times that are later than the actual time of appearance of the conscious intention to move. But the subjects confirmed that for W reports they concentrated on noting their *earliest* awareness of any urge/intention to move. They further stated that their mental set for 'observing' W time was also different from that for M time.

Furthermore, the available evidence indicates that the subjects' experience with attending to the awareness of actually moving (M) may have induced them to report W times that were somewhat more, not less, negative relative to EMG zero time (*see* Results, end of Section I). The mean W values were on the average about 50 ms more negative when, in a given session, the W series was performed after instead of before an M series (*see* Table 2D, etc.).

Awareness of 'actually moving' (M) preceded the EMG. Mean M values were generally negative to EMG zero time for most subjects, and consistently though slightly negative (average about −40 ms) relative to S values for all subjects. Timing of M so as to precede the activation of muscle contraction indicates that M was not reflecting awareness of proprioceptive sensory impulses elicited by the movement. It suggests the possibility that M reflected an awareness associated more immediately with initiation of efferent cerebral output for the movement. Components that follow the main negative RP shift are recordable just prior to movement, including a negative 'motor potential' that begins about 50 ms before the EMG (*see* Deecke *et al.*, 1976; Shibasaki *et al.*, 1980).

### *Unconscious and Conscious Initiation or Control of Voluntary Acts*

Since onset of RP regularly begins at least several hundreds of milliseconds before the appearance of a reportable time for awareness of any subjective intention or wish to act, it would appear that some neuronal activity associated with the eventual performance of the act has started well before any (recallable) conscious initiation or intervention could be possible. Put another way, the brain evidently 'decides' to initiate or, at the least, prepare to initiate the act at a time before there is any reportable subjective awareness that such a decision has taken place. It is concluded that cerebral initiation even of a spontaneous voluntary act, of the kind studied here, can and usually does begin *unconsciously*. The term 'unconscious' refers here simply to all processes that are not expressed as a conscious experience; this may include and does not distinguish among preconscious, subconscious or other possible nonreportable unconscious processes.

A general hypothesis had already been proposed that some substantial time period of appropriate cerebral activity may be required for eliciting all specific conscious experiences (Libet, 1965). This developed out of experimentally based findings that cortical activities must persist for up to 500 ms or more before 'neuronal adequacy' for a conscious sensory experience is achieved (*see* Libet, 1966, 1973, 1981*a*; Libet *et al.*, 1972). In that hypothesis, those cerebral activities that did not persist sufficiently would remain at unconscious levels. The present evidence appears to provide support for that more general hypothesis. It suggests that a similar substantial period of cerebral activity may also be required to achieve 'neuronal adequacy' for an experience of conscious intention or desire to perform a voluntary act.

The present evidence for the unconscious initiation of a voluntary act of course applies to one very limited form of such acts. However, the simple voluntary motor act studied here has in fact often been regarded as an incontrovertible and ideal example of a fully endogenous and 'freely voluntary' act. The absence of any larger meaning in the simple quick flexion of hand or fingers, and the possibility of performing it with capriciously whimsical timings, appear to exclude external psychological or other factors as controlling agents. It thus invites the extrapolation that other relatively 'spontaneous' voluntary acts, performed without conscious deliberation or planning, may also be initiated by cerebral activities proceeding unconsciously.

These considerations would appear to introduce certain constraints on the potential of the individual for exerting conscious initiation and control over his voluntary acts. However, accepting our conclusion that spontaneous voluntary acts can be initiated unconsciously, there would remain at least two types of conditions in which conscious control could be operative. (1) There could be a conscious 'veto' that aborts the performance even of the type of 'spontaneous' self-initiated act under study here. This remains possible because reportable conscious intention, even though it appeared distinctly later than onset of RP, did appear a substantial time (about 150 to 200 ms) before the beginning of the movement as signalled by the EMG. Even in our present experiments, subjects have reported that some recallable conscious urges to act were 'aborted' or inhibited before any actual movement occurred; in such cases the subject simply waited for another urge to appear which, when consummated, constituted the actual event whose RP was recorded (Libet *et al.*, 1982). (2) In those voluntary actions that are not 'spontaneous' and quickly performed, that is, in those in which conscious deliberation (of whether to act or of what alternative choice of action to take) precedes the act, the possibilities for conscious initiation and control would not be excluded by the present evidence.

## ACKNOWLEDGEMENTS

We are indebted to an anonymous editorial reviewer of the paper for helpful comments. This work was supported in part by the Research Support Program of the Mount Zion Hospital and Medical Center, San Francisco.

## REFERENCES

AUDLEY R J (1973) Some observations on theories of choice reaction time: tutorial review. In: *Attention and Performance IV*. Edited by S. Kornblum. New York and London: Academic Press.

BORING E G (1957) *A History of Experimental Psychology*. Second edition. New York: Appleton-Century-Crofts, pp. 146-147.

CHURCH J D, COBB E B (1971) Nonparametric estimation of the mean using quantal response data. *Annals of Institute of Statistical Mathematics*, **23**, 105-117.

CREUTZFELDT O D, RAGER G (1978) Brain mechanisms and the phenomenology of conscious experience. In: *Cerebral Correlates of Conscious Experience*. Edited by P. A. Buser and A. Rougeul-Buser. Amsterdam: Elsevier/North Holland Publishing Company, pp. 311-318.

DEECKE L, GRÖZINGER B, KORNHUBER H H (1976) Voluntary finger movement in man: cerebral potentials and theory. *Biological Cybernetics*, **23**, 99-119.

DEECKE L, KORNHUBER H H (1978) An electrical sign of participation of the mesial 'supplementary' motor cortex in human voluntary finger movement. *Brain Research, Amsterdam*, **159**, 473-476.

ECCLES J C (1982) The initiation of voluntary movements by the supplementary motor area. *Archiv für Psychiatrie und Nervenkrankheiten*, **231**, 423-441.

EFRON R (1973) An invariant characteristic of perceptual systems in the time domain. In: *Attention and Performance IV*. Edited by S. Kornblum. New York and London: Academic Press, pp. 713-736.

GARNER W R (1954) Context effects and the validity of loudness scales. *Journal of Experimental Psychology*, **48**, 218-224.

GILDEN L, VAUGHAN H G JR, COSTA L D (1966) Summated human EEG potentials with voluntary movement. *Electroencephalography and Clinical Neurophysiology*, **20**, 433-438.

GREEN D M, LUCE R D (1973) Speed-accuracy trade off in auditory detection. In: *Attention and Performance IV*. Edited by S. Kornblum. New York and London: Academic Press, pp. 547-569.

GROLL-KNAPP E, GANGLBERGER J A, HAIDER M (1977) Voluntary movement-related slow potentials in cortex and thalamus in man. In: *Attention, Voluntary Contraction and Event-Related Cerebral Potentials. Progress in Clinical Neurophysiology*, Volume 1. Edited by J. E. Desmedt. Basel: Karger.

KORNHUBER H H, DEECKE L (1965) Hirnpotentialänderungen bei Willkürbewegungen und passiven Bewegungen des Menschen: Bereitschaftspotential und reafferente Potentiale. *Pflügers Archiv für Gesamte Physiologie*, **284**, 1-17.

LIBET B (1965) Cortical activation in conscious and unconscious experience. *Perspectives in Biology and Medicine*, **9**, 77-86.

LIBET B (1966) Brain stimulation and the threshold of conscious experience. In: *Brain and Conscious Experience*. Edited by J. C. Eccles. Berlin: Springer-Verlag, pp. 165-181.

LIBET B (1973) Electrical stimulation of cortex in human subjects, and conscious sensory aspects. In: *Handbook of Sensory Physiology*. Edited by A. Iggo. Heidelberg: Springer-Verlag, Volume 2, pp. 743-790.

LIBET B (1981*a*) The experimental evidence for subjective referral of a sensory experience backwards in time: reply to P. S. Churchland. *Philosophy of Science*, **48**, 182-197.

LIBET B (1981*b*) ERPs and conscious awareness; neurons and glia as generators. In: *Electrophysiological Approaches to Human Cognitive Processing. NRP Bulletin*, Volume 20. Edited by R. Galambos and S. A. Hillyard. Cambridge, Mass. The MIT Press Journals, pp. 171-175, 226-227.

LIBET B, ALBERTS W W, WRIGHT E W, FEINSTEIN B (1972) Cortical and thalamic activation in conscious sensory experience. In: *Neurophysiology Studied in Man*. Edited by G. G. Somjen. Amsterdam: Excerpta Medica, pp. 157-168.

LIBET B, WRIGHT E W JR, FEINSTEIN B, PEARL D K (1979) Subjective referral of the timing for a conscious sensory experience. A functional role for the somatosensory specific projection system in man. *Brain*, **102**, 193-224.

LIBET B, WRIGHT E W JR, GLEASON C A (1982) Readiness-potentials preceding unrestricted 'spontaneous' *vs* pre-planned voluntary acts. *Electroencephalography and Clinical Neurophysiology*, **54**, 322-335.

NAGEL T (1974) What is it like to be a bat? *Philosophical Review*, **83**, 435-450.

PIEPER C F, GOLDRING S, JENNY A B, MCMAHON J P (1980) Comparative study of cerebral cortical potentials associated with voluntary movements in monkey and man. *Electroencephalography and Clinical Neurophysiology*, **48**, 266-292.

SHIBASAKI H, BARRETT G, HALLIDAY E, HALLIDAY A M (1980) Components of the movement-related cortical potential and their scalp topography. *Electroencephalography and Clinical Neurophysiology*, **49**, 213-226.

STERNBERG S, KNOLL R L (1973) The perception of temporal order: fundamental issues and a general model. In: *Attention and Performance IV*. Edited by S. Kornblum. New York and London: Academic Press, pp. 629-685.

*(Received July 20, 1982. Revised December 14, 1982)*

THE BEHAVIORAL AND BRAIN SCIENCES (1985) 8, 529–566
*Printed in the United States of America*

# Unconscious cerebral initiative and the role of conscious will in voluntary action

**Benjamin Libet**
*Department of Physiology, School of Medicine, University of California, San Francisco, Calif. 94143*

**Abstract:** Voluntary acts are preceded by electrophysiological "readiness potentials" (RPs). With spontaneous acts involving no preplanning, the main negative RP shift begins at about −550 ms. Such RPs were used to indicate the minimum onset times for the cerebral activity that precedes a fully endogenous voluntary act. The time of conscious intention to act was obtained from the subject's recall of the spatial clock position of a revolving spot at the time of his initial awareness of intending or wanting to move (W). W occurred at about −200 ms. Control experiments, in which a skin stimulus was timed (S), helped evaluate each subject's error in reporting the clock times for awareness of any perceived event.

For spontaneous voluntary acts, RP onset preceded the uncorrected Ws by about 350 ms and the Ws corrected for S by about 400 ms. The direction of this difference was consistent and significant throughout, regardless of which of several measures of RP onset or W were used. It was concluded that cerebral initiation of a spontaneous voluntary act begins unconsciously. However, it was found that the final decision to act could still be consciously controlled during the 150 ms or so remaining after the specific conscious intention appears. Subjects can in fact "veto" motor performance during a 100–200-ms period before a prearranged time to act.

The role of conscious will would be not to initiate a specific voluntary act but rather to select and control volitional outcome. It is proposed that conscious will can function in a permissive fashion, either to permit or to prevent the motor implementation of the intention to act that arises unconsciously. Alternatively, there may be the need for a conscious activation or triggering, without which the final motor output would not follow the unconscious cerebral initiating and preparatory processes.

**Keywords:** conscious volition; event-related chronometry; free will; mental timing; motor organization; readiness potentials; unconscious processes; voluntary action

One of the mysteries in the mind–brain relationship is expressed in the question: How does a voluntary act arise in relation to the cerebral processes that mediate it? The discovery of the "readiness potential" (RP) opened up possibilities for experimentally addressing a crucial feature of this question. The RP is a scalp-recorded slow negative shift in electrical potential generated by the brain and beginning up to a second or more before a self-paced, apparently voluntary motor act (Deecke, Grözinger & Kornhuber 1976; Gilden, Vaughan & Costa 1966; Kornhuber & Deecke 1965). The long time interval (averaging about 800 ms) by which RP onset preceded a self-paced act raises the crucial question whether the conscious awareness of the voluntary urge to act likewise appears so far in advance. If a conscious intention or decision to act actually initiates a voluntary event, then the subjective experience of this intention should precede or at least coincide with the onset of the specific cerebral processes that mediate the act.

This issue has recently been subjected to experimental tests and analyses, which I shall review briefly (Libet, Gleason, Wright & Pearl 1983; Libet, Wright & Gleason 1982; 1983). The experimental findings led us to the conclusion that voluntary acts can be initiated by unconscious cerebral processes before conscious intention appears but that conscious control over the actual motor performance of the acts remains possible. I shall discuss these conclusions and their implications for concepts of "the unconscious" and of conscious voluntary action. I propose the thesis that conscious volitional control may operate not to initiate the volitional process but to select and control it, either by permitting or triggering the final motor outcome of the unconsciously initiated process or by vetoing the progression to actual motor activation. (The reader is referred to our original cited research papers for the full details of the experimental techniques and observations together with their evaluation, etc.)

## 1. Definitions of voluntary action and will

Since the meanings assigned to the terms "voluntary action" and "will" can be quite complicated and are often related to one's philosophical biases, I shall attempt to clarify their usage here. In this experimental investigation and its analysis an act is regarded as voluntary and a function of the subject's will when (a) it arises endogenously, not in direct response to an external stimulus or cue; (b) there are no externally imposed restrictions or compulsions that directly or immediately control subjects' initiation and performance of the act; and (c) most important, subjects *feel* introspectively that they are

 0140-525X/85/040529-38/$06.00 

performing the act on their own initiative and that they are *free* to start or not to start the act as they wish. The significance of point (c) is sharply illustrated in the case of stimulating the motor cortex (precentral gyrus) in awake human subjects. As described by Penfield (1958) and noted by others, under these conditions each subject regarded the motor action resulting from cortical stimulation as something done *to* him by some external force; every subject felt that, in contrast to his normal voluntary activities, "he," as a self-conscious entity, had not initiated or controlled the cortically stimulated act.

The technical requirements of experiments do impose limits on the kinds of voluntary choices and settings available to the subject. The nature of the acts must be prescribed by the experimenter. In the studies to be discussed here the acts were to consist uniformly of a quick flexion of the fingers or wrist of the right hand; this yielded a sharply rising electromyogram (EMG) in the appropriate muscle to serve as a trigger for 0-reference time. The subjects were free, however, to choose to perform this act at any time the desire, urge, decision, and will should arise in them. (They were also free *not* to act out any given urge or initial decision to act; and each subject indeed reported frequent instances of such aborted intentions.) The freedom of the subject to act at the time of his choosing actually provides the crucial element in this study. The objective was in fact to compare the time of onset of the conscious intention to act and the time of onset of associated cerebral processes. The specific choice of what act to perform was not material to the question being asked.

Volitional processes may operate at various levels of organization and timing relative to the voluntary act. These may include consciously deliberating alternative choices as to what to do and when, whether or not to act, whether or not to comply with external orders or instructions to act, and so on. If any of these processes are to result in the motor performance of a voluntary act, they must somehow work their way into a "final common motor activation pathway" in the brain. Without an overt motor performance any volitional deliberation, choosing, or planning may be interesting for its mental or psychological content, but it does not constitute *voluntary action*. It is specifically this overt performance of the act that was experimentally studied by us.

In the present experimental paradigm subjects agree to comply with a variety of instructions from the experimenter. One of these is an expectation that the subject is to perform the prescribed motor act at some time after the start of each trial; another is that he should pay close introspective attention to the instant of the onset of the urge, desire, or decision to perform each such act and to the correlated spatial position of a revolving spot on a clock face (indicating "clock time"). The subject is also instructed to allow each such act to arise "spontaneously," without deliberately planning or paying attention to the "prospect" of acting in advance. The subjects did indeed report that the inclination for each act appeared spontaneously ("out of nowhere"), that they were consciously aware of their urge or decision to act before each act, that they felt in conscious control of whether or not to act, and that they felt no external or psychological pressures that affected the time when they decided to act (Libet et al. 1982; Libet, Gleason, Wright & Pearl 1983).

Thus, in spite of the experimental requirements, the basic conditions set out above for a voluntary act were met. Conditions for the subject's decision as to when to act were designated to represent those one could associate with a conscious, endogenously willed motor action, so that one could study the cerebral processes involved in such an act without confusing them with deliberative or preparatory features that do not necessarily result in action.

Finally, one should note that the voluntary action studied was defined operationally, including appropriate and reliable reports of introspective experiences. The definition is not committed to or dependent upon any specific philosophical view of the mind–brain relationship. However, some implications that are relevant to mind–brain theories will be drawn from the findings.

## 2. Cerebral processes precede conscious intention

Two experimental issues have to be resolved in order to obtain a relevant answer to the questions about the relative timing of conscious intentions and cerebral processes in the performance of voluntary acts: (1) Is the RP a valid indicator of cerebral processes that mediate voluntary acts? (2) How can one meaningfully measure the onset of the conscious intention, urge, or will to perform a specific voluntary motor act?

### 2.1. RPs in voluntary acts

Self-paced acts were used in the discovery of RPs (Gilden et al. 1966; Kornhuber & Deecke 1965) and in subsequent RP studies (e.g., Deecke et al. 1976; Shibasaki, Barrett, Halliday & Halliday 1980; Vaughan, Costa & Ritter 1968). Such acts have features that may compromise the exercise of free volition or confuse its interpretation: (a) Recording an RP requires averaging many events. When these self-paced acts are repeated in a continuous series, with irregular intervening intervals of 3–6 sec as selected by the subject, they become boring and may come to be performed in a stereotyped and almost automatic way, with no assurance that conscious control could be exercised in each trial. (b) Since subjects were asked to act within an allotted time interval, they may be under pressure consciously or unconsciously to plan to act within the time limit; that is, the subject's voluntary choice of when to act may be compromised by an external requirement. (c) Subjects are required not to blink until just after each act. The need to blink may impel the subject to act, thus serving as an external controlling factor.

In a study of what we termed "self-initiated" acts, these external forces were minimized or eliminated (Libet et al. 1982). Each trial in an averaging series of 40 trials was initiated as a separate independent event after a flexible delay determined by each subject's own readiness to proceed; there was no limit on the time in which subjects were to act; they were given the option to blink if necessary. For each trial, subjects were asked to perform a simple quick flexion of the wrist or fingers at any time they felt the "urge" or desire to do so; timing was to be entirely "ad lib," that is, spontaneous and fully endoge-

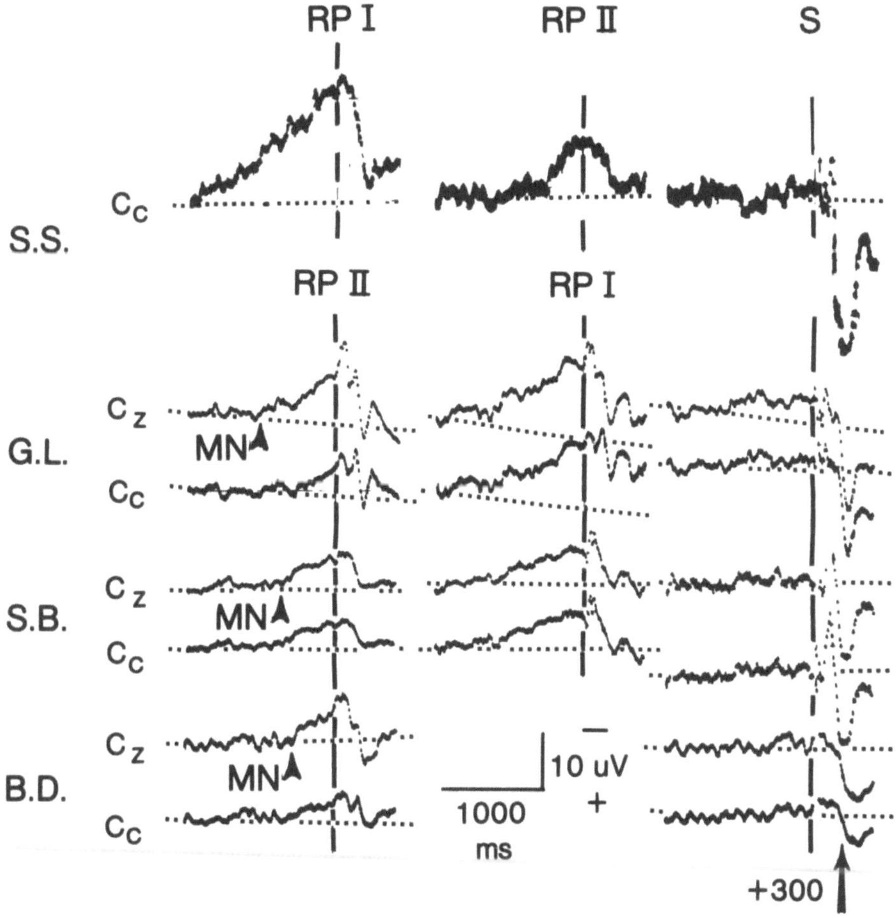


Figure 1. Readiness potentials (RP) preceding self-initiated voluntary acts. Each horizontal row is the computer-averaged potential for 40 trials, recorded by a DC system with an active electrode on the scalp, either at the midline-vertex ($C_z$) or on the left side (contralateral to the performing right hand) approximately over the motor/premotor cortical area that controls the hand ($C_c$).

When every self-initiated quick flexion of the right hand (fingers or wrist) in the series of 40 trials was (reported as having been) subjectively experienced to originate spontaneously and with no preplanning by the subject, RPs labeled type II were found in association. (Arrowheads labeled MN indicate onset of the "main negative" phase of the vertex recorded type II RPs in this figure; see Libet et al. 1982. Onsets were also measured for 90% of the total area of RP; see Table 1B). When an awareness of a general intention or preplanning to act some time within the next second or so was reported to have occurred before some of the 40 acts in the series, type I RPs were recorded (Libet et al. 1982). In the last column, labeled S, a near-threshold skin stimulus was applied in each of the 40 trials at a randomized time unknown to the subject, with no motor act performed; the subject was asked to recall and report the time when he became aware of each stimulus in the same way he reported the time of awareness of wanting to move in the case of self-initiated motor acts.

The solid vertical line through each column represents 0 time, at which the electromyogram (EMG) of the activated muscle begins in the case of RP series, or at which the stimulus was actually delivered in the case of S series. The dashed horizontal line represents the DC baseline drift.

For subject S.S., the first RP (type I) was recorded before the instruction "to let the urge come on its own, spontaneously" was introduced; the second RP (type II) was obtained after giving this instruction in the same session as the first. For subjects G.L., S.B., and B.D., this instruction was given at the start of all sessions. Nevertheless, each of these subjects reported some experiences of loose preplanning in some of the 40-trial series; those series exhibited type I RPs rather than type II. Note that a slow negative shift in scalp potential that precedes EMGs of self-initiated acts (RP) does not precede the skin stimulus in S series. However, evoked potentials following the stimulus are seen regularly to exhibit a large positive component with a peak close to +300 ms (arrow indicates this time); this P300 event-related potential had been shown by others to be associated with decisions about uncertain events (in this case, the time of the randomly delivered stimulus), and it also indicates that the subject is attending well to the experimental conditions. (Modified from Libet et al. 1982.)

nous. (For full technical details see Libet et al. 1982; Libet, Gleason, Wright & Pearl 1983.) Subjects reported that they were aware of the urge or intention to move before every act in the series; that is, the acts were not automatic or involuntary "tics." The absence of any larger meaning in this act appears to exclude external psychological or other factors as controlling agents. Acts of this kind may thus be taken as paradigmatic examples of unrestricted volition, at least in regard to choosing when to act. The basic initiating process for these simpler volitional acts may be the same as that for the actual motor expression of other, more complex forms of voluntary action, since the latter are manifested behaviorally only when final decisions to move have been made.

These self-initiated, endogenous acts were indeed found to be preceded by RPs (Libet et al. 1982). When all 40 self-initiated acts in an averaging series were performed with this spontaneous ad lib timing, with no reports of specific preplanning to act, the recordable averaged RP generally had an onset for its main negative rise at about 550 (±150) ms before the motor act began; these were called "type II" RPs (see Figure 1). (As is customary, the beginning of the muscle activity is signaled by the onset of the electromyogram, EMG, recorded at an appropriate muscle. This provides the "0-time" trigger for averaging the preceding scalp potential at the vertex and for other timing features.)

In some trials, subjects did report experiencing some general preplanning or preparation to act in the near future a few seconds before the act, despite the encouragement to be completely spontaneous. These occurrences were reported during the "debriefing" conducted at the end of each series of 40 trials. In those series that included even a small number of such reported experiences, a ramplike RP with onset at about −1050 ms (±175) was typically recorded (the "type I" RPs, Figure 1); these RPs were called type I because they resembled those RPs previously described for self-paced acts (e.g., Deecke et al. 1976). However, subjects all insisted that the more specific urge or intention to perform the actual movement was still experienced just before each act in a type I series, just as in the type II series; and they clearly distinguished this urge or intention from any advance feelings of preplanning to move within the next few seconds. In other experiments that required deliberate preplanning by instructing the subject to act at a preset time, there appeared a large ramplike RP that resembled the type I RP of our self-initiated acts. We concluded, therefore, that the RP component that starts at about −550 ms, the one that predominates in type II RPs recorded when all acts in a 40-trial series are spontaneous, is the one uniquely associated with an exclusively endogenous volitional process. The latter process is distinguished from a looser preintentionality or general preparation-to-act-soon that is not necessarily endogenous (Libet et al. 1982).

### 2.2. Timing the conscious intention to act

It presented a difficult challenge to devise the operational criteria for determining the time at which the subjects become aware of wanting or deciding to act. One begins with the premise that this subjective event is only accessible introspectively to the subject himself; some kind of report of this by the subject is therefore a requirement (Libet 1966; 1973; 1981b). Conscious subjective experience, in this case an awareness of the endogenous urge or intention to move, is a primary phenomenon; it cannot be defined in an a priori way by recourse to any externally observable physical event, including any behavioral action not directly representative of the subject's introspective report (Beloff 1962; Creutzfeldt & Rager 1978; Eccles 1980; Libet 1965; 1966; 1981a; 1981b; Nagel 1979: Popper & Eccles 1977; Thorpe 1974). The report, whether a verbal one or some other motor indication (e.g., pressing an answer key), *cannot* be an immediate one made as soon as the conscious experience has occurred: (a) Cerebral preparations for the motor action of reporting might introduce some confusing RPs of their own. (b) There could be a substantial delay for neurally organizing and achieving the motor actions required to make the report. (c) When a premium is put on the speediness of a response, as in measuring reaction time to a stimulus, there is no assurance that the motor response directly indicates when an actual subjective experience has occurred. The fast response to a stimulus can represent an unconscious mental process; but when the subject becomes consciously aware of the stimulus some hundreds of ms later (Libet 1965; 1966; 1973), the experience can be subjectively referred backward in time to an early neural signal (Libet 1981a; 1982; Libet, Wright, Feinstein & Pearl 1979).

For present purposes the experience of the time of the first awareness of wanting to move ("W") was related by the subject to his observation of the "clock position" of a spot of light revolving in a circle on the face of a cathode ray oscilloscope (CRO); the subject subsequently recalled and reported this position of the spot. (For technical details see Libet, Gleason, Wright & Pearl 1983.) Thus, the timing of this experience was converted to a reportable, visually related spatial image, analogous to reading and later recalling the clock time for any experience. This indicator of the time of first awareness of the intention to move could then be compared to (a) the actual time of the voluntary motor act, as indicated by the EMG recorded from the appropriate muscle, and (b) the time of appearance of the simultaneously recorded RP that is generated by the brain in advance of each act. For all self-initiated acts studied, the actual mean Ws for each series of 40 acts averaged about −200 ms (Table 1); that is, subjects reported becoming consciously aware of the urge to move 200 ms before the activation of the muscle (EMG) (Libet, Gleason, Wright & Pearl 1983).

### 2.3. Difference between RP onset and reported time of conscious intention, W

The RP onset time was found to be consistently in advance of W, the time of initial awareness of wanting to move (Table 1). For all of the series in which all 40 acts were experienced as fully spontaneous and unplanned, the average RP onset of (type II, described above) was about −535 ms relative to the initiation of muscle action (as indicated by the EMG). Reported times of conscious intention to act (W) in these same series with type II RPs averaged about −190 ms. The average onset of these RPs therefore *precedes* average W by about 345 ms. (For the significance of the even larger discrepancy in series ex-

Table 1. *Average times (ms) of reported awareness and recorded readiness potentials (RP) for all experimental series on 5 subjects, in 6 or more separate sessions for each subject. Each series consisted of 40 trials in which subjects reported only W or M or S times in that entire series. (Modified from Libet, Gleason, Wright & Pearl 1983.)*

*A.* Reported awareness times (ms) relative to recorded muscle activation (EMG).

| | W[a] | | (W-S)[b] | | M[c] | | (M-S) | |
|---|---|---|---|---|---|---|---|---|
| Subject | n[d] | $\bar{X}$ | n | $\bar{X}$ | n | $\bar{X}$ | n | $\bar{X}$ |
| S.B. | 8 | −125 | 5 | −123 | 4 | −59 | 4 | −19 |
| G.L. | 8 | −282 | 5 | −136 | 4 | −202 | 4 | −60 |
| B.D. | 7 | −152 | 4 | −249 | 4 | +51 | 4 | −32 |
| S.S. | 7 | −246 | 4 | −145 | 4 | −118 | 4 | −7 |
| C.M. | 8 | −227 | 4 | −165 | 4 | −103 | 4 | −20 |
| Grand averages | 38 | −207 | 22 | −160 | 20 | −86 | 20 | −28 |

*B.* reported time of conscious intention (W) related to recorded RP onset, separated for type I and II (see text).

| Type of RP, | Reported awareness times | | Onset of RP (in W series) | | (Onset RP) *minus* (W) using onset of: | | (Onset RP) *minus* (W-S), using onset | | |
|---|---|---|---|---|---|---|---|---|---|
| for W series | n[d] | W | $RP^{e}_{MN}$ | $RP_{90\%}$ | $RP_{MN}$ | $RP_{90\%}$ | n | $RP_{MN}$ | $RP_{90\%}$ |
| II | 20 | −192 | −535 | −527 | −343 | −333 | 14 | −366 | −323 |
| I | 12 | −233 | −1025 | −784 | −825 | −522 | 6 | −950 | −585 |

[a]W = time of first awareness of wanting to move (see text). [b]S was based on reported time of awareness of the sensation elicited by a near-threshold electrical stimulus pulse to the hand, delivered at a randomly irregular time in each trial. The attentive and other conditions (subject's observing and recalling "clock time" for each S) closely paralleled those for the W and M series, except that the event was an externally induced sensory one instead of a self-initiated motor one. The difference (S) between reported and actual stimulus times may be regarded as a measure of the subject's error or "bias" when observing and reporting under the experimental conditions employed (see text and Libet et al. 1982; Libet, Gleason, Wright & Pearl 1983). Almost all subjects exhibited a negative net bias for S (except for B.D.). For (W-S) values, the S bias exhibited by each subject is subtracted from the W values available in the same sessions. [c]M was time reported for subjects' awareness that they were actually moving, instead of wanting to move as for W. The consistently negative though smaller values for M suggest that it reflects the time of initiation of the final motor cortical output, i.e., the endogenous "command to move" (McCloskey et al. 1983), rather than the awareness of proprioceptive sensory impulses evoked after onset of the movement (see text). [d]n = number of series, each of 40 trials. Each average or $\bar{X}$ value for n series is the mean of the mean Ws (or mean Ms), each of which was determined for each series of 40 trials (see Libet, Gleason, Wright & Pearl 1983). [e]Onsets of RP, relative to EMG (electromyogram indicating that the activation of the muscle has started), are given for both the "main negative shift" (MN), as estimated by eye, and for the time at which the last 90% of the total area under the RP tracing begins.

hibiting type I RPs, those recorded when some acts were preplanned, see Libet, Gleason, Wright & Pearl 1983.)

This timing relationship, with the "physical" (cerebral process) preceding the "mental" (conscious intention), held not just for average values of all series but for each individual series of 40 self-initiated acts in which RP and W were recorded simultaneously. Although RPs of 40 events were averaged to produce the recorded RP, statistical and mathematical evaluation of the experimental data strongly supported the view that each individual RP precedes each conscious urge (see Libet, Gleason, Wright & Pearl 1983). The timing relationship also held regardless of which of the available parameters was used either to measure the onset of the RP (for the onset of its main negative component or for 90% of its area), or for W (using either the "actual" or the "order" mode of recall of the clock position of the revolving spot at the time of conscious intention; see section 2.4.3). Confidence in the significance of the difference between RP onset and W is further raised by the fact that it was almost invariably large in all the individual series when compared to the standard error of the mean value for W in each respective series. In addition, the individual W time reported for each act in a series of 40 trials was almost never negative to (timed in advance of) the onset of the averaged RP recorded for that series. In view of the foregoing considerations (and additional methodological checks listed in Libet, Gleason, Wright & Pearl 1983), the substantial interval by which RP onset precedes W appears sufficiently reliable. Questions about the validity and meaning of the values must still be considered.

### 2.4. Validity of criteria for the time of a conscious intention to act

Because subjective experiences are not directly accessible to an external observer, it may be logically impossible for the external observer to determine directly any feature of the experience (Creutzfeld & Rager 1978; Libet et al. 1979; Nagel 1979). This restriction applies also to the actual time of a subjective experience (Harnad, unpublished; Libet et al. 1979). We do not normally apply the criterion of logical impossibility to the validity of introspective reports by the people around us in everyday life although we do attempt to evaluate the accuracy of these reports. I do not know of any serious believer in Berkeleyan solipsism, even though that position may be logically unassailable. (On the other hand, the descriptions even of externally observable physical events cannot be regarded as having an absolute validity; they have been appropriately viewed as mental representations or constructs elicited by or developed from the available sensory experiences, e.g., Margenau 1984.)

One is always faced, then, with the unacceptable alternative of not attempting to study a primary phenomenological aspect of our human existence in relation to brain function because of the logical impossibility of direct verification by an external observer. Or one can attempt to evaluate the accuracy of the introspective report and gain confidence in its validity by applying indirect controls, tests and converging operations. In the present study we rely on the subject's ability to associate his introspective awareness (of the urge or decision to move) with the (later reported) position of a visually observed revolving spot, the "clock time." The crucial experimental question thus becomes: Is there any convincing way of estimating what might be the discrepancy between actual and reported times (for the subject's introspective experience of the urge to move)? The several independent types of control evidence discussed below provide confidence that the accuracy of the reported clock times is sufficient for present purposes (i.e., for determining the significance of the difference between RP onset and time of conscious intention).

**2.4.1. Comparisons of simultaneous events.** Our method requires that the subject observe simultaneously, for later report, the conscious urge or intention to move and a visual experience of "clock position" for the revolving spot on the CRO. Subjective timing comparisons of simultaneous but disparate events are known to be subject to potential errors (see Boring 1957; Efron 1973; Sternberg & Knoll 1973). However, we introduced a control series in each experimental session to help measure such an error. For this, a skin stimulus was delivered at an irregular, randomized time after the start of each trial and the subject reported the time of his awareness of that stimulus. All procedures were otherwise the same as in series of self-initiated acts (except that awareness of the stimulus replaced awareness of the urge to move). The actual time of the stimulus in the control series was later known to the investigator, and the discrepancy between the subject's reported timing and the actual stimulus time could be objectively determined. To the extent that simultaneous observation of visual clock time and awareness of skin sensation shares similar processes and difficulties with simultaneous observation of clock time and awareness of urge to move, one may regard any measured "error" in reports of stimulus time as an estimate of the potential error in reports of W (time of awareness of wanting to move). Skin sensations were commonly reported to occur somewhat in advance of (negative to) the actual delivery time, reminiscent of the prior entry effect (e.g., Allan; 1978; Boring 1957). However, the amount of the error found in the stimulus series did not qualitatively alter the difference between onset of RP and W; in fact, it generally enlarged the difference (Table 1).

**2.4.2. Judging onset time of an endogenous mental event.** It might be proposed that subjects do not judge the onset of an endogenous mental event such as conscious intention the same way they judge the onset of an experience induced externally by a skin stimulus. In relation to such a suggestion we note:

a. Each subject was instructed to "watch for" and report the earliest appearance of the awareness in question, and subjects did not raise any difficulties about doing this.

b. The onset time even of an intracerebrally generated event of some complexity, although admittedly induced by an applied stimulus, can be reported with no significant delays. In earlier work (Libet et al. 1979), onset time of a vaguely perceived near-threshold sensation elicited by a stimulus to a cerebral somatosensory structure (medial lemniscus) was judged subjectively to differ by only a few tens of ms from the sharper sensation elicited by a skin stimulus. In addition, both the medial lemniscus and the sensory cortex required repetition of stimulus pulses (at 20 per sec) for at least 200 ms, to elicit any subjective sensory experience at all in those experiments. Yet the subjects could consistently report a different onset time for each; they reported that the medial lemniscus–induced sensation began with no significant delay relative to the sensation elicited by a single pulse stimulus to the skin, whereas onset of the cortical sensation was delayed by the amount of the required stimulus duration (Libet et al. 1979).

c. For two different though related endogenous mental events related to the same voluntary act, the subjects consistently reported different onset times with an appropriate direction of difference. Under the identical experimental conditions for studying the self-initiated acts, the subjects were asked to report the clock time for their awareness of actually moving (M) instead of for awareness of wanting to move (W). M values were, unexpectedly, negative to EMG-0 time and slightly but consistently negative to reported times for awareness of skin stimulus (S) in which no movement was involved (see Table 1A).

Because M times were slightly before actual movement, this suggested that M may reflect awareness associated with the immediate initiation of cerebral motor outflow (Libet, Gleason, Wright & Pearl 1983). This would be in accord with the findings by McCloskey, Colebatch, Potter & Burke (1983) that subjective timing of one's own "command to move" preceded the EMG by up to 100 ms; a sensation of having already moved, elicited by input from peripheral sensory sources, was found to be separately reportable with an appropriately delayed time. M thus appears to be an endogenous mental event, different from but related to W. Nevertheless, the subjects did not confuse their reports of onset times for M with those of W; reports of W times (for awareness of wanting to move) were consistently negative to (in advance of) M times (for awareness of actually committing the movement), by about 120 ms on the average.

**2.4.3. Modes of reporting.** One way to test and improve confidence in the validity of the reported timings lies in using different and independent but converging modes of observing and reporting. Two quite different modes were used for reporting the "clock positions" of the CRO spot at the time of awareness: (a) absolute readings and (b) order relative to final stopping positions of the CRO spot, varied randomly (see Libet, Gleason, Wright & Pearl 1983). Yet both modes produced values for W that were essentially indistinguishable. (When reporting in the "order" mode, subjects had to recall the position of the moving spot [at the time of initial awareness of the urge to act] only with respect to a final resting position of the spot that was varied randomly in different trials. Subjects needed to make judgments about whether the CRO spot came to rest at a clock position that was "earlier" or "later" than the recalled position of the revolving spot when they were aware of the urge; they did not have to specify an absolute clock position of the moving spot associated with W [Libet, Gleason, Wright & Pearl 1983]. See also McCloskey et al. [1983] for an analogous order method for timing judgments.)

**2.4.4. Nonrecallable initial awareness of conscious intention?** It might be argued that a nonrecallable phase of a conscious urge exists, so that the reported time would apply only to a later, recallable phase of awareness. However, one should note that to report W time, the subject need recall only the clock position of the revolving spot at the time he first becomes aware of the urge or intention to move and not necessarily the initial awareness itself. In any case, there is no evidence for a nonrecallable initial awareness. But, like some other conceivable hypothetical uncertainties in timing an endogenous mental event, such a hypothesis cannot be excluded since it is presently not experimentally testable.

### 2.5. *RP as indicator of cerebral initiation*

For the experimental question about the initiation of a voluntary act, one must also consider whether the onset of recorded RP is a valid indicator of the time when cerebral processes begin to produce the act. The precise role of the cerebral activity represented by the RP in the initiation of the voluntary process is yet to be determined. It appears likely that the component of the RP associated with volitional preparation to act is generated in the supplementary motor area, a portion of the cerebral cortex located on the mesial surface of each hemisphere facing the midline (Deecke & Kornhuber 1978; Eccles 1982a; Libet et al. 1982). RPs associated with spontaneous self-initiated acts (type II) are indeed distinctly maximal at the vertex of the head (Libet et al. 1982), a scalp site that is above and adjacent to the supplementary motor areas. It has been proposed that the initial neuronal events in all voluntary movements arise in the supplementary motor areas (Eccles 1982). However, for present purposes it is not necessary that the full role of the supplementary motor area of the RP processes be established. It is only necessary to accept the RP as a valid *indicator* of *minimum* onset times for cerebral processes that initiate the voluntary act, even if these processes should be initiated elsewhere in the brain.

It might be proposed that the RP does not indicate directly or indirectly the specific initiation of the voluntary act. Rather, the RP might represent preprogramming processes that develop periodically without signifying a volitional function. The actual initiation of a given voluntary act would then depend on conscious activation or triggering of one of these preparatory sequences so as to generate an actual motor discharge. Such a proposal would seem to be an ad hoc speculation not supported by the experimental evidence. (a) The proposal would predict that endogenous RPs appear repeatedly without any associated subjective awareness developing and with no actual voluntary movements occurring. This has not been experimentally demonstrated and would seem to be untestable with present techniques. The RP that precedes an individual voluntary act is not clearly discernible from the background rhythmic activity; averaging of the pre-EMG periods (1.4 sec) for 40 acts gave us a usable though still noisy RP shift at the vertex. However, one should note that individual spontaneous negative and positive slow potential (SP) shifts have been successfully recorded during 5-sec periods preceding a choice reaction test and found to be related to proficiency of performance (Born, Whipple & Stamm 1982). These interesting spontaneous SPs were apparently maximal at frontal rather than vertex sites and they were either negative or positive in polarity; they presumably reflect processes different from those of the negative RP that is maximal at the vertex and obtained in a different mental context. (b) The recorded RPs in self-initiated acts do not exhibit any special electrophysiological event that might signal introduction of an activating process at the reported time of about −200 msec for the conscious urge (Libet et al. 1982; Libet, Gleason, Wright & Pearl 1983). (For RPs in self-paced acts see also Deecke et al. 1976; Shibasaki et al. 1980.) (c) The available evidence suggests that an RP precedes every voluntary act as well as the conscious awareness of the urge to perform each act (Libet et al. 1982; Libet, Gleason, Wright & Pearl 1983). Consequently, the proposal against RP initiation of the act would at best result in a two-stage mediation; "preparatory" cerebral processes would still unconsciously initiate the volitional sequence but consummation of the actual motor action would depend on a conscious control function. This sort of role for the conscious function is compatible with the thesis being advocated in this paper.

Is it possible that the subject's introspective observation of his conscious intention for each act would itself introduce a cerebral process that affects the recorded RP (a question raised by an anonymous editorial reviewer)? In a small number of experiments RPs were recorded for series of 40 self-initiated movements in which no reports of awareness time were requested from or made by the subjects. The RPs of these "no-report" series were similar in form and onset times to RPs of the "report" series (Libet et al. 1982; Libet, Gleason, Wright & Pearl 1983). Furthermore, reporting the time of awareness of a sensory stimulus delivered at a randomly irregular time ("S" series) required the same kind of attention and introspection by the subjects as did the reporting in self-initiated acts; yet there were no significant pre-event potentials at all in association with the stimulation experiments (e.g., Figure 1; Libet et al. 1982; Libet, Wright & Gleason 1983). One may conclude that the "introspective process" did not affect the RPs in any manner significant to the conclusions in the study, and that if there were any electrophysiological correlates of introspective observation or of the attentive state required for it, they are not manifested in the scalp recordings of RPs at the vertex.

## 3. Unconscious initiation of voluntary acts

Onsets of RPs regularly begin at least several hundred ms before reported times for awareness of any intention to act in the case of acts performed ad lib. It would appear, therefore, that some neuronal activity associated with the eventual performance of the act has started well before any (recallable) conscious initiation or intervention is possible. This leads to the conclusion that cerebral initiation even of a spontaneous voluntary act of the kind studied here can and usually does begin *unconsciously*. (The term "unconscious" refers here simply to all processes that are not expressed as a conscious experience; this may include and does not distinguish among preconscious, subconscious, or other possible nonreportable unconscious processes.) Put another way, the brain "decides" to initiate or, at least, to prepare to initiate the act before there is any reportable subjective awareness that such a decision has taken place.

It might be argued that unconscious initiation applies to the kind of spontaneous but perhaps impulsive voluntary act studied here, but not to acts involving slower conscious deliberation of choices of action. The possible role of unconscious cerebral activities in conscious deliberation is itself a difficult and open question. In any case, after a deliberate course of action has been consciously selected, the specific voluntary execution of that action, i.e., the cerebral activation and implementation of the actual motor deed, may well be related to that for the ad lib kind of act we have studied. Even when a more loosely defined conscious preplanning has appeared a few seconds before a self-initiated act, the usual specific conscious intention to perform the act was consistently reported as having been experienced separately just prior to each act by all subjects (Libet et al. 1982; Libet, Gleason, Wright & Pearl 1983). This leads me to propose that the performance of every conscious voluntary act is preceded by special unconscious cerebral processes that begin about 500 ms or so before the act.

### 3.1. Cerebral basis of unconscious mental functions

A role for "the unconscious" in modifying and controlling volitional decisions and actions was advocated long ago (e.g., Freud 1955; Whyte 1960). This role was inferred from analyses of strong but indirect psychological evidence. The present experimental findings provide direct evidence that unconscious processes can and do initiate voluntary action and point to a definable cerebral basis for this unconscious function.

In addition, these findings are in accord with a previous general hypothesis that dealt with the question of how the subjective conscious experience of each individual is related to his cerebral processes and what distinguishes this from unconscious processes. That hypothesis proposed that some substantial time period of appropriate cerebral activity lasting hundreds of ms may be required for eliciting many forms of specific conscious experiences (Libet 1965). The hypothesis developed out of experimental findings that cortical activities must persist for up to 500 ms or more before "neuronal adequacy" for a conscious sensory experience is achieved (Libet 1966; 1973; 1981a; 1982; Libet et al. 1979). This led to the further inference, supported by evidence, that those cerebral activities which did not persist sufficiently long would remain at unconscious levels. The present evidence suggests that a similar substantial period of cerebral activity may also be required to achieve "neuronal adequacy" for an experience of conscious intention or desire to perform a voluntary act. The experience of the conscious intention to act would, in these terms, arise as a secondary outcome of the prior unconscious initiating process; nevertheless, it could still have a role either in completing the initiating process ("conscious trigger") or in blocking its progression ("veto").

## 4. The conscious function in voluntary action

If the brain can initiate a voluntary act before the appearance of conscious intention, that is, if the initiation of the specific performance of the act is by unconscious processes, is there any role for the conscious function? It is of course possible to believe that active conscious intervention to affect or control a cerebral outcome does not exist and that the subjective experience of conscious control is an illusion (e.g., Harnad 1982). However, such a belief is not required even by a monist, determinist theory, as seen in Sperry's (1980) formulation of an emergent consciousness that can interact with and affect neuronal activity; and the theoretical physicist Margenau (1984) has claimed that conscious intervention in brain function can occur without any expenditure of energy or violation of the known physical laws. In any case, the potentialities for conscious control may be considered at a phenomenological level; that is, we can for the present discuss operational possibilities for conscious control at a level which does not require a commitment to any specific philosophical alternatives for mind–brain interaction, whether these be determinism versus free will or epiphenomenalism versus mental intervention.

I propose that conscious control can be exerted before the final motor outflow to select or control volitional outcome. The volitional process, initiated unconsciously, can either be consciously permitted to proceed to consummation in the motor act or be consciously "vetoed." In a veto, the later phase of cerebral motor processing would be blocked, so that actual activation of the motoneurons to the muscles would not occur. Such a role is feasible since conscious intention is reported to appear about 150 to 200 ms before the beginning of muscle activation (signaled by the EMG), even though it occurs several hundred ms later than the cerebral initiating processes. The late cerebral processes thought to lead more directly to descending discharge in the pyramidal cells may be reflected in the so-called final motor potential (MP) component near the end of the RP shortly before the muscle activation. An MP that is generated in the premotor/motor cortex contralateral to the activated hand begins about 50 ms (Deecke et al. 1976) or perhaps as little as 10 ms (Shibasaki et al. 1980) before the muscle EMG. There would remain a net period of about 100 to 200 ms in which conscious control could block the onset of the MP. A "premotion positivity" (PMP) may also develop about 90 ms (Deecke et al. 1976) or about 50 ms (Shibasaki et al. 1980) before the EMG. The significance of this component is still unclear. But even if the PMP is

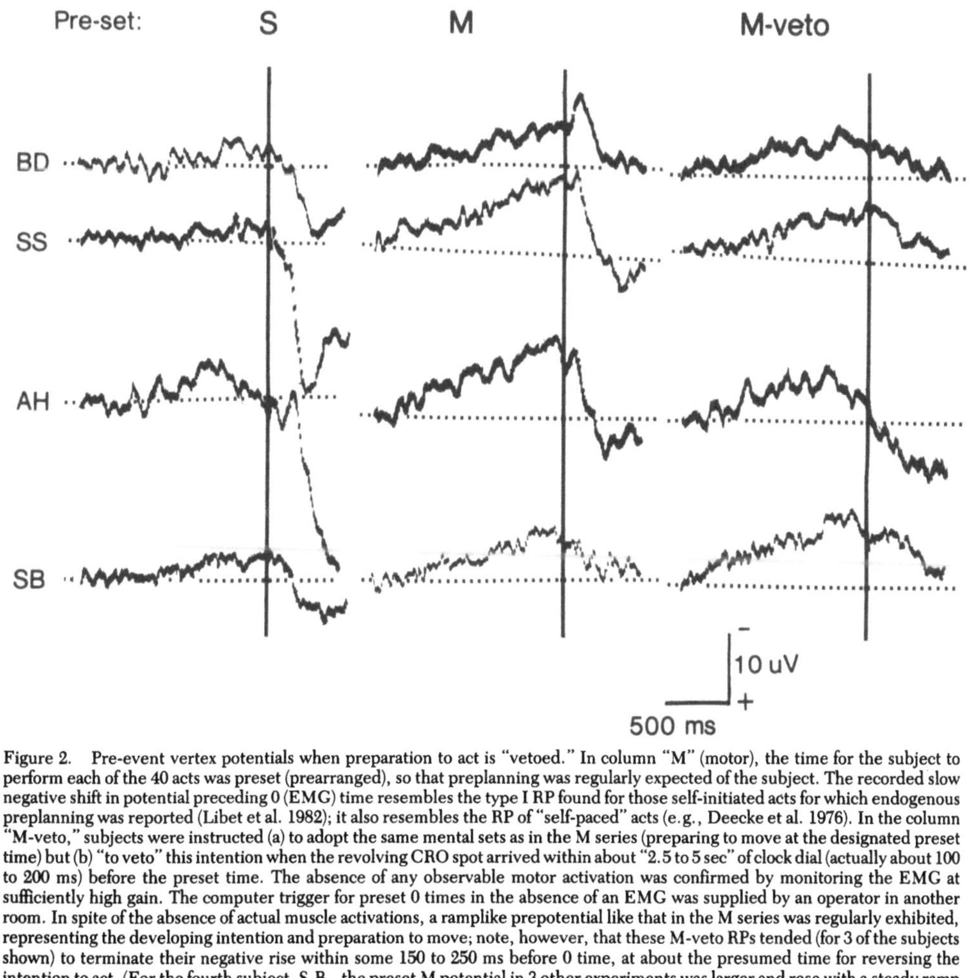


Figure 2. Pre-event vertex potentials when preparation to act is "vetoed." In column "M" (motor), the time for the subject to perform each of the 40 acts was preset (prearranged), so that preplanning was regularly expected of the subject. The recorded slow negative shift in potential preceding 0 (EMG) time resembles the type I RP found for those self-initiated acts for which endogenous preplanning was reported (Libet et al. 1982); it also resembles the RP of "self-paced" acts (e.g., Deecke et al. 1976). In the column "M-veto," subjects were instructed (a) to adopt the same mental sets as in the M series (preparing to move at the designated preset time) but (b) "to veto" this intention when the revolving CRO spot arrived within about "2.5 to 5 sec" of clock dial (actually about 100 to 200 ms) before the preset time. The absence of any observable motor activation was confirmed by monitoring the EMG at sufficiently high gain. The computer trigger for preset 0 times in the absence of an EMG was supplied by an operator in another room. In spite of the absence of actual muscle activations, a ramplike prepotential like that in the M series was regularly exhibited, representing the developing intention and preparation to move; note, however, that these M-veto RPs tended (for 3 of the subjects shown) to terminate their negative rise within some 150 to 250 ms before 0 time, at about the presumed time for reversing the intention to act. (For the fourth subject, S.B., the preset M potential in 3 other experiments was larger and rose with a steady ramp form until at least 50 to 100 ms before 0 time, unlike the M recorded in the session shown here; see Libet et al. 1982.) In column "S," a skin stimulus delivered at similar preset times replaced the preparation to act. Pre-event potentials were absent or relatively insignificant in the S series, in spite of attention and anticipation for each event being similar to those in M and M-veto series, in that the subject had to watch for and report those events in which the stimulus was omitted at the preset time. (Modified from Libet, Wright & Gleason 1983.)

assumed to reflect cortical motor activation just preceding the efferent discharge (Deecke et al. 1976) there would still remain about 60 to 100 ms after the "corrected" time of conscious intention, or 110 to 150 ms after the uncorrected time, in which conscious control could affect the PMP process.

#### 4.1. Evidence for "veto" control

The evidence for conscious veto is of two kinds: (a) Subjects in our study of RPs and conscious timings reported that during some of the trials a recallable conscious urge to act appeared but was "aborted" or somehow suppressed before any actual movement occurred; in such cases the subject simply waited for another urge to appear, which, when consummated, constituted the actual event whose RP was recorded (Libet et al. 1982). However, there is presently no technique available for recording and analyzing any RPs that may be associated with such spontaneous, irregularly appearing conscious urges to act that do not lead to an actual motor event. (b) In series of acts to be performed at prearranged times, subjects were instructed in advance to veto the developing intention/preparation to act and to do this about 100 to 200 ms before the prearranged clock time at which they were otherwise supposed to act. In these series a ramplike pre-event potential was still recorded during >1 sec before the preset time (Figure 2, "M-veto"), even though no actual muscle activation occurred (Libet, Wright & Gleason 1983). This resembles the RP of self-initiated acts when preplanning is present (Libet et al. 1982, type I RP). The form of the "veto" RP differed (in most but not all cases) from those "preset" RPs that were followed by actual movements; the main negative potential tended to alter in direction (flattening or reversing) at about 150–250 ms before the preset time (Libet, Wright & Gleason 1983). This difference suggests that the conscious veto interfered with the final development of RP processes leading to action. (Whether the above-mentioned MP or PMP components of RP are specifically eliminated by such a conscious veto remains to be analyzed.) In any case, the preparatory cerebral processes associated with an RP can and do develop even when intended motor action is vetoed at approximately the time that conscious intention would normally appear before a voluntary act.

The veto findings suggest that preparatory cerebral processes can be blocked consciously just prior to their consummation in actual motor outflow. As an alternative study, we might have randomly presented an external signal at which the subject would veto the prearranged or preset act. (External signaling to veto an act after a given *self-initiated* RP has begun is not technically feasible, since the individual RPs are not sufficiently discernible from the background EEG activity.) However, an externally signaled veto would not be an endogenous conscious process; as a quick reaction to a sensory signal it could even be generated unconsciously. It would of course be even more desirable to study the uninstructed veto of a spontaneous, self-initiated act, but, as mentioned, this is not presently possible technically because an objective trigger time for averaging RPs would not be available.

#### 4.2. Conscious "trigger" versus "veto"

An alternative mode of conscious control might lie in a requirement that a conscious "trigger" finally impel the unconsciously initiated cerebral processes to achieve the actual motor act. Conscious control would then have an active role in completing or consummating the volitional process; the absence of a positive conscious trigger would mean no actual motor act occurs. If one grants the availability of the veto process, then an active trigger role becomes a redundant and unnecessary means of achieving conscious control. On the other hand, it is conceivable that both modes of control, active trigger and veto blockage, are available. Whether by active positive triggering or by vetoing the completion of the volitional process, the conscious function may be thought of as selecting from among the possible acts developed by the unconscious initiating processes.

Would the appearance of a conscious trigger or veto also require its own period of prior neuronal activity, as is postulated for the development of the conscious urge or intention to act and for a conscious sensory experience? Such a requirement would imply that conscious control of the volitional outcome, whether by veto or by an activating trigger, is itself initiated unconsciously. For *control* of the volitional process to be exerted as a *conscious initiative*, it would indeed seem necessary to postulate that conscious control functions can appear without prior initiation by unconscious cerebral processes, in a context in which conscious awareness of intention to act has already developed. Such a postulate can be in accord either with a monist view, in which a conscious control function could be an ongoing feature of an already emergent conscious awareness (Margenau 1984; Sperry 1980), or with a dualist interactionist view (Popper & Eccles 1977).

### 5. Free will and individual responsibility

This is not the place to debate the issue of free will versus determinism in connection with an apparently endogenous voluntary action that one experiences subjectively as freely willed and self-controllable (see Eccles 1980; Hook 1960; Nagel 1979; Popper & Eccles 1977). However, it is important to emphasize that the present experimental findings and analysis do not exclude the potential for "philosophically real" individual responsibility and free will. Although the volitional process may be initiated by unconscious cerebral activities, conscious control of the actual motor performance of voluntary acts definitely remains possible. The findings should therefore be taken not as being antagonistic to free will but rather as affecting the view of how free will might operate. Processes associated with individual responsibility and free will would "operate" not to initiate a voluntary act but to select and control volitional outcomes. (Voluntary action and responsibility operating behaviorally within a deterministic view would, of course, be subject to analogous restrictions.)

Some may view responsibility and free will as operative only when voluntary acts follow slower conscious deliberation of alternative choices of action. But, as already noted above, any volitional choice does not become a voluntary action until the person moves. In the present study, the subjects reported that the same conscious urge or decision to move that they experienced just before each voluntary act was present and that it was similar whether or not any additional experience of general preplanning had already been going on. Indeed, the reported times for awareness of wanting to move were essentially the same for fully spontaneous acts and those with some preplanning (Libet, Gleason, Wright & Pearl 1983). One might therefore speculate that the actual motor execution even of a deliberately preselected voluntary act may well involve processes similar to those for the spontaneously voluntary acts studied by us. The urge or intention actually to perform the voluntary act would then still be initiated unconsciously, regardless of the preceding kinds of deliberative processes.

The concept of conscious veto or blockade of the motor performance of specific intentions to act is in general accord with certain religious and humanistic views of ethical behavior and individual responsibility. "Self-control" of the acting out of one's intentions is commonly advocated; in the present terms this would operate by conscious selection or control of whether the unconsciously initiated final volitional process will be implemented in action. Many ethical strictures, such as most of the Ten Commandments, are injunctions not to act in certain ways. On the other hand, if the final intention to act arises unconsciously, the mere appearance of an intention could not consciously be prevented, even though its consummation in a motor act could be controlled consciously. It would not be surprising, therefore, if religious and philosophical systems were to create insurmountable moral and psychological difficulties when they castigate individuals for simply having a mental intention or impulse to do something unacceptable, even when this is not acted out (e.g., Kaufmann 1961).

ACKNOWLEDGMENTS
This paper is based on a presentation at a conference, "Cerebral Events in Voluntary Movement," held at Castle Ringberg in West Germany November 14–19, 1983, organized by J. C. Eccles, O. D. Creutzfeldt, and M. Wiesendanger, under the auspices of the Max Planck Society (abstract in *Experimental Brain Research*, 1985). I thank Moreen Libet for helpful comments on an earlier draft of the paper.

# Open Peer Commentary

*Commentaries submitted by the qualified professional readership of this journal will be considered for publication in a later issue as Continuing Commentary on this article. Integrative overviews and syntheses are especially encouraged.*

## Problems with the psychophysics of intention

Bruno G. Breitmeyer
*Department of Psychology, University of Houston, Houston, Texas. 77004*

Several methodological and conceptual problems come to mind after a reading of Libet's article. For one, the timing of all consciously apprehended events under investigation was measured relative to the "clock position" of a dot revolving in a circle. Similar timing methods plagued by several problems have been used for over 100 years. Using a revolving dial, Wundt (1904) noted that the perceived time of a sensory event relative to the simultaneously visually perceived position of the rotating dial depended crucially on the angular rate of the dial's rotation and the other sense being stimulated. Libet's work is based on a single angular dot velocity; hence, despite acceptance of his particular implementation of the procedure by refereed journals, there is a very strong possibility that his measures are idiosyncratic.

Moreover, the timing of S, the awareness of a tactile stimulus, does not serve as a clear control that allows one to regard any timing "error" here as an indication of the potential error found in timing W, the awareness of the intent to act. First, judgments of intermodal sensory simultaneity depend on the particular senses investigated and the stimuli used. Besides the prior entry effect noted by Libet, intrinsic latency and processing rate differences among senses as well as latency differences introduced extrinsically by use of a near-threshold tactile stimulus relative to a clearly suprathreshold visual dot stimulus (Libet, Gleason, Wright & Pearl 1983) render use of any one estimate of timing error arbitrary and suspect. Second, attending to W may not be equivalent to attending to S, as Libet assumes. Indeed, one can voluntarily allocate attention to endogenously produced cognitive/mental processes as well as to mental processes produced exogenously by sensory stimuli. However, in the latter case a compulsory, stimulus-evoked allocation of attention is typically also engaged, as illustrated by Remington's (1980) and Jonides's 1981) studies of attention to brief suprathreshold visual stimuli. Insofar as Libet's near-threshold tactile stimuli were above threshold, their presentation would also evoke such an obligatory or nonvoluntary attention.

Even if one were to pass over these pertinent methodological problems, several concerns of a more conceptual nature need addressing. First, in what sense can the voluntary acts as operationally defined by Libet be paradigmatic of volitional action generally, particularly when he draws certain weighty religio-ethical implications from his findings? As Libet admits, his experimentally reduced acts of finger/wrist flexion occur in the absence of any larger meaning. Hence they are as limited in application to our understanding of volitional action as use of nonsense syllables is to our understanding of memory. By what rules do we proceed from these experimental findings to human volitional action (or memory) occurring inextricably within a rich, varied, and meaningful context? William James (1950) held that a *strictly* voluntary act must be guided throughout its whole course not only by volition but also by idea and perception. Moreover, he observed that consciousness, besides being primarily a selective, intentional process, is more or less intense depending on action's being more or less significant and hesitant (nonhabitual), that is, where indecision is present to a greater or lesser degree. Consequently, one might at least require that subjects choose freely among several actions, each of which carries some practical consequence (cost and benefit) rather than merely choosing to act or not in some stereotyped and inconsequential way.

To counter the requirement that a strictly voluntary act be characterized by slow conscious deliberation and existential alternatives of action, Libet notes that no volitional choice becomes voluntary action until the person moves. The implication is that Libet's paradigmatic acts tap this final, effective conscious intent, which invariably appears approximately 350 ms after an RP is generated but 200 ms before one actually moves. It should be noted that the actions investigated by Libet have been performed (by myself and several of my colleagues) without awareness of intent to act. By requiring subjects to attend to awareness of intent, Libet may have imposed intention artificially and in a way that is not comparable with more ecologically and existentially valid voluntary and intentional acts.

To illustrate, up to this point I was not consciously aware of intending to write down these thoughts. Yet a prior intention to write a critique occurred days ago. In fact, however, I could have chosen to intentionally write out my critique word by word, that is, with clear awareness of each intent to write each work just prior to writing it. Yet this or Libet's "hyperintention" brought about by self- or by experimental instruction in no way represents my voluntary actions in general. At best the hyper-awareness of intention functions as a monitor retrospecting on my much earlier plan, decision, or intention to write rather than as an instigator, motivator, or modulator of writing activity. In this view, the awareness of intent, though it falls just after the onset of RP and just before the onset of movement, poses neither a scientific nor a philosophical problem and has little if any bearing on issues of free will and responsibility.

Finally, even if one admits the legitimacy of Libet's procedure and interpretation, Libet hedges on and skirts around an important issue. Libet would have it that one can discuss the operational possibilities of conscious control of action on purely phenomenological grounds without commitment to specific philosophical alternatives such as determinism versus free will or epiphenomenalism versus mental intervention. Such a phenomenological bracketing is well-nigh impossible since it asks one to suspend any thesis of reality including the metaphysical assumptions hidden behind the very scientific enterprise being undertaken by Libet. In the context of his work, how can one talk of possibilities of conscious control, and not turn this talk into idle chatter, without taking a stand in particular on epiphenomenalism versus mental (conscious) intervention? On the one hand, if the conscious permissive "trigger" or restrictive "veto" is preceded by causally efficacious yet unconscious neural activity just as in the case of the consciously experienced intent to move (Harnad 1982), then that consciousness is mere afterthought, a reflection on events outside its causal control and, therefore, epiphenomenal. On the other hand, consciousness is a fact to each of us. Insofar as its existence is undeniable, it is a troublesome and abiding enigma, particularly to any accepted version of natural evolution. For to have evolved it must be as causally efficacious as is the hand that writes these words. Hence consciousness, including any conscious "trigger" or "veto," calls for some form of mental intervention. As scientists, we cannot stand on the sidelines and suspend or bracket the thesis of natural evolution. To do so would further mystify consciousness to a degree warranting silence.

## Free will and the functions of consciousness

Bruce Bridgeman

*Zentrum für Interdisziplinäre Forschung, Universität Bielefeld, 48 Bielefeld 1, Federal Republic of Germany*

Libet attempts nothing less than a beginning of the physiology of free will, an area where philosophical work previously has enjoyed a total lack of empirical restraint. The philosophical issues won't go away yet, however, and they remain important to interpreting the experiments. Two problems deserve special comment: the demand characteristics of the experiment and the generalization from millisecond-level operations to long-term behavioral planning.

A careful analysis of the experimental conditions reveals that the subjects' wills were not as free as the Libet article implies, for the small, sharp movements that they were instructed to make were not freely willed but were requested by the experimenter. The will of a subject was no more free in this design than in reaction-time experiments; the only difference between this experiment and the latter paradigms is that the instruction and the movement are decoupled in time. While performing the task, the subjects do nothing more than obey the instructions. The acts are a step removed from the instructions, and the issue of the source of timing for the irregularly repeated acts is an important one, but the behaviors should not be confused with instances of free will. It is even possible that free will, like the mind–body problem, will disappear as our understanding of the physiology of experience increases.

In a sense the subjects in the Libet experiments are asked to behave as though they had free will, whether such a thing really exists or not. Under these circumstances it is not clear whether we are seeing some fundamental property of the human nervous system or merely the program that the subject has set into play. To give another example of this process, consider a subject in a psychophysical experiment who is asked to draw boxes on pieces of paper. The psychologist could study the box-drawing machine as though it were designed only for this task, and the dynamics of the behavior, its physiological concomitants, and so on could be studied in detail. Box-drawing centers could be found in the brain, box-detecting circuits could be described in the visual system, and the prebox potentials could be analyzed. The artificiality of the task, though, would not be apparent no matter how detailed the analysis; in fact, the more detailed the analysis the less likely it is that the results will be interpreted as specialized operations of a more general-purpose machine. The subject has programmed himself to behave as if he were a box drawer and nothing else. Similarly, Libet's RPs may have characteristics unique to the rather specialized and unusual tasks required of his subjects. This is not to say that Libet's paradigms are invalid but only that they should be interpreted with caution.

The temptation to overgeneralize a specific task with its unique demand characteristics may also be related to the generalization of the veto principle at the end of Libet's article. The Bible's injunction not to commit adultery, we may expect, will be handled very differently from Libet's injunction not to move the fingers on a given trial. The confusion of levels is an error that I have called "Uttalism"after Uttal's (1971) injunction that properties of single-cell receptive fields cannot automatically be applied to behaviors of the whole organism. This problem has arisen in visual masking, where neurophysiologically based models, whether computer simulations (Bridgeman 1971; 1978; Weisstein 1972) or qualitative theories (Breitmeyer & Ganz 1976), rely on mechanisms too limited to reflect the subleties of real human behavior. No amount of tinkering with these theories will deal with practice and attention effects, for example, nor will they explain strong effects of rather small differences in stimulus patterns on masking. Similarly, the Libet data, important as they are, should not be confused with physiological studies of self-control in human behavior.

The finding that consciousness enters after the beginning of an identifiable set of neurological events can be viewed in the context of consciousness as a neurological system like any other, with specific jobs that help the organism to function effectively. Its jobs include handling situations that are difficult, dangerous, or novel (Norman & Shallice 1980), and it serves among other things to establish action schemata, order their priorities, and monitor their progress. Thus consciousness must be involved when a behavior is about to be executed, if that behavior might interfere with other ongoing schemata. In Libet's special case the only ongoing task is to sit still. Here, that stage of organizing a behavior that first requires access to consciousness can occur only a few hundred milliseconds before the behavior begins. We do not yet know what happens in the more general case, when other action programs are being executed at the same time.

## Consciousness and motor control

Arthur C. Danto

*Department of Philosophy, Columbia University, New York, N.Y. 10027*

It is a truth universally acknowledged that a physiologist in possession of a metaphysical prejudice must be in want of philosophical help. It is inconceivable save with reference to some such prejudice that Libet would find it necessary at the end of his paper, to postulate functions whose existence would be incompatible with everything he had up to that point been at pains to show. These are "conscious control functions," which "can appear without prior initiation by unconscious cerebral processes." But everything up to then would have disposed us to believe that motor acts are the consequence of exactly such initiating processes, revealed to the consciousness of the agent about 350 ms after onset, with the motor act itself taking place about 150 ms thereafter, barring endogenous intervention. But then, in that last fateful interval, abruptly and without experimental motivation, between the intention and the act falls the shadow of alien ideas. These are the "conscious control functions" that "trigger" or "veto" the act and that spring, cerebrally unsummoned, into being. Freud famously said the hysterical symptom seems to have no knowledge of anatomy. When a physiologist relaxes his laboratory scruples in favor of what must be physiologically mysterious, he is to be diagnosed as in the grip of a kind of metaphysical hysteria.

Surely conscious control functions have some physiological substance if they have physiological effects. And surely it should be an empirical matter whether or not their occurrence be cerebrally initiated through that kind of neuronal activity which precedes the occurrence of subjectively experienced intentions or "wantings to act." So why should it seem necessary to postulate them as thus unpreceded unless one believes precedent unconscious activity must queer some theory held dear by the writer – perhaps a position on the free will question? If Libet is right that "the present experimental findings do not exclude the potential for 'philosophically real' individual responsibility and free will," why should he act as though they did exclude that by postulating what he feels must be in place in order that responsibility and freedom have application? Philosophy must learn to live with scientific truth.

It seems to me that the existence of free will does not have as close a connection with "conscious deliberation of alternative choices of action" as Libet supposes. Choosings between alternative courses of action, in the preponderance of motor acts we perform, occur as the outcome of deliberations of which we are barely conscious, if at all. A slow-motion film of Matisse shows the artist making countless decisions with his fingers that at normal speed looks like a single confident chalk stroke defining the edge of a leaf. He may or may not have been conscious of each decision, but I suspect that he was conscious only of drawing a leaf. Consciousness, in moral theory, plays its role only in connection with premeditation, for which there is neither time nor occasion in the sort of spontaneous choosings we do in life and in the sort of laboratory Libet's work presumes. Happily, we are so wired that deliberation may occur without the mediation of consciousness at all. Consciousness is evolution's gift to us for rather special deliberative employment having to do, as responsibility and free will have to do, with *courses* of action – with *projects* – rather than the basic sorts of acts involving the simple flexion of a muscle or the moving of a hand to no further purpose.

Suppose one were to designate as intentions the entire cerebral processes that eventuate in motor acts, rather than restricting the intention to that fragment of the cerebral process which becomes conscious? The concept of intention was framed well before there was knowledge of cerebral process, but once it is accepted that much of deliberative action transpires without becoming conscious to the agent – because its being conscious would reduce our efficiency as agents – the concept might easily be extended to cover more than would have been necessary in periods when the mental and the conscious were closely identified. We might indeed think, in those cases in which some segment of the intention becomes conscious, of the preceding segment as *preconscious* intention. Then, in the standard case, this is what happens: The intentional is formed; some milliseconds later the agent becomes conscious of his intention; some milliseconds later the motor act occurs as intended. Why do we need an extra "trigger" since there is no empirical basis for its existence but only a "necessary postulation"? It would be like requiring a trigger in mechanics in order to explain the fact that a body, moving in a straight line with uniform velocity, continues to move in a straight line with uniform velocity, when in fact all we need is an explanation of acceleration, or change in direction and velocity. Why should not the intention be enough to trigger the movement? I surmise that Libet thinks that simply allowing to take place what is already in process is too passive a role for conscious intention if freedom is to be robust enough for our moral vision of ourselves. In my view, all we need to explain is *changes* in intention. But these can be well under way before we are conscious of the change, with the entire cerebral process, including the fragment of it that is conscious, as the veto of the previous intention. There is plenty of time to abort the action if the intention arises before consciousness of veto.

In brief, instead of the conscious control functions playing the special on–off role of metaphysical switches, we have the play of cerebral processes, in which consciousness informs us of what we have decided to do. Whether these decisions themselves are free belongs to a different topic, but my claim is that freedom and consciousness have less to do with each other, and certainly so in the execution of simple behaviors, than Libet supposes. Once he realizes that it is only because he believes that they have much more to do with each other than the data he presents justifies, he may drop from the inventory these curious operations that owe their existence in his article to an insufficiently self-conscious agenda.

Knowing what we are embarked upon need not be a causally inert fact about ourselves when in fact we are embarked upon projects with horizons wider than the circumscribed boundaries of the laboratory. In these straitened confines, the projects to which responsibility and freedom have application scarcely can flourish. Commonly we do not simply move our hands; we do so with larger purposes in mind–to wave away a canapé, to signal the death of a gladiator, to stifle by gesture the cackle of subordinates, to set up perturbations for the distraction of a wasp, or to express some agitation or other through the language of the body. Our minds bent upon these, consciousness simply assures us we are in contact with ourselves.

## The time course of conscious processing: Vetoes by the uninformed?

Robert W. Doty

*Center for Brain Research, University of Rochester Medical Center, Rochester, N.Y. 14642*

Perhaps the most important feature of this latest in the series of ingenious experiments by Libet and his colleagues is the demonstration it provides that the neurophysiological basis of conscious awareness can be subjected to meaningful analysis. This has profound philosophical import, the more so since it adds further evidence for the probable uniqueness of the neural processes accessible to or directly producing conscious experience.

It has long been apparent that many, indeed probably most, neural transactions are utterly devoid of or incapable of an element of consciousness–for example, autonomic regulation, hormonal release, adaptations in visuomotor control, cerebellar activity, and all neuronal discharge during most of a night's sleep (see Doty 1975). A particularly dramatic example is the loss of visual sensation despite demonstrably continuing retinal input when one is viewing a Ganzfeld (Bolanowski & Doty 1982) or absolutely fixated image (Rozhkova, Nickolayev & Shchadrin 1982); the same is probably true for the disappearance of stimuli rotating about a fixed locus in the peripheral visual field (Hunzelmann & Spillmann 1984). On the other hand, in these instances the absence of a direct conscious concomitant to the neuronal activity in the forebrain clearly does not mean that such activity is inaccessible to consciousness. Rather, these phenomena of visual loss are probably an extreme example of the workings of that still mysterious tool of consciousness, selective attention. Thus, it is apparent that there are neural processes that lie forever outside the domain of conscious experience and that there are others for which a conscious concomitant is elective.

In still other instances it seems that information garnered from sensorial processes lacking an experiential component can nevertheless be incorporated into the guidance of movements consciously controlled. These issues have previously been well discussed in these pages in relation to the phenomena of afferent discharge from muscle spindles (Roland 1978) or blindsight (Campion, Latto & Smith 1983). However, the fact that unconscious neuronal activity is constantly in play during movement seems well recognized, as in the common inability to perform properly a habitual, rapid movement while endeavoring to exert conscious control over all its components. (Try intellectually constructing and planning the motions of your fingers in tying your shoes!)

Now, perhaps Libet's experiments are detecting this, the unconscious components of an organized movement. There is a voluntary initiation of these components, just as there can apparently be a voluntary cancellation (veto) of them. The actual decision to release the movement occurs only against the background of readiness, the point at which the subconscious set of the neuronal program, possibly being arranged in striated-cerebellar circuitry, is acceptably complete. The unconscious part, just as in tying one's shoe, proceeds pari passu with, and apparently slightly ahead of, the overt and consciously released movement; but this does not mean that the unconscious components proceed or arise independently of conscious control. After all, the neurons for each are all embedded and intertwined within the same brain; and one does not know yet whether the neuronal transactions resulting in conscious perception are a manifestation of a special type of neuron or a special form of activity within groups of neurons of diverse form and chemistry.

It seems to me that this is a much more satisfactory explanation of Libet's fascinating observations, that an aura of unconscious preparation for movement perpetually surrounds the ever-moving focus of consciousness, and that the aptly named "readiness potential" (Kornhuber & Deecke 1965), which Libet records prior to the "decision" to actually perform the movement, is a manifestation of this process. The alternative, which he seems to favor, is that "the brain" proceeds independently of conscious control to prepare movements, which can then be either consciously allowed or consciously "vetoed." The great flaw in this interpretation is that, if the preparatory movement is wholly outside conscious control, how could a conscious process then "know" what will ensue if it fails to veto the brain's proposal? In this scheme, consciousness is relegated to an intuitive process of guessing what it may be that "the brain" is up to and being ever on the alert that the demons of the unconscious do not set in motion some act inappropriate to the conscious plan. While such views of brain processes may be permissible in the poetic fantasy of Freudian psychology, they are not neurophysiologically convincing.

## Mental summation: The timing of voluntary intentions by cortical activity

John C. Eccles

*Max-Planck-Institut für Biophysische Chemie, Göttingen, Federal Republic of Germany*

My commentary starts with an acceptance of the extraordinary findings reported by Libet. With great ingenuity he has been able to train subjects to report retrospectively the timing of their voluntary intention to make a simple sharp movement. I am not concerned with the subtle distinctions he makes between types of conscious endogenously willed motor actions, for example, whether or not the subject was cognizant of planning in advance. For me the decisive discovery is that the subjectively experienced onset of intention to move is about 200 ms before the muscle activation and about 350 ms *after* the onset of the readiness potential (RP), which provides some integrated signal of the cortical activity preceding the movement.

To simplify my hypothesis, I will assume that the voluntary intention to move acts on the supplementary motor area (SMA) [see Goldberg; "Supplementary Motor Area Structure and Function," this issue] and thence through the various pathways to the motor cortex and so by the pyramidal tract to bring about the movement (cf. Eccles 1982b). It is very tempting to follow Libet in interpreting these findings as establishing that cortical activity (of the SMA, for example) initiates not only the voluntary movement but also, after some hundreds of milliseconds, the introspective experience of having initiated the movement, which thus becomes an illusory experience. I shall consider later Libet's veto hypothesis, by which he attempts to preserve the responsibility of the conscious self by means of its power to veto the ongoing cortical activities that would otherwise lead to the movement.

I now present a hypothesis that accepts all of Libet's experimental observations but that nevertheless preserves fully the role of conscious intention in initiating the movement. The hypothesis has several components.

(1) It is proposed that there is a fluctuating background of activity in the cerebral cortex and in the SMA that can in part be generated by the reticular activating system and that was proposed by Oshima (1983), possibly to involve a "set" for movements.

(2) As discovered by Libet, the mental intentions reported by subjects begin about 200 ms before the movement. The hypothesis is that *these intentions tend to be timed unconsciously* by the subjects so as to take advantage of the spontaneous fluctuations in the cortical activity ((1) above). Since the RP as observed is formed by the averaging of a large number (fifty to hundreds) of recordings of scalp potentials with zero time given by the onset of the electromyogram, it is a mistake to assume tacitly that the averaging eliminates the random fluctuations. If there is a tendency for the initiation of the movements to occur during the excitatory phases of the random spontaneous activity, the earlier phase of the RP may be no more than the averaging of the premonitory spontaneous activity. If that is so, the RP does not signify that cortical activity initiates the movement. Instead, the hypothesis is that the spontaneous fluctuations of cortical activity merely adjust the phase of the conscious initiation to the intention some 200 ms before the movement.

(3) It is further postulated that this timing of the intention in relation to the phases of cortical activity is a learned phenomenon having the advantage that it secures opportunistically the most effective occasions for initiating voluntary actions. The lower right corner of Figure 1 illustrates the hypothesis. It is to be noted that the activities of the SMA are reciprocally related to the mental intentions, the arrows being directed both ways across the frontier between mind and brain.

(4) In the further development of the hypothesis we have to consider how the mental event of an intention can cause changes in the neuronal responses of the SMA. Let us first focus attention on a single synaptic bouton, which may be, for example, on a pyramidal cell of SMA. As shown for very diverse central synapses by Jack, Redman, and Wong (1981) and by Korn and Faver (1985), a presynaptic impulse evokes the liberation from the bouton of a single synaptic vesicle probabilistically, the probability factor being usually less than 1 in 2. This probability can be increased or decreased with consequent changes in synaptic effectiveness. As described by Akert, Peper, and Sandri (1975), each bouton has a single paracrystalline structure, the presynaptic vesicular grid that holds about 50 synaptic

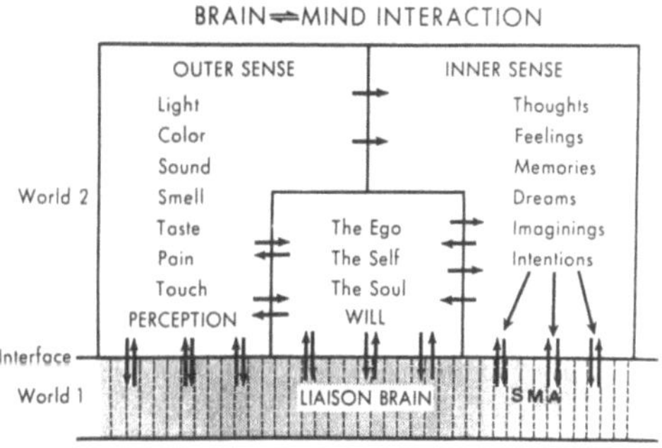


Figure 1. (Eccles). Information flow diagram for brain–mind interaction in human brain. The three components of World 2 – outer sense, inner sense, and psyche or self – are diagrammed with their communication shown by arrows. Also shown are the lines of communication across the interface between World 1 and World 2 – that is, from the liaison brain to and from these World 2 components. The liaison brain is the columnar arrangement indicated by the vertical broken lines.

vesicles, and somehow it controls the probability of their emission. The hypothesis is that the immaterial mental event of intention acts analogously to a probability field of quantum mechanics, as proposed by Margenau (1984), and modifies the probability of emission of a synaptic vesicle by a presynaptic impulse. Thus an intention is effective only insofar as there is an adequate quota of presynaptic impulses; hence the necessity for the learned timing of intentions in relation to the fluctuating waves of SMA background activity.

(5) Any effect of a mental intention in altering probabilities of quantal emission from a bouton is orders of magnitude too small to cause the sequence of neuronal actions leading to an effective discharge of motor pyramidal cells. It is conjectured that there has to be an immense collusive action of the mental intention on the multitude of boutons on one neuron and on a large assemblage of similarly acting neurons. This is in accord with the findings of Brinkman and Porter (1979) that, when a monkey is carrying out a voluntary act, there is excitation of many similarly acting neurons in the supplementary motor area 100 to 200 ms before the onset of the electromyogram.

(6) Furthermore, according to the hypothesis there is also a reverse flow of information (Figure 1), the SMA activity being subconsciously "sensed" when a mental intention is being initiated. This is the most obscure component of the hypothesis. Yet it is generally recognized that in the perceptual areas of the cortex much activity can occur subconsciously, as in the refined experiments of Libet (1973) on somatosensory perception, where weak repetitive stimulation of the somatosensory cortex may have to continue for 0.5 sec before the cortical activity reaches the threshold for conscious perception.

The veto experiments of Libet are very ingenious and offer further evidence of mental control of cortical activity with the late flattening of the RP.

In conclusion, the hypothesis here presented offers a general explanation of the findings of Libet while preserving the essential character of dualist interactionisms. The early phase of the RP may be no more than an artifact arising from the technique of averaging. There is no scientific basis for the belief that the introspective experience of initiating a voluntary action is illusory.

NOTE

**Commentator's mailing address: CH 6611 Contra (TI), Switzerland**

## Brain mechanisms of conscious experience and voluntary action

Herbert H. Jasper
*University of Montreal and the Montreal Neurological Institute, McGill University, Montreal, Quebec, Canada H3Z 1E7*

For many years Libet has been carrying out carefully controlled crucial electrophysiological experiments on the relation between electrical stimulation and responses in sensory cortex and pathways in the conscious human brain and verbal reports of conscious awareness with the surprising result that it seems to require considerable time (about 500 ms) for activity in sensory systems to reach the threshold of conscious awareness. The precise neuronal mechanisms involved in this delay have not been specified. It has long been known from experiments carried out under light barbiturate anaesthesia or natural sleep that evoked potentials and unitary responses from single cells in sensory cortex (somatic, visual, or auditory) are preserved, even including the complex information processing involved in feature detection in visual cortex as studied by Hubel and Wiesel, in states that probably preclude conscious awareness (light barbiturate anaesthesia).

Libet now uses the "readiness potential" (RP) to time unspecified cortical events that precede an ad libitum voluntary motor act as compared to the timing of the subject's conscious awareness of intention to move, with the surprising conclusion that willed voluntary movements arise out of brain mechanisms that precede conscious awareness of the intention to move and must therefore be subconscious. Controls on the reliability of subjective reports of the timing of conscious awareness of intention to move depend on the accuracy of memory, introducing another important factor that in my opinion has not given adequate consideration. Is it not possible that brain mechanisms underlying awareness may occur without those which make possible the recall of this awareness in memory afterward? Patients with epileptic automatisms, for example, may carry out many apparently intentional complex motor acts, often remarkably appropriate ones (such as driving in traffic), without being able to recall having done so afterward. A similar state of apparently "automatic" behavior may occur with certain drugs such as scopolamine. I realize that it may be impossible to dissociate mechanisms of awareness from those of memory recall under the conditions of these experiments, but there is a problem here that should be given serious consideration.

Concerning the more philosophical implications of these studies, Libet should be commended for his ingenious and precise experiments, which have clarified, if not solved, the age-old problem of mind–brain relationships. I agree that mental events can be considered scientific data even though they are difficult to measure, and that they may well play a most important role in the direction of behavior and consequently of the brain mechanisms underlying this behavior, while at the same time mental events must depend upon highly integrative brain functions (i.e., interactionism rather than dualism). It may well be that there are specialized neuronal systems extending throughout cortical and subcortical structures but separate from specific afferent and efferent pathways to cerebral cortex, which mediate mechanisms of conscious awareness, analogous to the outworn hypothesis of the reticular system or the "centrencephalic system" of Penfield.

Libet has provided us with important temporal constraints on two aspects of this problem: the temporal summation required for conscious awareness and the delay in awareness of conscious intention of voluntary movement. I would suggest that he now direct more of his attention to brain circuits separate from the primary sensory or motor pathways in the search for mechanisms more closely related to mechanisms of consciousness, as originally suggested by Hughlings Jackson in his search for brain mechanisms of "highest level seizures."

## Voluntary intention and conscious selection in complex learned action

Richard Jung

*Department of Neurophysiology, University of Freiburg, D-7800 Freiburg, Federal Republic of Germany*

Libet's experiments are limited to the recording of readiness potentials (Kornhuber & Deecke 1965), which precede the decision to make or veto brief finger flexions. These simple movements are made voluntary, but the will acts here only as a trigger. Willed intention is more important in goal-directed and complex movements such as writing. These also contain many unconscious mechanisms and become partly automatized by learning. Slow brain potentials recorded during action may give additional information complementing the analysis of readiness potentials that appear before movement.

I agree with Libet that the conscious will mainly selects and controls our action and that unconscious preparatory cerebral mechanisms are important. I doubt Libet's assertion, however, that the subject's will does not consciously initiate specific voluntary acts. It is true in complex and learned movement too that several more or less unconscious motivations contribute to the action. In man, however, even emotional or instinctive actions and skilled movements can be voluntarily initiated, directed, and set for their duration, as they can be inhibited and blocked by will. Willed intention is normally related to consciousness.

***Cerebral correlates of intention.*** The interaction of instinctive, willed, and learned factors in human decisions to act can be demonstrated by skilled movements and mental activity such as language and calculation tasks. Cerebral correlates of these conscious acts have been recorded in man (Jung 1984).

The electrophysiological correlates of goal-directed and writing movements are large surface negative potentials that appear as an increase of the readiness potentials at the precentral and parietal cortex (Grünewald-Zuberbier et al. 1978; Jung et al. 1982). The aiming potentials terminate in a positive shift when the goal is reached (Figure 1A). The preparatory body posture and balance accompanying the consciously steered goal-directed movement become unconscious after the primarily will-controlled movement is trained (Jung 1981; 1982). The aiming potentials are probably related to the willed performance of goal-directed movements and to their programming. Normally, our consciousness is concerned only with the goal and not with the automatic and learned mechanisms of action involved in its pursuit. Owing to the limits of conscious information content, conscious intention is only a small part of the whole action program.

***Limited capacity of consciousness.*** In conscious perception and voluntary action the information flow of the human nervous system is extremely reduced from the input of $10^7$ to about 20–50 bits/sec (Küpfmüller 1971). This narrow range of consciousness necessitates selective processing and automatized programs for all voluntary skilled movements (Jung 1981). Such unconscious motor programs are acquired by learning.

Let me explain the selective and restricted role of the conscious contribution to complex action by the experience of goal-directed movements and other tasks. As a subject in the experiment shown in Figure 1A, I was consciously aware of my aiming intention during the action and of two other intentions that were in the background and less salient. The first intention, to direct the object to the goal, began with the readiness potential of 1 sec duration and continued for 3 sec. The second, to fixate the target and not to look to my hand, was less conscious, and the third, to suppress blinking by staring, was sometimes interrupted by involuntary blinks. Of course, special activation of arm muscles, needed during the task, was not conscious. Hence, the voluntary conscious intention to reach the target was combined with a negative veto to avoid eye movements and associated with automatized hand movements.

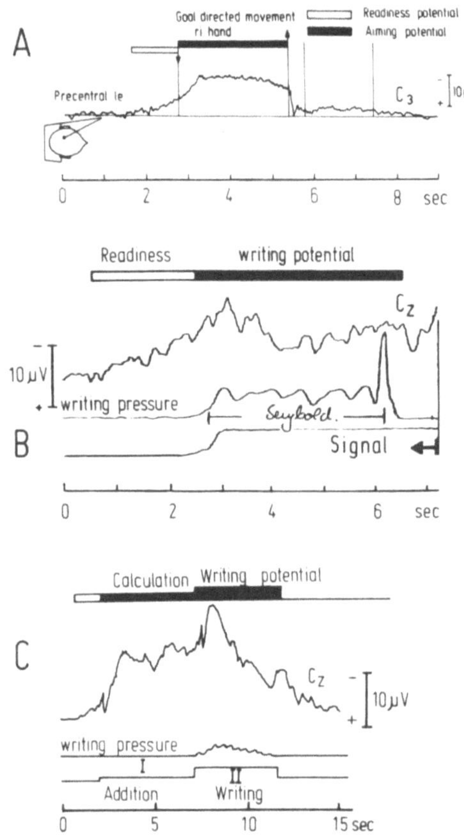


Figure 1. (Jung). Readiness, aiming, writing, and calculation potentials in normal right-handers. Reference: A, mastoids; B, C, earlobes. *A.* Aiming potential (***Zielbewegungspotential***) in left precentral] region ($C_3$) during goal-directed movements of right hand (from ↓ to ↑). Backward averaging of 34 trials. The readiness potential begins 1 sec before movement starts, approximately with the conscious decision to reach the target. Adapted from Grünewald-Zuberbier et al. (1978) and Jung (1981). *B.* Writing potential at vertex ($C_z$) while subject writes his name. The readiness potential begins 2 sec before movement starts, together with the intention to write. Hand movement is recorded as writing pressure. Backward averaging of 32 trials. Adapted from Jung et al. (1982). *C.* Calculation potential during addition of 2 two-digit numbers followed by writing potential. Both are triggered by acoustic signals. During the initial period of expectation and ocular fixation a small readiness potential arises. A large negative calculation potential follows in period I after hearing the task numbers. Writing down the results causes the second largest peak in period II. The triggering signals elicit evoked potentials that precede the slower potentials. A conscious intention to calculate and write follows the perception of these signals. Forward averaging of 32 trials. Experiment no. MSV 135/1.

Writing one's name, as shown in Figure 1B, was preceded by a conscious decision to start each writing sequence. However, the subject was not aware of the detailed performance of finger movements since repeated name writing had become automatized. In 1C the subject consciously intended the mental calculations of the perceived numbers in period I and the writing down of the result in period II. The special mode used in problem solving, however, often *failed* to be consciously experienced. Such acts had been automatized by years of learning writing and calculation.

***Variations of conscious timing.*** For the writing act shown in Figure 1B, Libet's final time of decision, said to arise about 200 ms before action, would imply an extremely long "unconscious initiative" of several seconds. The readiness potentials before repeated word writing may last 2–3 sec; that is, they can be six times longer than the readiness potentials recorded before simple movements (Schreiber et al. 1983). We interpret this as a sign of more complex and thus more time-consuming preprogramming. It appears improbable that such cerebral potentials of 3 sec duration are initiated "unconsciously" without willed intention. Our subjects tell us that they experience a first impulse to act well before writing begins. This preparatory intention, however, is rather vague, and we are not conscious of the learned complex cerebral programs of writing after they become automatized during 8 to 10 years of learning in school.

During training of skilled movements, which before learning had been guided by conscious control, a progressive reduction of conscious intention occurs, thus leading to automatization. The willed intention to start such trained motion programs becomes restricted to voluntary triggering and timing.

In summary, I do not deny unconscious elements in voluntary movements. Rather, I stress the importance of preconscious motivation, learned and automatized mechanisms, that is, of unconscious programs that contribute to voluntary action. I doubt only that Libet's experiments can prove the unconscious initiation of all self-paced voluntary acts. His results may be explained by the small information capacity of conscious introspection and of recall during the combined observation of the clock and the intention to move.

## Consciousness as an experimental variable: Problems of definition, practice, and interpretation

Richard Latto

*Department of Psychology, University of Liverpool, Liverpool L69 3BX, England*

The traditional role of the electrophysiologist has been to relate electrical events in the brain to sensory inputs and motor outputs. More recently there have been attempts to relate electrophysiological activity to the cognitive processes, such as expectancy (Grey Walter, Cooper, Aldridge McCallum & Winter 1964) and selective attention (Hillyard, Simpson, Woods, Van Voorhis & Munte 1984), that lie between these inputs and outputs. It is normally assumed that aspects of these processes may be part of our conscious experience and that we may sometimes draw on evidence from conscious experiences to suggest how they operate, but as theoretical constructs they are essentially neutral in their relation to consciousness. Expectancy and selective attention can be identified and discussed without having to specify what parts of the processes are conscious and what parts are unconscious. Libet is now extending the paradigm still further by attempting to relate electrophysiological activity directly to conscious experiences. He is well aware of the enormous problems raised by doing this and he addresses most of them impressively and effectively. But it is the function of *BBS* Commentaries to raise questions and dwell on perceived weaknesses. This can be done under three headings.

***1. Problems of definition.*** The three conditions laid down in section 1 of the target article as necessary for establishing voluntariness of action seem reasonable. But condition (c), that the subject *feels* he is acting voluntarily, is not easy to handle operationally and is therefore of doubtful usefulness. How do we know the subjects' self-reports are accurate? Feelings of voluntariness are labile and difficult to tie down, as the attempts to investigate the voluntariness of actions performed under posthypnotic suggestion (Wagstaff 1981) or of subjects' obedience to authority (Milgram 1974) make clear. The problem here is the compliance, which may be operating on the feelings of voluntariness or on the report of that feeling, or both. Supposing Penfield (1958) had stimulated some point higher up in the chain of motor control than the motor cortex and found not only that consistent and repeatable movements were produced but that the subjects also reported that they felt that they performed the act on their own initiative and that they were free to start or stop the act as they wished. Now suppose the patients were shown exactly what was being done to them and then asked again whether their movements were voluntary. If the explanation had been made properly, they would be bound to say they were not acting voluntarily. Which self-report would be the correct one? Maybe it does not matter too much to Libet whether or not the report is accurate in the context of voluntariness, but this difficulty of reliability applies to all reports of conscious experience, including the report of conscious intentions, which is the central variable in Libet's experiments. It must matter if these are not reliable.

***2. Practical problems.*** There are two difficulties here. The first concerns the reliability of self-reports of awareness and is really a more operational restatement of the problem raised at the end of the last section. The subjects are being asked to report when they are first aware of the intention to act. Assuming there is a gradual development of awareness, this is equivalent to making a threshold judgment using an ascending method of limits (for a more detailed account of threshold determination procedures, see Haber & Hershenson, 1980, chap. 2). As a technique, such judgments are open to criterion shifts, and therefore threshold shifts, when the subject changes his biases and expectations. The technique would also normally be expected to overestimate the threshold, leading in Libet's experiments to a time-of-occurrence estimate later than if it had been possible to use more reliable threshold techniques. Similar difficulties arise in all attempts to use awareness as an experimental variable, for example, in studies of blindsight, discriminative behavior elicited by stimuli of which the patient is unaware as the result of damage to his visual cortex (Campion & Latto 1985; Campion, Latto & Smith 1983), and in attempts to demonstrate semantic activation without conscious identification (Holender 1986; Latto & Campion 1986). The fundamental problem is that it is not possible to use adequate, criterion-free signal-detection procedures in a situation where the independent variable of signal or stimulus strength is the subject's conscious experience and is therefore not only not under experimental control but is also unknown to the experimenter.

The second practical difficulty for Libet is in the attempt to estimate relative timing. He has shown in other experiments (Libet, Wright, Feinstein & Pearl 1979) that the experience of a discrete stimulus may be subjectively referred back in time to an earlier neural signal closer to (though still later in time than) the stimulus. That is, the event was perceived as happening about 200 ms before the neural processes necessary to produce its perception were complete. It is therefore likely that in the present experiments the perceived time at which the moving spot is at a certain position would also be referred back. Now if the perceived time at which the awareness of the intention to act occurs is also referred back by exactly the same amount, there is no problem. The two events, awareness and the perception of the spot, will be in synchrony and the timing of the latter will be an objectively accurate guide to the timing of the former. But if the perceived time at which the awareness of intention occurs is not referred back, in the way that the perceived time of direct cortical stimulation is apparently not referred back (Libet et al. 1979), or if it is referred back by a different amount, then the timing procedure is invalidated. There is no evidence either way on this. It is worth noting, however, that if awareness of intention is not referred back at all, then the timing procedure would give a time for its occurrence that was late by the amount the perception of the spot was being referred back.

There are therefore two possible reasons why Libet's time-of-occurrence estimates for the awareness of intention might appear late relative to the more objective measurements of the time-of-occurrence of the readiness potential, in addition to Libet's own explanation that the readiness potential does actually develop before conscious awareness.

***3. Problems of interpretation.*** If we accept Libet's evidence and conclude that the initiation of a voluntary act is unconscious, at least for his experimental situation of deciding when to make a voluntary movement, then we should surely also accept that the whole process of voluntary action might sometimes be unconscious. How would Libet's condition (c) (see section 1) apply in such a case? It would require the passive and retrospective reporting of events that at the time they occurred were not open to conscious experience. So, applying condition (c), at the time it occurs the action is involuntary, but the subsequent reporting of a feeling of voluntariness retrospectively converts it into a voluntary act. Alternatively, we have to conclude that condition (c), and consciousness, are irrelevant to the question of whether or not to act is voluntary.

Libet suggests that the reason why conscious awareness does eventually develop in his situation is in order that there may be a conscious veto on action. He does not make it clear why this veto has to be conscious. If the initiation of a voluntary act can be unconscious, why could not the subsequent veto also be unconscious? Nor is the evidence for a veto presented in section 4.1 very strong. His subjects' anecdotes could be interpreted in other ways, for example, as the occurrence of mistaken feelings that an act had been initiated. And the findings (Libet, Wright & Gleason 1983) from the experiment with preset responses that were not therefore voluntary according to Libet's own criteria but that could be voluntarily suppressed by the subject when instructed to do so by the experimenter are using a paradigm so different from the central one that it is difficult to generalise between them.

If the veto is set aside as a role for consciousness, we are left with consciousness as a passive process taking a few hundred milliseconds to develop, both for external stimuli and for internal decision-making processes of the kind described here. Even without all the other difficulties outlined above, this rather barren conclusion should be enough to suggest that the electrophysiological investigation of conscious awareness is not yet a fruitful branch of science. The slow negative potential over the frontal lobes was first described by Grey Walter et al. (1964). Perhaps the siren call of its currently fashionable name, the readiness potential, with its implied association with consciousness, should be rejected in favour of a return to Grey Walter's original and far more neutral label, the contingent negative variation.

## Do we "control" our brains?

Donald M. MacKay

*Department of Communication and Neuroscience, University of Keele, Staffordshire ST55BG, England*

Libet is to be congratulated on the care and ingenuity with which he has articulated his position – a challenge to the rest of us to articulate our views with similar care. Given the presupposition that our conscious initiative means "*control* of the volitional process," it may (as he suggests) be "necessary to postulate that conscious control functions can appear without prior initiation by unconscious cerebral processes." This presupposition, however, seems to me unwarranted and arguably mistaken. There is, I think, an alternative way of looking at the relation between conscious control and brain activity that would wholeheartedly support Libet's emphasis on human responsibility without requiring the dubious postulate he thinks necessary.

We all know what it means to control the movement of a car by using our limbs; we also speak of consciously controlling, for example, a skilled finger movement, and people have even learned to control (i.e., regulate) the firing of their own motor neurones under suitable feedback. What makes these cases of "control" is that we have criteria in terms with which to *evaluate* what happens. Mere outward causal linkages are not sufficient. What we cannot in principle evaluate, we cannot control.

I see no reason to hold that in this sense we normally "control" (or should have any wish to control) our most central brain processes. That our conscious thinking, valuing, and choosing (sometimes) *determines the form* of our action is, I believe, a fact of daily experience. That such conscious mental activities have direct correlates in our brain activity seems a well-founded hypothesis, especially if the correlated brain activity is thought of in informational/stochastic rather than physical/energetic categories. That we consciously *control* these correlates, however, does not follow.

To see the logical non sequitur here, consider first an inanimate example. The autopilot in an aircraft in a clear sense evaluates and controls the plane's altitude, speed, and the like. It does so in and through an internal computational network of physical processes, which are ultimately linked to receptors and effectors in the aircraft. But does it in the same sense "control" these internal processes? Surely not; these are processes that it has no means of evaluating, for it is in them that it has its own being as an evaluative controller. Its evaluative and other computational processes certainly *determine the form* of their physical embodiment; but it would be a confusion of categories to say that they *control* it.

Now of course we are conscious agents, while autopilots (we believe) are not; but the same distinctions between categorical levels of analysis must clearly be recognised in the human case. If, as I suggest, we think of our conscious agent as *embodied in* our physical brain activity, then some (though not all) of that activity will have its form determined by our conscious thinking, valuing, and deciding. Motor acts casually dependent on such activity may then be under conscious control; but it would make no sense – it would involve a confusion of category levels – to conclude that we must therefore be able to "control" the cerebral correlates of our own thinking or that the cerebral correlates of a conscious decision must appear without causal initiation in prior cerebral processes.

From this perspective there is a clear distinction between those cerebral processes which are, and those which are not, direct *correlates* of conscious experience; but it would be inept to apply the category conscious/unconscious to any cerebral process as such. If, as I have argued elsewhere (MacKay 1951; 1966; 1982), the direct correlate of conscious experience is cerebral activity at a self-supervisory evaluative level, Libet's data have a simple and instructive interpretation. Far from suggesting that we have no conscious control over acts whose cerebral causes antedate our awareness of the urge to act, they would merely indicate that conscious volition is embodied in a stochastic cerebral process, in which the setting of evaluative criteria is not triggered until some prior physical process reaches a critical threshold. Whether or not "vetoing" is possible thereafter, the action is ours, because it is in that very same stochastic process that we have our being as conscious and deliberative agents. On this view the link between conscious decision and action is even more intimate than that between cause and effect (MacKay 1965; 1980). More firmly than any interactionist hypothesis, it pins to our own door responsibility for all we consciously choose to do.

## Toward a psychophysics of intention

Lawrence E. Marks
*John B. Pierce Foundation Laboratory and Yale University, New Haven, Conn. 06519*

Consider the following *Gedanken* experiment. The experiment generally follows Libet's design – a subject generates a simple and well-defined, self-initiated act, such as wrist flexion, at various points in times wholly of the subject's own choosing – but it contains some modifications. The first two modifications, though they make my paradigm technically impossible, nevertheless remain true to the spirit of Libet's. First, I shall assume that we are able to measure the "readiness potential" or RP on each and every occasion on which it occurs. Second, I shall assume that if we otherwise follow Libet's paradigm, we will find that every occurrence of an RP leads, about 400 ms later, to a report of conscious intent to perform the motor act (and is unrelated to any possible decision within the next 150 ms or so to "veto" the performance of the act). In other words, I am assuming that the RP is not irregular in appearance or an artifact of averaging but is regularly and reliably related to a subsequent awareness of conscious intent.

Now, however, I wish to make a more profound change – a "*Gedankener*" change – in the experimental paradigm. Assume it were possible within the course of an experimental session to "stop action," both physiological and mental, to halt the proceedings at a point, say, 200 ms after the main negative shift of the RP, which would be about 200 ms before the subject would report, under normal circumstances, becoming aware of conscious intent. With the sequence of events on "hold" in this wholly imaginary experiment, I would like to be able to query the subject, "Do you think you are likely to want to move your wrist within the next few tenths of a second?" Of course, as a control measure I would ask the same question of the subject at other, randomly chosen, points in time, points at which no RP had been in evidence in recent moments. Were all of this possible, I strongly suspect that subjects would be much more likely to acknowledge an intent to act at "test" moments – that is, during the supposed "unconscious interval" between the occurrence of the readiness potential and the first awareness of an intent to act – than at "controls."

This suspicion is largely represented by an analogy I would like to draw between intentions to perform voluntary acts and perceptions of weak signals. Consider now a parallel experiment on signal detection: On a given trial a weakly luminous light may be flashed, and the subject is asked to respond either "yes," a light was detected, or "no," it was not. In fact, the light is flashed on half of the trials; the other half contains "blanks." Under a given set of conditions (instructions, rewards or punishments for various types of correct or incorrect responses) and with a suitably selected light intensity, a subject will correctly detect (respond "yes" to) a certain fraction of the presentations of the light stimulus (hits) but will also incorrectly identify (respond "yes" to) some smaller proportion of the nonstimulus or blank presentations (false alarms). As instructions, rewards, and so forth are manipulated but the intensity of the light is held constant, the percentages of hits and false alarms rise and fall in tandem.

To construct this example, I have freely borrowed from the enormous literature on the theory of signal detectability (see, for example, Green & Swets 1966; Swets, Tanner & Burdsall 1961), a theory that interprets these kinds of findings as follows: Every presentation of the sensory stimulus produces an internal response, which adds to the always present, ineluctable noisy background, a background that is continuously fluctuating. To the subject, then, the task of detecting a stimulus becomes one of distinguishing stimulus-plus-noisy-background from noisy background alone. The subject tries to maximize performance by setting up a cutoff along the dimension of the underlying internal sensory response: When that response exceeds the cutoff, the subject response "yes"; otherwise, the subject responds "no." The overt response is therefore jointly a function of the internal sensory reaction to the stimulus and the particular cutoff or criterion. In principle, some information is available on every stimulus trial. A telling finding is, for instance, that in a multiple-choice paradigm (a signal is presented to one of four possible locations), when first guesses are incorrect, second guesses can be correct at a frequency above chance.

May an analogous "criterion" exist for the reporting of conscious intention to perform motor acts? In a sense what I am proposing is that up to some point near (within 100–200 ms of) the projected time of voluntary motor activation, "awareness" of the intention is as much a function of the "criterion for reporting" as it is of the strength of the underlying intention itself. Just as many a subject in a psychophysical "threshold" experiment will set a high criterion, avoiding "false alarms" but at the cost of "missing" many stimulus trials, so too may some time interval following an RP be one of high criterion for reporting an awareness to act. It follows from this analysis that the final brief moment before action, or veto, is one in which either the criterion drops to a level sufficiently low that intention is dramatically evident or the "intensity" of the underlying intent increases markedly, for it is likely that the "intensity" of intentions themselves can fluctuate, can differ from occasion to occasion or over time.

But the most significant point, I think, is the possibility that there may be some very general processes or mechanisms governing the transition from nonawareness to awareness, from nonperceived to perceived, across the so-called threshold of consciousness. Might we seek a unified theory of conscious elements (percepts, intentions, et al)? Rather than dichotomize between not aware and aware, I would suggest a probabilistically determined continuum. Long ago, Leibniz (1916) argued both for the existence of "unperceived perceptions" and for a continuity in the gradations or qualities of consciousness. The conscious entities he identifies as monads. Although Leibniz denied that temporality applies to monads, I propose the opposite–that there is a temporal continuity in which I would call the "potential awareness" of voluntary acts, and that it is precisely this temporal continuity to which a psychophysical model applies.

## Conscious and unconscious processes: Same or different?

Philip M. Merikle and Jim Cheesman
*Department of Psychology, University of Waterloo, Waterloo, Ontario, Canada N2L 3G1*

Being cognitive psychologists interested in the study of unconscious perceptual processes, we read Libet's review of his research program with great interest to see whether it provided new insights into the relationship between conscious and unconscious processes. Unfortunately, we were disappointed. From our point of view, Libet's research only documents an implicit assumption made by many cognitive psychologists, namely, that self-reports of conscious awareness are based on underlying brain processes. Furthermore, although some of Libet's conclusions concerning the relationship between unconscious brain processes and conscious awareness are interesting, these ideas lack empirical support, because they are based on speculations that are untestable using his methodology.

To understand our conclusions, it is useful to consider the approach adopted by cognitive psychologists to study unconscious perceptual processes. Two basic questions have guided research in this area: (1) Does perceptual information for which there is no conscious awareness influence behavior? (2) Do conscious and unconscious perceptual processes lead to distinguishable behavioral consequences? In order to answer these questions, cognitive psychologists have used research designs that allow the potentially distinctive effects of conscious and unconscious perceptual processes to be compared or contrasted. The experiments, in their simplest form, involve two separate conditions: One condition involves the assessment of performance following the presentation of unconscious perceptual information, while the second condition involves an evaluation of performance when the same perceptual information leads to conscious awareness.

In contrast to the approach adopted by cognitive psychologists, Libet's critical empirical findings are derived from observations made within a single experimental condition. The key finding underlying the entire paper is that a readiness potential (RP) *always* precedes a self-report of conscious awareness, which, in turn, *always* precedes a voluntary action. Thus, in Libet's experiments, RPs, conscious awareness, and voluntary action are *perfectly* correlated. Given this perfect correlation, it is impossible to distinguish the potentially distinctive behavioral consequences of brain processes that do and do not lead to conscious awareness. Furthermore, without evidence to indicate that RPs in the *absence* of reported awareness also precede behavioral acts, there is no empirical support for Libet's critical assumption that RPs and self-reports of conscious awareness reflect unconscious and conscious processes, respectively. In fact, given the perfect correlation between the two measures, there is simply no need to distinguish between these measures theoretically.

The only conclusion that can be made with confidence on the basis of Libet's findings is that conscious awareness of an impending voluntary action is always preceded by specific brain processes. This empirical observation, although interesting, only confirms a generally held implicit assumption. As long as it is assumed that conscious awareness is based on underlying brain processes, an assumption consistent with the views of most cognitive psychologists, then it is not surprising that certain brain processes occur prior to self-reports of conscious awareness. In fact, how could it be otherwise? For example, in our studies of perceptual awareness for visual stimuli (e.g., Cheesman & Merikle 1984; in press), activity in the optic nerve must logically occur prior to conscious awareness of the stimuli. Thus, by establishing that brain processes always precede self-reports of conscious awareness, Libet has only confirmed a necessary implicit assumption made by most cognitive psychologists. In our opinion, if RPs always precede reports of conscious awareness, then this entire sequence of brain and behavioral responses should be viewed as reflecting conscious activity.

Finally, Libet's findings do not address the questions that cognitive psychologists find most interesting. These questions concern the separate or distinctive roles of unconscious and conscious processes in determining voluntary action. Because of inherent limitations it is not possible to use Libet's methodology to investigate either the distinctive contributions of conscious and unconscious processes or the interactions that may occur between these two types of processes. Thus, even though Libet discusses a number of interesting ideas concerning how conscious and unconscious processes may interact, his empirical findings do not provide any support for these speculations, since his results demonstrate only that voluntary actions are preceded by two perfectly correlated events: RPs and self-reports of conscious awareness. It is this correlation that must be eliminated before the distinctive roles of conscious and unconscious processes can be established.

## Conscious decisions

Chris Mortensen

*Department of Philosophy, University of Adelaide, North Terrace, Adelaide, South Australia 5001, Australia*

Libet distinguishes two possible functional roles for the urge to move specifically coming to consciousness: veto and trigger. The difference is that a conscious veto is something whose absence leads to the action's occurring; a conscious trigger, on the other hand, is something necessary for an action, so that without the trigger the action does not occur.

Libet, however, draws attention to earlier work according to which a short (300-ms) period of appropriate cerebral activity is required to achieve the "neuronal adequacy" necessary for a conscious experience. He proposes that the same is true of the relationship between the readiness potential, RP, and the decision to act at about −200 ms. This suggests a third role, a more functionally epiphenomenal one, according to which the apparent conscious decision is neither a veto nor a trigger but *merely* the coming to consciousness of an unconscious process already in progress and indicated by the RP.

Libet's arguments for the veto hypothesis amount to arguments for the functional efficacy of the conscious experience of a decision. First, subjects not infrequently report the conscious experience of an intention to act followed by the acts being aborted. Second, in different experiments with instructions to veto at preset times when the decision becomes conscious, the same RP occurs but flattens out at about the time (−200 ms) at which the intention becomes conscious and veto reportedly occurs. Libet notes that such evidence must be indirect because of the fact that the individual RP spike is not detectable above background noise and must hence be derived by averaging. Both of these arguments are, however, consistent with the veto itself's arising from a prior unconscious veto process and only "incidentally" later coming to consciousness, as the former point about neuronal adequacy would also suggest. It must be conceded that the second argument adds the feature that conscious instructions by the experimenter are a causally relevant factor that it is not unreasonable to suppose operates via a conscious veto mechanism. But this is not logically forced on us. Indeed, in all these experiments, conscious instructions by the experimenter at the beginning are (partly) responsible for the unconscious RP when it occurs, if not its exact timing.

There is one standard argument for the functional efficacy of the conscious experience of intention: Whatever seems real to consciousness (even if it is an illusion) needs explanation and so is not functionally epiphenomenal. But to concede that consciousness has a function is not to say that the function is specifically *veto or trigger*. Furthermore, the implied evolutionary argument here cuts both ways: Since we are focusing on the last-minute decision to move, rather than "diffuse" preplanning (which presumably has whatever "higher" function conscious deliberation has), then the mechanisms involved may well be older and evolutionarily prior to conscious decisions.

For all this, Libet's case for a causally relevant conscious decision is persuasive. However, there is a specific problem with the conscious-trigger hypothesis, namely, that the conditions of the experiments are artificial to the extent that much normal movement occurs when our minds are very much on other things. Conceivably a trigger occurring later than the onset of the RP still invariably occurs, unconscious but sufficiently like a conscious decision to be worth retaining that term for. But aside from the difficulty of verifying such a trigger, it seems neater to opt for a positive veto function. This also fits better with the efficacy of instructions: If instructions to veto operated to hold back a trigger, then this looks like an internal veto mechanism anyway. It is harder to agree with Libet's suggestion that both trigger and veto functions might be independently present in the conscious experience. Remember that before the conscious trigger would operate, there has been a rising RP for 300 ms already. But a true trigger is *necessary* for action. In its absence the action would not occur, irrespective of a veto function, consigning the latter to an epiphenomenal role. Similarly for vetos: In the absence of a veto, action proceeds. The only way to have both, it would seem, is to have them acting in series, which would be a more complicated conscious mechanism. All of this, of course, accepts Libet's assumption that vetos and triggers are not trivial converses of one another (with "trigger" defined as "absence of veto" and vice versa). The trivial converse manoeuvre is a way to have both together at merely linguistic cost, but there does seem to be a genuine difference between vetos and triggers that such linguistic legislation would obscure. Again, the fact that much action occurs with one's mind on other things suggests that conscious vetos have an (occasional) role but that conscious triggering does not. A better role for a trigger would be whatever *unconscious mechanism sets off the RP rise.*

Finally, the moral implications are, I suspect, not what Libet proposes, even though I am sympathetic with Libet's general position at this point. Libet seems to be operating on some sort of *moral responsibility* theory, according to which one can be held responsible only for one's exercisings of conscious control. While such exercisings obviously are morally significant, it seems better to base moral prescriptions on those which will be efficacious in moral education. Remember that we are dealing with initially unconscious final decisions to act, subject to last-minute conscious veto or trigger. At least if the veto model is correct, such final unconscious decisions are exceedingly dangerous beasts, well deserving of castigation and the attention of moral educators. (We have already argued that prior instructions affect the arising of the RP.) The same point would apply on the third, consciousness-as-functionally-epiphenomenal suggestion. Only on the conscious-trigger model would the preceding unconscious process be serving a less-than-morally-vital function.

## Brain physiology and the unconscious initiation of movements

R. Näätänen

*Department of Psychology, University of Helsinki, 00170 Helsinki, Finland*

The new technology introducing a variety of means to monitor different aspects of brain activity has made it possible to obtain a many-sided and detailed picture of brain processes occurring during various subjective experiences and behavior. However, although each subjective experience may have its unique brain state, the reverse is not true: Data from diverse sources are rapidly accumulating to suggest that a number of brain processes occur in response to sensory stimuli and underlie essential aspects of information processing but have no representation in conscious experience (e.g., Näätänen in press; Näätänen et al. 1978; 1980; 1982; in press). Libet and his associates' insightful research on the initiation of movement, which in an interesting way links physiological and subjective data, appears to provide a particularly important case of brain processes with no simultaneous subjective counterpart. This is because, as Libet claims, these brain processes precede the conscious experience of the intention and decision to initiate a movement. Such brain processes are usually regarded as conscious if the movement occurs in the absence of abrupt environmental change or stimuli.

This appears miraculous, and we should therefore examine very carefully the situation and data giving rise to such a radical claim. Libet's central thesis is that well before (by some hundreds of milliseconds) we consciously decide (or experience an intention) to perform a motor act, the movement-related slow potential called the readiness potential (RP) or *Bereitschaftspotential*, discovered by Kornhuber and Deecke (1965), starts to develop in our brains. This means that if we could monitor the readiness potential on-line on a single-trial basis (i.e., if the signal-to-noise ratio were good enough to make it possible to determine the presence of the RP in the raw unaveraged EEG, we would be able to see in advance when the subject was going to experience an intention to perform the instructed movement.

First of all, I am convinced of the soundness of this data-base from some of my own pilot work of over a decade ago. Puzzled by the long duration of the RP before the actual movement compared to the fact that even unwarned motor responses in reaction-time experiments occur within a much shorter time from stimulus onset (see, e.g., Näätänen 1971; Näätänen & Koskinen 1975; Näätänen & Merisalo 1977), I, in pilot experiments with T. Järvilehto, tried to "fool" the cerebral RP generator by concentrating on reading a book and suddenly, acting on movement decisions occurring "out of nowhere" by pressing a response switch. In this way we tried to produce a movement with no preceding RP or with only a very short one. Nevertheless, much to our surprise, RPs of quite a long duration were still there although the subject felt he had (immediately) followed a sudden, spontaneous urge to press the switch.

Although Libet's data-base is unassailable, his conclusions can be questioned. He seems to ignore the fact that the specific nature of the movement was determined in detail by the instructions, practice, and preceding repetitions, and that hence the only decision of the subject involved the *timing* of this *preplanned* movement. Moreover, even the decision to perform this movement can be regarded as already having been made (consciously) by him at the beginning of the experiment: The subject knows and has agreed that he is going to produce quite a large number of these movements sooner or later, within some reasonable time, before he can leave (and receive his payment), and that it is only the timing of each single movement of this specified type that is under his control – and even that not fully but within certain quite wide limits. Consequently it appears to be somewhat questionable to describe this motor act as "spontaneous" or "fully endogenous" and occurring with "no preplanning." It is accordingly not possible to agree with Libet's main conclusion that "cerebral initiation of a spontaneous voluntary act begins unconsciously." This conclusion means (and was intended to mean – judging from the author's discussion of free will) that even the *type* of motor act to be performed is unconsciously chosen (a veto of a conscious decision is also regarded as possible, however). Perhaps, but this cannot be concluded from the present data, since the type of motor act and whether it would be repeatedly performed during the session was *consciously* decided by the subject on receiving the experimental instructions. Consequently, the discussion of the possible implications of Libet and his associates' results for the issue of free will involves an unnecessary expression of concern. If I decide to go to a liquour store – regarded by some in this country as an immoral decision – I am sure there is no RP preceding this decision, whereas an RP might precede the conscious experience of deciding to initiate the chain of muscular events leading to this end.

Nevertheless, the brain's deciding *when* to perform a preplanned motor act well before the mind decides this is certainly of sufficient interest to warrant discussion in these respected pages. This specified motor act is, presumably, in some state of facilitation for reasons discussed above, that is, there is some central, and perhaps even peripheral, facilitation of this particular motor pattern that might contribute to the dissociation between RP development and its "subjective counterpart." Moreover, after each instance of performing this movement, there might be a subtle conscious decision with regard to the moment of the next movement in the sequence, which might then trigger the RP onset with this predetermined delay. (The distribution of the intermovement intervals might be highly informative here: Some deadline rather than the Bernoulli type of distribution might be reflected in it.)

In any case, Libet and his associates' work has provided a model case of the ingenious application of available physiological and psychological measures to understanding the mind-body relationship in the initiation of a preplanned (and repeated) motor act with spontaneous (within certain limits) timing.

ACKNOWLEDGMENT

This work was supported by the Academy of Finland.

## Libet's dualism

R. J. Nelson
*Department of Philosophy, Case Western Reserve University, Cleveland, Ohio 44106*

Libet presents two principal theses: Given the experimental setting and his findings concerning volitional processes,

(i) A subject's spontaneous, conscious urge to act is initiated by an unconscious cerebral process signaled by RPs.
(ii) The act itself, "conscious voluntary action . . . may operate . . . to select and control" the volitional process.

Except for possible doubts about the veridicality of the introspection and reporting of urges and possible questions about timing, it seems to me the experiments combined with the cited supporting experiments do indeed establish (i) and (ii).

There remains, however, a dangling question whether some mediating cerebral process precedes the occurrence of conscious control (ii). According to Libet, "there is presently no technique available for recording and analyzing any RPs that may be associated" with conscious *vetoes* of an act. And presumably there are no techniques for identifying the role of cerebral processes, if any, in positive *triggering* of an act, although this is not very clear from what Libet says. Of course, absence of adequate technique does not imply that there are no such underlying processes.

Nevertheless, Libet concludes that it would indeed be *necessary* to postulate:

(iii) Conscious control functions (ii) can appear without prior initiation by unconscious cerebral processes.

Thus conscious voluntary control is autonomous with respect to the brain.

From what he says in the last section of the paper I suspect that Libet, in suggesting this postulate, wants to make room for a kind of scientific warrant for the proposition that human beings have voluntary control of at least some of their actions, in the straightforward sense of popular psychology and ethics. In our nonphilosophical moments most of us feel that conscious actions are not merely part and parcel of a purely physical, cerebral stream of events. As agents we *cause* actions. So lacking evidence as to physical causal factors in conscious voluntary control, the postulate provides an appealing sop to our ordinary intuitions. It even has some scientific warrant inasmuch as it enjoys support, according to Libet, from Margenau and Popper and Eccles, as indicated below.

I believe, however, that the following considerations indicate that Libet might better have suspended judgment, awaiting either further experiments from which something more definite could be concluded about the presence or absence of RPs (or other laboratory indicators) preceding conscious voluntary activity, or developments in cognitive psychology that could afford better clues as to the role of actions in intentional life.

Notice that the postulate is *not* grounded in Libet's definition of "voluntary action," which stipulates (a) that the action must arise endogenously, (b) that it must be without external constraints, and (c) that the subject must feel free to act if he wishes. For a conscious event (e.g., an urge [i]) could arise endogenously and be initiated by a cerebral process, and the subject could feel free (satisfying [c]) without being free, which is hardly news. Moreover, the proposition is not supported in any way by the experiment, since reporting the feeling of freely performing an act does not entail that the conscious control function can appear without initiation by a cerebral process. That the control function, as postulated, can occur withour prior initiation by brain processes is not supported by the experiment, by the principal results (i) and (ii) that derive from it, or by the underlying definitional concepts.

However, the postulate (iii) is not inconsistent with (i) and (ii), although as a set they are curiously *incoherent*. Why is one mental event – the urge – initiated by a brain process, while the other – the conscious voluntary act – is not (at least, why is it postulated not to be so initiated)?

This incoherent mix betrays a strange sort of double dualism. Before explaining what I mean, let me clear up in advance a possible misunderstanding about the term "initiate." This can be understood in a direct, empirical way within the context of experiment. As initiating event, the RP-signaled process is the head of a uniform, regular sequence, with the felt urge being the contiguous second element. (This is slightly reminiscent of Hume; however, Hume's analysis of "cause" does not mix putatively physical with phenomenological events – ideas, impressions – so one hesitates to say on Humean grounds that "initiate" means "cause"). But Libet must mean in these interpretational passages more than empirical regularity. I suspect there are all sorts of deep cerebral processes that regularly and uniformly precede conscious voluntary control but are wholly without influence on action. So I suggest that Libet must mean by "process A initiates event B" that B would *not normally occur* without A, that is to say, that A *causes* B. It is not easy to grasp the force of (iii) unless he means "cause" by "initiate," which of course loads an added philosophical burden on an otherwise neutral experimental term.

If this is right, a certain type of cerebral process causes an urge to act, whereas, by the postulate, conscious control functions can occur without being caused by forerunning or concomitant cerebral processes. The *double* dualism is this: (1) some mental events, that is, urges, are caused by physical events. This is a form of interactionism (not of substantive interactionism, as no claim has been made that there is a substantial mind being influenced by a material brain); on the other hand, (2) other mental events, that is, conscious active control, might be features of an emergent conscious awareness that has "already developed" (cf. Margenau 1984). (As an alternative Libet suggests that the postulate "can be in accord with a dualistic interactionist view" [Popper & Eccles 1977]. But this idea is unintelligible. If conscious control is uncaused by unconscious cerebral process it certainly cannot be the result of interaction with the brain, unless the interaction is something the subject is *conscious* of, or else there is some kind of extracerebral bodily process that causes it – both of which are extremely unlikely.) So conscious voluntary control is part of a conscious stream parallel to, but not interacting with, cerebral process.

Adoption of Libet's suggested postulate leads to an interpretation of the experiment having the incredible consequence that there is an *interactive* dualism of physical and mental events, as in the case of urges, and yet a parallel *noninteractive* dualism, as instanced by voluntary actions. The remedy to this confusion is to drop the postulate and pursue the elusive connection via further experiment, possibly within the conceptual framework of a strictly materialistic view of mind and brain.

## Timing volition: Questions of what and when about W

James L. Ringo
*Center for Brain Research, University of Rochester Medical Center, Rochester, N.Y. 14642*

Libet is to be congratulated on having both the courage to seek experimental answers to deep and difficult questions about mind, brain, and conscious control of voluntary movement and for finding such ingenious scientific methods for attacking such resilient questions. Conclusions drawn from work in this area are likely to be monumental ones and as such will demand the most solid foundations possible. As Libet is clearly aware, one of the more difficult points in his work is the self-report of the "urge," W. He has tried to assess the timing of this report by at least partly independent methods and hence to increase its reliability. Aside from the question of when W occurs there may also be some question of what the subjects are reporting. In the main task the subject is asked to initiate a movement at a self-chosen and pseudorandom time. There must be some way in which this point in time is chosen, that is, there must be some initiation. The mechanism responsible for initiation must be mostly in a state (or in a mode, or at a level) that does not cause initiation and on pseudorandom occasions must go to a state that does. One possibility for such a mechanism is a fluctuating potential occasionally crossing some threshold and producing an initiation. Upon back averaging from the result of the initiation (the electromyogram, EMG) one might find something very much like the RPs recorded. In a sense there would then indeed be an unconscious initiation of the movement when the pseudorandom fluctuation crossed threshold but one that was fully set up by the "consciousness." What is being suggested here is that the instructions to produce spontaneous movements may cause the subjects to create an unusual mental state in which brain potentials trigger a previously willed decision.

The possibility that the subject is essentially monitoring some brain potential (or some correlate thereof) and initiating a movement when this potential exceeds some criterion may be open to experimental test. The distinction to be made is between a potential associated with movement and a potential associated with the requirement of spontaneous initiation. The experiment is as follows. In circumstances that are otherwise the same, subjects are asked to choose (and later report) a clock position on a pseudorandom, spontaneous basis. That is, just as in Libet's main experiment, the urge to "act" should come out of nowhere, but in this case the "action" would be simply to note the clock time. With recording of the EEG and the clock position an average could be constructed later by back-averaging from the reported clock time (of the urge). The discovery of a potential preceding the urge would suggest that the type II RP stems from the requirements of spontaneous initiation, while a failure to find a potential would strengthen the interpretation of the type II RP as the harbinger of the motor act. Such an experiment might at least help determine whether the recorded potentials are more clearly associated with the voluntary act (physical) or the decision (mental).

A second and less testable point is that the subjects may be reporting the "peak" of an urge that actually has an extent in time. That is, perhaps we should not imagine the production of an instantaneous urge that is then sent out to the appropriate motor control areas and generates activity (from which idea we would expect the urge to precede the RP); instead, the urge may have a start, a rise, and a peak. If for the moment we think of the urge as having a physical source and form, it may be that the urge is produced by areas or cell groups connected to the areas that produce RPs; the start of an urge would start an RP, the rise of an urge would produce the rise of the RP, and so on. Such a system might produce an (unrecorded) "urge waveform" that precedes the RP by a few tens of milliseconds. This early RP might reflect the motor system's being "readied" in an effort to anticipate as well as possible the outcome of the "will's" decision and hence to save time (this is somewhat analogous to "look-ahead" computer methods). Since in the experimental situation the likely motor act is quite predictable and only the time is unknown, readying the system for the motor performance is not unreasonable, so the very beginning of the "urge waveform" might very well begin the production of the RP. When asked to report an instant in which the urge occurred, however, the subjects may be choosing the peak of the "urge waveform" (which follows the beginning of the RP) instead of the beginning of the "urge waveform" (which leads the RP). Perhaps if the subjects could be instructed to choose between two (or more) movements as well as to choose a time, all in a spontaneous matter, then an anticipatory RP would be less likely since the desired movement would be less predictable.

## Sensory events with variable central latencies provide inaccurate clocks

Gary B. Rollman

*Department of Psychology, University of Western Ontario, London, Ontario, Canada N6A 5C2*

Libet's earlier analyses of central timing processes for sensory experiences have been cogent and clever, his views on complementary experiences associated with motor acts also are often insightful. However, unless I misunderstand Libet's methodology and rationale, a serious logical flaw exists in his determination of the absolute times of conscious intention to act (W), awareness of actual movement (M), and awareness of a tactile sensation (S). If so, alterations in the interpretation of Libet's absolute values are required, although the relative times between some of these events may still be generally correct.

Libet measures the time of the first awareness of wanting to move (W) by having the subject report, retrospectively, his observation of the "clock position" of a spot of light revolving on an oscilloscope screen when such an experience occurred. By relating this to the clock time when the actual motor act began, using a record of the electromyogram (EMG) from the appropriate muscle, Libet claims to have determined that subjects become "consciously aware of the urge to move 200 ms before the activation of the muscle."

The perceived position of the clock at the time a subject experienced awareness appears to be confused with the actual time when the awareness took place. Such readings do not occur instantaneously. Sensory events are registered centrally only after a latency of up to several hundred milliseconds. A clock value of "0 ms" is transduced, coded, and transmitted through the retina, optic pathways, and subcortical and cortical regions before it can be "read" as stating "0 ms." By that time, of course, the face of the physical clock tells a very different time, "N ms."

Consequently, the clock time described by the subject as occurring simultaneously with his intention to move is a central representation of an event that occurred N msec earlier. The actual time of the occurrence that Libet wants to measure is N ms later than the value the observer reported. It is difficult to estimate the value of N. Fitts and Deininger (1954) found reaction time in a clock-reading task to be about 400 ms, a value that must include sensory, motor, and decisional components. If N is as long as 300, then the subject's awareness of the urge to move does not occur 200 msec *before* activation of the muscle, as Libet proposes. Rather, it occurs $-200 + N$ or 100 ms *following* the movement. If N is 100 ms, the awareness occurs $-200 + N$ or 100 ms *before* the beginning of the EMG. Clearly both positive and negative times are possible because of the lability of the central latency and uncertainty whether early or late components of the neural response are involved.

The determination of M, the "clock time for the awareness of actually moving," suffers from the same defect. Libet notes that "M values were, unexpectedly, negative to EMG −0 time." Again, consider that the time described by the subject was the time on the clock N ms before that reading was actually perceived. Real time is N ms later. If N is 300 ms, the true value of M changes from Libet's reported −86 ms to −86 + 300 or +214 ms. The perception of movement occurs *subsequent* to actual movement, and the "negative" value that emerges from Libet's method is not unexpected.

The earlier reinterpretation offered above suggested that W could really be +100 ms (if N is 300 ms). This implies that the subject does not become conscious of the urge to move until 100 ms after the movement has occurred. Before dismissing this as counterintuitive, consider that the second part of the reinterpretation suggests that M, the time when the movement is perceived, is +214 ms, indicating that the movement itself is not perceived until 214 ms after it has taken place. The urge to move is perceived 114 ms before the movement is perceived in both Libet's analysis [−200 − (−86) = −114] and my own (100

− 214 = −114). Libet's values may reflect the relative times of the critical events, but they do not correctly reflect either their absolute value or their sign. Since N is unknown, no accurate values of W, M, or S can be obtained.

A further complication arises in the proper determination of M. As Libet indicates, a judgment regarding the occurrence of movement may accompany either the motor command or feedback from the movement. If it is the latter, the latency of the appropriate reafferent signal must also be considered in determining the relationship between recall clock times and true latency between critical central events.

Likewise, the value reported for S, the time when a skin stimulus is perceived, is subject to additional problems. To determine W, Libet compares a peripherally initiated event (visual examination of the clock) with a central event (intention to move). In measuring S, he compares two peripherally initiated events, those triggered by clock movement and skin stimulation. Both of them will require a considerable latency (almost certainly different) before they are perceived.

Those latencies are influenced by both stimulus characteristics and task demands. If conduction time were equivalent in the visual and somatosensory systems for one set of parameters, adjustment of intensity for either signal could tip the balance in one direction or the other (Rollman 1974). Given that the tactile task involves simply detecting the presence of a stimulus on the skin while the visual task requires discrimination of clock position, latency for the second judgment is likely to be considerably greater. If the decision about the time of touch onset occurs when the neural representations of the tactile pulse and the clock position jointly reach some central locus, the longer-latency visual event must have taken place prior to the presentation of the tactile signal. Under such conditions a negative value for S must occur (it was about −50 ms for Libet's parameters).

This outcome follows from the differential transmission times for the two stimuli; Libet's footnote to Table 1 labels it "error" or "bias." The wide potential variability in the value of S as a consequence of changes in stimulus parameters, plus the fact that a tactile pulse is a peripheral event whereas the intention to move arises centrally, negates taking Libet's S as "a measure of the potential error in reports of W."

Libet has wrestled admirably with the complexities underlying the timing of conscious intention to act. Unfortunately, the situation seems even more complex than he anticipated.

## Are the origins of any mental process available to introspection?

Michael D. Rugg

*Psychological Laboratory, University of St. Andrews, St. Andrews, Fife KY16 9JU, Scotland*

Putting to one side questions of methodology and the issue of how a special causal role for a "conscious" process can be established, I shall argue that there are a number of logical and conceptual problems with Libet's thesis. The thesis is that the initiation of a voluntary motor act is under the control of a system or systems whose activity is not accessible to conscious introspection, at least until some time after it has begun, while the processes causing a modification of such an act are closely associated in time with introspectively derived feelings of control over it.

First, this thesis depends crucially upon the assumption that there is a necessary relationship between the execution of a voluntary "willed" action, such as a finger movement, and the prior existence of the variable chosen by Libet to index the onset of the processes leading up to the action, the readiness potential (RP). Thus, it would, for example, be necessary to plausibly rule out the existence of individuals in whom, as a result of, say, a brain lesion, RPs have been abolished, but not the capacity for voluntary action. To my knowledge, no such study has yet been carried out, and in the absence of any relevant data pertaining to this issue the proposition that a necessary relationship exists between RPs and voluntary movement is at least questionable. In addition, in the absence of any knowledge as to the precise functions with which the RP is associated, it seems premature to propose that the emergence of an RP indicates the onset of processes leading to a *specific* voluntary act, as opposed to the beginnings of some more general "arousal" or "priming" process serving as the precursor to a wide range of potential acts. The choice of the specific act to be performed may indeed be associated with the very process giving rise to the introspectively experienced "will" to perform that act. Inasmuch as the emergence of the RP prior to the time of this feeling of an urge to act is associated with exclusively nonspecific aspects of motor output, a crucial role would indeed exist in the initiation of an act for the processes associated with its conscious "willing" (but see below). Although denied as such by Libet, this position seems significantly at variance with the essence of the thesis advanced in the target article.

A further difficulty concerns the limited scope given to the notion of an act or action. Within the framework of contemporary cognitive psychology it is not uncommon for there to be no hard and fast conceptual distinction between overt motor acts and their covert, mental analogues (see, for example, Posner, 1980, for such an exposition with respect to mechanisms of visual search and attention). In this vein, I argue that it is quite reasonable to consider a covert mental event such as a "consciously" taken decision to be a type of voluntary act. This being so, one might reasonably question whether the precursors of such an act are any more amenable to conscious introspection than those associated with an overt action such as a finger movement. A relevant example in the present context is the decision to "veto" a previously initiated finger movement. This is considered by Libet to be an example of the role of conscious control in motor function: specifically, to "select or control volitional outcome." On the basis of the above arguments, the precursors of the "veto" decision might themselves have origins that are as inaccessible to introspection as those associated with the original decision to initiate the act in question. One is therefore forced to the conclusion that there is no evidence for the conscious control of the initiation of *any* definable overt or covert act; the origins of all behaviour, whether this is ultimately expressed in an observable motor act or not, and irrespective of whether any aspect of its precursors eventually enters consciousness, may arise from processes to which we have no introspective access.

Thus the distinction drawn by Libet between the intention to act and the fulfillment of that intention, in terms of the former being outside an individual's "control" and the latter within it, ceases to be meaningful. Although it may be reasonable to argue that a necessary component of any "voluntary" act is an introspective awareness of an intention to execute it, this is not the same as arguing that this awareness itself has a special causal status. To reiterate, the origins of this awareness, and of any modifications to it, may always precede and thus determine its contents.

## Conscious intention is a mental fiat

Eckart Scheerer

*Department of Psychology, University of Oldenburg, D-2900 Oldenburg, Federal Republic of Germany*

Libet jumps from neurophysiology straight to philosophy as if there were no psychology in between. Contemporary psychology indeed has little to say about the "conscious will," but the will was a standard topic for earlier psychologists who took introspection seriously without neglecting physiology. To Americans, William James is the "classical" psychologist par excellence, and he included a long chapter on the "will" in his *Principles of Psychology* (James 1890). Another classical psychologist who spent his entire scientific life on "the analysis of the will" was Narziss Ach (1935). Libet shares with classical psychology a reliance on introspection and even some of its technical procedures. It is therefore appropriate to relate his work to the viewpoints maintained by psychologists such as James and Ach.

Libet finds it surprising that the "conscious will" does not have the function to "initiate a specific voluntary act" but only serves to "select and control volitional outcome." But a similar conclusion was reached by the classical psychologists. In Ach's analysis of the will, the concept of "determination" is central. It refers to the fact that once a certain task has been adopted by a subject, the selection and control of subsidiary mental processes is performed at an unconscious level. However, this does not mean that the execution of an overt voluntary act has no conscious antecedents except for a general "planning" stage (i.e., Ach's determination). As long as an act is not automatized, there will be "intentional sensations" representing the act or its outcome, and there will be some kind of mental consent to the occurrence of the act (Ach 1935, p. 122). The same thought had been expressed by William James (1890, vol. 2, p. 501): "An anticipatory image . . . of the consequences of a movement, plus (on certain occasions) the fiat that these consequences become actual, is the only psychic state which introspection lets us discern as the forerunner of our voluntary acts." James thought that the anticipatory image was obligatory, while the fiat was needed only when inhibitory influences had to be overcome. But when both are involved, the anticipatory image precedes the fiat.

How can we relate these concepts of classical psychology to the events in Libet's experiment? The critical event reported by Libet's subjects receives somewhat different names: "endogenous urge or intention to move"; "wanting to move"; "conscious intention to act." A distinction was made between acts that "were experienced as fully spontaneous and unplanned" and acts where "some general preplanning or preparation" was experienced. On the basis of Libet's description, one can identify the "general preplanning" with Ach's determination, or rather with its conscious equivalent, and the "intention to act" to the "consent" or "fiat" of both Ach and James. Because Libet's experiments involved a choice between two incompatible acts (responding and not responding), William James would have consented that the fiat was necessary.

So far, then, Libet's results are well in line with the viewpoints of classical psychology and provide them with a physiological underpinning. A general determination precedes the consent to allow to happen a specific act selected by the determination; the determination is occasionally reinstated in conscious form, and when this happens, it has a specific cortical correlate ("type I" RPs); however, the selection of the specific act occurs at an unconscious level, and it is noted only after the fact, at the stage of consent or fiat.

But what about James's anticipatory image or Ach's intentional sensations? Libet mentions the possibility that a "nonrecallable phase of a conscious urge exists," but he rejects this possibility as untestable. However, perhaps the question is not so much whether or not a certain event is "recallable" but whether or not it will be noticed at all. And here we should accept the premise that introspection works selectively, that in introspection we find only those events that we have been led to expect. Libet's subjects were apparently instructed to observe events related to the "volitional" aspect of mental activity as envisaged by everyday psychology; the occurrence of anticipatory images was never reported, perhaps because it was never asked for. The "expert observer" of introspective psychology should be reintroduced, and his attention should be directed to the possibility that a movement might be imaged before one "wants" or "intends" to act. If such judgments can indeed be made, they too can be timed with Libet's methods, and such timing might result in a coincidence with "type II" RPs. This would constitute an alternative interpretation of these RPs, removing much of the mystery with which they are surrounded in Libet's account. The fiat would then be preceded by a neural event having an immediate conscious correlate.

Another concept that bears some demystification is the "conscious veto." Its existence was demonstrated, long ago, in a situation remarkably similar to the indicator stimulus paradigm used by Libet for timing internal events. In one variant of the "complication experiment," the observer is asked to synchronize his response to the moment when a moving dot crosses a line. After some practice, such synchronization can be done with an accuracy of around 20 ms. The observer can then be instructed not to respond when the dot stops before crossing the line. The time needed for the "inhibition of a prepared voluntary act" (Hammer 1914) was found by Flachsbart-Kraft (1930) to be in the region of around 150 ms relative to the anticipated transit of the moving dot, and Woodworth (1938, p. 301) briefly mentions this work. The task used by the old investigators is the same as that used by Libet, and so are the "inhibition times." Ach (1935, p. 115) noted that the "inhibition time" was equivalent to simple visual reaction time, and from this he deduced that the inhibition of voluntary impulses and the time needed for it is a special case of the inhibition exerted by one antagonistic response on another. Thus, the "flashlike counter-command" (Ach 1935) consists in the replacement of one prepared voluntary act by another; there is nothing mysterious or even ethically relevant about the "conscious veto."

ACKNOWLEDGMENT
This paper was prepared during my stay at the Zentrum für Interdisziplinäre Forschung der Universität Bielefeld.

## The uncertainty principle in psychology

John S. Stamm
*Department of Psychology, State University of New York, Stony Brook, N.Y. 11794*

The arguments in Libet's target article are based on two time measures: RP onset and awareness of wanting to move (W). These are signs of, respectively, physiological and psychological processes. The reliability and significance of these two measures therefore, require careful consideration.

The RP onset is determined by the appearance of a negative deflection from baseline in the averaged EEG recording. Unfortunately, the stability of the pre-RP baseline is an averaging artifact, since ECG (electrocorticogram) recordings with implanted electrodes in monkeys (Stamm & Gillespie 1980) show continual baseline fluctuations at frequencies of several seconds. Similar pre-event baseline flucutations have been reported during single-trial scalp recordings from human subjects (Bauer & Nirnberger 1981; Born, Whipple & Stamm 1982). Furthermore, there is convincing evidence from physiological and psychological investigations that these fluctuations represent changes in neuronal excitability, with surface negativity indicative of heightened excitability, or cortical arousal. In Libet's experiments, the procedure suggests fairly long intervals between successive acts (report of "half-day sessions"), during which the subject is presumably relaxed and probably bored. It is accordingly conceivable that the subject's mental and behavioral processes tend to start during a period of heightened cortical excitability, that is, during the negative phase of the endogenous baseline fluctuation. This baseline shift would be obscured in the averaged EEG recording because of the considerable variability in RP onset of single events. Even a modest negative bias before RP onset would result in considerably shorter RP durations than those obtained from averaged recordings, which show a very slow rate of increased negativity. This argument can be experimentally examined only with single-trial recordings, which are now technically feasible.

The significance of the W event is difficult to understand, despite Libet's arguments that this is a valid index for timing the subject's awareness. He supports the short W latency with the control experiment of reported awareness of a skin stimulus. Unfortunately, this paper does not present the data for the times between reported and actual application of the stimulus, but an earlier publication (Libet, Wright & Gleason 1982) reports the mean times for six subjects as between −167 ms and +83 ms. This wide range, with some awareness times seemingly preceding the actual stimulus delivery, raise further questions about the timing of the mental processes. Assessments for the durations of mental activity have been obtained with reaction time paradigms that indicate response latencies of several hundred milliseconds for simple reactions and of 1 sec or more when a choice response is required (Born, Whipple & Stamm 1982). The attentive demands placed on Libet's subjects are severe, with instructions to: relax, gaze at the rapidly sweeping dot on the CRO, avoid eye blinks, monitor both one's internal state (intention) and the external "clock," execute the finger movements, and remember the dot position. While many of these functions are processed in parallel channels, they will certainly interact and prolong each other. Consequently, Libet's arguments for near simultaneity between the subject's internal state and his report is not well substantiated. The constraints for temporal assessments of internal states may be designated as the analogue of the Heisenberg uncertainty principle in physics, that is, that self-monitoring of an internal process interferes with that process, so that its precise measurement is impossible. According to these considerations, the subject's intention for a finger movement occurs at a considerable time before the W measure.

My arguments for a later onset of the true RP and earlier intent for the act than the times reported by Libet would lead to a reversal of the temporal sequence for these events. Certainly the assignments for these quite fragile measures in terms of unconscious and conscious functions is at best premature and does not contribute to our understanding of mental processes.

## Mind before matter?

**Geoffrey Underwood[a] and Pekka Niemi[b]**

*[a]Department of Psychology, University of Nottingham, Nottingham NG7 2RD, England and [b]Department of Psychology, University of Turku, 20500 Turku 50, Finland*

Awareness of an intention appears to occur after a physiological activity (the "readiness potential") otherwise associated with preparation for muscular contraction. Libet suggests that we can therefore conclude that the conscious will lags behind a decision to act that is itself physiologically based. We find this idea unacceptable on both conceptual and empirical grounds. In what follows, we shall reply to Libet's arguments by first pointing out the absence of conceptually acceptable hypotheses. We shall also argue that a number of mental operations intervene between a physiological correlate of a mental process and our awareness of experiencing that mental process. In effect, we suggest that Libet and his collaborators have undertaken a test for which there is no other outcome than the one they found.

We assume that it is not possible for a conscious intention to be formulated without any underlying physiological activity; this variety of monism, known as emergent materialism, is implicit in many contemporary cognitive theories. When we need to make assumptions about the relationships between cognitive processes and their underlying physical substrate – when identifying the possible psychological consequences of a deficit such as acquired dyslexia, for example – the physiological data are in agreement with the notion that normal psychological abilities are dependent on the normal operation of a normal brain. Unless the brain is intact, for instance, it will not operate appropriately, and this confirms the position that psychological performance is dependent on physiological competence. This does not imply that psychological experiences in some way *follow* the activity of the brain, but simply that the brain is necessary for psychological processes to be possible.

Our working assumption is that the activity of the brain, which is theoretically observable to a physiologist, is responsible for the experiences of the owner of the brain. Without those physiological activities, the experiences would not be possible. Accordingly, the question arises as to whether Libet and his colleagues *could* have found reports of intentions prior to the observation of the putative "readiness potentials." The answer is a very clear "no" and for the same reasons that music cannot be heard from a gramophone record that is not being operated in a specific way. The record is the physical substrate for the music, and its operation is correlated in time with the generation of the music. If the mind is the product of the physiological activation of the brain, as we shall suppose, then awareness could never precede the observation of such activation. Since becoming aware of an intention is but one of a series of mental processes associated with volitional movement, it is quite likely that awareness will follow after the intention itself. We are distinguishing here between conscious intention and the subsequent awareness of having intended to take some action. In other words, observers' reports can suggest only temporal contiguity, not simultaneity. Given that Libet's experiments could not, in principle, observe awareness of intention prior to the "readiness potential," we can now turn to the question of why they appear to show that physiological processes occur prior to the associated psychological processes.

To determine the time of onset of conscious volition, the experiments use a task that necessarily incorporates delays in the self-reports elicited. It may not be possible to avoid observing delays without using indirect inferential procedures, for self-reports must rely on experiential data collected some time in the psychological past. The task required that participants judge the position of a rapidly moving spot of light as soon as they become aware of having had an intention to move (Libet, Gleason, Wright & Pearl 1983): the problem is that making this judgment requires the use of mental processes that are limited resources (sometimes identified with "consciousness") and are necessary for volitional motor planning.

If limited mental resources are dedicated to the command of motor actions, what is then left to make an accurate judgment about the relative simultaneity between consciousness and a sensory event? In more detail now, the minimum necessary processes are as follows: The participant intends to move a hand, and at some time after this may become aware of having intended to move a hand. Here we have the conscious intention followed by awareness of that intention, and there is no suggestion of any unconscious initiation. The distinction between the two processes is the first source of delay in Libet's procedure. Awareness of having had an intention does not necessarily follow, and with certain overpracticed actions the performer need not be aware of any intention or planning. Examples include tying one's shoelaces when dressing, shifting gears when driving a car, or holding a racquet at the approriate angle when playing tennis. These are cases of automatized actions (Reason 1979; Underwood 1982), in which the presence of an environmental calling pattern is in itself sufficient to initiate an action sequence. These cases are possible when the relationship between environmental conditions and their appropriate actions are invariant and can be overlearned. Lack of awareness cannot be taken as evidence of lack of conscious intention, however, and the two must be seen as separate processes.

After becoming aware of having had an intention the participants must then refocus their attention on a moving spot of light. This is not the same as redirecting their gaze, of course, for it is quite possible to gaze at some point in space while attending to one's thoughts of future motor actions. To return attention from the thoughts to the visual world would require time, however, and this is the second source of delay in Libet's procedure. The extent of the delay attributable to attention switching is something which is not established (Broadbent 1971).

The position of the spot of light must then be judged and remembered. These final two processes are common to both the experimental task and the control task, and so their importance can be neglected here. However, the first two processes are not present in the control task and therefore allow us to dismiss its use: becoming aware of having had an intention ("metavolition"?) and redirecting attention from this thought process to a point in space. If it can be demonstrated that the time required to become aware of having had an intention is of negligible duration, and that the time to switch attention between a cutaneous sensation and visual space (the control task) is the same as that to switch from a thought to visual space, then the data would be more convincing.

We are also curious to know how Libet is able to distinguish between a volitional intention to act, which is said to be unconscious, and an intention to veto an act, which is said to be conscious. Does a veto require an intention, and why should it differ from other intentions by being conscious?

## Nineteenth-century psychology and twentieth-century electrophysiology do not mix

C. H. Vanderwolf

*Department of Psychology, University of Western Ontario, London, Ontario, Canada N6A 5C2*

Libet has summarized a curious research program aimed at the identification of the time of occurrence of a conscious mental process that leads to the generation of a voluntary motor act. At the heart of this program is the assumption that people are directly aware (by introspection) of some sort of endogenous brain process that controls voluntary movement. This is essential if consciousness is to have the regulatory role that Libet's hypothesis proposes. However, the assumption appears to be incorrect. Laszlo (1966) has shown that following compression of the upper arm with an inflatable cuff, active kinesthesis disappears well before a loss of motor ability. Subjects are able to squeeze something or tap their fingers even though they deny that any movement is occurring and refuse to believe that the anesthetized hand is moving until they are permitted to verify this visually. This demonstrates that humans have no direct awareness of the brain processes that generate hand movement. Not only the initiation but the entire process of generating a movement is unavailable to introspection. One knows that one's hand has moved only as a result of kinesthetic feedback from it. Further support for this conclusion can be found in a report by Melzack and Bromage (1973) that a feeling of being able to produce voluntary movement in a phantom limb (produced experimentally by injection of a local anesthetic into the region of the brachial plexus) is completely dependent on the preservation of residual electromyogram (EMG) activity (and, presumably, feedback from it) in the affected limb.

These results raise an interesting question. If people are aware of their own voluntary movement only as a result of sensory feedback, how can Libet's subjects tell that they "want" to move nearly 200 ms prior to the onset of recorded EMG activity? This apparent problem may be due to nothing more than a failure of Libet's recording technique to detect peripheral neuromuscular changes that herald the onset of a voluntary movement. It is well known that there are changes in the excitability of monosynaptic spinal reflexes in humans well before a voluntary movement (e.g., Papakostopoulos & Cooper 1973) and early researches by Jacobson (1930a; 1930b) and others demonstrated that low-amplitude EMG bursts in the relevant muscles accompany imagining a movement or thinking about it. Thus, it is probable that when Libet's subjects detect that a movement is imminent they are reacting to a peripheral sensory event, that is, to changes in their muscles, rather than to an endogenous mental or brain event.

The foregoing results are consistent with the general conclusion that humans have little or no direct awareness of the central processes that cause their own behavior. Libet appears to be unaware of the history of attempts to investigate the mind by introspection. This was a serious scientific endeavor in the period of (approximately) 1880–1910. As a result of this work it became apparent that "mental" processes are generally not open to direct examination. We are aware of physical events in the external world and those inside our own bodies and of very little else. Such knowledge as we do have of the causes of our own behavior is the result of inference rather than direct awareness (Hebb 1980; Skinner 1974).

Libet's conceptual approach to his work is an excellent illustration of the low level of communication that generally exists between the behavioral sciences and mainstream neuroscience. The greatest advances in behavioral research in this century have been made by the Lorenz–Tinbergen school in Europe and the operant conditioning school, which has been associated particularly with B. F. Skinner in America. Both schools have found it advantageous to abandon the introspective mentalistic approach to behavior that has been an integral part of Western philosophy for centuries. Leading cognitive psychologists have also recognized that mental activity is largely unavailable to introspection (e.g., Pylyshyn 1973; Nisbett & Wilson 1977). Unfortunately, most neuroscientists are unaware of the conceptual advances made in behavioral work or else do not understand how to apply these insights to their own work. Consequently they tend to accept the philosophical and psychological hypotheses of the seventeenth to nineteenth centuries as self-evident truths. As a result of this weak conceptual basis, attempts to relate brain electrophysiology to mental processes have generally been unsuccessful. However, if one attempts to relate brain electrophysiology to *behavior*, there may be a greater prospect of success, as I have attempted to show in previous papers (Vanderwolf & Robinson 1981; Vanderwolf, 1983a; 1983b).

## Conscious wants and self-awareness

Robert Van Gulick

*Department of Philosophy, Syracuse University, Syracuse, N.Y. 13210*

Professor Libet's ingenious experimental procedures provide remarkably detailed information about the temporal structure of cerebral events preceding voluntary movement. However, clarifying the notion of a conscious mental state may help in assessing the relevance of his data to issues concerning the role of consciousness in the production of voluntary action.

Libet takes his data to show (or at least to suggest strongly) that the initiating causes of voluntary movement (readiness potentials, or RPs) are not conscious mental states. But the notion of a conscious mental state is ambiguous in a relevant and theoretically important way. A *conscious mental state* may be either *a mental state of which one is conscious* (i.e., a mental state that is an object of self-awareness) or *a state of being conscious of some mental state* (i.e., a state of self-awareness which has a mental state as its object). It is the latter notion that Libet seems to have in mind. For in his view, it is only the state (W) that occurs 300 to 400 ms after the onset of the RP state that counts as a conscious mental state.

However, the initiating (RP) state may well be a conscious mental state in the former sense, and there is some basis for holding that it is the first sense that is relevant to the case at hand. When we speak of a conscious want, urge, or intention we normally mean a *want of which we are conscious* or aware, whereas in both psychoanalytic and everyday parlance, an unconscious want or desire is one of which we are not conscious or of which we have no awareness. Urges, wants, and desires are not likely to be conscious states in our second sense (i.e., states of self-awareness) since they are not normally states of awareness at all. They are motivational states, which should not be confused with the awareness we may have of them.

Thus if what Libet's subject becomes aware of when he becomes aware of his intending or wanting to move (W) is just the causally initiating RP state, then that RP state will count as a conscious want or intention in our first sense, since it is a mental state of which the subject is self-aware.

However, given the time delay between the onset of RP and the onset of W, it might seem that the RP is not a conscious want at the time when it initiates the causal production of a movement but only becomes one 300 to 400 ms later. The significance of this time delay depends on further causal and temporal facts about the brain. If consciousness is a sort of self-monitoring or self-scanning process, there will always be some time lag between the onset of a cerebral state and awareness of that state. Libet indeed gives some indirect support for such a view when he notes that 200 ms of stimulation is required to produce subjective awareness of a cerebral event. If no mental state ever becomes a conscious state (i.e., an object of self-awareness) until several hundred milliseconds after its onset, then RP states would be no less conscious states than any other mental states.

Restated in light of our distinction, Libet's claim is still an interesting and important one. His data appear to show that under his experimental conditions the event initiating a voluntary movement is not a state of self-awareness. However, that result may still be compatible with RPs being conscious wants or intentions in every respect in which wants or intentions are ever conscious (i.e., they become wants or intentions of which we are conscious in as short an interval as the self-scanning process of the brain allows).

## Neural/mental chronometry and chronotheology

**Gerald S. Wasserman**

*Department of Psychological Sciences, Purdue University, West Lafayette, Ind. 47907*

Given enough commitment to a cause, anyone can fail to take account of otherwise obvious basic principles. Such is the case for Libet, whose target article overlooks fundamental measurement concepts and also pays no attention to relevant empirical findings of psychology. Libet is thereby led to make two egregious errors:

1. He fails to distinguish between a measuring operation and the thing being measured; these do not have to be coincident or synchronous. A classic example would be the determination of a star's velocity by measuring its red shift. The spectral measuring operation and the star being measured are separated by vast amounts of space and time. This general metrical caveat applies equally well to the brain/mind problem. It specifically applies to Libet's attempt to use objective phenomena (spots of light, skin shocks, and electrophysiological potentials) to measure neural/mental timing. Such an attempt cannot succeed without establishing the temporal relations between these objective phenomena and the neural/mental activities they purport to measure.

This caution would apply even if no one had ever done any empirical research on the brain/mind. But of course such research has been done, and it has taught that bioelectric signals take time to propagate through the brain. It has similarly taught that mental propagation takes time as well. Indeed, the elementary fact that a sensation is delayed with respect to the stimulus that evoked it has been known for centuries. Elsewhere, I have noted the ancient origins of this idea in the work of Ptolemy and Francis Bacon (Wasserman 1978).

2. Libet also fails to consider that information does more than just propagate between the environment and the brain/mind. Information also has to be processed. There is no reason to assume that complex neural/mental information-processing operations do not take time. Instead there is a copious literature that indicates that information processing does take time. Elsewhere, a review of some of that literature has been given (Wasserman & Kong 1979). The work cited in that review shows that mental chronometry is a serious discipline. This is a literature with which Libet appears to be unfamiliar. It also suggests a different interpretation of Libet's findings that can be brought out by a careful examination of the details of Libet's own experiments.

A subject voluntarily chooses to move, and his movement is measured by means of the EMG (electromyogram). Some hundreds of milliseconds before the EMG-defined movement appears, a set of externally measurable readiness potentials (RPs) appears on the scalp. So far, so good. The RP and EMG are both objectively measurable with adequate precision. But the subject also reports when he became aware of his intention to move by observing a rotating spot and reporting as a clock coordinate the position the spot was in when the intention began. This clock position is converted into objective time measurements designated as W. Here is where the trouble begins: The time when the external objective spot occupies a given clock position can be determined easily, but this is not the desired result. What is needed is the time of occurrence of the internal brain/mind representation of the spot. Libet does not recognize this problem and concludes that subjects begin to make voluntary movements without being aware of what they are doing. The quantity RP − W is offered as a measure of the interval of "unconscious initiative"; it is claimed to be more than 300 ms.

It is easy to show, however, that RP − W cannot be accepted as a valid measuring tool. Metrical principles permit the possibility of a delay (D) between the neural/mental representation of the spot-clock relative to the objective spot-clock. And ample data exist to show that D is not zero. Hence RP − W must be in error by an amount equal to D that would increase the correct value. This error might be discounted because accounting for it would only make Libet's claimed effect larger. But that would be too narrow an approach; a proper approach would recognize that this particular error is merely one undeniable exemplar of a class of metrical problems that afflict Libet's argument.

Further examination reveals these other problems: RP − W + D would be a fair measure only if it took zero time to process the neural/mental representation of the display in order to determine the position of the spot. The underlying assumption that produced this crucial proposition can be readily demonstrated by critically examining the asymmetry in Libet's treatment of the sensory and motor parts of his experiment:

Consider how Libet views the motor task that requires the subject to flick his fingers or hand. This is about as easy a motor task as one could imagine; no specific flick is demanded as long as the flick exceeds a minimum amount. (Note that use of the EMG removes any ballistic delays due to limb inertia. The quantity RP − EMG is a measure only of the difference in time of two bioelectric potentials, one in the arm and the other in the head.) The quantity RP − EMG estimates the minimum time required for the neural/mental processing of the motor flick; this minimum quantity comes out at about 500 ms. Most of this delay is not due to simple axonal conduction delays. There is no reason to be surprised that processing a simple motor task takes this much time; there are many comparable results.

But there is also a sensory task to be done. This sensory analysis must be initiated by the same voluntary initiative that initiates the motor task. Logically, there is no requirement for both tasks to start together even though the instructions would seem to call for joint onset. But whether they do or do not start simultaneously can be determined only by research, not by assumptions. What we do know is that the sensory task is at least as complicated as the motor task; the subject must analyse the information in the clock-spot representation and decide where the spot is. Is it possible that the senory analysis takes 0 ms while the motor programming takes 500 ms? The greatest problem of this line of work is that no attempt was made by Libet to consider the problem of this analysis time. This failure is central, for if the clock-spot processing time were only a few hundred milliseconds, then the effect claimed by Libet would vanish.

Libet does offer a putative control in the form of a separate sensory experiment in which the subject relates the visual clock-spot (W) to the somesthetic sensation evoked by an electrical stimulus delivered to the skin (S). But this experiment is no control at all; it is afflicted by the same metrical problems that affect the main experiment. For the external S and W are not of interest. Rather, it is their internal representations that matter. Both S and W have to propagate into the brain/mind, so both will have delayed representations. How likely is it that both will be subject to identical propagation delays? Furthermore, the representations of both S and W have to be processed. Is it likely that both will take the same amount of time to be processed? Finally, the processing of W in the control experiment takes place under conditions different from those of the processing of W in the main experiment. Is it likely that the W-processing time is the same in both experiments? These questions are not addressed by Libet. Instead, all we have is the fact that S can be computed (from the grand averages for W and for W − S) to differ by 47 ms from W. This just means that some differential delay exists. In order for the two stimuli to seem to be simultaneous, one has to precede the other by 47 ms to overcome the differential delay. But the putative control experiment does not give any basis for using this simple objective measurement to determine absolutely any of the multiple internal delays described above.

Libet's research has provided several exemplars of the metrical problems that affect neural/mental chronometry. The shock-spot sensory experiment and the finger-flick motor experiment both confirm the existence of neural/mental delays. The actual experiments themselves are not original. It is only Libet's interpretation of these commonplace data that is striking, and this interpretation founders when its basis is examined.

## Pardon, your dualism is showing

Charles C. Wood

*Neuropsychology Laboratory, VA Medical Center, West Haven, Conn. 06516 and Departments of Neurology and Psychology, Yale University, New Haven, Conn. 06510*

Libet's intriguing experiments on electrical stimulation of human cortex and their implications for the mind–body problem (Libet 1966; 1973; Libet et al. 1979) have provoked considerable controversy (e.g., Churchland, 1981a, 1981b; Popper & Eccles 1977; Libet 1981a), and the target article promises to continue in that tradition. Perhaps more than any other investigator, Libet has ingeniously combined subjective and objective variables in his experiments in a way that consistently rubs our noses in one aspect or another of the mind-body problem.

Other commentators will no doubt wish to quibble with aspects of the experimental procedures, data analysis, the possible role of "prior entry effects" and other judgment biases, the distinction between type I and type II RPs, and the magnitude of the within- and between-subject variability (see Tables 1 and 2 in Libet et al. 1983). Instead, I will accept for purposes of discussion Libet's major finding that RPs begin 350–400 ms before subjects report the "initial awareness of intending or wanting to move (W)" in order to concentrate upon his fundamental assumption that W judgments must either precede or coincide with RPs in order for conscious intention to initiate voluntary movements.

My title is intended as gentle encouragement for Libet to make explicit his tacit assumptions regarding conscious experience and the mind–body problem because I believe they have caused him to overlook alternative explanations that pose no difficulty for the concept of conscious initiation of voluntary movements.[1] In my opinion, Libet's fundamental assumption about the temporal relationship that should exist between RPs and W judgments is decidedly (substance) dualist in character. He assumes: "If a conscious intention or decision to act actually initiates a voluntary event, then the subjective experience of this intention should precede or at least coincide with the onset of the specific cerebral processes that mediate the act." I characterize this assumption as dualist because it makes sense only if one believes that "conscious intention" is not mediated by a physical process or processes in the brain but by something else. That is, it makes sense only if we assume that conscious intention to move is not part of "the specific cerebral processes that mediate the act."

In contrast, if we assume that conscious intention is one of the many brain processes that contribute to the initiation of a voluntary movement, then the explanation for the obtained results is straightforward. According to this view, the brain process(s) that mediate the conscious intention to act *must* begin before subjects can report that they are aware of that intention. To assume otherwise is to assume that conscious intention arises full-blown, out of nothing, instantaneously ("the Devil made me do it"), a prospect that is decidedly dualistic. Unless conscious experience is totally unlike every physical process we know anything about, it must have a nonzero time course; if its time course is anything like that of other brain processes, then tens or hundreds of milliseconds is certainly reasonable.

If it seems strange to suggest that some of the neural events that contribute to conscious experience should be detectable before the completion of the process(es) that mediate that experience, consider the same question applied to a multiuser computer operating system that allocates computer resources as a function of the number and priority of competing tasks (I am not, of course, suggesting that conscious experience is analogous to an operating system in any deep sense; I simply assume that both are complex processes exhibited by suitably organized physical systems). An important element in such an operating system is a scheduling routine that intermittently examines the competing tasks and the available resources in order to determine which tasks will receive which resources next. The key point is that the scheduler is itself a program that takes time to execute. Consequently, there is a time period during which the scheduler is executing but is not yet complete so that no scheduling "decision" has yet been made (on the scheduler's current pass). In a similar manner, if conscious intention is mediated by a physical process in the brain, then the neural events that mediate subjects' conscious intention to act must necessarily begin before subjects become consciously aware of them. We have little difficulty dealing with the fact that it takes time to become aware of external sensory stimuli (the current S judgments notwithstanding).[2] Indeed, Libet's own cortical stimulation studies (Libet et al. 1979) emphasize just how much time may be necessary. It is therefore surprising that Libet has difficulty with the possibility that similar time intervals would be required to become aware of internal states such as those upon which W judgments are based.

There is an alternative reading of Libet's assumption that W judgments should concide with or precede RP onset that does not commit him to such a strong dualist position but that is inconsistent with other parts of the target article as well as with other RP data. He might be assuming that conscious intention is indeed a physical brain process but that RPs reflect exclusively motor activity and hence must be preceded by subjects' W judgments. The difficulties with this interpretation are: (a) RPs as defined and measured by Libet are generally interpreted as reflecting various preparatory processes assumed to occur in advance of actual motor activity, which is thought to be reflected in scalp recordings only in the last 50–100 ms before EMG onset (for review, see Deecke et al. 1984); (b) the neural generators of RPs and other premovement potentials have not been fully determined, although other structures as well as primary motor cortex appear to be involved (e.g., Arezzo & Vaughan 1975; Gemba et al. 1980; Hashimoto et al. 1980); and (c) even if RPs exclusively reflected motor activity, Libet would need to explain why motor-related activity would be evident in scalp recordings and preceding activity associated with the intention to move would not.

The possibility that at least some of the activity that contributes to RPs preceding voluntary movements may be generated by neurons that contribute to conscious intention raises interesting suggestions concerning the functional role of the neural system(s) that generate RPs (e.g., Deecke et al. 1976; Popper & Eccles 1977). However, as Libet correctly notes, the onset of RPs should be interpreted only as an indicator of the "minimal onset times for cerebral processes that initiate the voluntary act" since even earlier activity could be present and not evident in scalp recordings. This is because RPs and other surface electrical potentials are aggregate, incomplete measures of the neural events occurring at a particular time (see Vaughan 1982; Wood & Allison 1981). The neurons that generate RPs may not themselves help to mediate conscious intention but may lie "downstream" from them.

I have tried to suggest how the obtained temporal relationship between RP onset and W judgments can be explained without resorting either to a nonphysical basis for conscious experience (i.e., substance dualism) or to Libet's conclusion that all so-called voluntary actions are "unconsciously initiated." Neither of these (to me) undesirable conclusions is required if conscious experience (both of external stimuli and internal states) is mediated by physical processes in the brain that take time to operate. According to this hypothesis, Libet's conclusion regarding unconscious initiative is correct only in the restricted sense that components of the neural system that mediates conscious experience cannot themselves mediate that experience in the same way that components of an operating system's scheduler cannot themselves mediate scheduling. Thus, although the components of the system that mediates conscious experience are themselves unconscious, this does not mean that conscious intention must be limited to a subsequent "veto" role over "unconscious initiative," as Libet suggests.[3] Self-reference and part–whole relationships are among the reasons why conscious experience is the perplexing philosophical and scientific question that it is. There are plenty of reasons to be concerned about the role of unconscious processes in cognition and behavior (e.g., Dennett 1978; Fodor 1983; Freud 1925), but the possibility that RPs precede W judgments should not be among them.

ACKNOWLEDGMENT

This commentary was supported by the Veterans Administration and NIMH Grant MH-05286. I am grateful to T. Allison and G. McCarthy for helpful discussion.

NOTES

1. Lest I be guilty of hiding my own assumptions and biases, I briefly summarize them here. So little is known about the properties and mechanisms of conscious experience that adopting any position is risky business. Nevertheless, I believe that it is more reasonable as a provisional hypothesis to assume that conscious experience is an as-yet-unknown property or capacity of a suitably organized physical system (i.e., one that obeys the laws of physics as we know them) than it is to assume that it involves some substance or phenomenon that lies outside the confines of physical law. Because this is a working hypothesis, I am eager to entertain logical arguments that it is incoherent, empirical evidence that it is incorrect, or data/theories suggesting that the relevant physical laws are seriously flawed – in this respect, I'm from Missouri.

2. That S judgments preceded the sensory stimulus by approximately 50 ms illustrates the type of "prior entry effects" and other judgment biases that can occur even in temporal-order tasks much simpler than those employed here (see Sternberg & Knoll 1973). As Libet notes, however, the direction of the bias in the S condition is opposite to that required to explain away the fact that RPs preceded W judgments (assuming that similar judgment biases occur in the S and W tasks), and the error in the M condition is similar to that reported by McCloskey et al. (1983).

3. Here again the scheduler analogy can be helpful. At the level of the individual instructions of the scheduling routine, the scheduler is a fixed, deterministic process. However, at the level of the scheduler as a whole and its interaction with the remainder of the system, the outcome of each execution of the scheduler is not fixed or deterministic because it depends on the competing tasks and available resources at the time (i.e., on the environment in which it executes). For additional discussion of how rigid, "dumb" processes at one level can underlie what appear to be flexible, "smart" processes at another, see Hofstadter (1979) and Dennett (1984).

# Author's Response

## Theory and evidence relating cerebral processes to conscious will

Benjamin Libet

*Department of Physiology, School of Medicine, University of California, San Francisco, San Francisco, Calif. 94143*

Not unexpectedly, the commentators have raised a number of controversial issues. These center on the validity and meaning of the experimental observations and on alternative interpretations of their implications. Commentators had many different kinds of arguments to make on the same general issue. This made it difficult not only to draw together all comments on a related issue but also to cite every relevant commentary; I hope I will be forgiven for any such omissions of citation. I thank the commentators for their conscientious efforts and am gratified that many of them find merit in our experimental questions, design, and observations, even when they do not fully agree with my conclusions and proposals about volitional processes.

### 1. Validity and meaning of the experimental observations

***1.1. Time of conscious intention (W).*** Our experimental values for W were subjected to a variety of critical comments regarding validity, reliability, and quantitative significance. (Many of these criticisms were already anticipated and discussed in the target article [TA] section

2.4.) Much of the criticism appears to reflect differences in experimental approaches to investigating conscious events. My own approach, perhaps influenced by my being a physiologist, has been to accept direct observations (in this case reported times of awareness) as primary evidence; the meaning of such evidence should not be altered unless it is necessitated by other directly relevant observations. A number of the psychologists among the commentators appear to bring with them outlooks conditioned either by behaviorist methodology and philosophy (overt and covert) or by a history of attempts to conceptualize conscious perceptual and volitional processes that are based on observations not directly relevant to the issue of introspective awareness and its timing.

The distinction between a subjective experience (which is only introspectively accessible to the individual), and some externally observable physical or "behavioral" event still seems to elude some commentators. The sterility and irrelevance of behavioristic studies for the mind–brain issue have been recognized even by many former traditional behaviorists (including the late David Krech, 1969, as expressed in his William James lecture of 1967 before the American Psychological Association). **Vanderwolf** adheres to a "classical" behaviorism. His insistence that mental activity is largely unavailable to introspection and that we are only aware of physical (i.e., sensory) events is unrealistic; he would appear to be denying that he is aware of his own thoughts. He argues that reports of W [time of awareness of intention (*wanting*)] appearing before muscle activation must really be due to detection of sensory signals from unrecorded premovements of muscles. This ad hoc construction is required by his philosophy but is without any experimental basis; our EMG (electromyogram) recording was sensitive enough to pick up single motor unit potentials. Vanderwolf's citations on volition and kinesthesia are misleading: Laszlo (1966) reported that speed of key tapping but *not* volitional power was affected by kinesthetic loss; Melzack and Bromage (1973) dealt with the feeling of actually being able to move a "phantom" (not normal) limb; they in no way indicated that subjects were unable to generate a conscious intention or wish to move.

The possibility of discrepancy between actual and reported times for the subject's awareness of wanting to move was discussed in TA 2.4, but this possibility was raised in different ways by a number of commentators as still providing a serious challenge to my use of reported W values in establishing the relationship of conscious intention to the initiation of the voluntary act.

Some commentators (**Latto, Marks, Ringo**) propose that awareness (of the urge to move) must arise in a graded manner and reach some peak or "threshold" before the subject can or will report it. If this is correct, some degree of awareness would actually be present before reported W. It would accordingly reduce or eliminate the temporal difference between RP onset and conscious intention. The experimental basis for this view appears to lie in signal-detection studies, from which Marks argues that subjects would set a high criterion in selecting a threshold level of awareness to be reported. Even if such a detection theory were applicable here, there is no basis for assuming that our subjects wanted to "avoid false alarms" and thus waited for a "strong signal." There was no test for the correctness of their report; any W report time was completely acceptable. Hence there was no reason for reluctance to report *any* awareness. But, more fundamentally, signal-detection studies are based on forced choices; their results are not directly applicable to studies of awareness processes (see Libet 1979; 1981a; 1981b). In our present study, subjects were asked to associate "clock time" with their earliest awareness of the urge to move. They did not report being aware of any preceding graded intention or urge (except for the different and separate awareness of preplanning when that occurred with type I RPs; see Libet et al., 1982). When a subject reports that he feels or is aware of absolutely nothing, whether as here in the period before W or in experiments with stimulation of sensory cortex (e.g., Libet 1973; 1982), I regard it as a distortion of the primary evidence for an investigator to insist, on the basis of a (possibly misapplied) theory, that the subject really was aware of something.

Some commentators propose that there are various other cerebral time factors that could make the reported time of conscious intention (W) significantly different from its actual time (**Breitmeyer, Latto, Rollman, Stamm, Underwood, Wasserman**). The difficulty centers chiefly on the potential for delays in becoming visually aware of the position of the revolving CRO (cathode ray oscilloscope) spot (clock time). Such delays could affect the temporal relationship between the reported clock time (W) and the actual introspective awareness of wanting to move. That there is probably a substantial delay (in hundreds of milliseconds) for becoming aware of a sensory stimulus would indeed follow from our own earlier direct experimental studies of this issue (Libet 1965; 1966; 1973; 1981a, 1982). The existence of such a delay is therefore conceded; even though the reasons for delay offered by most commentators were either speculative or based on irrelevant data. (For example, long reaction times [RTs] are cited by Rollman and by Stamm as evidence for lengthy "mental processing." But RTs do not measure or depend on awareness and cannot be used as primary indicators of when sensory awareness is achieved – e.g., Libet 1973; 1981a. Incidentally, I was not unaware of Wasserman's views about mental chronometry; see Libet 1979.)

One should not confuse *what* is reported by the subject with *when* he may become introspectively aware of what he is reporting. As described in TA 2.4.2(b), our earlier studies provided direct evidence for this distinction and for a subjective referral backward in time. The latter automatically "corrects" one's conscious perception to coincide with the real time of the stimulus (Libet et al. 1979; Libet 1981a; 1982). This can explain, for example, why a runner in a race can take off within 50–100 ms after the starting gun, presumably well before he becomes introspectively aware of the stimulus, but later reports that he heard the gun *before* taking off. Now, although it may take substantial time for cerebral processes to develop the introspective awareness of the urge to move, as I indeed postulate, there is no basis for expecting a subjective backward referral of its perceived timing. Backward referral has only been found in the timing of an external sensory stimulus, and even then it specifically requires the primary cortical response to the fast sensory projection pathway for its mediation (see TA 2.4.2).

**Latto** and **Rollman** assume that the appearance of conscious intention must coincide with the *delayed* awareness of visual clock spot position for both to be regarded as simultaneous by the subject. The backward referral for the visual spot would then lead to an incorrect (earlier) report of clock time for conscious intention. But one should recognize that the subject was required only to associate conscious intention with a visual signal (the position of the revolving spot) whose *content* he would report some seconds after the event. He did not have to be concurrently *aware* of the visual signal in order to associate it correctly with conscious intention; this associated visual position was recalled later, after awareness of it. This is analogous to making fairly correct subjective observations and appropriate associations with respect to diverse sensory stimuli and endogenous experiences in general, even when cortical delays in actual awareness may differ (see Libet et al. 1979; Libet 1981a). The more appropriate inference, from the existence of variable though substantial cerebral delays in awareness of sensory stimuli, is that sensory signals can be meaningfully identified well before introspective awareness of them develops (Libet 1978; Libet 1981a, 1982).

The validity of our skin-stimulus experiment as a control for error in reporting simultaneous events was considered in TA 2.4.1 but was further called into question by some commentators (**Breitmeyer, Stamm, Underwood**). Stamm cites the broad range of our observed skin timings for all subjects (−167 to +83 ms) as a significant uncertainty in timing such mental processes. But data in our original report (Libet, Gleason, Wright & Pearl 1983) showed that the mean timing for each individual subject consistently exhibited small standard errors (SEs) (not far from 20 ms in all cases) and was thus characteristic for each. When each subject's characteristic timing of skin stimuli was subtracted from his reported W times for conscious intention, the "net" W times still followed the onset of RP in the same subject by intervals close to those obtained for the grand average of W times. That is, the actual errors in timing skin stimuli, regardless of the individual, did not appreciably affect the crucial difference between RP onset and W times. It should also be noted that the reliability of W reports, which worried some commentators (e.g., **Latto**) was very good. SE values for each series of 40 trials were typically not far from 20 ms and had no statistical impact on the significance of mean W values.

Suggestions concerning hypothetical differences between times needed for attention to a skin stimulus compared to W (**Breitmeyer, Stamm, Underwood & Neimi**) do not seem to be applicable to our studies. Our subjects were asked to attend continuously to the revolving clock spot and to wait for the appearance of the conscious urge to move (in the W series) or of the conscious sensory experience (in the skin-stimulus series, S). There is no operational reason to believe there was a significant time difference between attentional factors in these two associations, W with spot versus S with spot.

In any case, having subjects associate awareness of a skin stimulus (instead of an urge to move) with the clock position of the revolving spot provided the best available control experiment for assessing the error under the specific conditions used when obtaining reports of W times. The measured timing errors with skin stimuli were not large enough to affect the significance of W timings (relative either to RP [readiness potential] onset or EMG onset), and there is presently no definitive experimental basis for believing that the error in associating conscious intention with clock position would be so much larger as to affect the significance of W. Even if the potentially relevant but speculative errors in W timings proposed by commentators were valid, they would probably not be large enough to affect the significance of the RP-W-EMG temporal relationship.

Regarding unreported awareness, **Scheerer** suggests that an additional component of introspective intention may precede the one reported by our subjects. This is attributed to William James (an "anticipating image") and N. Ach ("intentional sensations"). There is no basis for believing that such a hypothetical component was "missed" by our subjects (Ss). They already had a good image of the anticipated act well before each trial. The free volitional feature was purely one of choosing *when* to act. Also, they were asked to report any introspective feelings that might have preceded the reported earliest awareness of W. The only additional reported awareness was the one for preplanning to act some time within the next few seconds (associated only with some series and a different, "type I" RP; Libet et al., 1982). Ss consistently distinguished this more occasional preplanning awareness from the consistent conscious urge immediately associated with each act.

The possibility that some conscious awareness might develop earlier than W, but without any associated memory processes, and hence without being recallable (**Jasper**), does present a problem, but was not experimentally testable. Examples of "automatic" complex behavior that is inaccessible to recall (such as that during certain epileptic seizures) could be regarded as unconscious manifestations rather than as actions associated with conscious awareness without memory, just as many actions and reactions of normal people appear to be accomplished unconsciously with no specific awareness of them. In any case, as I noted in TA 2.4.4, our subjects did not actually have to recall *any* awareness to make a W report; they only had to be able to associate the clock position of the revolving spot with the first awareness of an urge to move, and later to recall and report that associated spatial image. As indicated above, it would not be necessary to be immediately aware of the associated visual signal in order to recall its appropriate content later.

**1.2. What does the recorded RP represent?** The spontaneity of the voluntary acts under study was questioned directly by **Näätänen** and indirectly by **Ringo.** I would reiterate that each trial was conducted as a separate event, at the subject's convenience, with no set intertrial interval, and that in those series associated with a type II RP, subjects reported experiencing full spontaneity with no preplanning in every trial (Libet et al. 1982). Our distinction between type I and II RPs, associated with the presence and absence respectively of preplanning experiences, has more recently been confirmed by Goldberg, Kwan, Borrett, and Murphy (1984).

**Latto** suggests that the subjects' reports of *feeling* they are acting voluntarily may represent a compliance with what is expected rather than an endogenous process. Even if, as in Latto's hypothetical experiment, a cerebral stimulus site could be found to produce a movement associated with a feeling of volition, it does not follow that the subject would, as Latto predicts, report not acting voluntarily when shown that his movement was instigated by the stimulus. Based on our own extensive experience with subjects reporting conscious sensory responses to cerebral stimuli (Libet 1973), I am certain that the subjects would still report what they *felt* in Latto's hypothetical case, namely, a feeling of wanting to move, even though they would also recognize that the actual instigator was external. There is no necessary conflict for the subject when he reports a *feeling* whose nature he can distinguish from what he observes externally as a physical occurrence; the subject knows what his own experience was and is always encouraged to describe that.

Several commentators (**Eccles, Ringo, Stamm**) propose that the averaged RP, as recorded over 40 events, is actually masking spontaneous or random fluctuations in slow pre-event potentials and that the onset of the meaningful RP is later and perhaps coincides with the time of conscious intention. In such a case there would be no reason to conclude that the cerebral processes initiating the voluntary act precede the appearance of conscious intention. (This issue was already considered in part in TA 2.5).

a. Regardless of the circumstances under which any prepotentials arise, they must be contributing a regular component to the recorded average RP. If they were so irregular and random as not to contribute to the RP whose onset is measured by us, they would have been canceled out by the averaging.

b. No change in contour or components of the averaged RP appears at the reported time of conscious intention (W), that is, at about 150 to 200 ms before muscle action (EMG), whether the RPs represent spontaneous voluntary acts (Libet et al. 1982) or self-paced ones (Deecke et al. 1976). There is therefore no electrophysiological evidence for a distinction between early, random fluctuations (that lead to or allow the appearance of the actual initiating process) and a late potential associated with conscious intention and actual "decision" to initiate the act. Note also that "motor potentials" associated with the development of final outflow from contralateral motor cortex appear well after W time (see Deecke et al. 1976; Shibasaki et al. 1980). They do not account for the main negative rise even in the type II RP that at W time (−200 ms) is still maximal at the vertex (near the supplementary motor area).

c. The vertex recorded RP is associated only with preparation for actual movement, not with processes of attention, expectancy, or any other possible fluctuations that may play some role in volition (e.g., Libet et al. 1982; Libet, Wright & Gleason 1983). There is no basis for regarding the early RP as nonspecific, as **Rugg** suggests. The whole RP, whose onset provides the basis for our thesis, is directly related to an impending voluntary motor act. That is, if the recorded averaged RP reflects individually variable and random fluctuations, each of which develops into a full volitional act, such fluctuations would have to originate in the motor preparatory structure(s) responsible for the whole recorded RP. If other processes or fluctuations do precede and help initiate the recorded RP, their bioelectric counterparts are not appreciably recordable at the vertex, at least under our experimental conditions. (Incidentally, the slow component of the contingent negative variation may be in a special subgroup of RPs–see Libet, Wright, & Gleason 1983–rather than the converse, as proposed by **Latto.**

d. Even if (despite the foregoing) one were to assume that the early portion of the averaged RP does represent spontaneous or random rises in cerebral "excitability," the putatively more definitive and intentional initiating process would have to await the development of each such excitability rise to a level that permits the initiating process to proceed (also argued by **Rugg**). Such a mechanism would still impose a limitation on *when* a specific voluntary initiating process could arise. The conscious function could not then itself decide when to move; it could only select which adequate but randomly appearing fluctuation to proceed with. The conscious initiating process would become the "trigger" that follows an unconscious though nonspecific preprocess. As **Eccles** puts it, conscious intention would itself be *timed* unconsciously. Such a view would differ from my proposal for conscious control only in regarding the preprocesses as not constituting specific initiators of the act.

**Ringo** suggests an experimental test that he believes may establish whether the type II RP (recorded with spontaneous voluntary acts) represents some pseudorandom fluctuation which, when it crosses some threshold level, leads to a conscious initiation of the movement. He proposes that the subject be "asked to choose (and later report) a clock-position on a pseudorandom spontaneous basis." Although such a perceptual choosing "action" might indeed be a spontaneous endogenous event, it cannot be regarded as the equivalent of an urge to move, since no motor act is expected or contemplated. On the evidence that RPs are associated only with preparation or intention to move (Libet, Wright & Gleason, 1983), one would expect no similar *pre*-event potential with the choosing of a visual signal. On the other hand, the evidence already indicates that RPs can appear even when the intention to act is not consummated in an actual motor event (as in our "veto" experiment). That in itself would not settle the issue of whether the RP represents "nonspecific" fluctuations that may lead to conscious initiation or a "specific" initiating process that precedes appearance of conscious intention. Other considerations are required, as discussed previously.

**Rugg** argues that my thesis requires that there be a "necessary" relationship between a voluntary action and the RP processes that precede the act. As indicated in TA 2.5, the RP need only reflect the fact that some cerebral process consistently begins well before W times; this would be true even if it should turn out that the RP process is only directly related to the volitional preparatory process and could be dispensed with without losing the potential for voluntary action. The consistent and regular onset of RP before the voluntary acts studied must mean that RP processes, regardless of how they are causally related to the voluntary act, have been set into motion in relation to the impending act. Surely Rugg is not suggesting that the RPs were simply chance occurrences unrelated to impending voluntary acts.

**Merikle & Cheesman** argue that one must demonstrate that RPs precede behavioral acts even without conscious intention in order to accept our thesis that RP onset before W signifies an unconscious motor initiating process. But the whole RP has in fact been shown to appear only in association with a preparation to move. As indicated above, to show that relevant cerebral processes start before W, RP processes need not themselves be in the direct cerebral path leading to the voluntary act. On the other hand, evidence does suggest that the RP processes (probably occurring in the supplementary motor area) are directly involved in preparations to act. This, together with our present evidence that RP onset substantially precedes conscious intention, does suggest that RP provides a significant physiological indicator of those unconscious behavioral actions that involve preparation (in contrast to motor reactions to unsignaled stimuli); but, contrary to Merikle & Cheesman's view, such additional studies are not crucial to my present thesis about RP and conscious intention.

On the other hand, **Van Gulick** proposes that RP processes which precede W should not be regarded as unconscious but rather as a state of conscious intention, of which the subject later becomes "self-aware" (at W time). This appears to impose a semantic play of words on the actual findings. Since there is no operational manifestation of any awareness of intention until W, one should at most refer to the preceding RP processes as developers, but not direct representations, of a state of conscious intention.

**1.3. How is our experimental act related to "normal" voluntary action?** There are several concerns about the significance of the act we studied, a spontaneously initiated quick flexion of fingers or wrist, in relation to voluntary actions in general (**Breitmeyer, Bridgeman, Danto, Jung, Latto**). We wanted our measurements of relative *timings* (for the onsets of RP and W) to be quantitative and operationally definable, without reliance on intuitive impressions or speculations. Such an objective is much more difficult, if not impossible, to achieve with any of the more common voluntary actions recommended by the commentators for study. Even in our paradigm, the not infrequent appearance of an experience of preplanning within some seconds before the act had to be (and fortunately could be) clearly distinguished from the experience of the conscious intention that more immediately preceded each spontaneous voluntary act. The earlier, temporally looser awareness of preplanning can be regarded as a more deliberative form of conscious intention, but it was not possible to time its onset relative to the onset of the accompanying "type I" RP (which arose considerably earlier than type II RPs in series of acts devoid of preplanning; Libet et al. 1982).

On the other hand, characteristics of preplanning experiences and other considerations led me to propose that even more deliberate voluntary actions, when they finally reach the condition of performing the actual motor act, include processes with characteristics similar to those studied in our simpler, spontaneous voluntary act (see TA 1 and 3). None of the commentaries appear to me to present any convincing evidence or argument against such a proposal. RPs are of course not unique to the special acts we studied, contrary to **Bridgeman**'s concern. The experiments described by **Jung** provide important studies of how RPs can be detected in and used for analyzing the more complex actions of writing, aiming, and so on. RPs have also been associated with the onset of speaking (Grözinger, Kornhuber & Kriebel 1977). Our conclusion – that the 350–400 ms by which RP onset precedes W indicates a period of unconscious initiative for the acts we studied – should not be taken to imply that the seconds-long RPs before repeated word writing (in Jung's experiment) indicate a correspondingly long unconscious initiative. As I have already noted, conscious intention in a more deliberative, preplanning situation is distinguishable from our "W."

I accept **Jung**'s proposal that in a fully learned skill like writing conscious intention is concerned primarily with the "goal and not with the automatic and learned mechanisms of action." As **Breitmeyer** notes, one need not be aware of intending to write immediately before each word. However, it does not follow that nonautomatic acts, that is, those in which conscious intention precedes each act (as in our studies), cannot be preceded by an unconscious cerebral process. Obviously, many deliberately planned intentions involve acts that are not automatic or overlearned but rather are each immediately preceded by an intention to act. Motor acts that have become "automatic" and are not accompanied by an experience of intention before each act are not of interest when one is studying the nature of conscious intention and control, even though they are set into motion by a general intention. Indeed, the available evidence suggests that an automatic act, even though it follows a general intention, is not preceded by any substantial RP (e.g., Libet et al. 1982).

## 2. Conscious control and the mind–brain relationship

Given our experimentally based conclusion that cerebral processes initiating or leading to a voluntary act are initially unconscious, I looked for a way in which the appearance of conscious intention (at 350–400 ms after RP onset but 150–200 ms before muscle activation, EMG) might still play a role in determining the outcome of the unconsciously initiated process. For this, I proposed two possibilities: (a) Conscious intention could signify a conscious triggering process, without which the volitional process would not be consummated; this would agree in principle with the process postulated by **Eccles,** although we differ on how to interpret the prior portion of the RP. (b) Alternatively, when conscious awareness of the intention to move has appeared, an ensuing conscious function might veto or block the consummation into a motor act. The veto alternative was the more attractive one to me, as well as to **Mortensen.** (I agree with Mortensen that the trigger and veto functions would probably have to operate in series if both modes of control were to be independently present.) **Scheerer**'s suggestion that our conscious veto experiment was equivalent to a simple visual reaction-time paradigm misses some crucial distinctions. In the earlier work he cites, subjects were instructed not to respond if a moving dot stopped before crossing a line that otherwise signaled them to respond. Such an inhibition or veto would indeed measure a reaction-time response to a sensory signal and could in principle be accomplished even unconsciously. In our experiment the subject consciously knew in advance that he was about to veto and was not reacting to a signal having an unknown incidence.

**2.1. Are prior cerebral processes required for conscious control?** There is nothing in our new evidence to entail that a conscious veto or trigger is not itself initiated by preceding cerebral processes, as correctly noted by a number of commentators (**Danto, Doty, Latto, Mortensen, Nelson, Rugg, Underwood & Niemi, Wood**; a related argument was made by Harnad, 1982). With such prior processes, any conscious control would itself be initiated unconsciously, as in the case of conscious awareness of intention to move. That is a viable proposition and could perhaps lead to certain testable inferences. Indeed, it would be in accord with my own general hypothesis that a substantial period of cortical activity is in general required in order to elicit a conscious experience (e.g., Libet 1965; 1981a, 1982). However, one must remain open about the applicability of such a general hypothesis to all forms of conscious experience, particularly in the area of intention for and control of voluntary acts. After all, it must also be noted that there is presently no directly applicable evidence *against* the appearance of a conscious control function without prior unconscious cerebral processes.

The arguments for and against the necessity of prior unconscious processes in conscious control really concern matters of philosophical viewpoint, rather than matters of scientific substance. The view that consciousness cannot be primary seems to be based on the widely held premise that some form of identity theory correctly describes the mind–brain relationship (i.e., conscious experience is assumed to be a property or introspectively observable aspect of the underlying neural activities: (**Danto, MacKay, Merikle & Cheesman, Underwood & Niemi, Van Gulick, Wood**). The argument is then that since a conscious experience is based on neural activities, it would require prior causation by cerebral processes. In the general assumption that there always has to be an appropriate ongoing background of cerebral function (to make any mental or conscious manifestations possible), all modern theories of the mind–brain relationship would be in agreement. But we are considering the sufficient, not merely the necessary, conditions, that is, which specific neural activities are uniquely involved in the direct and immediate development or appearance of the conscious function?

I would argue that my proposal of a conscious control that would not itself be initiated unconsciously is compatible with any mind–brain theory. There is no logical imperative in any mind–brain theory that requires specific neural activity to precede the appearance of a conscious event or function. Such a condition cannot be established by a priori arguments and must be experimentally shown to exist, as has been done for a conscious sensory experience (Libet 1973; 1982) and for conscious intention (Libet, Gleason, Wright & Pearl 1983). Even identity theory would be compatible with the occurrence of sudden, spontaneous neural patterns that were immediately associated with conscious events. The issue of prior neural processes would therefore not be primarily one of monism versus dualism, as explicitly suggested by **Nelson.** For similar reasons, and contrary to the view of **Wood,** dualism is not necessarily present or implied in the proposition we experimentally tested, namely, that awareness of the intention to act should precede or coincide with the onset of the RP *if* a conscious intention initiates a voluntary act. On any mind–brain theory, even a determinist one, there could be no a priori assurance that conscious intention (whatever its underlying nature) would follow the onset of neural sequences specifically generating a voluntary act; indeed, many scientists and philosophers have tended to write and speak as if the reverse were true.

The foregoing considerations also bear on the suggestion of **Doty** (who has contributed greatly to our understanding of conscious processes) that conscious control must form part of the same process that unconsciously initiated the conscious intention to move. This is an acceptable argument, as indicated above, as long as one is not concerned to provide a mechanism for conscious control as a spontaneous initiative not developed out of prior unconscious processes. According to Doty, the flaw in having the conscious veto arise separately after conscious intention is that the conscious process would not "know" what motor act will ensue if it fails to veto. But even after the appearance of conscious awareness of intention to act, there remains 100–150 ms in which the conscious function could "evaluate" and decide on whether to veto that intention. We are still far from being able to say with any confidence how quickly a conscious function could evaluate and block the processes leading to an act once conscious intention has appeared.

**2.2. Responsibility and free will.** An appropriate caution is recommended by **Bridgeman** against too facile a transference from our results, based on simple spontaneous voluntary acts, to the larger issues of voluntary behavior in general, self-control, and free will. However, he misses the mark in viewing our subject's acts as not freely willed and as equivalent to reaction-time responses. When a subject is acting at times that he experiences as having himself chosen spontaneously, it seems ad hoc and unsupported to regard his acts as unwilled, programmed responses to special instructions. I have already indicated why our overall findings do suggest some fundamental characteristics of the simpler acts that may be applicable to all consciously intended acts and even to responsibility and free will (if the latter do exist). Scientific progress has almost always depended on discoveries made with simpler, controllable experimental paradigms which then provide the basis for larger inferences. The problems of brain function in relation to conscious voluntary action may also require such an experimental analysis. Speculations and theories not based on experimental data directly relevant to the experience of conscious intention have thus far provided little more than representations of personal philosophical viewpoints.

There can of course be different ways to interpret the significance of the results for the issue of individual responsibility. **MacKay** carefully sets forth a reasoned argument based on a form of identity theory, an argument related to that made by **Doty** and some others. MacKay argues that "what we cannot in principle evaluate, we cannot control." That is, the conscious function would have to be able to evaluate the outcome of impending neural actions in order to control them. But subjective experience is in a phenomenological category that does not include any of the externally observable physical systems that provide the reference analogies in MacKay's argument, and it may be a mistake to argue that "the same distinctions between categorical levels of analysis" must apply. We should not yet presume to know a priori the rules that describe how the conscious function must operate. In any case, I have already argued above that even such a philosophy as MacKay's does not necessarily or logically exclude the appearance of conscious control without specific prior processes.

**MacKay** also argues that if our conscious agency is embodied in our physical brain activity the forms of cerebral activity associated with conscious intention and control must be developing out of and inherent in the whole sequence of processes. He argues that this kind of identity between conscious decision and neural action should serve firmly to "[pin] to our own door responsibility for all we consciously choose to do." However, those for whom MacKay's view of responsibility may not represent a convincingly active process can legitimately turn to an interactionist approach, whether monist emergent (Sperry 1980; **Jasper**) or dualist (Popper & Eccles 1977). The available scientific evidence does not discriminate in favor of one or the other of these views whether interactionist or not.

**Eccles** sets forth, in a systematic and straightforward fashion, a hypothesis of how a conscious entity operating within a dualist interactionist framework might work to initiate voluntary acts and implicitly exert control and responsibility. Eccles has ingeniously adapted his philosophical view to the opportunities and constraints presented by our observations; one would hope for a similar kind of impact from our experimental findings on other mind–brain models that are likewise compatible with the data. However, I must repeat that my own interpretation of the meaning of the RP processes that precede conscious intention differs from that of Eccles (see above, section 1.2). A consideration of all known features of RPs has led me to postulate a more specific initiating role for the unconscious processes that precede conscious intention. In my proposal, the conscious function selects among the unconsciously initiated volitional motor impulses by either triggering one to completion or preferably (see TA section 4.2) by vetoing it. In Eccles's proposal, conscious intention is the specific initiator, but it is nevertheless timed unconsciously to occur when a nonspecific cortical change becomes favorable for proceeding with a motor act.

Finally, I accept **Mortensen**'s contention that the unconscious initiations of conscious intentions are deserving of "moral education," to the extent that this can be efficacious in affecting the tendency to or the context in which our unconscious initiating impulses to act arise. However, when any such moral education takes the form of imparting feelings of guilt, shame, or malevolence for an unconsciously initiated process, it would seem to be demanding responsibility for something not directly manageable at the conscious level. Actual motor performance of the act is both consciously controllable (in my thesis) and ethically meaningful, since it is the motor act that has a real impact on one's fellow man. Moral constraints on actual voluntary motor actions, rather than on the having conscious intentions or urges to act, would thus be based upon realistically achievable goals of responsibility.

## References


Ach, N. (1935) *Analyse des Willens*. Urban & Schwarzenberg. [ES]

Akert, K., Peper, K. & Sandri, C. (1975) Structural organization of motor end plate and central synapses. In: *Cholinergic mechanisms*, ed. P. G. Waser. Raven Press. [JCE]

Allan, L. G. (1978) The attention switching model: Implications for research in schizophrenia. *Journal of Psychiatric Research* 14:195–202. [taBL]

Arezzo, J. & Vaughan, H. G., Jr. (1975) Cortical potentials associated with voluntary movements in the monkey. *Brain Research* 88:99–104. [CCW]

Bauer, H. & Nirnberger, G. (1981) Concept identification as a function of preceding negative or positive spontaneous shifts in slow brain potentials. *Psychophysiology* 18:466–69. [JSS]

Beloff, J. (1962) *The existence of mind*. MacGibbon and Kee. [taBL]

Bolanowski, S. J., Jr. & Doty, R. W. (1982) Monocular loss, binocular maintenance of perception in Ganzfeld. *Abstracts, Society for Neuroscience* 8:675. [RWD]

Boring, E. G. (1957) *A history of experimental psychology*. Appleton-Century-Crofts. [taBL]

Born, J., Whipple, S. C. & Stamm, J. (1982) Spontaneous cortical slow-potential shifts and choice reaction time performance. *Electroencephalography and Clinical Neurophysiology* 54:668–76. [taBL, JSS]

Breitmeyer, B. & Ganz, L. (1976) Implications of sustained and transient channels for theories of visual pattern masking, saccadic suppression, and information processing. *Psychological Review* 87:52–69 [BB]

Bridgeman, B. (1971) Metacontrast and lateral inhibition. *Psychological Review* 78:528–39. [BB]

(1978) Distributed sensory coding applied to simulations of iconic storage and metacontrast. *Bulletin of Mathematical Biology* 40: 605–23. [BB]

Brinkman, C. & Porter, R. (1979) Supplementary motor area in the monkey: Activity of neurons during performance of a learned motor task. *Journal of Neurophysiology* 42:681–709. [JCE]

Broadbent, D. E. (1971) *Decision and stress*. Academic Press. [GU]

Campion, J. & Latto, R. (1985) What is a blindsight? *Behavioral and Brain Sciences* 8:755–757. [RL]

Campion, J., Latto, R. & Smith, Y. M. (1983) Is blindsight an effect of scattered light, spared cortex, and near-threshold vision? *Behavioral and Brain Sciences* 6:423–86. [RWD, RL]

Cheesman, J. & Merikle, P. M. (1984) Priming with and without awareness. *Perception & Psychophysics* 36:387–95. [PMM]

(in press) Word recognition and consciousness. In: *Reading research: Advances in theory and practice*, vol. 5, ed. D. Besner, T. G. Waller & G. E. MacKinnon. Academic Press. [PMM]

Churchland, P. S. (1981a) The timing of sensations: Reply to Libet. *Philosophy of Science* 48:492–97. [CCW]

(1981b) On the alleged backwards referral of experiences and its relevance to the mind–body problem. *Philosophy of Science* 48:165–81. [CCW]

Creutzfeldt, O. D. & Rager, G. (1978) In: *Cerebral correlates of conscious experience*, ed. P. A. Buser & A. Rougeul-Buser. Elsevier/North Holland. [taBL]

Deecke, L., Bashore, T., Brunia, C. H. M., Grünewald-Zuberbier, E., Grünewald, G. & Kristeva, R. (1984) Movement-associated potentials and motor control. *Annals of the New York Academy of Sciences* 425:398–428. [CCW]

Deecke, L., Grözinger, B. & Kornhuber, H. H. (1976) Voluntary finger movement in man: Cerebral potentials and theory. *Biological Cybernetics* 23:99–119. [taBL, CCW]

Deecke, L. & Kornhuber, H. H. (1978) An electrical sign of participation of the mesial "supplementary" motor cortex in human voluntary finger movement. *Brain Research* 159: 473–76. [taBL]

Dennett, D. C. (1978) *Brainstorms: Philosophical essays on mind and psychology*. Bradford Books. [CCW]

(1984) *Elbow room: The varieties of free will worth wanting*. Bradford/MIT Press. [CCW]

Doty, R. W. (1975) consciousness from neurons. *Acta Neurobiologiae Experimentalis* 35:791–804. [RWD]

Eccles, J. C. (1980) *The human psyche*. Springer. [taBL]

(1982a) The initiation of voluntary movements by the supplementary motor areas. *Archiv für Psychiatrie und Nervenkrankheiten* 231:423–41. [taBL]

(1982b) How the self acts on the brain. *Psychoneuroendocrinology* 7:271–83. [JCE]

Efron, R. (1973) An invariant characteristic of perceptual systems in the time domain. In: *Attention and performance*, vol. 4, ed. S. Kornblum. Academic Press. [taBL]

Fitts, P. M. & Deininger, R. L. (1954) S-R compatibility and information reduction. *Journal of Experimental Psychology* 48:483–92. [GBR]

Flachsbart-Kraft, F. (1930) Messung von Hemmungszeiten. *Zeitschrift für Psychologie* 117:73–145. [ES]

Fodor, J. (1983) *The modularity of mind*. Bradford/MIT Press. [CCW]

Freud, S. (1925) *Collected papers*. Hogarth Press. [CCW]

(1915/1955) *The unconscious*. Standard Edition. Hogarth Press. [taBL]

Gemba, H., Sasaki, S. & Hashimoto, S. (1980) Distribution of premovement slow cortical potentials associated with self-paced hand movements in monkeys. *Neuroscience Letters* 20:159–63. [CCW]

Gilden, L., Vaughan, H. G., Jr. & Costa, L. D. (1966) Summated human EEG potentials with voluntary movement. *Electroencephalography and Clinical Neurophysiology* 20:433–38. [taBL]

Goldberg, G., Kwan, H. C., Borrett, D. & Murphy, J. T. (1984) Differential topography of the movement-associated scalp potential with internal vs. external dependence of movement timing. *Archives of Physical Medicine and Rehabilitation* 65:630. [rBL]

Green, D. M. & Swets, J. (1966) *Signal detection theory and psychophysics*. Wiley. [LEM]

Grey Walter, W., Cooper, R., Aldridge, V. J., McCallum, W. C. & Winter, A. L. (1964) Contingent negative variation: An electric sign of sensorimotor association and expectancy in the human brain. *Nature* 203:380–84. [RL]

Grözinger, B., Kornhuber, H. H. & Kriebel, J. (1977) Human cerebral potentials preceding speech production, phonation and movements of the mouth and tongue with reference to respiratory and extracerebral potentials. In: *Progress in clinical neurophysiology*, vol. 3, *Language and hemispheric specialization in man: Cerebral event-related potentials*, ed. J. E. Desmedt. Karger. [rBL]

Grünewald-Zuberbier, E., Grünewald, G. & Jung, R. (1978) Slow potentials of the human precentral and parietal cortex during goal-directed movements (Zielbewegungspotentiale). *Journal of Physiology* (London) 284:181–82P. [RJ]

Haber, R. N. & Hershenson, M. (1980) *The psychology of visual perception*, 2d ed. Holt, Rinehart & Winston. [RL]

Hammer, A. (1914) Untersuchung der Hemmung einer vorbereiteten Willenshandlung. *Psychologische Studien* 9:321–65. [ES]

Harnad, S. (1982) Consciousness: An afterthought. *Cognition and Brain Theory* 5:29–47. [BGB, tarBL]

(unpublished) Conscious events cannot be localized in real time. [taBL]

Hashimoto, S., Gemba, H. & Sasaki, K. (1980) Premovement slow cortical potentials and required muscle force in self-paced hand movements in the monkey. *Brain Research* 197:415–23. [CCW]

Hebb, D. O. (1980) *Essay on mind*. Erlbaum. [CHV]

Hillyard, S. A., Simpson, G. V., Woods, D. L., Van Voorhis, S. & Munte, T. F. (1984) Event-related brain potentials and selective attention to different modalities. In: *Cortical integration*, ed. F. Reinoso-Suarez & C. Ajmone-Marsan. Raven Press. [RL]

Hofstadler, D. R. (1979) *Gödel, Escher, Bach: An eternal golden braid*. Basic Books. [CCW]

Holender, D. (1986) Semantic activation without conscious identification in dichotic listening, parafoveal vision, and visual masking: A survey and appraisal. *Behavioral and Brain Sciences*, in press. [RL]

Hook, S., ed. (1960). *Dimensions of mind*. New York University Press. [taBL]

Hunzelmann, N. & Spillmann, L. (1984) Movement adaptation in the peripheral retina. *Vision Research* 24:1765–69. [RWD]

Jack, J. J. B., Redman, S. J. & Wong, K. (1981) The components of synaptic potentials evoked in cat spinal motoneurones by impulses in single group Ia afferents. *Journal of Physiology* 321:65–96. [JCE]

Jacobson, E. (1930a) Electrical measurements of neuromuscular states during mental activities. 1. Imagination of movement involving skeletal muscle. *American Journal of Physiology* 91:567–608. [CHV]

(1930b) Electrical measurements of neuromuscular states during mental activities. 2. Imagination and recollection of various muscular acts. *American Journal of Physiology* 94:22–34. [CHV]

James, W. (1890) *The principles of psychology*. Holt. Dover repr., 1950. [BGB, ES]

Jonides, J. (1981) Voluntary versus automatic control over the mind's eye's movement. In: *Attention and performance*, vol. 9, ed. J. B. Long & A. D. Baddeley. Erlbaum [BGB]

Jung, R. (1981) Perception and action. In: *Advances in physiological sciences, Regulatory functions of the CNS: Motion and organization principles*, ed. J. Szentágothai, M. Palkovits, & J. Hámori. Pergamon Press; Adadémiai Kiadó. [RJ]

(1982) Postural support of goal-directed movements: The preparation and guidance of voluntary action in man. *Acta Biologica Academiae Scientiarum Hungaricae* 33:201–13. [RJ]

(1984) Electrophysiological cues of the language dominant hemisphere in man: Slow brain potentials during language processing and writing. *Experimental Brain Research* Suppl. 9:430–50. [RJ]

Jung, R., Altenmüller, E. & Natsch, B. (1984) Zur Hemisphärendominanz für Sprache and Rechnen: Elektrophysiologische Korrelate einer Linksdominanz bei Linkshändern. *Neuropsychologia* 22:755–75. [RJ]

Jung, R., Hufschmidt, A. & Moschallski, W. (1982) Langsame Hirnpotentiale beim Schreiben: Die Wechselwirking von Schreibhand und Sprachdominanz bei Rechtshändern. *Archiv für Psychiatrie und Nervenkrankheiten* 232:305–24. [RJ]

Kaufmann, W. (1961) *The faith of a heretic*. Doubleday. [taBL]

Korn, H. & Faber, D. S. (1985) Regulation and significance of probabilistic release mechanisms at central synapses. In: *New insights into synaptic function*, ed. G. M. Edelman, W. E. Gall & W. M. Cowan, Neurosciences Research Foundation. Wiley. In press. [JCE]

Kornhuber, H. H. & Deecke, L. (1965) Hirnpotentialänderungen bei Willkürbewegungen und passiven Bewegungen des Menschen: Bereitschaftspotential und reafferente Potentiale. *Pflügers Archiv für Gesamte Physiologie* 284:1–17. [RWD, RJ, taBL, RN]

Krech, D. (1969) Does behavior really need a brain? In: *William James: Unfinished business*, ed. R. B. McLeod. American Psychological Association. [rBL]

Küpfmüller, K. (1971) Grundlagen der Informationstheorie und der Kybernetik. In: *Physiologie des Menschen, vol. 10, Allgemeine Neurophysiologie*, ed. O. H. Gauer, K. Kramer, & R. Jung. Urban & Schwarzenberg. [RJ]

Laszlo, J. (1966) The performance of a simple motor task with kinaesthetic sense loss. *Quarterly Journal of Experimental Psychology* 18:1–8. [rBL, CHV]

Latto, R. & Campion, J. (1986) Approaches to consciousness: Psychophysics or philosophy? *Behavioral and Brain Sciences*, in press. [RL]

Leibniz, G. W. F. von (1704/1916). *New essays concerning human understanding*. Macmillan. [LEM]

Libet, B. (1965) Cortical activation in conscious and unconscious experience. *Perspectives in Biology and Medicine* 9:77–86. [tarBL]

(1966) Brain stimulation and the threshold of conscious experience. In: *Brain and conscious experience*, ed. J. C. Eccles. Springer. [tarBL, CCW]

(1973) Electrical stimulation of cortex in human subjects, and conscious sensory aspects. In: *Handbook of sensory physiology*, ed. A. Iggo. Springer. [JCE, tarBL, CCW]

(1978) Neuronal vs. subjective timing for a conscious sensory experience. In: *Cerebral correlates of conscious experience*, ed. P. A. Buser & A. Rougeul-Buser. Elsevier/North Holland Biomedical Press. [rBL]

(1979) Can a theory based on some cell properties define the timing of mental activities? *Behavioral and Brain Sciences* 2:270–71. [rBL]

(1981a) The experimental evidence for subjective referral of a sensory experience backwards in time. *Philosophy of Science* 48:182–97. [tarBL, CCW]

(1981b) ERPs and conscious awareness; neurons and glia as generators. In: *Electrophysiological approaches to human cognitive processing. NRP Bulletin*, vol. 20, ed. R. Galambos & S. A. Hillyard, MIT Press. [tarBL]

(1982) Brain stimulation in the study of neuronal functions for conscious sensory experiences. *Human Neurobiology* 1:235–42. [tarBL]

Libet, B., Gleason, C. A., Wright, E. W. & Pearl, D. K. (1983) Time of conscious intention to act in relation to onset of cerebral activities (readiness-potential); the unconscious initiation of a freely voluntary act. *Brain* 106:623–42. [BGB, tarBL, GU, CCW]

Libet, B., Wright, E. W., Jr., Feinstein, B. & Pearl, D. K. (1979) Subjective referral of the timing for a conscious sensory experience. A functional role for the somatosensory specific projection system in man. *Brain* 102:193–224. [RL, tarBL, CCW]

Libet, B., Wright, E. W., Jr. & Gleason, C. A. (1982) Readiness-potentials preceding unrestricted "spontaneous" vs. pre-planned voluntary acts. *Electroencephalography and Clinical Neurophysiology* 54:322–35. [tarBL, JSS]

(1983) Preparation- or intention-to-act, in relation to pre-event potentials recorded at the vertex. *Electroencephalography and Clinical Neurophysiology* 56:367–72. [RL, tarBL]

McCloskey, D. I., Colebatch, J. G., Potter, E. K. & Burke, D. (1983) Judgements about onset of rapid voluntary movements in man. *Journal of Neurophysiology* 49:851–63. [taBL, CCW]

MacKay, D. M. (1951) Mindlike behaviour in artefacts. *British Journal for the Philosophy of Science* 2:105–121. [DMM]
(1965) From mechanism to mind. In: *Brain and mind*, ed. J. R. Smythies. Routledge and Kegan Paul. [DMM]
(1966) Cerebral organizaton and the conscious control of action. In: *Brain and conscious experience*, ed. J. C. Eccles. Springer. [DMM]
(1980) The interdependence of mind and brain. *Neuroscience* 5:1389–91. [DMM]
(1982) Ourselves and our brains: Duality without dualism. *Psychoneuroendocrinology* 7:285–94. [DMM]
Margenau, H. (1984) *The miracle of existence*. Ox Bow Press. [JCE, taBL, RJN]
Melzack, R. & Bromage, P. R. (1973) Experimental phantom limbs. *Experimental Neurology* 39:261–69. [rBL, CHV]
Milgram, S. (1974) *Obedience to authority: An experimental view*. Harper & Row. [RL]
Nagel, T. (1979). *Mortal questions*. Cambridge University Press. [taBL]
Näätänen, R. (1971) Non-aging fore-periods and simple reaction time. *Acta Psychologica* 35:316–27. [RN]
(in press) Selective attention and stimulus processing: Reflections in evoked potentials, magnetoencephalograms, and regional cerebral blood flow. In: *Attention and performance*, vol. 11, ed. M. I. Posner & O. S. Marin. Erlbaum. [RN]
Näätänen, R., Gaillard, A. W. K. & Mäntysalo, S. (1978) Early selective-attention effect on evoked potential reinterpreted. *Acta Psychologica* 42:313–29. [RN]
(1980) Brain potential correlates of voluntary and involuntary attention. In: *Motivation, motor and sensory processes of the brain: Electrical potentials, behaviour and clinical use*, ed. H. H. Kornhuber & L. Deecke. Elsevier. [RN]
Näätänen, R. & Koskinen, P. (1975) Simple reaction times with very small imperative-stimulus probabilities. *Acta Psychologica* 39:43–50. [RN]
Näätänen, R. & Merisalo, A. (1977) Expectancy and preparation in simple reaction time. In: *Attention and performance*, vol. 6, ed. S. Dornic. Erlbaum. [RN]
Nääntänen, R., Sams. M. & Alho, K. (in press) The mismatch negativity: The ERP sign of a cerebral mismatch process. *Electroencephalography and Clinical Neurophysiology*, Supplement. [RN]
Näätänen, R., Simpson, M. & Loveless, N. E. (1982) Stimulus deviance and evoked potentials. *Biological Psychology* 14:53–98. [RN]
Nisbett, R. E. & Wilson, R. E. (1977) Telling more than we can know: Verbal reports on mental processes. *Psychological Review* 84:231–59. [CHV]
Norman, D. A. & Shallice, R. (1980) Attention to action: Willed and automatic control of behavior. Technical report, Center for Human Information Processing, University of California, San Diego. [BB]
Oshima, T. (1983) Intracortical organization of arousal as a model of dynamic neuronal processes that may involve a set for movements. In: *Motor control mechanisms in health and disease*, ed. J. E. Desmedt. Raven Press. [JCE]
Papakostopoulos, D. & Cooper, R. (1973) The contingent negative variation and the excitability of the spinal monosynaptic reflex. *Journal of Neurology, Neurosurgery and Psychiatry* 36:1003–10. [CHV]
Penfield, W. (1958) *The excitable cortex in conscious man*. Liverpool University Press. [RL, taBL]
Popper, K. R. & Eccles, J. C. (1977) *The self and its brain*. Springer [taBL, RJN, CCW]
Posner, M. I. (1980) The orienting of attention. *Quarterly Journal of Experimental Psychology* 32:3–25. [MDR]
Pylyshyn, Z. (1973) What the mind's eye tells the mind's brain: A critique of mental imagery. *Psychological Bulletin* 80:1–24. [CHV]
Reason, J. (1979) Actions not as planned: The effects of automatisation. In: *Aspects of consciousness*, vol. 1, ed. G. Underwood & R. Stevens. Academic Press. [GU]
Remington, R. (1980) Attention and saccadic eye movements. *Journal of Experiental Psychology: Human Perception and Performance* 6:726–44. [BGB]
Roland, P. E. (1978) Sensory feedback to the cerebral cortex during voluntary movement in man. *Behavioral and Brain Sciences* 1:129–71. [RWD]
Rollman, G. B. (1974) Electrocutaneous stimulation. In: *Cutaneous communication systems and devices*, ed. F. A. Geldard. Psychonomic Society. [GBR]
Rozhkova, G. I., Nickolayev, P. P. & Shchadrin, V. E. (1982) Perception of stabilized retinal stimuli in dichoptic viewing conditions. *Vision Research* 22:293–302. [RWD]
Schreiber, H., Lang, M., Lang, W., Kornhuber, A., Heise, B., Keidel, M., Deecke, L. & Kornhuber, H. H. (1983) Frontal hemispheric differences of the Bereitschaftspotential associated with writing and drawing. *Human Neurobiology* 2:197–202. [RJ]
Shibasaki, H., Barrett, G., Halliday, E. & Halliday, A. M. (1980) Components of the movement-related cortical potential and their scalp topography. *Electroencephalography and Clinical Neurophysiology* 49:213–26. [taBL]
Skinner, B. F. (1974) *About behaviorism*. Alfred A. Knopf. [CHV]
Sperry, R. W. (1980) Mind–brain interaction: Mentalism, yes; dualism, no. *Neuroscience* 5:195–206. [tarBL]
Stamm, J. S. & Gillespie, O. (1980) Task acquisition with feedback of steady potential shifts from monkeys prefrontal cortex. In: *Motivation, motor and sensory processes of the brain: Electrical potentials, behavior and clinical use*, ed. H. H. Kornhuber & L. Deecke. Elsevier. [JSS]
Sternberg, S. & Knoll, R. L. (1973) The perception of temporal order: Fundamental issues and a general model. In: *Attention and performance*, vol. 4, ed. S. Kornblum. Academic Press. [taBL, CCW]
Swets, J. A., Tanner, W. P. & Birdsall, T. G. (1961). Decision processes in perception. *Psychological Review* 68:301–40. [LEM]
Thorpe, W. H. (1974) *Animal nature and human nature*. Methuen. [taBL]
Underwood, G. (1982) Attention and awareness in cognitive and motor skills. In: *Aspects of consciousness*, vol. 3, ed. G. Underwood. Academic Press. [GU]
Uttal, W. R. (1971) The psychobiological silly season, or what happens when neurophysiological data become psychological theories. *Journal of General Psychology* 84:151–66. [BB]
Vanderwolf, C. H. (1983a) The influence of psychological concepts on brain-behavior research. In: *Behavioral approaches to brain research*, ed. T. E. Robinson. Oxford University Press. [CHV]
(1983b) The role of the cerebral cortex and ascending activating systems in the control of behavior. In: *Handbook of behavioral neurobiology*, vol. 6, *Motivation*, ed. E. Satinoff & P. Teitelbaum. Plenum Press. [CHV]
Vanderwolf, C. H. & Robinson, T. E. (1981) Reticulo-cortical activity and behavior: A critique of the arousal theory and a new synthesis. *Behavioral and Brain Sciences* 4:459–514. [CHV]
Vaughan, H. G., Jr. (1982) The neural origins of human event-related potentials. *Annals of the New York Academy of Sciences* 388:125–38. [CCW]
Vaughan, H. G., Jr., Costa, L. D. & Ritter, W. (1968) Topography of the human motor potential. *Electroencephalography and Clinical Neurophysiology* 25:1–10. [taBL]
Wagstaff, G. F. (1981). *Hypnosis, compliance and belief*. St. Martin's Press. [RL]
Wasserman, G. S. (1978) Limulus psychophysics: Temporal summation in the ventral eye. *Journal of Experimental Psychology: General* 107:276–86. [GSW]
Wasserman, G. S. & Kong, K.-L. (1979) Absolute timing of mental activities. *Behavioral and Brain Sciences* 2:243–304. [GSW]
Weisstein, N. (1972) Metacontrast. In: *Handbook of sensory physiology*, vol. 7(4). *Visual Psychophysics*, ed. D. Jameson & L. Hurvich. Springer. [BB]
Whyte, L. L. (1960) *The unconscious before Freud*. Basic Books. [taBL]
Wood, C. C. & Allison, T. (1981) Interpretation of evoked potentials: A neurophysiological perspective. *Canadian Journal of Psychology* 35: 113–35. [CCW]
Woodworth, R. S. (1938) *Experimental psychology*. Holt. [ES]
Wundt, W. (1874/1904). *Principles of physiological psychology*, vol. 2. Macmillan. [BGB]

# 17

# Are the Mental Experiences of Will and Self-Control Significant for the Performance of a Voluntary Act?

## Response to Commentaries by L. Deecke and by R. E. Hoffman and R. E. Kravitz

Reprinted from *The Behavioral and Brain Sciences*
Vol. 10, No. 4, pp 781–786, 1987

Behavioral and Brain Science 10 (4): 781-786

*Commentary on* **Benjamin Libet (1985) Unconscious cerebral initiative and the role of conscious will in voluntary action. BBS 8:529–566.**

**Abstract of the original article:** Voluntary acts are preceded by electrophysiological "readiness potentials" (RPs). With spontaneous acts involving no preplanning, the main negative RP shift begins at about −550 ms. Such RPs were used to indicate the minimum onset times for the cerebral activity that precedes a fully endogenous voluntary act. The time of conscious intention to act was obtained from the subject's recall of the spatial clock position of a revolving spot at the time of his initial awareness of intending or wanting to move (W). W occurred at about −200 ms. Control experiments, in which a skin stimulus was timed (S), helped evaluate each subject's error in reporting the clock times for awareness of any perceived event.

For spontaneous voluntary acts, RP onset preceded the uncorrected Ws by about 350 ms and the Ws corrected for S by about 400 ms. The direction of this difference was consistent and significant throughout, regardless of which of several measures of RP onset or W were used. It was concluded that cerebral initiation of a spontaneous voluntary act begins unconsciously. However, it was found that the final decision to act could still be consciously controlled during the 150 ms or so remaining after the specific conscious intention appears. Subjects can in fact "veto" motor performance during a 100–200-ms period before a prearranged time to act.

The role of conscious will would be not to initiate a specific voluntary act but rather to select and control volitional outcome. It is proposed that conscious will can function in a permissive fashion, either to permit or to prevent the motor implementation of the intention to act that arises unconsciously. Alternatively, there may be the need for a conscious activation or triggering, without which the final motor output would not follow the unconscious cerebral initiating and preparatory processes.

## The natural explanation for the two components of the readiness potential

Lüder Deecke
*Neurological University Clinic Vienna, Lazarettgasse 14, A - 1097 Vienna, Austria*

Libet is not the first who distinguishes between two components of the readiness potential or Bereitschaftspotential (BP, Kornhuber & Deecke 1964; 1965). We have long maintained the view that there is an early and a late BP component (cf. Deecke et al. 1976; 1984). The existence of these components, however, does not depend on constructs so remote from experimental verifiability that they abandon the firm ground of natural science. The most simple explanation of the two BP components is that two different cortical sources are responsible for the generation of the BP.

Although the BP is widespread and can be recorded over both hemispheres and the midline depending on the task it precedes, two principal generators seem to prevail. These are the supplementary motor area (SMA), which generates the early symmetrical component, and the rolandic motor cortex (MI), which generates the late asymmetric (i.e., contralateral) component preceding finger movement.

We have further substantiated this interpretation in our commentary (Kornhuber & Deecke 1985) on Goldberg's (1985) *BBS* target article. There is not only electroencephalographic (Deecke et al. 1983) but also magnetoencephalographic (Deecke et al. 1985) evidence for such distinction. Topographical recordings are needed to distinguish between the two components; the few electrodes used by Libet are insufficient. Since the SMA component of the BP leads in the temporal chain of premovement events, we believe that the SMA has a key role in initiating voluntary movement (and in every sequence of multisequential actions). A "preconscious" appearance, if there is any, of the SMA BP, does not particularly disturb the neurologist, who is familiar with the various infraconscious brain operations and controls, who knows the agnosias and neglect syndromes and asks himself why phylogenesis may have invented consciousness: for the sake of data reduction. That's why the method of introspection is limited. Introspection may fail, but this does not mean that all that is not accessible to it is supernatural.

## Feedforward action regulation and the experience of will

Ralph E. Hoffman and Richard E. Kravitz

*Department of Psychiatry, Yale University School of Medicine, New Haven, Conn. 06520*

A *BBS* target article by Libet (1985) indicates that for simple actions such as finger flexion, a physiologically generated readiness potential (RP) precedes subjects' initial awareness of intending-to-act (referred to as "wanting" or W), which in turn precedes motor behavior. Libet argues that W signifies a process whereby subjects consciously choose whether or not prior biologically generated action initiatives receive motor implementation; his experimental paradigm is assumed to reflect intentional action in general.

Wood's (1985) commentary notes the dualism of this model, which includes as different stages of a single process brain events as well as decisions by human agents. The following is an alternative, nondualistic account of W based on the phenomenology of the experience of action, recent experimental studies of the regulation of motor behavior, and the self-reports of Tourette's patients. This model suggests that the intentionality of actions is inferred rather than directly experienced.

What are the contents of our consciousness? First we experience images as percepts, representing what we construe to be a world outside ourselves. Second, we experience images independent of any external stimuli which seem to arise within ourselves. Like external images, internal images can combine all sensory modes; our consciousness is commonly populated by internal images that are visual and/or auditory, the latter typically (but not exclusively) being "talking to ourselves."[1] Third, we experience feelings and emotions that can reflect a wide range of shades and hues. Fourth, we consciously register our body and its actions; this mode of awareness provides a broad range of information, including proprioceptive and/or visual awareness of our bodily displacements such as limb, head, lip and eye movements, and auditory perception of our utterances. The latter three forms of awareness compose the different aspects of the conscious of Self (to borrow a phrase from William James 1890/1983, ch. 10).

W, as an experience, is not construed to reflect the world outside ourselves. Thus W must either be an internal image, affective state, or awareness of action. As the initial "urge-to-act," W connotes a kind of tension; this quality suggests that W belongs to the feeling and emotional class of experience. On closer scrutiny, however, there are three reasons to doubt that W, as a purely affective experience, can connote the intentionality of particular actions:

(A) Feelings and emotions cannot be directly "willed" (Gordon 1986). We cannot decide to feel sad, anxious, tense, and so on, and then directly experience that feeling. We are able, at times, to induce certain feelings and emotions, but this requires prior generation of images, attending to certain external perceptions, or undertaking certain actions which, singly or in combination, provoke the desired feeling. One characteristic of human action is that it can be directly "willed"; therefore, the initial conscious antecedent of any such intentional action cannot be (or cannot merely be) a feeling or emotion.

(B) Feelings and emotions lack particularity (cf. Wittgenstein 1980, sect. 148). When I am sad/happy/anxious/tense the feeling does not specify what I am sad/happy/anxious/tense about. The job of keeping me informed of the circumstances of these feelings is performed, again, by my internal images, external perceptions, and that which is revealed by my actions. Thus, W, as a pure feeling, can represent a "generic" urgency but cannot specify a particular act.

(C) The absence of W does not cause the corresponding action to be experienced as "involuntary" (see commentary by Breitmeyer 1985). For example, a routine activity such as knotting my tie need not be accompanied by any W; nonetheless, this action is not experienced as involuntary in the sense of being out of my control.

Thus, to connote the intentional initiation of particular actions, W, as an experience, must reflect something other (or more) than a feeling. Since it occurs prior to manifest action, W cannot be the experience of the action itself. In light of our outline of the three experiential forms of "self-awareness," internal images remain, for W, as the only other possibility. The phenomenology of internal images is in two relevant aspects opposite to that of feelings and emotions.

(A*) Images can be "willed" without mediating experiences: I can decide to have an image of the Empire State Building or the last movement of Beethoven's Ninth Symphony and then directly see/hear the corresponding image (cf. Wittgenstein 1980, sect. 80, 88, 111).

(B*) Images possess particularity. If I picture the act of hammering a nail, the act itself is represented more or less unambiguously (cf. Wittgenstein 1980, sect. 85).

These considerations suggest that internal images could constitute the conscious experience of initiating an action.

Viewing W as having an image component is by no means a new idea. William James described "anticipatory images" that were visual and/or proprioceptive memories of earlier actions which, when reactivated, correspond to the conscious decision to act (James 1890/1983, ch. 26). Scheerer (1985), in his commentary on Libet's target article, postulated that an anticipatory image of this sort coincides temporally with the RP. Although this move preserves the notion that we can consciously initiate our own actions, Scheerer's suggestion requires two distinct anticipatory experiences: the first an imagery experience coinciding with RP and a second W experience, which, according to Libet's data, occurs somewhat later than the RP.

Empirical studies by Kelso (1982) suggest a link between anticipatory images and the second of two feedforward regulatory processes for motor behavior.[2] The first, *efferent copy*, is a replication of the motor command that modifies the perception of the action as it unfolds; for example, information about eye muscle activity is used to rectify the apparent movement of the visual scene induced by ocular movement. The second, *corollary discharge*, consists of a representation of expected sensory consequences of actions generated prior to the action. If the sensory experience of an action deviates from this anticipatory representation, rapid peripheral adjustments can be made without needing to return to central motor initiation programs. If they were conscious, corollary discharge representations would take the form of images that could be proprioceptive, tactile, visual, or in the case of, say, speech generation, auditory. A number of experimental studies have highlighted the robust effects of conscious imagery in facilitating and modifying motor activity (Finke 1979; Greenwald 1970a; 1970b), thus supporting the hypothesis that corollary discharge representations could take the form of conscious images. [see also Roland: "Sensory Feedback to the Cerebral Cortex During Voluntary Movement in Man" *BBS* 1(1) 1978; and Stein "What Muscle Variable(s) Does the Nervous System Control in Limb Movements?" *BBS* 5(4) 1982.]

A perspicuous account of W immediately follows: An unconsciously generated action initiative (signaled by the RP) brings about a corollary discharge image of the action in consciousness at a somewhat later time, which in turn provokes a secondary affective "urge" to match motor behavior to the image. The image and urge *together* constitute W; their onset is predicted to occur between the RP and actual motor behavior as observed by Libet (1985). According to this model, however, W is a conscious correlate of feedforward regulatory processes rather than an intentional decision by a human agent.

This raises the question of whether the W experience yields an adequate paradigm for intentional behavior. Consider the example of Tourette's syndrome, a classic disorder of *involuntary* action involving motor tics and automatic vocalizations. Electroencephalograms that are time-locked to the tics of Tourette's patients do not demonstrate any RP, even though a robust RP is induced when these same subjects mimic their own tics (Obeso et al. 1981). Libet's model, predicting that an intentional action requires an RP-signaled action initiative to "gate," is consistent with these data.

But if one looks closely at the phenomenology of Tourette's stereotypies an interesting discrepancy emerges. Certain patients, upon introspection, report that their ticcing is actually preceded by a mounting urge or tension, which is in turn induced by a phantom sensation or image, frequently with tactile or proprioceptive features: The resulting motor behavior relieves the mounting affective tension (Bliss 1980; Cohen et al. 1984). If the patient is aware that the motor behavior is initiated to relieve the prior affective "urge," he can experience the tic as *voluntary*.

One of us (Hoffman 1986) has recently argued that the "voluntariness" of an action is not directly experienced at all, but rather is inferred as a function of whether the action is concordant with current, consciously accessible goals. In fact, most action is experienced as neither voluntary nor involuntary (cf. the tie-knotting example in C above, and the commentary by Breitmeyer 1985). But if we need to determine the voluntariness of tie-knotting, we can optionally do so by accessing a goal representation (such as visually "seeing" the completed tie knot, or saying to oneself something like, "I've got to look good for work today") and thereby take the action to be voluntary. Even the extreme example of a tic not initiated by an RP can still be associated with a consciously accessible goal – namely tension relief – and can therefore be inferred as voluntary.

Insofar as W can precede the action, the illusion that the former is the causal antecedent of the latter is maintained. However, this temporal ordering simply reflects the emergence of regulatory processes which mold the form of the action by priming sensory systems prior to it. To claim that W causes the action would be erroneous in the same sense that it would be incorrect to assert that activation of a radar system causes the airplane to take off just because the pilot generally turns on the radar, for guidance purposes, prior to takeoff. Libet's hypothesis that mental decisions control biological processes adds an unnecessary complication which does not account for other data. The belief that a self, indeed a my-self, is the initiating agent reflects an optional interpretation of the action that occurs post hoc (cf. Harnad 1982). W is neither sufficient nor necessary for this inferential process.

NOTES

1. Internal images can represent concepts but should not be equated with them; when we think of a tree, we generally "see" a "picture" of a tree or say to ourselves the word "tree." These visual or auditory images do not "contain" the concept of a tree (including, for instance, a tree's relationship to plants, dependence on light, water, etc.) but at best connote the availability of such information for other cognitive operations.

2. These important ideas were pointed out to us by Gail Zivin (1986).

# Author's Response

## Are the mental experiences of will and self-control significant for the performance of a voluntary act?

Benjamin Libet

*Department of Physiology, School of Medicine, University of California, San Francisco, Calif. 94143*

In their critique of my target article (Libet 1985) and in making the case for their model of will and intentionality, **Hoffman & Kravitz** raise several fundamental issues which I shall try to sort out for separate but interrelated discussion. In our experimental study, the subject's report of the initial awareness of the intention (or desire or decision) to act (i.e., to perform the act now, as distinguished from more distant planning to act) was indeed referred to as W ("wanting"). However, contrary to Hoffman & Kravitz's account, I did *not* argue that such an awareness (W) "signifies a process whereby subjects consciously choose whether or not" to act. Such a conscious control process was postulated by me to arise subsequent to the appearance of W; the hypothesis of conscious control (and the evidence for it) should not be confused with the operationally observed phenomenon of the initial awareness of the intention to act. W was reported to appear 350 or 400 msec after the onset of specific cerebral processes (the readiness potential or RP) associated with spontaneous voluntary acts; consequently I argued that (without making further assumptions) the volitional initiative to act could not have been generated consciously. I assume that Hoffman & Kravitz are chiefly contesting my thesis that there could be conscious mental control of the final motor implementation of the unconsciously initiated intention to act. However, they devote much of their commentary to their view of the nature of W as an experience of will, and I shall first discuss that view.

**1. The experience of will.** After listing what they regard as the four classes of our conscious experience, Hoffman & Kravitz suggest that "W belongs to the feeling and emotional class." On their further argument that feelings and emotions lack particularity, they develop the view that W must also represent an experience of internal images; "images can be 'willed'" and they "possess particularity" (for the intended specific act). Hoffman & Kravitz accordingly suggest that "internal images could constitute the conscious experience of initiating an action." Let's examine the evidence offered in support of this proposal:

(a) As already indicated in my original response (Libet 1985r) to Scheerer's commentary (Scheerer 1985), we found no experimental basis for the existence of "anticipatory images," which were originally proposed by William James and are now hypothesized by Scheerer as experiences that coincide with the RP phase that precedes W. Although asked to report any introspective experiences that preceded their initial awareness of wanting to move, subjects did not report any anticipatory sensory images. In the Hoffman & Kravitz account the "image and urge *together* constitute W," but the conscious image of the action is regarded as the first of the two to appear and it "in turn provokes a secondary affective 'urge' to match motor behavior to the image." This introduction of an additional conscious experience of an image is an ad hoc requirement of Hoffman & Kravitz's hypothesis, not based on actual observation under the conditions in question. Are we to accept the primary evidence of the subjects' introspective report (as I do), or are we going to insist that the subject had a conscious experience which he himself does not report and would even deny having had? The unitary W experience reported by the subject includes the wish or urge to perform a particular act; its alleged lack of particularity rests only on Hoffman & Kravitz's assumption that this urge must be regarded as a pure diffuse feeling.

(b) The experimental studies referred to (Finke 1979 and Kelso 1982) do not appear to provide direct evidence on the proposed role for anticipatory images in conscious voluntary action. Any ability of conscious imagery to modify motor outcomes or patterns does not directly address the hypothesis that anticipatory images actually constitute the experience of initiating an action.

**2. Does the W experience yield an adequate paradigm for intentionality?** Hoffman & Kravitz raise this question because in their model "W is a conscious correlate of feedforward regulatory processes rather than an intentional decision by a human agent." As already noted above, in my view the feature of intentional decision by a human agent consists of conscious control subsequent to W (W being the reported time of awareness of an intention initiated unconsciously well before W).

Hoffman & Kravitz consider the example of Tourette's syndrome in relation to this question. First they agree that the motor tics of Tourette's patients are generally regarded as involuntary actions. Especially significant is the evidence that no RP precedes such tics, even though a robust RP does appear before a voluntary act in the same subjects when they voluntarily mimic their own tics (Obeso et al. 1981). However, many Tourette's patients report introspective experiences of phantom sensations or images, which induce a localized affective tension that is finally relieved by a tic or vocalization (Bliss 1980; Cohen et al. 1984). It seems to escape the notice of Hoffman & Kravitz that these features of Tourette phenomenology provide strong evidence to *contradict* their proposal that internal sensory images constitute the conscious experience of initiating a voluntary action. Tourette's patients have intense internal images before ostensibly *involuntary* movements!

Hoffman & Kravitz may feel that this difficulty is resolved by suggesting that the Tourette motor behavior may really be a voluntary type of experience, the patient being aware that the motor act is initiated to relieve the intense urge or tension induced by the sensory images. Without attempting to settle the issue of how to use the label "voluntary" under such conditions, it is at least clear that the initiation of a Tourette tic differs from that of normal voluntary acts; only the latter are preceded by an RP and are associated with a feeling or experience of independent intention to act rather than a compulsion to act that is driven by abnormally and spontaneously arising sensory images.

Having proposed their "anticipatory image hypothesis" to account for the experience of voluntariness, Hoffman & Kravitz then go on to suggest that voluntariness "is not directly experienced at all, but rather is inferred" (by the subject) in association with "a consciously accessible goal." To support this view they use (a) the debatable conclusion that Tourette tics are experienced as voluntary in the normal sense (already discussed above) and (b) the argument that "most action is experienced as neither voluntary nor involuntary." In support of the latter point they cite the case of well-learned, automatic actions (such as tie-knotting) that may be optionally inferred to be voluntary by the subject by consciously "accessing a goal representation." That many if not most behavioral actions may not be experienced as voluntary does not address the nature of those actions which are experienced as voluntary (even if they should be in the numerical minority).

In my original response (Libet 1985r, sect. 1.3) I had agreed with Jung's argument (Jung 1985) that in a fully learned skill (such as writing, in his case) conscious intention is concerned primarily with the goal; that is, there need not be a conscious awareness of intending to write that immediately precedes writing each word (as also noted by Breitmeyer 1985). But motor acts which have individually become automatic are not the kinds that are relevant to the issue of conscious intention to perform a given act, when such intention does occur (see Libet 1985r for further discussion on this point). A subject might infer (after the fact, as Hoffman & Kravitz suggest) that a series of automatically accomplished acts were in accord with his goals and therefore not involuntary. But a phenomenological *experience* of the conscious intention to knot one's tie would only have occurred if and when one had been aware of the specific wish to start the whole tie-knotting process; and even then, the experience would bear only on the initiating movement for the sequence of automatic acts. Indeed, if even the initiation of a tie-knotting sequence were performed automatically (without a specific conscious intention, as an unconscious response to externally or internally generated time cues, etc.), there would be no basis for assigning an experience of *conscious* intention to the whole process. This would hold even if the subject were aware of what he had done after the fact. Hoffman & Kravitz seem to be confusing (1) "determining," after the action, whether one's acts were consistent with one's general wishes or goals, and (2) the conscious intention (and control) that one experiences (and reports) occurring before the act and not post hoc.

**3. Monism, dualism, and epiphenomenalism.** I did indeed propose that conscious control of the motor outcome of the unconsciously initiated volitional process could be operative, either positively (by permitting the active triggered process to go to completion) or negatively (by vetoing, that is, blocking the final motor consummation of the process). I noted that such a hypothesis, and the associated evidence, leaves open a *potentiality* for free will (in the sense that decisions by a mental self could affect neuronal processes). But I was carefully considering conscious control at a phenomenological level, as an experience; that is, we could "discuss operational possibilities for conscious control at a level which does not require commitment to any specific philosophical alternatives for mind-brain interaction, whether these be determinism versus free will or epiphenomenalism versus mental intervention" (Libet 1985, p. 536). I elaborated this further (Libet 1985r, p. 563), stating that "a conscious control that would not itself be initiated unconsciously is compatible with any mind–brain theory." This of course leaves each of us free to apply whichever mind-brain theory appeals to us, as recognized, for example, by MacKay (1985).

In one strictly deterministic view, as set out by Hoffman & Kravitz, both W and conscious control may be regarded as simple epiphenomena, correlates of "feedforward regulatory processes rather than [representing] an intentional decision by a human agent"; there is of course no proof of the validity or invalidity of such a belief. But when Hoffman & Kravitz assert that a hypothesis like mine, involving conscious control of neuronal processes, adds unnecessary complications (in comparison to their belief system), they seem to be in error both on general philosophical grounds and in their specific arguments. Their proposal, that the experience of voluntarism is simply a post-hoc inference, is treated in their last paragraph as if it were a fact, from which they then naturally conclude that conscious control is neither sufficient nor necessary! But they do not actually know that a subject's experience of (or, as they put it, belief about) self-conscious control in fact occurs post hoc (after the action). Indeed, our experimental evidence, in which W and mental veto are reported before an action, contradicts their post-hoc hypothesis.

Hoffman & Kravitz's analogy to a radar system, serving to guide rather than cause an airplane to take off, is not appropriate or relevant to the case of conscious control of an act. First, having argued that W and the experience of conscious control are post-hoc epiphenomena, Hoffman & Kravitz should find it difficult to assign to these epiphenomana a "guidance" function. More important, it is their pilot who turns on the radar; but our question is, who or what turns on the pilot! On Hoffman & Kravitz's model, feedforward regulatory (neuronal) processes turn the pilot on with no need for conscious intention and no role for conscious decisions or controls. Hoffman & Kravitz claim that a hypothesis of conscious mental control does not account for "other data" (data they do not spell out); but in fact it is their view that faces problems from other data. For example, how do they account for the empirical existence of the conscious correlates and the intuitive impact of these on one's decisions to act, since it is these conscious phenomena that become an "unnecessary complication" for their system?

This question can in fact be answered even within the terms of a monistic brain/mind identity theory (see Libet 1985r, p. 563), when this is not so narrowly constrained as in the approaches of Hoffman & Kravitz and presumably also that of Harnad (1982). Sperry (1976; 1980) has elaborated a monistic interactionist answer, in which the conscious mind is not a passive correlate but "instead an essential constituent of higher brain processing – and an active causal determinant in brain activity" (1976, p. 12). Other philosophical difficulties or arguments aside, dualist interactionism also deals readily with this issue (Eccles 1980; Popper & Eccles 1977). I have noted that "the available scientific evidence does not [crucially] discriminate in favor of one or the other" of the brain/mind theories (Libet 1985r, p. 564), although it may affect how convincing a given theory appears to any given individual.

**4. Types of readiness potential (RP) are related to "natural" differences in conscious intentions to act.** Deecke's commentary seems to be misdirected on the issue of RP (readiness potential) components, and philosophically flawed in his view of conscious experience and its investigation.

Our study was not primarily concerned with components in an RP or with their precise neuronal sources (Libet et al. 1982; Libet 1985). What we found was that different RPs accompanying similar, self-initiated acts could be categorized into two different types, for virtually all subjects. The Type I RP starts earlier and is ramplike; it appears when subjects report experiencing some endogenous general preparation to act within a few seconds before the act. (Type I RPs resemble the RPs previously discovered by Kornhuber and Deecke, 1965, for self-paced acts, in which volition was not entirely free of external constraints; see Libet 1985, p. 530.) The Type II RP starts more abruptly at about −550 msec (before the EMG [electromyogram] signals the initial activation of the muscle); it appears when subjects report that all their acts have been performed with ad lib timing, with no experiences of preplanning before each spontaneous inclination to act.

Our discovery of two different types of RPs and their relation to the subject's endogenous pattern of conscious intention to act had no direct relation to components in RPs for self-paced acts as analyzed by Deecke et al. (1976). The second or late component discussed by **Deecke** does not start to be significant until −150 msec, and it is predominantly contralateral (rolandic motor cortex in origin). The later of our two types of RP (i.e., II) starts at about −550 msec (400 msec before the start of Deecke's "later" component); it is decidedly maximal at the vertex, that is, it is probably mostly from the SMA (supplementary motor area), although a contribution from contralateral rolandic area is also likely (see Libet et al. 1982). Another component in self-paced RPs was described by Shibasaki et al. (1980); their so-called NS' component starts at about −500 msec and may be related in part to our type II RP (Libet et al. 1982).

Explanations of RPs at a neuronal level do not in themselves establish how RPs are related to the appearance of conscious intention. The two approaches (relating RPs to neuronal and to subjective volition) are complementary, not mutually exclusive. **Deecke** states "The existence of these (RP) components, however, does not depend on constructs so remote from experimental verifiability that they abandon the firm ground of natural science." The reports by subjects about their different experiences of preplanning versus spontaneity (associated with type I and II RPs respectively) were not theoretical constructs but empirical observations. The reliability of the correlation between the type of subjective report and the type of RP was statistically established as verifiable and significant (Libet et al. 1982; 1983). Indeed, another group of investigators has since confirmed this relationship (Goldberg et al. 1984). The method of introspection may have its limitations, but it can be used appropriately within the framework of natural science, and it is absolutely essential if one is trying to get some experimental data on the mind-brain problem. Although Deecke's neurologist may not have been "particularly disturbed" by an unconscious initiation of a voluntary act, he had no experimentally direct and quantitative knowledge of its existence until our report (Libet et al. 1983; Libet 1985); some neurologists have in fact argued against our conclusion (e.g., Jung 1985, pp. 544–45).

## References


Bliss, J. (1980) Sensory experiences of Gilles de la Tourette syndrome. *Archives of General Psychiatry* 37:1343–47. [REH, BL]

Boschert, J., Hink, R. F. & Deecke, L. (1983) Finger movement versus toe movement-related potentials: Further evidence for supplementary motor area (SMA) participation prior to voluntary action. *Experimental Brain Research* 55:73–80. [LD]

Breitmeyer, B. G. (1985) Problems with the psychophysics of intention. *Behavioral and Brain Sciences* 8:539–40. [REH, BL]

Cohen, D. J., Riddle, M. A., Leckman, J. F., Ort, S. & Shaywitz, B. A. (1984) Tourette's syndrome. In: *Neuropsychiatric movement disorders*, ed. D. V. Jeste & R. J. Wyatt. American Psychiatric Press. [REH, BL]

Deecke, L., Bashore, T., Brunia, C., Grünewald-Zuberbier, E., Grünewald, G. & Kristeva, R. (1984) Movement associated potentials and motor control. In: *Brain and information: Event-related potentials*, ed. R. Karrer, J. Cohen & P. Tueting. Annals of the New York Academy of Sciences 425:398–428. [LD]

Deecke, L., Boschert, J., Brickett, P. & Weinberg, H (1985) Magnetoencephalographic evidence for possible supplementary motor area participation in human voluntary movement. In: *Biomagnetism: Applications and theory*, ed. H. Weinberg, G. Stroink & T. Katila. Pergamon. [LD]

Deecke, L., Grözinger, B. & Kornhuber, H. H. (1976) Voluntary finger movement in man: Cerebral potentials and theory. *Biological Cybernetics* 23:99–119. [LD, BL]

Eccles, J. C. (1980) *The human psyche*. Springer. [BL]

Finke, R. (1979) The functional equivalence of mental images and errors in movement. *Cognitive Psychology* 11:235–64. [REH, BL]

Goldberg, G., Kwan, H. C., Borrett, D. & Murphy, J. T. (1984) Differential topography of the movement-associated scalp potential with internal vs. external dependence of movement timing. *Archives of Physical Medicine and Rehabilitation* 65:630. [LD, BL]

Gordon, R. M. (1986) The passivity of emotions. *Philosophical Review* 95:371–92. [REH]

Greenwald, A. (1970a) A double stimulation test of ideomotor theory with implications for selective attention. *Journal of Experimental Psychology* 84:392–98. [REH]

(1970b) A choice reaction time test of ideomotor theory. *Journal of Experimental Psychology* 86:20–25. [REH]

Harnad, S. (1982) Consciousness: An afterthought. *Cognition and Brain Theory* 5:29–47. [REH, BL]

Hoffman, R. E. (1986) Verbal hallucinations and language production processes in schizophrenia. *Behavioral and Brain Sciences* 9:503–48. [REH]

James, W. (1890/1983) *The principles of psychology*. Reprinted by Harvard University Press. [REH]

Jung, R. (1985) Voluntary intention and conscious selection in complex learned action. *Behavioral and Brain Sciences* 8:544–45. [BL]

Kelso, J. A. (1982) Concepts and issues in human motor behavior: Coming to grips with the jargon. In: *Human motor behavior*, ed. J. A. Kelso. Erlbaum. [REH, BL]

Kornhuber, H. H. & Deecke, L.: (1964) Hirnpotentialänderungen beim Menschen vor und nach Willkürbewegungen, dargestellt mit Magnetbandspeicherung und Rückwärtsanalyse. Pflügers Arch. 281:52. [LD, BL]

Kornhuber, H. H. & Deecke, L. (1965) Hirnpotentialänderungen bei Willkürbewegungen und passiven Bewegungen des Menschen: Bereitschaftspotential und reafferente Potentiale. *Pflügers Archiv für Gesamte Physiologie* 284:1–17. [LD, BL]

Libet, B. (1985) Unconscious cerebral intiative and the role of conscious will in voluntary action. *Behavioral and Brain Sciences* 8:529–39, 529–66.

(1985r) Theory and evidence relating cerebral processes to conscious will. *Behavioral and Brain Sciences* 8:558–66. [REH, BL]

Libet, B., Gleason, C. A., Wright, E. W. & Pearl, D. K. (1983) Time of conscious intention to act in relation to onset of cerebral activities (readiness-potential); the unconscious initiation of a freely voluntary act. *Brain* 106:623–42. [BL]

Libet, B., Wright, E. W. Jr. & Gleason, C. A. (1982) Readiness-potentials preceding unrestricted "spontaneous" vs. preplanned voluntary acts. *Electroencephalography and Clinical Neurophysiology* 54:322–35. [BL]

MacKay, D. M. (1985) Do we "control" our brains? *Behavioral and Brain Sciences* 8:546. [BL]

Obeso, J. A., Rothwell, J. C. & Marsden, C. D. (1981) Simple tics in Gilles de la Tourette's syndrome are not prefaced by a normal premovement EEG potential. *Journal of Neurology, Neurosurgery, and Psychiatry* 44:735–38. [REH, BL]

Popper, K. R. & Eccles, J. C. (1977) *The self and its brain*. Springer. [BL]

Scheerer, E. (1985) Conscious intention is a mental fiat. *Behavioral and Brain Sciences* 8:552–53. [REH, BL]

Shibasaki, H., Barrett, G., Halliday, E. & Halliday, A. M. (1980) Components of the movement-related cortical potential and their scalp topography. *Electroencephalography and Clinical Neurophysiology* 49:213–26. [BL]

Sperry, R. W. (1976) Changing concepts of consciousness and free will. *Perspectives in Biology and Medicine* 20:9–19. [BL]

(1980) Mind-brain interaction: Mentalism, yes; dualism, no. *Neuroscience* 5:195–206. [BL]

Wittgenstein, L. (1980) *Remarks on the philosophy of psychology*, ed. G. H. von Wright & H. Nyman. University of Chicago Press. [REH]

Wood, C. C. (1985) Pardon, your dualism is showing. *Behavioral and Brain Sciences* 8:557–58. [REH]

Zivin, G. (1986) Image or neural coding of inner speech and agency? *Behavioral and Brain Sciences* 9:534–35. [REH]

# Encyclopedia of Neuroscience

## Volume I

Edited by
George Adelman

Foreword by Francis O. Schmitt

With 239 Figures

## Consciousness: Conscious, Subjective Experience

*Benjamin Libet*

*Consciousness* as a neuroscientific concept has been loosely employed to encompass several different meanings or aspects of cerebral function in human and nonhuman animals. The term is often applied to states of responsiveness to the environment—being conscious or in a coma, awake or asleep, and being alert or aroused within the waking state. These states can be described behaviorally by observing the human or animal. Here we restrict ourselves to the meaning of consciousness as one of subjective awareness and experience, whether it be sensory experiences of our environment, external and internal, or subjective experiences of our feelings and thoughts, or simply awareness of our own existing self and presence in

A *Pro Scientia Viva* Title

BIRKHÄUSER
Boston · Basel · Stuttgart
1987

the world. Our own subjective inner life, including sensory experiences, feelings, thoughts, volitional choices, and decisions, is what really matters to us as human beings. And it is the cerebral, neuronal basis of our subjective experiences that is at issue in the problem of the mind-brain relationship.

**Basic definitions and concepts**

Conscious or subjective experience or awareness is clearly accessible only introspectively to the subject having the experience. It is a primary phenomenon whose nature cannot be defined by any other externally observable event, whether molecular or behavioral. There is considerable evidence that an intimate and lawful relationship exists between neural processes and conscious experience. But even a complete knowledge of the observable neural processes in the brain of another individual would not in itself tell us what that subject is experiencing or feeling.

There are no a priori rules governing the relationship between conscious mental events and brain events; the rules must be discovered. The phenomenon of subjective referral of a sensory experience in the spatial dimension illustrates this principle. The spatial configuration of a subjectively experienced visual image, for example, is considerably different from the configuration of neuronal activities in the brain that accompanies and represents the sensory input giving rise to the visual images; what you see, subjectively, is quite different from the neuronal pattern for that image. The cerebral neuronal representation is actually a spatial distortion of the original image coming into the eye; but subjectively the image is "referred" to the original source of the input in a way that corrects the neuronal distortion. Direct electrical stimulation of primary sensory cortex provides an even more obvious demonstration of this; a stimulus applied to the lateral portion of the postcentral gyrus produces a sensation that is felt to be located not at that stimulated site of the brain but is rather subjectively referred to, and felt in, the hand or arm and in a direction reversed from the postcentral representation. Clearly, a complete knowledge of the neuronal representation would not, without validation by the subject's report, tell us where or what sensation is being subjectively experienced. Similarly, the experience of color cannot be described by finding that certain neural units may respond to specific frequency bands of the light spectrum.

Although each individual has access to and can be certain about only his own conscious experiences and feelings, we do commonly concede and accept the premise that other human beings have their subjective experiences (except for those of us who want to adopt Bishop Berkeley's sollipsism as a serious view of life). We are also confident that one person can communicate something about his subjective experiences to another. But the validity of what is communicated depends upon the degree to which both individuals have had similar or related experiences. A congenitally completely blind person can never share the conscious experience of a visual image, regardless of how detailed a verbal description he is given by a sighted individual. The same limitation applies to all experiences in less dramatic, more subtle ways. For example, electrical stimulation of somatosensory cortex can produce sensations related to but sufficiently different from those generated by normal sensory input, so that the subjects could only relate some roughly understandable approximation of these experiences to the experimental observer, in whom similar modes of sensory generation had never been employed.

**Neuroscientific study of conscious experience**

One of the chief long-range goals of neuroscience is to understand the neuronal basis of conscious experience, i.e., the mind-brain relationship. To investigate this problem we must be sure to adopt valid operational criteria for studying the mental, the conscious experience side of the relationship; i.e., we must directly study the phenomenon of subjective experience if we are to obtain meaningful answers to the problem.

In the absence of direct observational access to the subject's conscious experience, we must rely on the subject's introspective report of experience. The observer must ask the subject the appropriate question about the latter's conscious experience and be confident that the subject understands the question. The report is often most conveniently verbal, but it may be nonverbal (e.g., a sign made by a finger) if the latter clearly represents the subject's introspective experience. Any report should be made only after sufficient time for introspection and not as part of any speedy reaction-time procedure. The issue of the accuracy and reliability of the subject's communication of introspective experience must be dealt with in each study individually. Suitable control tests can in fact be devised, and satisfactory reliability of reports is obtainable.

*Reports of conscious experience vs. behavioral detection.* There is an imperative corollary of the foregoing operational definition of conscious experience: Any measured indicator that can be dissociated from or independent of subjective awareness would be invalid, not acceptable as a primary indicator of subjective experience. This would even include behavioral responses that depend on cognitive and decision-making processes, unless these were validated by the primary evidence of the subject's introspective reports. For example, we have all had the experience of driving an automobile and becoming subjectively engrossed in thoughts not related to the mechanics of driving; nevertheless, the driving proceeds while all sorts of sensory signals are being properly recognized, evaluated and acted upon (usually successfully), without any consciously introspective awareness or later recallability. Among experimental paradigms, signal detection studies provide sophisticated examples of a generally unacceptable approach to conscious experience. The forced-choice responses in such studies could be made independent of introspective awareness of the signal, although they may be excellent indicators of whether some type of detection has occurred. Even when subjects are giving confidence ratings of their responses in a signal detection study, they are rating their forced choices and are not necessarily directly reporting their subjective experiences. There is, in fact, evidence that signal detection can occur with signals that are distinctly below the threshold required for any conscious awareness of the signal. Indeed, most sensory signals probably do not reach conscious awareness; but many of them lead to modified responses and behaviors, as in simple everyday postural and walking activities, and have therefore clearly been detected and utilized in complex brain functions.

**Do animals have conscious experience?**

There is no way to answer this in any absolute sense. But it should be clear that it is difficult if not impossible to carry out valid studies of conscious experience in nonhuman animals. Obviously, we cannot meaningfully ask an animal to report about a shared introspective experience in the validity of which we have confidence. Second, as just noted, complicated cognitive and purposeful behaviors can proceed even in human beings without introspective awareness of them, and so we cannot safely assume that such behaviors in animals are expressing subjective experience. Some investigators and writers have proposed that adaptive, problem-solving behaviors in animals indicate conscious thought and experience. But even in humans, the most complex and even mathematical problem solving can and often does proceed at unconscious levels, as has been repeatedly described by many creative thinkers, artists, and others. Again, one must be careful to hold to the primary criterion of conscious experience per se, as a phenomenon that cannot automatically be described by any nonvalidated behavioral expression.

### Are computers conscious?

Some enthusiasts of machines with artificial intelligence have tended to equate complexity of abstract processing with the existence of conscious experience. As we have noted, there is no such necessary connection even in the human brain. Again, there is the fundamental error in assuming that conscious experience can be identified on purely behavioral criteria, whether for human or machine. It has been argued that a machine might eventually be constructed with such sophistication that an external observer could not distinguish its behavior and responses from those of a human being. Even if we grant such a possibility, for the sake of argument, such an apparent identity of behavioral responses does not require that the two systems, machine and human brain, are alike in every respect. In fact they are constituted differently and operate with different mechanisms. The unique makeup of the brain, and indeed certain special neuronal actions in it (see below), may be essential correlates of what we recognize as conscious subjective experience.

### Memory of conscious experience; dreams

Reporting about an introspective awareness requires at least some short-term memory and recallability of the experience. This presents no experimental difficulty, except to exclude as subjects those patients with a serious deficiency in this regard (e.g., Alzheimer's disease). Furthermore, it is the remembered and recallable body of our conscious experiences which is most meaningful to our own continuum of self-identity, although admittedly past unconscious processes presumably have a powerful influence on present outlooks, emotions, urges to act, etc.

However, there remains the question of whether conscious experience can be dissociated from the memory of it; i.e., can there be conscious awarenesses or experiences which are ephemeral and not remembered? The answer appears to be yes, based on the features of dream states; there is presently no experimental way to test this issue in the waking state. As is now well known, a person awakened during or very shortly after any period of rapid eye movement (REM) sleep (as defined by the electroencephalograph pattern, eye movements, etc.) usually can report having just had some dream experiences. If the person is awakened later, he cannot recall any dream. Judging from REM sleep episodes, we have five or more dreams each night but retain no awareness of these except for those followed by awakening. Dream experiences are clearly conscious, and sometimes all too vividly so. It appears that the memory consolidation processes are operating so weakly during dream states that the experiences are not recallable, unless the memory process is activated by awakening and by attentively dwelling on the dream experiences. Dreaming can thus represent conscious experiences not necessarily tied to memory and recall processes.

### Are there unique cerebral neuronal actions for conscious experience?

A normally functioning brain is a necessary but not a sufficient condition for the appearance of a conscious experience, as already suggested above. What specific neuronal actions may be required for a conscious experience to occur? The relating of subjective experience to cerebral activities requires responsive human subjects, in whom most experimental approaches have necessarily been limited to extracranial techniques. An ability to study and manipulate neuronal activity more directly, by intracranial techniques, can be more instructive especially in relation to questions of causality. These techniques have included anatomical lesions, made surgically or pathologically, studies of local cerebral circulation and metabolism by recently developed noninvasive methods, effects of pharmacologic

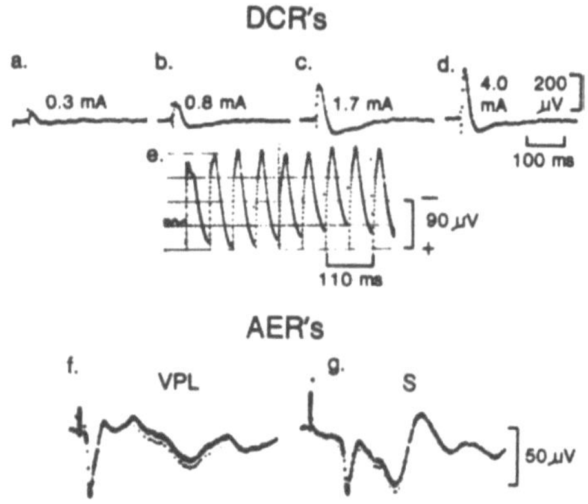


**Figure 1.** Electrophysiological cortical responses to stimuli, accompanied and not accompanied by conscious experience. Direct cortical responses (DCRs) of human somatosensory (SI) cortex of awake and conscious parkinsonian patient, evoked by stimulation (with 0.3 msec pulses) of an adjacent site a few millimeters away. Each tracing in a-d is the average of 18 responses; stimulus pulses, 1/2 sec, peak currents as indicated. Subject reported he did not feel any of these single-pulse stimuli. In e, the liminal I pulses at 0.8 mA (see b) were delivered at 20/sec for 0.5 sec; each of the 10 separate trains averaged for e did elicit a conscious sensory response, with utilization TD = 0.5 sec (see Fig. 2). Recordings were made with dc amplifier system, positive downward. Calibrations in e differ from those for a-d. Average evoked responses (AERs) recorded on SI cortex in response either to ipsilateral thalamic (VPL) or contralateral skin (S) stimuli, in an awake and conscious patient with heredofamilial tremor. Each tracing is the average of 250 responses at 1.8/sec; total length of trace is 125 msec. In f, stimuli applied in VPL (ventroposterolateral nucleus); subject reported not feeling any of these stimuli, even though peak currents were 6 times the liminal intensity that was adequate for sensory experience when a train of pulses 60/sec lasting > 0.5 sec was applied. In g, stimuli S (skin of back of hand); peak currents were at 2 times threshold for subjectively feeling a single pulse, and all stimuli were felt. Note that initial surface positive cortical evoked potential was actually larger with a single VPL than with S stimulus, but VPL was not felt. a-d, f-g. From Libet et al., 1967, reprinted by courtesy of *Science*. e, From Libet et al., 1972, with permission of Excerpta Medica.

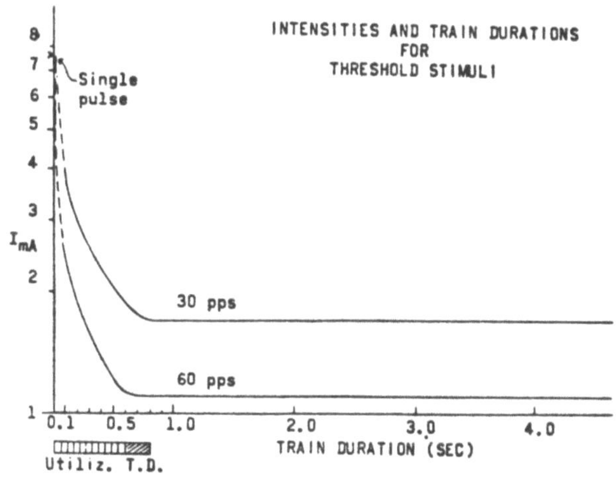


**Figure 2.** Temporal requirement for stimulation of sensory cortex in humans. The curves were drawn through points for each intensity/train duration (TD) combination for repetitive stimulus pulses just adequate to elicit a threshold conscious experience of somatic sensation, when applied to primary somatosensory cortex SI (postcentral gyrus). Curves are presented for two different pulse repetition frequencies, employing rectangular pulses of 0.5 msec duration. Bar for utilization TD indicates the minimum train duration required (or utilized) in order to elicit any conscious sensation when intensity is at a minimum effective level (liminal I). Note that liminal I remains constant even if TD is increased above the utilization TD. From Libet, in Eccles, 1966.

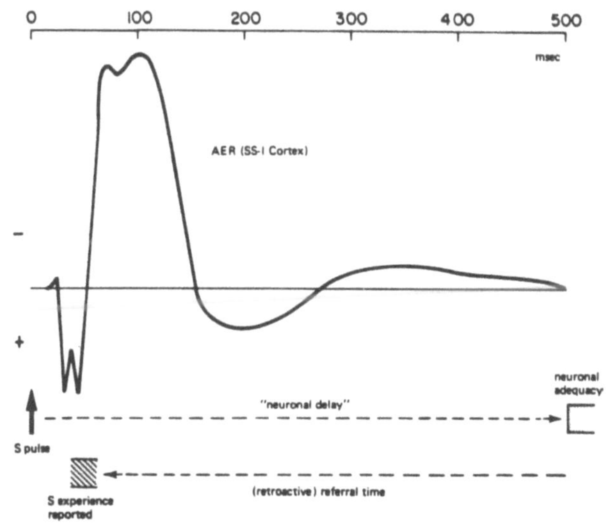


**Figure 3.** Diagram of hypothesis for subjective referral of a sensory experience backward in time. The average evoked response (AER) recorded at SI cortex was evoked by pulses just suprathreshold for sensation (at about 1/sec, 256 averaged responses) delivered to skin of contralateral hand. Below the AER, the first line shows the approximate delay in achieving the state of neuronal adequacy that appears (on the basis of other evidence) to be necessary for eliciting the sensory experience. The second line shows the postulated retroactive referral of the subjective timing of the experience, from the time of neuronal adequacy backward to some time associated with the primary surface-positive component of the evoked potential. The primary component of AER is relatively highly localized to an area on the contralateral postcentral gyrus in these awake human subjects. The secondary or later components, especially those following the surface negative component after the initial 100 to 150 msec of the AER, are wider in distribution over the cortex and more variable in form even when recorded subdurally (see, for example, Libet et al., 1975). It should be clear, therefore, that the present diagram is not meant to indicate that the state of neuronal adequacy for eliciting conscious sensation is restricted to neurons in primary SI cortex of postcentral gyrus; on the other hand, the primary component or "timing signal" for retroactive referral of the sensory experience would be a function more strictly of this SI cortical area. (The later components of the AER shown here are small compared to what could be obtained if the stimulus repetition rate were lower than 1/sec and if the subjects had been asked to perform some discriminatory task related to the stimuli, as seen, for example, in Desmedt and Robertson, 1977.) From Libet et al., 1979, by permission of *Brain*.

agents, and direct electrical stimulation of, as well as electrophysiological recording from, cerebral tissue. The intracranial electrophysiological methods permit the most discrete spatial and temporal studies, including controlled alterations of neuronal activity. We have used them in direct experimental investigations of the specific issue in question, using awake and responsive human subjects, and shall briefly summarize our findings.

Many neural responses to stimulation can develop without leading to any conscious experience. For example, stimulating the cortical surface can produce large electrophysiological direct cortical responses (DCR) with no awareness by the subject (Fig. 1a-d). Similarly, the primary evoked potential elicited by a peripheral sensory stimulus leads to no conscious sensation if the appropriate later event-related potentials do not accompany the primary one (Fig. 1f,g). Unique kinds and durations of neuronal activities appear to be required to elicit a conscious sensory experience, but not unconscious forms of stimuli detection.

*Time factors*. Substantial duration of specific cerebral activations, of up to about 500 msec, turned up experimentally as one of the most interesting neuronal requirements (Fig. 1e; Fig. 2). This implies there is a substantial delay before neuronal adequacy for a sensory experience is achieved. However, further evidence indicated there is a subjective referral of the experience back to the time of the initial sensory signal that normally arrives at the cerebral cortex within 15–25 msec after a sensory stimulus; the primary evoked potential represents the response to this early signal, which comes up by way of the fast, specific projection system (Fig. 3). Subjectively, then, the skin sensation would appear to have no significant delay, even though the experience is not neuronally elicitable until some hundreds of milliseconds later. Such a subjective referral in the temporal dimension is analogous to that in the spatial dimension. Both referrals serve to project the subjective image closer to the spatial and temporal features of the real stimuli, even though the adequate neuronal representation distorts both the spatial pattern and the timing of the process.

More recently, it was found that a substantial period (several hundred milliseconds) of cerebral activity also precedes the reportable time of awareness of wanting or deciding to perform a voluntary action. A general hypothesis was proposed that many if not most conscious experiences require a substantial minimum period of cortical activation; and that shorter periods of such cortical activation may elicit unconscious mental operations. A major cerebral determinant of the difference between unconscious and conscious mental events would be the duration of the appropriate neuronal activities. This would allow most operations to proceed unconsciously. It also provides an opportunity for the unconscious modification of the content of a subjective experience or even for its total (Freudian) repression, during the time in which the conscious event is "developing."

**See also** Activation, Arousal, Alertness, Attention; Dreaming; Imagery, Mental; Mind, Psychobiology of; Sleep

## Further reading

Eccles JC, ed (1966): *Brain and Conscious Experience*. New York: Springer-Verlag

Eccles JC (1980): *The Human Psyche*. New York: Springer-Verlag

Libet B (1973): Electrical stimulation of cortex in human subjects, and conscious sensory aspects. In: *Handbook of Sensory Physiology*, Iggo A, ed. Berlin: Springer-Verlag

Libet B (1982): Brain stimulation in the study of neuronal functions for conscious sensory experiences. *Human Neurobiol* 1:235–242

Libet B, Gleason, CA, Wright EW, Pearl DK (1983): Time of conscious intention to act in relation to onset of cerebral activities (readiness-potential); the unconscious initiation of a freely voluntary act. *Brain* 106:623–642

Nagel T (1979): *Mortal Questions*. Cambridge: Cambridge University Press, Chapters 11, 12, 14

Penfield W (1958): *The Excitable Cortex in Conscious Man*. Liverpool: Liverpool University Press

# 19

# The Timing of a Subjective Experience
## Response to a Commentary by D. Salter

Reprinted from *The Behavioral and Brain Sciences*
Vol. 12, No. 1, pp 181–187, 1989

BEHAVIORAL AND BRAIN SCIENCES (1989) 12, 181–187

# Continuing Commentary

*Commentary on* **Benjamin Libet (1985) Unconscious cerebral initiative and the role of conscious will in voluntary action. BBS 8:529–566.**

**Abstract of the original article:** Voluntary acts are preceded by electrophysiological "readiness potentials" (RPs). With spontaneous acts involving no preplanning, the main negative RP shift begins at about −550 ms. Such RPs were used to indicate the minimum onset times for the cerebral activity that precedes a fully endogenous voluntary act. The time of conscious intention to act was obtained from the subject's recall of the spatial clock position of a revolving spot at the time of his initial awareness of intending or wanting to move (W). W occurred at about −200 ms. Control experiments, in which a skin stimulus was timed (S), helped evaluate each subject's error in reporting the clock times for awareness of any perceived event.

For spontaneous voluntary acts, RP onset preceded the uncorrected Ws by about 350 ms and the Ws corrected for S by about 400 ms. The direction of this difference was consistent and significant throughout, regardless of which of several measures of RP onset or W were used. It was concluded that cerebral initiation of a spontaneous voluntary act begins unconsciously. However, it was found that the final decision to act could still be consciously controlled during the 150 ms or so remaining after the specific conscious intention appears. Subjects can in fact "veto" motor performance during a 100–200-ms period before a prearranged time to act.

The role of conscious will would be not to initiate a specific voluntary act but rather to select and control volitional outcome. It is proposed that conscious will can function in a permissive fashion, either to permit or to prevent the motor implementation of the intention to act that arises unconsciously. Alternatively, there may be the need for a conscious activation or triggering, without which the final motor output would not follow the unconscious cerebral initiating and preparatory processes.

## Voluntary process and the readiness potential: Asking the right questions

David Salter

*Department of Psychology, University of Newcastle-upon-Tyne, England NE1 7RU*

Following Libet's own remark that "the precise role of the cerebral activity represented by the RP [Readiness Potential] in the initiation of the voluntary process is yet to be determined" (1985a, p. 535), this commentary re-examines the role of the readiness potential in the context of earlier investigations (Libet 1979; 1982). Libet (1985a) makes three major claims. (1) Unconscious cerebral activity begins about 400 msecs before the conscious voluntary act. (2) Results from these studies of impulsive, spontaneous action provide *direct* evidence about the causal role of unconscious cerebral activity in initiating volitional actions. Libet concludes that (3) the subject's awareness of an intention to move occurs only after this preparatory activity, with the direct implication that conscious control is secondary to the unconscious initiation of the act.

These conclusions are based upon evidence that the onset of cerebral RP processes appears to occur before the subject becomes aware of wanting to move the finger (W), which itself precedes the option to veto the intended movement. Crucial to our accepting this evidence are two aspects of the experimental procedure: *the objective time check* (the psychological and methodological factors involved in cross checking the subjective sensation with an accurate, external time reference), and *the subjective report* (the subject's own introspective monitoring of the sensation of wanting to move).

***The objective time check: Division of attention.*** Establishing an objective time check for W raises two problems: the potential psychological difficulty of monitoring two independent sensations simultaneously, and the methodological appropriateness of using a skin stimulus, S, as a control to check on the timing of W. The time difference between RP onset and W is open to question on the grounds that limited attentional resources might not allow the subject to monitor the urge-to-move while accurately perceiving the clock at the same time. The theoretical assumption of simultaneity at this point is critical because the addition of a few hundred miliseconds in registering the clock time for W, which if it can be shown to be artefactual, would undermine the solidity of the case for unconscious initiation. Assuming for the moment that divided attention does cause the subject to distribute priority in processing between these two information sources, the "urge-to-move" or the clock face, the implications can be made explicit. Of the two, the most probable attention priority is to monitor the gradual emergence of an urge-to-move (since this was emphasized by the experimental instructions). Only after this sensation has been unequivocally identified will attention be switched to the clock face to register its time. In the less likely (though possible) situation, attention is directed primarily toward the clock face. Now the urge-to-move must become sufficiently powerful as it accumulates to distract attention involuntarily from clock monitoring. In retrospect, we cannot establish which alternative was followed by the subjects, but in either case both outcomes lead to a similar result: a time registration for W, which is objectively delayed.

Independent evidence from psychological research has shown that behavioral changes usually arise with divided attention when the information of the combined sources exceeds capacity. It is an empirical question whether the information of the combined tasks in Libet's procedure is sufficiently great after extended practice to make this dual task a resource-limited one (Norman & Bobrow 1975; Navon & Gopher 1980). Selective attention is thus one possible source of an artefactual delay in timing event W. A more substantive argument for the time difference between RP onset and W takes up the question of validating the timing of W by the times obtained from S.

***The use of S as a timing control for W.*** Using a skin stimulus, S, which the subject reports as a clock time, seems a straightforward way to check on the subject's ability and reliability in making simultaneous judgments because the delivery time of the skin stimulus is objectively known. Hence the margin of error can be readily calculated. Libet calculated the error margin for S and applied it to W working on the assumption that discriminating an endogenous mental event W, the urge-to-move, and discriminating the skin stimulus S, involve similar processes. But this assumption does not seem well founded because the subjective timing for the skin stimulus S involves retroactive referral compared with the endogenous mental event W.

***Retroactive referral.*** Libet et al. (1979) reported two categories of stimuli that required repetitive pulses for periods in excess of 200 msecs for any subjective experience to be elicited at all. This cumulative activity was termed *neuronal adequacy*: a product of the intensity of the pulse stimulation and the pulse train duration. Separate investigations (Libet 1982; Libet et al. 1983) have also noted that skin stimulation is special in that the introspective report of the time of the sensory experience compensates retroactively for the elapsed delay caused by neuronal adequacy. The time is referred back to a localized, primary evoked potential, which registers at the cortex some 15 msecs after the stimulus pulse is delivered to the finger. The phenomenal perception of the sensory event is therefore experienced as occurring at the very beginning of the neural activity necessary for its subsequent perception because the surface positive potential spike acts as an event marker to anchor the mental experience in time.

Suppose one now applies these findings to the control skin stimulus on the reasonable assumption that S is characterized by a similar volley in the ascending specific (lemniscal) projection to register as a surface positive spike. There are two consequences: First, the ascending volley would evoke involuntary attention, and concurrent perceptual processing of the clock face would be momentarily disrupted so that the specific clock time registered in working memory until the report stage would be the time immediately prior to the onset of the skin stimulus. Libet's time data for the control stimulus S support this interpretation by confirming that subjects showed a consistent bias and paradoxically reported times in advance of skin stimulus onset by some 50 msecs on average. An even more relevant observation is that because S is characterized by a surface positive spike, its report would be characterized by retroactive referral, unlike W, the endogenous, emergent stimulus. This characteristic, along with the following observations, make S an unsatisfactory timing control for W, which is now considered.

***The subjective report: The endogenous mental event W.*** The fact that subjects monitored a cumulative activity that they were eventually able to identify as W raises the technical difficulty of establishing exactly when identification occurs with an endogenous signal like W, which arises gradually, compared with a stimulus that can be induced more abruptly. The report of W therefore has all the attendant problems that beset establishing a "threshold" judgment for the detection of a signal-to-noise ratio ($d'$) combined with the further psychological factor of an individual caution criterion ($\beta$) for report. Libet argues that signal detection analysis is an inappropriate paradigm because it usually involves forced-choice and the avoidance of false alarms, neither of which is directly applicable in this case. The latter is debatable. Notwithstanding this rebuttal, signal detection theory seems to capture the changing signal-to-noise ratio of the underlying processes which are the focus of introspective attention and identification in W. If the caution criterion further delayed the conscious report of the then identified urge-to-move, the objective time gap between RP onset and W would be increased further, giving rise to an erroneous conclusion that initiation was unconscious. A residual question is exactly how long this hold-up might be.

In many respects the mental event, W, is like the subjective awareness of an induced cortical stimulus, C, with its period of neuronal adequacy. But rather than involving negative referencing in time as S does, C-type stimulation has the opposite effect: to quote Libet et al. (1979, p. 197) "a less sharp onset for cerebrally induced sensory experience might tend to bias the reported timing order in the direction of this experience starting relatively later than its actual time of onset might warrant." Similarly, medial lemniscal (LM) stimulation "resembles the cortical stimulus in requiring similarly long durations of up to about 500 msecs; unlike the skin stimulus, a single LM pulse is completely ineffective in producing a conscious sensory experience" (Libet 1982, p. 239). Given that both cortical and medial lemniscal stimulation with *induced* onset require neuronal adequacy of this order, it is to be expected that W arising more gradually will require similar durations before conscious perception. This means that even when subjects are instructed to report their wanting to move at the earliest moment possible, they will be physically unable to do so before some 300–350 msecs have elapsed. Such a delay would explain the time differential between RP onset as measured at the vertex and report times for W.

Making the processes reported in Libet (1985a) explicit thus leads to a more circumscribed conclusion that cerebral activity which has not yet achieved neuronal adequacy can be measured in advance of attentive introspection and report, bearing in mind that the subsequent report of the phenomenal experience with some types of stimulation such as S is subject to retroactive timing.

***A paradigm framed to answer the wrong question.*** Attempting to polarize either consciousness or unconsciousness as the sole initiator of action is almost certainly misleading. The subjects in this experimental procedure participated as the experimenters instructed them; it was a necessary condition that they accede to this request. We may therefore assume that the knowledge and intention to act in the prescribed manner was represented in some form of physical brain state. We may further assume that subjects were conscious of their ongoing motivation to act because this conscious intention was an essential ingredient of their participation. Given these reasonable assumptions, the experimental conclusions seem fatally compromised and unable to provide any decisive evidence one way or the other about the question of sole initiation.

What the independent evidence about neuronal adequacy and retroactive referral seems to suggest is that questions of timing, or strict order, across levels of parallel operations cannot provide answers about moral responsibility. Because the two levels of activity operate on different time bases, this can cause major difficulties when we attempt to interpret the data. It makes better operational sense to frame questions about temporal ordering in terms of one level of activity – say, motor versus somatosensory – although even within one level of operation, isolating the origin of initiating activity may not be possible. For example, if action leading to awareness derives from a motor-sensory loop, then delays of the order of 300–400 msecs might be expected during which the average evoked response establishes neuronal adequacy. Evidence from the RP readings is compatible with such a build-up of activity. Yet if voluntary action could be traced back to a particular site in the brain, this would be likely to prove more significant to our conceptual understanding because it operates at an appropriate scale and avoids the pitfalls which attend questions or explanations framed in terms of Cartesian dualities. My hunch is that relevant facts about the role of consciousness are more likely to "fall out" from judicious indirect methods that keep levels of procedural investigation separate (and thereby avoid metaphysical problems) than from attempting to tackle the problem head on.

## Editorial Commentary

The problem of the objective timing of subjective events and their neural substrates may be more serious than considerations about divided attention and "neuronal adequacy" suggest. Subjective "time" and objective time may simply be objectively incommensurable. Both neural and muscular responses can be objectively timed. So can introspective reports (in the sense that we can know when they occur). But what could possibly be the objective timer for the *contents* of an introspective report ("X occurred now" or "at t")?

One is tempted to say that responses (motor or neural) at least set outer limits on the timing of subjective events: Responses are effects, hence their subjective causes must have preceded them. But the causal status of subjective events is itself one of the questions under investigation in mental timing research. To presuppose that they must come first is to beg the question. With latencies as short as the ones at issue here, the report that "X occurred at t" is compatible with the true time (of the subjective experience of X, and its concurrent neural substrate) occurring at t, before t, or even after t.

The problem of incommensurability is most prominent for the absolute timing of subjective events, but relative timing is not free of it either. Except if we again beg the question by granting the contents of subjective reports face validity, introspection can only tell us when an event *seemed* to occur, or which of two events *seemed* to occur first. There is no independent way of confirming that the real timing was indeed as it seemed.

Incommensurability is a methodological problem, not a metaphysical one. The subjective event X may indeed be identical with a neural event X. There is simply no way to pinpoint objectively *which* neural event X is.

# Author's Response

## The timing of a subjective experience

Benjamin Libet
*Department of Physiology, School of Medicine, University of California, San Francisco, CA 94143*

**Salter's** beginning quotation of my remark about "the precise role of . . . the RP . . ." is misleadingly out of context and irrelevant here. My remark conveyed the point that we don't yet know either where the actual initiating processes for a voluntary act begin or precisely how the RP reflects the mediation of this. I emphasized, however, that this in no way diminishes our ability to use the RP as an indicator of the minimum advanced *timing* for the onset of cerebral preparations to perform a voluntary act (Libet 1985a, pp. 535–36). Salter's primary concerns, as well as those of the EDITORIAL COMMENTARY, are with the possible difficulties in accepting our measurements of subjective timing. Salter's concern about "asking the right questions" is finally expressed in his concluding paragraph. Let us first consider the issues raised concerning subjective timing.

Salter proposes that a division of attention, when the subject monitors two independent experiences (that of wanting to move, W, and the position of the revolving "clock" spot) may introduce an artefactual delay in the reported clock-time of W. First, there is no evidence to indicate that the task exceeded the subjects' capacity to handle both experiences without difficulty or significant error. The argument that attention is completely restricted to a single mental event at a time and that there is a significant time delay in switching attention to another is an ad hoc construction in the present circumstances. Salter states, as if it were a fact, that the urge-to-move emerges gradually, and that this experience (he reveals his behavioristic bias by calling it a "sensation") has to be "unequivocally identified" before the subject will switch attention to the clock, and so on. As I have already pointed out (Libet 1985r, p. 559), the experimental evidence is against such an assumption.

Second, our control series, in which a skin stimulus (S) replaced W as the experience to be paired with clock-time, served to indicate the magnitude of any errors in the whole process for reporting the association of the two simultaneous events under our experimental conditions; such errors would include those for attention switching proposed by Salter. The objectively observed reporting errors for the timing of S averaged about −50 msec; this error did not significantly affect our conclusions and would in fact make the corrected reported timing of W have an even greater delay after onset of RP. Salter objects to this method because subjective timing of S involves retroactive referral (Libet et al. 1979), whereas that of the endogenous mental event W does not. This kind of discrepancy and proposed difficulty has already been fully discussed, and dismissed as not a serious problem (Libet 1985r, pp. 559–60).

Salter then finds a variety of faults with trying to obtain any meaningful subjective report of the timing of the endogenous mental event W. His "fact that subjects monitored a cumulative activity that they were eventually able to identify as W" is not a fact at all, but rather a hypothetical construction. Such a view stems from a signal-to-noise detection *theory*. This may serve well for the analysis of the detection of signals but need not be applicable to introspective reporting of conscious experiences (see further, Libet 1985r, p. 559). Indeed, we have recently carried out an experimental study (article in preparation) which confirms the cerebral and behavioral difference between forced-choice detection of a sensory signal and the conscious awareness of the signal. With respect to a "caution criterion" for report by the subject, I have already noted that subjects were instructed to attend to and report time for their earliest awareness of the urge or wish to move (Libet 1985a; 1985r); that is, they were cautioned *not* to wait for some development of intensity of that awareness.

Salter then goes on to restate, without realizing it, my own argument (Libet 1985a, p. 536, section 3.1) that the mental event W should be expected to develop after a delay of up to 500 msec for achieving "neuronal adequacy" for awareness, just as we had found for conscious sensory responses to the stimulation of somatosensory cortex. However, Salter appears to have the strange notion that this physiological requirement affects our conclusions, since the subjects "will be physically unable" to report any cerebral events that precede reported W time. Of course they are unable to do so, if the activity required to precede and develop the conscious awareness is all at an unconscious level, which the evidence shows! In the case when a cortical stimulus train of 500 msec elicited a conscious sensory experience, the subjects reported being aware of *nothing* when the stimulus train was cut down to 400 msec (Libet et al. 1979).

Salter then argues that our subjects already had a conscious intention to act, as an essential ingredient of their participation, and so "the experimental conclusions seem fatally compromised." Salter may not have carefully read the original target article (Libet 1985a, section 2) or my replies on this issue to other commentators (Libet 1985r, section 1.3; Libet 1987a; 1987b). To repeat, one must distinguish between a general intention to act and the specific intention "to act now." Indeed, the recorded RPs even for self-initiated "act now" events were of two types which themselves distinguished between these levels of intention. Type I RPs were associated with the subjects' reports of some conscious preplanning in the seconds before the act (related to the "general intention to act"), whereas type II RPs appeared with fully spontaneous acts without any conscious preplanning (except for the earlier general intention to act at some time of the subjects' own choosing). The whole objective of the study was to see what the temporal sequence of events in the brain was, relative to the specific conscious intention to "act now". In the study, an individual initiates the actual voluntary motor act at a time of his own choosing, well after he has formed a deliberate general intention to perform the act sometime in the impending future.

Salter would like to keep separate the investigation of the "two levels of activity," the neuronal and the subjective, "and thereby avoid metaphysical problems." His "hunch is that relevant facts about the role of consciousness are more likely to 'fall out' from judicious indirect methods." In my view, his hunch is itself fatally compromised if the objective is to study the actual relationship between neural-physical processes and conscious, subjective events. His bias appears to be closely related to that of traditional behaviorists, who do not accept the experimental study of subjective experience because it must rely on introspective reports rather than on externally observable behavior. But subjective experience is only accessible to the individual having it. It is in a phenomenological category that is independent of (i.e., cannot be described by) externally observable events. Therefore, although there are admittedly great difficulties in designing, performing, and interpreting reliable and meaningful experiments in which the subjective and associated physical events are coupled, that is the only clear way in which real information and discoveries about the relationship are going to be achieved (Libet 1987c).

The related EDITORIAL COMMENTARY focuses on incommensurability between timings of subjective and neural events as a virtually insurmountable methodological problem rather than a metaphysical one.

I have already conceded in previous publications (Libet et al. 1979; Libet 1985a, section 2.4) that it is logically impossible for the timing (or indeed any feature) of a subjective experience to be directly determined or directly verified by an external observer. But, rather than give up the possibility of studying the physiology of subjective experience, "one can attempt to evaluate the accuracy of the introspective report and gain confidence in its validity by applying indirect controls, tests and converging operations" (Libet 1985a, section 2.4; see also Libet 1987c).

I agree that there is no absolute objective timer for the *contents* of an introspective report and that we only have what *seemed* to the subject to be the timing of the event. The chief experimental source of potential doubt about the relation between the reported and actual timings of a subjective event comes from our own finding of a subjective referral backwards in time. We found that the reported timing of a sensory experience appears to be subjectively antedated to the initial incoming neural signal even though there is a neuronal delay (of up to 500 msec) before the experience could have arisen (Libet et al. 1979; Libet 1982). In such a case, the content of the introspective report does not accurately tell us when the subjective event actually occurred.

However, it should be recognized that this startling discrepancy between such timings appears to be a specific feature of sensory experiences in response to peripheral stimuli. Indeed, even in this class of experiences, the discrepancy (i.e., the subjective antedating) is present only when the fast, specific ascending projection pathway properly delivers its signal to the cerebral cortex. In contrast, in the stimulation of somatosensory cortex with a surface electrode the sensation elicited *did* exhibit a relative delay in the introspectively reported timing of the sensory experience; the delay roughly matched the minimum stimulus time required to elicit that experience. Surface cortical stimulation did not effectively excite the fast ascending nerve pathway (which enters from below the cortex), hence no subjective antedating occurred; on the other hand, stimulating the specific ascending pathway at a subcortical site in the brain (i.e., at medial lemniscus) did result in subjective antedating even though a minimum stimulus time of some hundreds of msec (similar to that for cortex) was required before any experience could have appeared (Libet et al. 1979; Libet 1982).

The subjective referral of a sensory experience in time is therefore a unique specific feature of this and presumably other normal sensory systems. There is no reason to believe that such a referral is occurring in other cerebral functions; the evidence with direct cortical stimulation supports this. Consequently, although we must concede that we do not know with certainty that the contents of introspectively reported timings accurately represent the actual timings of the subjective events, there is good reason to assume for the present that they do. In the specific case of W, the reported time of intention to act now, there is no known neural component present in the activity represented by the preceding readiness-potential that would be related to the specific sensory pathway; hence there is no basis for invoking any subjective referral in time in the content of W. We should therefore proceed with studies of these fundamental issues, tempering our conclusions with the appropriate cautions about the potential limitation raised by the EDITORIAL COMMENTARY.

## References


Libet, B. (1979) Can a theory based on some cell properties define the timing of mental activities? *Behavioral and Brain Sciences* 2:270–71. [DS]

(1982) Brain stimulation in the study of neuronal functions for conscious sensory experiences. *Human Neurobiology* 1:235–42. [rBL, DS]

(1985a) Unconscious cerebral initiative and the role of conscious will in voluntary action. *Behavioral and Brain Sciences* 8:529–58. [rBL, DS]

(1985r) Theory and evidence relating cerebral processes to conscious will. *Behavioral and Brain Sciences* 8:558–66. [rBL]

(1987a) Awarenesses of wanting to move and of moving: Response to Lawrence H. Davis. *Behavioral and Brain Sciences* 10:318–21. [rBL]

(1987b) Are the mental experiences of will and self-control significant to performance of a voluntary act? Response to Commentaries by L. Deecke and by R. E. Hoffman & R. E. Kravitz. *Behavioral and Brain Sciences* 10:781–86. [rBL]

(1987c) Consciousness: Conscious, subjective experience. In: *Encyclopedia of neuroscience*, vol. 1, ed. G. Adelman. Birkhauser. [rBL]

Libet, B., Gleason, C. A., Wright, E. W. & Pearl, D. K. (1983) Time of conscious intention to act in relation to onset of cerebral activity (readiness-potential). The unconscious initiation of a freely voluntary act. *Brain* 106:623–42. [DS]

Libet, B., Wright, E. W., Jr., Feinstein, B. & Pearl, D.K. (1979) Subjective referral of the timing for a conscious sensory experience: A functional role for the somatosensory specific projection system in man. *Brain* 102:191–222. [rBL, DS]

Navon, D. & Gopher, D. (1980) Task difficulty, resources, and dual task performance. In: *Attention and performance*, vol. 8, ed. R. S. Nickerson. Erlbaum. [DS]

Norman, D. A. & Bobrow, D. G. (1975) On data-limited and resource-limited processes. *Cognitive Psychology* 7:44–64. [DS]

Reprinted from *Models of brain function*
Ed. Rodney M. J. Cotterill

# CONSCIOUS SUBJECTIVE EXPERIENCE VS.UNCONSCIOUS MENTAL FUNCTIONS: A THEORY OF THE CEREBRAL PROCESSES INVOLVED.

BENJAMIN LIBET

Department of Physiology, University of California,

San Francisco, CA. 94143-0444, U.S.A.

How cerebral processes give rise to, or are related to, conscious subjective experience is a question regarded as perhaps the most fundamental and challenging that scientists would like to address. Yet this question is seldom addressed experimentally, and then rarely in a meaningful manner. This is in large part an outcome of that kind of logical positivism which accepts as valid only behavioral events directly observable and measurable by the experimenter. Additionally, the ability to conduct any experiments involving invasive, intracranial studies of brain function in the conscious human subject is obviously severely limited.

1. **EXPERIMENTAL GUIDELINES.** What guidelines should be followed for a valid and meaningful experimental study of this question? (See Libet, 1965; 1973; 1987.)

### 1.1 Introspective Reports

The first principle is that conscious subjective experience, awareness of some thing or some event, is directly accessible only to the individual having the experience, not to an external observer. Therefore, the only measurement of it that has primary validity requires an introspective report by the subject. The "transmission" of an experience by a subject to the observer can, of course, involve distortions and inaccuracies, particularly when emotional issues intervene. Study of the simplest kinds of experience, e.g., a somatosensory feeling, can minimize such potential invalidity. A corollary of this first principle is that externally observable events, behavioral or physiological (e.g., EEGs, ERPs), are not valid primary indicators of a subjective experience, unless they are a part of the subject's introspective report appropriately elicited. Evidence and theoretical models based on purposeful and successful behavioral responses, or on cognitive and decision-making events, or even on complex and abstract problem-solving, cannot by themselves provide answers to the question of brain and subjective experience since all of these can and often do proceed unconsciously/preconsciously without awareness by the subject.

### 1.2. No A Priori Rules

The second principle is that there are no *a priori* rules that describe the relationship between neural-brain events and subjective-mental events. The rules must be discovered and established by simultaneous observation of both phenomenological categories, the "physical" and the "subjective-mental". It also follows that even a complete knowledge of the neural-physical makeup and events would not, in itself, produce a description of any correlated subjective experience (e.g., Nagel, 1979).

### 1.3 Subjective Referral of Experiences

The phenomena of subjective referral of sensory representations can serve to illustrate the second principle. Subjective referral or "projection" in the spatial dimension was already well known. For example, the subjective visual image we experience has a form and quality not directly evident in the pattern of observed neural activations in the brain which are associated with the experience. There is also the subjective "filling in" of neurological blind spots, or even of whole blind visual fields. Subjective referral in the temporal dimension was only discovered more recently (Libet et al., 1979). (The experimental evidence for this phenomenon includes some that is basic to the ensuing part of this paper, and it will be presented briefly.) It should be noted that both types of subjective referral are mental functions introspectively reportable but not apparent as such in the related neural activities (Libet et al., 1979; Libet, 1982; see also Sherrington, 1940 and Eccles, 1979). These mental phenomena could not have been predicted from even a full knowledge of the underlying patterns of neural activity.

### 1.4 Neural Delay for Subjective Experience

In the initial study we found that a substantial time period of neuronal stimulation in the cerebral somatosensory system is required in order for the subject to have a reportable sensory experience. The required duration of such activation varied with intensity (I) of the cerebral stimulus; with the weakest liminal I (below which no sensation could be elicited even with long stimulus trains) a substantial minimum train duration was required (Fig.1) (Libet et al., 1964; Libet, 1973). This minimum at liminal I, termed the "utilization train-duration" or "U-TD", averaged about 500 msec. Similar U-TDs were observed regardless of pulse frequency (15/s to 120/s) and at all cerebral points in the specific sensory pathway (S-I cortex and its subcortex, ventrobasal thalamus and medial lemniscus).

On the other hand, a single pulse (at or near the absolute minimum I) could be sufficient when applied to skin, peripheral nerve or dorsal columns. In spite of this, we postulated that substantial times of activity were required to achieve cerebral "neuronal adequacy" for awareness following a single peripheral stimulus pulse, just as for cerebral stimuli. Such a

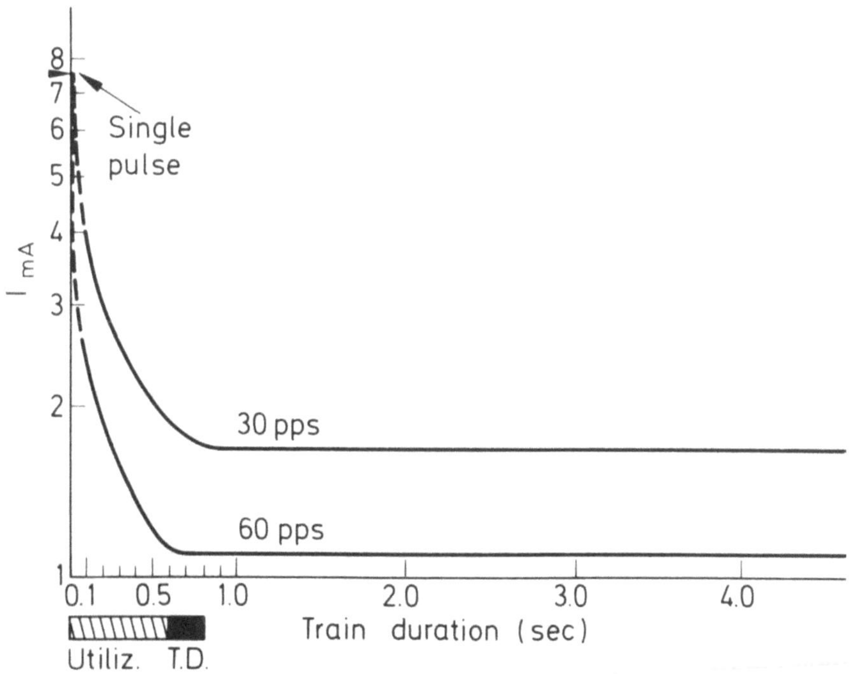


Fig. 1. Temporal requirement for stimulation of somatosensory (SI) cortex in human subjects. (The curves are diagrammatic representations of measurements in several different individuals, which were not sufficient in each case to produce a full curve individually.) Each point on a curve indicates the combination of intensity (I) and train duration (TD) for repetitive pulses that is just adequate to elicit a threshold conscious sensory experience. Separate curves shown for stimulation at 30 pps and 60 pps. At the liminal I (below which no sensation was elicited even with long TDs), a similar minimum TD of about 0.6sec ± was required for either pulse frequency. Such values for this "utilization TD" have since been confirmed in many ambulatory subjects with electrodes chronically implanted over SI cortex and in ventrobasal thalamus. (From Libet, 1973.)

single peripheral stimulus pulse elicits a series of event-related-potentials (ERPs) at the cerebral cortex, and the later components appear to represent necessary correlates for sensory experience (Libet et al., 1967, 1975). Several additional lines of evidence were developed in support of this postulate (Libet, 1973, 1978, 1981, 1982), but these will not be reviewed here.

The concept of a "neuronal delay" of *up to* about 500 msec, before a *sensory experience* can appear in response to a sensory input, should not be confused with or deemed inconsistent with the *ability to discriminate* among much briefer time intervals present in certain pulsatile stimuli. For example, one can discriminate among vibratory stimuli at relatively higher frequencies applied to the skin, or appreciate the difference in pitch for sound waves differing in frequencies, all containing wavelengths with durations much shorter than our "neuronal delays". The question we are addressing in such instances is -- *when* does one become *aware* of whatever temporal discrimination has been achieved?

### 1.5 Subjective Referral Backwards in Time.

If there is a substantial neuronal delay required before achieving a sensory experience or awareness, is there a corresponding delay in the *subjective* timing of the experience? In accordance with the second principle above, one could not answer this question solely from the neural knowledge available. In fact, an appropriate experimental test of this question indicated that there is no appreciable delay for the *subjective* timing of a normally arriving sensory input (Libet et al., 1979)! Experiments utilizing reports of relative order of subjective timings for stimuli at skin, medial lemniscus and S-I cortex enabled us to conclude the following: (l) After a delayed achievement of neural adequacy for awareness, there is an automatic subjective referral of the experience backwards in time, to approximately the delivery time of the stimulus. (2) The initial cortical response (primary evoked potential at SI cortex), to the fast specific (lemniscal) projection sensory message, serves as the timing signal for this backward referral (Fig. 2). The experience would thus be subjectively antedated and would appear to the subject to occur without the actual substantial neural delay required for its production.

Subjective referrals in space and in time both, interestingly, employ the same sensory specific projection system to the cerebral cortex, to supply the signals utilized in the referrals. Also both serve to "correct" (subjectively) the spatial and temporal distortions of the real stimuli introduced by their neuronal representations.

## 2. CONSCIOUS AND UNCONSCIOUS MENTAL FUNCTIONS.

It is generally agreed that much if not most mental functions can proceed unconsciously,

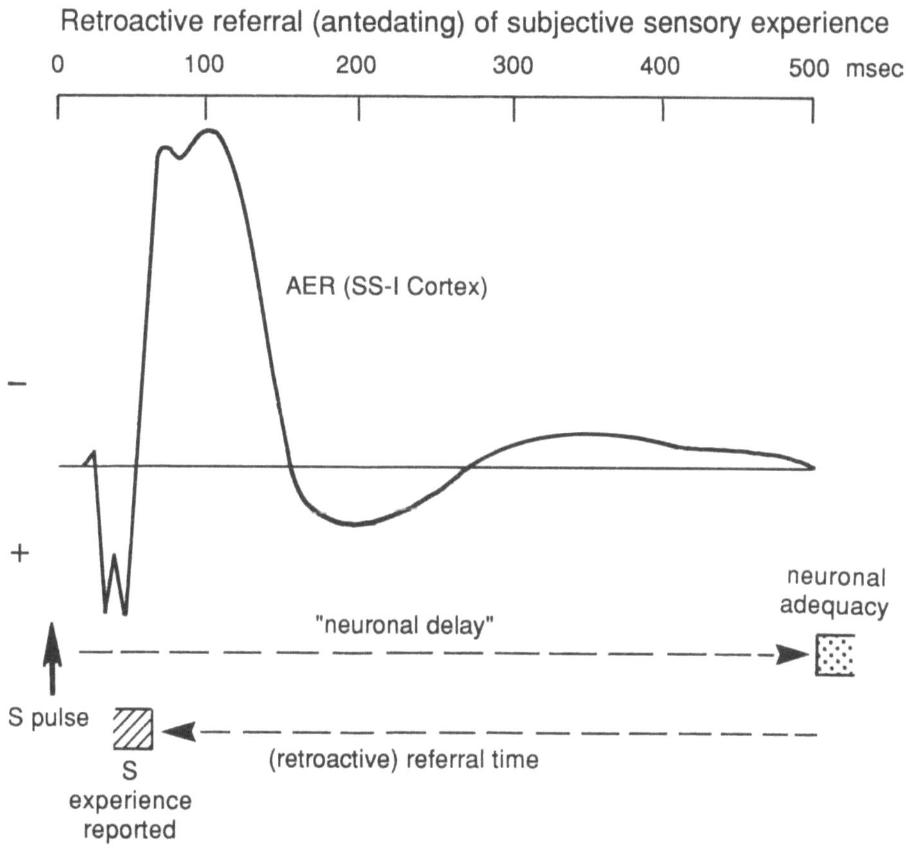


Fig. 2. Diagram of subjective referral of a sensory experience backward in time. The averaged evoked response (AER) recorded at SI cortex was evoked by single pulses just above threshold for sensation (at about 1/sec, 256 averaged responses) delivered to skin of contralateral hand. Below the AER, the first line shows the approximate delay in achieving the state of neuronal adequacy that appears (on the basis of other evidence) to be necessary for eliciting the sensory experience. The second line shows the apparent retroactive referral of the subjective timing of the experience (as reported by the subject) from the time of neuronal adequacy backward to some time associated with the primary surface-positive component of the evoked potential. This would explain the subject's reporting the sensations elicited by a single pulse to the skin and a 0.2 to 0.5 sec train of pulses to sensory cortex is reported as delayed.

The primary component of AER is relatively highly localized to an area on the contralateral postcentral gyrus in these awake human subjects. The secondary or later components, especially those following the surface negative component after the initial 100 to 150 msec of the AER, are wider in distribution over the cortex and more variable in form even when recorded subdurally (see, for example, Libet et al., 1975). It should be clear, therefore, that the present diagram is not meant to indicate that the state of neuronal adequacy for eliciting conscious sensation is restricted to neurons in primary SI cortex of postcentral gyrus; on the other hand, the primary component or "timing signal" for retroactive referral of the sensory experience would be a function more strictly of this SI cortical area. (The later components of the AER shown here are small compared to what could be obtained if the stimulus repetition rate were lower than 1/sec and if the subjects had been asked to perform some discriminatory task related to the stimuli.) (From Libet et al., 1979.)

and there is no need to document this proposition here. Mental functions, whether conscious or unconscious, would apply to all those recognized as psychological in nature, including cognitive, conative (decision-making), learning and recalling, thinking (including complex, abstract and creative thought), etc. The term "unconscious" is used here as a general operational one to cover all mental functions which are not reportable as introspective subjective experiences; it would include the more theoretical categories like subconscious, preconscious, etc., and is intended to cover a broad array of normal and abnormal mental processes, not limited to the usage in Freudian "repression", etc. To attempt an experimental study of brain functions which are necessary and sufficient for the appearance of a conscious experience would be almost impossibly broad and daunting at present. But attempting to specify some crucial *differences* between cerebral processes for conscious and those for unconscious mental functions, seemed experimentally more feasible.

### 2.1 A "Time-on" Theory

Based on our evidence for a neural delay factor in conscious experience, I proposed this theory to explain the difference between cerebral processes mediating a conscious mental function and those for an unconscious one (Libet, 1965, 1973, 1981, 1982, 1985a; Libet et al., 1983a). The theory states that the transition, from an unconscious mental function or event to one that reaches awareness and is consciously-subjectively experienced, can be a function simply of a sufficient increase in duration (or "time-on") of appropriate neural activities. That is, appropriate neural activities whose duration is below some minimum substantial duration (in 100s of msec) could mediate a mental function that remains unconscious; but when such activities persist for longer than that minimum time (as may be effected by influences from changes in attention, etc.), subjective awareness of the mental function can appear.

It should be clear that the theory does not exclude other important or even controlling distinctions between processes mediating conscious vs. unconscious mental functions. For example, the specific kinds of neural activity and/or specifically active sites in the brain *may* differ crucially. It is only proposed that the "time-on" factor is superimposed, as a critical requirement, upon any other differentiating factors. Indeed, one wants at present to avoid designing any specificity or mechanism into the theory other than the "time-on" factor *per se*. There are two basic propositions inherent in the theory, each of which is experimentally testable and falsifiable. (1) A minimum duration of appropriate neural activity, of up to about 500 msec depending on conditions, is required in order to elicit a conscious experience or awareness of an event. The available evidence already supports this proposition, as indicated above. (2) When appropriate neural activity has a duration briefer than that required for awareness, it may still mediate an unconscious mental function, without any subjective awareness of it. In addition to earlier indirect evidence (Libet, 1981, 1982), more recent studies have produced experimental evidence which directly supports this second proposition, as follows.

## 2.2 Time-on Theory and the Initiation of a Voluntary Act

The relevant experimental question here is -- does measurable cerebral activity start before or after the appearance of conscious, subjective intention to perform a fully voluntary act? If cerebral activity begins first, by a significant time margin, that would constitute a direct demonstration of unconscious cerebral mediation of an important mental function, before the neural activity becomes sufficient for awareness of this intention; such a situation is predicted by the "time-on" theory.

The readiness-potential (RP) is a slow ERP (event-related-potential) whose onset precedes a "self-paced" movement by 800 msec or more. After its discovery by Kornhuber and Deecke (1965; Deecke et al., 1976), we established that fully endogenous, spontaneous voluntary acts (not subject to some external constraints in the "self-paced studies) are also preceded by a type of RP with onset at about -550 msec (Libet et al., 1982; Fig. 3). The onset of such "self-initiated" RPs was taken to be an indicator of the minimum advance starting time for specific cerebral processes leading to a voluntary act.

Measurement of the timing of the associated subjective event, i.e., the time of appearance of the *awareness* of the intention or wish to act (W), was based on the subject's report (following each act) of the "clock-time" associated with the first such awareness in each trial (Libet et al., 1983). (The control observations to indicate the accuracy of such reporting, and the analyses by myself and others of the validity of such reporting of the subjective timing of awareness, have been presented fully elsewhere, - see Libet et al., 1983a; Libet, 1985a,b; 1989).

The results of that investigation ( Fig. 4) led to the following experimental conclusions: (i) The cerebral processes that precede a voluntary motor act begin at least 350 msec *before* the subject is aware of his/her intention or wish (W) to "act-now". (ii) But this awareness (W) still appears about 150 to 200 msec before activation (EMG) of the muscles involved. This evidence thus demonstrates an observable but unconscious cerebral process associated with an initiation of at least the preparation to perform a freely voluntary act - something normally regarded as a fundamental mental event. The evidence is also in accord with the proposal that the development of the associated subjective awareness of such an initiating intention requires a substantial period of neural activities. These conclusions depend on the assumption that the readiness-potential (RP) does represent cerebral processes that are

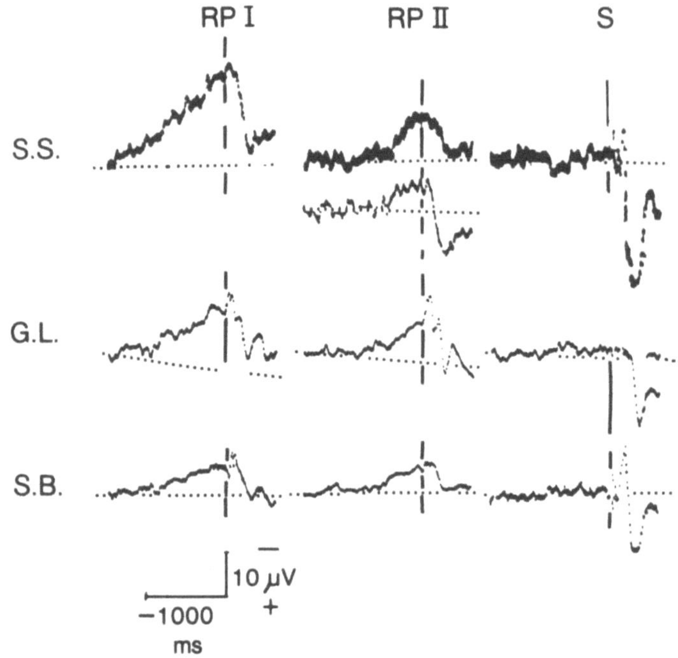


Fig. 3. Readiness potentials (RP) preceding self-initiated voluntary acts. Each horizontal row is the computer-averaged potential for 40 trials, recorded by a DC system with an active electrode on the scalp, either at the midline-vertex ($C_Z$) for subject G.L. and S.B., or on the left side (contralateral to the performing right hand) approximately over the motor/premotor cortical area that controls the hand ($C_3$) for S.S.

RPs labeled type II were found when every self-initiated quick flexion of the right hand (fingers or wrist) in the series of 40 trials was (reported as having been) subjectively experienced to originate spontaneously and with no preplanning by the subject. Type I RPs were recorded when an awareness of a general intention or preplanning to act some time within the next second or so was reported to have occurred before some of the 40 acts in the series (Libet et al., 1982). In the last column, labeled S, a near-threshold skin stimulus was applied in each of 40 trials at a randomized time unknown to the subject, with no motor act performed; the subject was asked to recall and report the time when he became aware of each stimulus in the same way he reported the time of awareness of wanting to move in the case of self-initiated motor acts. (No significant prepotential is seen before S stimuli, but a large $P_{300}$ERP follows S, indicating attention to and cognitive uncertainty for S here present.)

The solid vertical line through each column represents "0 time", at which the electromyogram (EMG) of the activated muscle begins in the case of RP series, or at which the stimulus was actually delivered in the case of S series. The dashed horizontal line represents the DC baseline drift.

For subject S.S., the first RP (type I) was recorded before the instruction "to let the urge come on its own, spontaneously" was introduced; the second RP (type II) was obtained after giving this instruction in the same session as the first. (The lower tracing shows another such RP II for S.S. in a later session.) For subjects G.L. and S.B., this instruction was given at the start of all sessions. Nevertheless, each of these subjects reported some experiences of loose preplanning in some of the 40-trial series; those series exhibited type I RPs rather than type II.

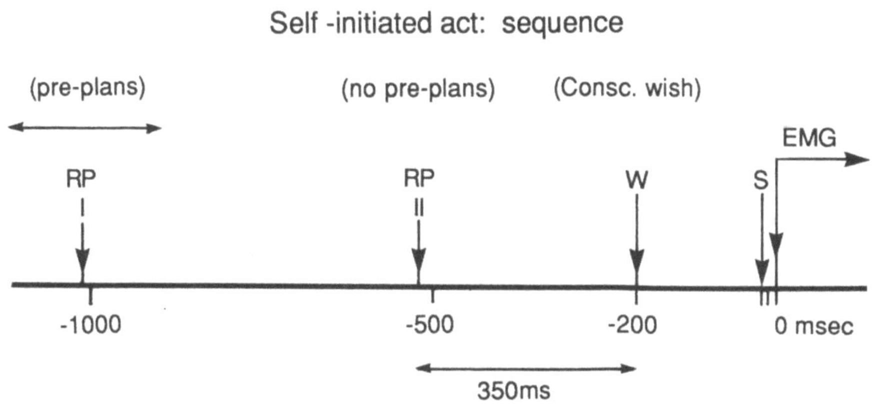


Fig. 4. Diagram of sequence of events, cerebral and subjective, that precede a fully self-initiated voluntary act. Relative to 0 time, signalled by the onset of EMG of the suddenly activated muscle, the RP (an indicator of related cerebral neuronal activities) begins first, - at about -1050 msec when some preplanning is reported (RP I) or about -550 msec with spontaneous acts lacking immediate preplanning (RP II). Subjective awareness of the wish to move (W) appears at about -200 msec, some 350 msec after onset even of RP II but well before the act (EMG). Subjective timings reported for awareness of the randomly delivered S (skin) stimulus averaged about -50 msec relative to actual delivery time.

actually involved in or meaningfully related to the initiation of the volitional preparation to "act now". An alternative view of the significance of the RP has been proposed by Eccles (1985) and also somewhat differently by Ringo (1985) and Stamm (1985). Although this alternative view cannot be simply ruled out, it requires extra ad hoc assumptions and does not fit the overall evidence as consistently or satisfyingly as the view we have proposed (Libet, 1985a,b; 1989).

**2.3 "Time-on" Theory and Sensory Detection Without Awareness.**

When stimulating ventrobasal thalamus (as well as SI cortex and medial lemniscus) in human subjects, we had found that a minimum train duration (TD) of pulses is required in order to elicit any reportable conscious sensation; with intensity level (peak current) near the absolutely liminal one, such minimum TDs averaged about 500 msec. With a TD below the required minimum, e.g., 300 msec when minimum was 400, subjects would report feeling no sensation. Although such a briefer stimulus train at the same intensity is inadequate for conscious sensation, it must clearly be exciting the same ascending axons that feed into the same cortical and other areas as does the longer train. The "time-on" theory would predict that such shorter lasting activations, even though they do not give rise to a reportable awareness, may be psychologically/mentally detected and lead to meaningful behavioral responses in an unconscious manner.

We are now concluding a direct experimental study of this proposition (Libet, Pearl, Morledge, Gleason, Hosobuchi, and Barbaro, unpublished). The subjects were patients with permanent implantations of electrodes in some portion of ventrobasal thalamus for the therapeutic purpose of controlling certain kinds of intractable, intolerable pain. Patients were completely ambulatory, coming in from home for the studies. The completely internalized subcutaneous receiver coil, which fed the stimulus pulses to the intrathalamic electrode tips, received the input from an externally overlying coil connected to an external stimulator box whose controls were modified so as to be run from our controlled computer-programmed source. At near liminal intensities, with a pulse frequency of 72 per sec, we confirmed our previous finding that a minimum TD of about 400 to 500 msec was required in order to elicit a reportable conscious sensation, typically a kind of localized paresthesia or "tingling". Evidence for behavioral detection of the stimulus, with or without awareness, was obtained in a simple forced choice paradigm. For this, the subject observed two 1 sec-long lights which went on in succession (separated by a 1 sec interval); the stimulus was delivered in one or the other of these two light-on periods in a randomly distributed manner for the different trials. The subject was asked to choose which light (#1 or #2) might have included something different, even if he felt or was aware of no sensation at all, and even if he had no consciously definable basis for his choice. The pulse numbers in each stimulus were also randomly varied in different trials, between 0 and 55 pulses (i.e., between TDs of 0 and approximately 750 msec), thus including some stimuli of which the subject was aware.

Although the full statistical analysis is still in progress, it is already quite clearly shown that correct behavioral detection of these signals, at levels substantially and significantly greater than the 50% pure chance result, did in fact occur with stimulus durations too brief to produce any reportable conscious sensations. Some detection without awareness occurred at values of about 10 pulses when sensory awareness required values of about 20 to 30 pulses . Such results would provide direct evidence (i) that detection without awareness is possible. (This also provides a conclusive experimental distinction between the meaning of purely behavioral detection and subject reports of awareness. It should be noted that this behavioral detection without awareness required some sort of cognitive and decision-making processes.) (ii) They also show, in relation to the theory, that the duration ("time-on") of neural activations can determine whether such detection remains at unconscious levels or is accompanied by a conscious subjective sensory awareness of the neural input.

## 3. SOME IMPLICATIONS OF THE "TIME-ON" THEORY.

### 3.1 Cerebral Representation

If the transition from an unconscious to a conscious mental function could be dependent simply on a suitable increase in duration of certain neural activities, then both kinds of mental functions could be represented by activity in the same cerebral areas. Such a possibility would be in accord with the fact that the constituents and processes involved in both functions are basically similar, except for the awareness quality. It is also in accord with the view that a broadly distributed neural activity pattern probably mediates both types of function. Separate cerebral sites for conscious vs. unconscious functions would thus not be necessary, although this is not excluded (as noted above).

### 3.2 "Filter" Function

The "time-on" requirement could serve as mechanism for the known condition that most inputs and cerebral activities do not reach awareness. This of course permits conscious-awareness to be uncluttered and to focus on one or a few specific issues at a time.

### 3.3 Quick Behavioral Responses Are Initially Unconscious

Responses in reaction time (R.T.) tests, for example, can be made within less than 100 msec, depending on the complexity of the signal, etc. R.T.s even with the simplest signal (e.g., one loud sound or a visual flash) involve cognitive and conative processing. On our theory, all this would often occur before awareness of the signal could develop. There is much indirect and anecdotal evidence to support this view of the initially unconscious nature of quick reactions. However, when delayed awareness does appear, there would ordinarily be a subjective antedating of its timing (Libet et al., 1979) - so that the subject believes he experienced the signal before reacting; for example, a racer may start within <100 msec after the starting gun, before he is consciously aware of the shot, but would later report having heard it before take-off.

### 3.4 Unconscious Mental Functions Proceed Quickly.

This would be so since they would not require the fuller "time-on" needed for a conscious one. This feature is obviously advantageous not only for quick meaningful reactions to signals but also in the ability to carry on creative, complex and intuitive thinking.

### 3.5 Permits Modulation of a Conscious Experience.

It is well known that the *content* of the introspectively reportable experience of an event may be modified considerably in relation to the content of the actual signal, whether this be an emotionally laden sensory image or endogenous mental event (which may even be fully repressed, in Freud's terms). In order for this to happen, some delay between the initiating event and the appearance of the conscious experience of it is essential. The "time-on" theory provides a built-in basis for the appropriate delays. We have produced some direct experimental evidence for such modulatory actions on the awareness of a simple sensory signal from the skin; in this, an appropriate cortical stimulus was begun up to 200 msec or more after the skin pulse, but could still either inhibit or enhance the sensory experience (Libet et al., 1972; Libet, 1978, 1982).

### 3.6 Conscious Control and Free Choice of Voluntary Action

As discussed above, an observable cerebral process (RP) regularly begins some 300 to 400 msec *before* the appearance of the conscious awareness of intention "to act now" (W). If a cerebral action unconsciously initiates (or, in the alternative view of Eccles and others, at least determines the available timing of) the process leading to performance of a voluntary act, that may raise a serious question about the role of conscious free choice or will in voluntary action. However, one must distinguish between the appearance of conscious intention (W) and that of a potential *conscious control* of the volitional outcome.

I suggest that conscious control differs from W not only in its appearance after W but in its "nature". Conscious control does not constitute another awareness of something. It is rather a different phenomenon which may impose a change after an awareness of intention (W) has already appeared (Libet et al. 1983b; Libet 1985a). Even if one assumes that the "time-on" theory applies to all kinds of subjective awarenesses (each preceded by an unconscious initiating process), the theory would not necessarily apply to the phenomenon of conscious *control*. On this basis, the theory would not require that conscious control is also initiated unconsciously. Thus, the potentiality for a form of free choice (in the classical sense) is not excluded by the theory, even if the theory is generalized to all awarenesses. Indeed, the experimental observations provide an opportunity for the occurrence of free choice, but apparently in the form of control rather than of initiation of an act (Libet, 1985a,b).

## ACKNOWLEDGMENT

The unpublished work cited here was supported by U.S. Public Health Grant NS-24298-01.

## REFERENCES

Deecke, L., Grozinger, B. and Kornhuber, H.H. (1976). 'Voluntary finger movement in man: Cerebral potentials and theory.' *Biol.Cybernetics*, **23**, 99-119.

Eccles, J.C. (1979). *The Human Mystery.* New York: Springer International, p. 225.

Eccles, J.C. (1985). 'Mental summation: The timing of voluntary intentions by cortical activity.' *Behav. and Brain Sci.*, **8**, 542-543.

Kornhuber, H.H. and Deecke, L. (1965). 'Hirnpotentialänderungen bei Willkürbewegungen und passiven Bewegungen des Menschen: Bereitschaftspotential und reafferente Potentiale.' *Pflügers Archiv für Gesamte Physiologie*, **284**, 1-17.

Libet, B. (1965). ' Cortical activation in conscious and unconscious experience.' *Perspect. Biol.Med.*, **9**, 77-86.

Libet, B. (1973). 'Electrical stimulation of cortex in human subjects, and conscious sensory aspects.' In *Handbook of Sensory Physiology*, Vol. II, Somatosensory System, ed. A. Iggo. New York: Springer, pp. 743-790.

Libet, B. (1978). 'Neuronal vs. subjective timing for a conscious sensory experience.' In *Cerebral Correlates of Conscious Experience* , eds. P.A. Buser and A. Rougeul-Buser. Amsterdam: Elsevier/North Holland Biomedical Press, pp. 69-82.

Libet, B. (1981). 'The experimental evidence for subjective referral of a sensory experience backwards in time: Reply to P.S. Churchland.' *Phil.of Sci.*, **48**, 182-197.

Libet, B. (1982). 'Brain stimulation in the study of neuronal functions for conscious sensory experience.' *Human Neurobiol.*,**1**, 235-242.

Libet, B. (1985a). 'Unconscious cerebral initiative and the role of conscious will in voluntary action.' *Behav. and Brain Sci.*, **8**, 529-539.

Libet, B. (1985b). 'Theory and evidence relating cerebral processes to conscious will.' *Behav. and Brain Sci.*, **8**, 558-566.

Libet, B. (1987). 'Consciousness: Conscious, Subjective Experience.' In *Encyclopedia of Neuroscience* , ed. G. Adelman. Birkhauser:Boston, pp. 271-275.

Libet, B. (1989). 'Cerebral processes that distinguish conscious experience from unconscious mental functions.' In *Principles of Design and Operation of the Brain*, eds. J.C.Eccles and M. Wiesendanger.*Exp.Brain Res., Suppl.*. (in press).

Libet, B., Alberts, W.W., Wright, E.W., De Lattre, L.D., Levin, G., and Feinstein, B. (1964). 'Production of threshold levels of conscious sensation by electrical stimulation of human somatosensory cortex.'*J. Neurophysiol.*, **27**, 546-578.

Libet, B., Alberts, W.W., Wright E.W. and Feinstein, B. (1967). 'Responses of human somatosensory cortex to stimuli below threshold for conscious sensation.' *Science*, **158**, 1597-1600.

Libet, B., Alberts, W.W., Wright, E.W., and Feinstein, B. (1972). 'Cortical and thalamic activation in conscious sensory experience.'In *Neurophysiology Studied in Man*, ed. G.G. Somjen. Amsterdam: Excerpta Medica, pp. 257-268.

Libet, B., Alberts, W.W., Wright, E.W., Lewis, M., and Feinstein, B. (1975). Cortical representation of evoked potentials relative to conscious sensory responses and of somatosensory qualities - in Man. In *The Somatosensory System*, ed. H.H. Kornhuber, Stuttgart: G. Thieme, pp. 291-308.

Libet, B., Wright, E.W., Jr. , Feinstein, B., and Pearl, D.K. (1979). 'Subjective referral of the timing for a conscious sensory experience: a functional role for the somatosensory specific projection system in man. ' *Brain*, **102**, 191-222.

Libet, B., Wright, E.W., and Gleason, C. (1982). 'Readiness-potentials preceding unrestricted "spontaneous" vs. pre-planned voluntary acts.' *Electroenceph.clin. Neurophysiol.*, **54**, 322-335.

Libet, B., Gleason, C.A., Wright, E.W., and Pearl, D.K. (1983a). 'Time of conscious intention to act in relation to onset of cerebral activities (readiness-potential); the unconscious initiation of a freely voluntary act.' *Brain* , **106**, 623-642.

Libet, B., Wright, E.W., Jr., and Gleason, C.A. (1983b). 'Preparation-or intention-to-act, in relation to pre-event potentials recorded at the vertex.' *Electroenceph.clin.Neurophysiol.*, **56**, 367-372.

Libet, B., Pearl, D.K., Morledge, D., Gleason, C.A., Hosobuchi, Y., and Barbaro, N. 'Signal detection without sensory awareness, produced by stimulation of somatosensory thalamus in man. ' (Unpublished, in preparation.).

Nagel, T.(1979). *Mortal questions,* Cambridge University Press.

Ringo, J.L. (1985). 'Timing volition: questions of what and when about W.' *Behav. and Brain Sci.*, **8**, 550-551.

Sherrington, C.S. (1940). *Man on his nature.* Cambridge University Press.

Stamm, J.S. ( 1985). 'The uncertainty principle in psychology.' *Behav. and Brain Sci.*, **8**, 553-554.

*Brain* (1991), **114**, 1731–1757

# CONTROL OF THE TRANSITION FROM SENSORY DETECTION TO SENSORY AWARENESS IN MAN BY THE DURATION OF A THALAMIC STIMULUS

## THE CEREBRAL 'TIME-ON' FACTOR

*by* BENJAMIN LIBET,[1,3] DENNIS K. PEARL,[1,4] DAVID E. MORLEDGE,[1,3] CURTIS A. GLEASON,[1] YOSHIO HOSOBUCHI[2] *and* NICHOLAS M. BARBARO[2]

*(From the Departments of [1]Physiology and [2]Neurosurgery and the [3]Coleman Memorial Laboratory in Otolaryngology, University of California, San Francisco, and the [4]Department of Statistics, Ohio State University, Columbus, Ohio, USA)*

## SUMMARY

A 'time-on' theory to explain the cerebral distinction between conscious and unconscious mental functions proposes that a substantial minimum duration ('time-on') of appropriate neuronal activations up to about 0.5 s is required to elicit conscious sensory experience, but that durations distinctly below that minimum can mediate sensory detection without awareness. A direct experimental test of this proposal is reported here.

Stimuli (72 pulses/s) above and below such minimum train durations (0–750 ms) were delivered to the ventrobasal thalamus via electrodes chronically implanted for the therapeutic control of intractable pain. Detection was measured by the subject's forced choice as to stimulus delivery in one of two intervals, regardless of any presence or absence of sensory awareness. Subjects also indicated their awareness level of any stimulus-induced sensation in each and every trial. The results show (1) that detection (correct > 50%) occurred even with stimulus durations too brief to elicit awareness, and (2) that to move from mere detection to even an uncertain and often questionable sensory awareness required a significantly larger additional duration of pulses. Thus simply increasing duration ('time-on') of the same repetitive inputs to cerebral cortex can convert an unconscious cognitive mental function (detection without awareness) to a conscious one (detection with awareness).

## INTRODUCTION

The existence of mental functions or operations that go on unconsciously, without reportable subjective awareness, is widely accepted. This is not the place to review the considerable clinical and experimental evidence that could be cited to support this view (*see*, e.g., Shevrin and Dickman, 1980; Marcel, 1983; Holender, 1986; Weiskrantz, 1986). The question of main interest here is what cerebral neuronal processes may control the transition between an unconscious and a conscious (subjectively experienced) mental function; that is, can a causal difference between (1) neuronal activity that supports/mediates an unconscious mental function and (2) neuronal activity that becomes sufficient to support/mediate a mental function carried out with conscious subjective awareness be identified?

Correspondence to: Dr B. Libet, University of California (S-762), San Francisco, CA 94143-0444, USA.

A 'time-on' theory has been proposed to provide one neural mechanism to distinguish between unconscious vs conscious mental functions (Libet, 1965, 1985, 1989). This entails two propositions. (1) A substantial minimum duration (or 'time-on') of appropriate neuronal activations is required to elicit a reportable conscious subjective experience. This duration can be up to 500 ms and more, when stimulation of a cerebral sensory system is at a liminal intensity for any sensory awareness. (2) Durations of activations which are less than the minimum required for awareness could mediate an unconscious mental function, one involving cognitive and conative responses to a sensory signal with no conscious sensory experience. The term 'unconscious' is used in a broad operational sense to mean a mental event which is not reportable as an introspective experience by the subjects (*see* Libet, 1987). This would include so-called preconscious and subconscious events. Considerable direct and indirect evidence for proposition (1) has already been developed (e.g., Libet *et al.*, 1979; Libet, 1981, 1982). Proposition (2), however, had thus far not been subjected to direct experimental testing of the kind that could potentially confirm or falsify it. Some indirect evidence was available from the timing of cerebral activity before a voluntary act (Libet, 1985, 1989).

The present study was designed to provide a direct test of the specific hypothesis that durations of repetitive inputs to sensory cortex too brief to elicit any sensory awareness can nevertheless be detected without awareness. Our previous study had shown that a stimulus of liminal intensity in the ventrobasal thalamus (or medial lemniscus) requires a minimum train duration of about 500 ms, independent of the pulse frequency, just as does S-I cortex (Libet *et al.*, 1964, 1979; Libet, 1973, 1982). In the present study train durations in a wide range above and below the minimum were delivered in different trials. Subjects were asked to give two kinds of responses after each stimulus delivery, one to test for psychological (mental) *detection* of the signal and another to report any *introspective experience or awareness* of a sensation. Detection was measured by a forced-choice response, as to which one of two time intervals may have been the one in which a stimulus was delivered, regardless of whether anything was felt by the subject. Thus propositions (1) and (2) of time-on theory were tested simultaneously in these direct comparisons of detection and awareness.

The design is based on our operational definition of conscious subjective awareness, that is, the introspective report of the subject that he/she has felt or experienced the simple sensation in question here. The case for the validity and reliability of such reports, when made under appropriate conditions, has been made elsewhere (e.g., Libet, 1987) and will be discussed further below. As will be shown below, an ambiguity of interpretation only arises when the subject reports that he is uncertain about whether or what he felt or experienced; how that category of subjective responses affects the overall conclusions will be considered in detail.

## METHODS

### *Subjects*

Subjects were drawn from a pool of patients in whom stimulating electrodes had been chronically implanted for the therapeutic relief of some forms of intractable pain (Hosobuchi, 1986; Levy *et al.*, 1987). The patients were ambulatory and for most study sessions they came from their homes. Each gave his/her fully informed consent in accordance with the Declaration of Helsinki, as prescribed by the Committee on Protection of Human Subjects of the University of California, San Francisco; the Committee also approved the entire experimental study in accordance with its own rigorous guidelines and those required by the National Institutes of Health. This included reimbursement of incidental expenses and $25 for each half-day study session. It was agreed that no risk was added by the study; the intensity and duration of the experimental stimuli were all far lower than the therapeutic stimuli routinely self-administered via the same electrodes and electronic devices. Further, if there was any indication of fatigue or loss of interest, or of any preference by the subject not to continue, the study session was to be terminated, without any prejudice implied to the patient; this happened in only a few instances out of the total of approximately 50 half-day (2 h) study sessions for all subjects.

There were 6 males and 3 females who underwent the full study; ages ranged from 44 to 59 yrs. An additional male subject was studied extensively in 5 sessions in an initial pilot experiment to test and refine the procedure, including gaining experience with the nature of a subject's responses about the presence or absence of sensory awareness of the stimuli. The chronic pain in these patients was apparently brought on by injury (blow to back; fall) in 5, including 1 with paraplegia due to a spinal cord transection at T6 plus some spinal cord injury at C5 (Case B.D.). In 1 (K.K.) the diagnosis was postherpetic neuralgia; in another (L.R.) it was lumbar arachnoiditis; and in another (H.C.) the pain syndrome was apparently chiefly a result of partial damage to ascending somatosensory tracts in the midbrain during a previous surgical procedure. In the ninth subject (S.D.) the diagnosis was 'thalamic pain syndrome', with some residual loss in pressure and proprioceptive sensibility on the right side. Thus in all of these 9 subjects, the pain syndrome was associated with damage to somatosensory pathways at some level.

Six subjects were taking daily medications for pain (in addition to using the thalamic stimulator). In 4 of these (K.D., J.S., J.P. and L.R.) this was acetaminophen (500 mg) plus codeine (30 or 60 mg) one to three times a day. In the fifth (M.P.) it was Percocet (1 tablet three times a day). A sixth (K.K.) was taking low doses of methadone (10 mg/day) and thorazine (50 mg/day). Of the 3 not taking pain medications, 1 (S.D.) regularly took phenytoin, 300 mg/day and 2 others (H.C. and B.D.) took L-DOPA. Although it is possible that some medications could have affected sensory perception, every subject appeared alert, attentive, responsive and articulate in the study sessions. In any case, it was not our objective to establish absolute levels of required stimulation, which might be sensitive to medications. The chief objective was to determine the *relative* train durations of stimuli required for detection without vs with sensory awareness, under given conditions of stimulus intensity and subjects' abilities in a given experimental session.

Many of the potential subjects experienced paraesthesiae (tingling, etc.) in the affected referral areas even at rest, that is, in the intervals between applying stimulation for relief of pain. Any substantial amount of such resting paraesthesiae makes it difficult if not impossible for the subject to report consistently the usually similar additional paraesthesiae experienced with a threshold stimulus. Consequently, we accepted for full study only those patients with few or no resting paraesthesiae, in whom near threshold sensations were discerned with consistency and without unmanageable difficulty.

*Stimulation*

The electrodes were bipolar platinum, with contact exposures of 1 mm and separation of 2 mm (Model 3380, Medtronics Inc., Minneapolis). The assembly had been implanted stereotaxically in the ventrobasal thalamus (n. VPL or VPM, depending on location of pain) with leads to a receiver coil which was implanted below the clavicles and internalized subcutaneously for permanent use. Location of the electrodes could not be verified by direct anatomical observation, but the physiological effects of stimulation were in accordance with the stereotaxic localizations. In any case, precise knowledge of electrode locations is not critical to the objectives of this study. A portable stimulator box (Models 3523, or 7520 or 3424, Medtronic, Inc.) supplied the electrical pulses, transmitted by another coil placed on the skin over the receiver coil. The initial pilot sessions showed that the battery supply in the box could not maintain a voltage sufficiently stable to enable us to achieve a relatively consistent threshold intensity during even 30–60 min of testing. Thus it was essential to replace the batteries with a constant voltage source at 9.0 V (Hewlett-Packard Power Supply, Model 6216A) during all experimental sessions. A stimulator box identical to that used therapeutically by patients was modified so that an external control could trigger the output of the box pulse by pulse, permitting us to control the frequency and number of pulses delivered by the box to its transmitter coil in each trial. Pulse shape and duration, 200 $\mu$s, remained a function of the box. Intensity (peak voltage) of pulses also remained regulatable only by the control knob on the box itself; since this proved to be too coarse to enable a sufficiently clear differentiation between a near-threshold (liminal) intensity and one giving a very strong sensation, we inserted an attentuator control in the circuit to the transmitter coil. The intensity could then be graded more finely to liminal levels after setting the coarser control on the stimulator box.

A pulse frequency of 72 pps was adopted for all experiments, to provide a sufficient range of pulse numbers for the different stimulus trials (*see below*). The pulse number (train duration) was controlled by a gating signal delivered to a Grass S-44 stimulator, set to deliver the requisite frequency and number of trigger pulses to the stimulation box. An isolation unit disconnected the patients' stimulator box from ground and uncoupled it from direct connection to the Grass stimulator. Each gating signal was delivered from a IBM-PC/AT with Data Translation, DT2821-F-8DI, D/A converter. The duration of the gating signal (calibrated in pulse numbers for the Grass stimulator's output) was controlled and varied as desired by a program control written in ASYST.

*Detection of the stimulus by the subject*

Detection was measured by a simple forced-choice response. Each stimulus was delivered during 1 of 2 lighted intervals, $L_1$ and $L_2$, independently randomized with equal probability in successive trials. $L_1$ and $L_2$ were indicated by two separate horizontally placed buttons; within each button a weak light went on for 1 s, first in $L_1$ (on the left) and after a 1 s interval, in $L_2$ (on the right). After $L_2$ light had turned on and off, the subjects had to choose that light ($L_1$ or $L_2$) during which the stimulus was delivered and to press the corresponding button for that choice. They were asked to make a choice even if they did not feel or experience any sensation either during the $L_1$ or the $L_2$ light-on time periods. In the absence of an associated sensory experience they were asked simply to give a best guess, letting this come spontaneously and quickly without trying to analyse their choice by deliberating about it. The use of 2 intervals in each trial, in which no stimulus is expected in 1 of the 2, allows the subject to compare *any* subjective sensory experience in 1 of the intervals with another putative blank interval. Also, the variation in the stimulus can be matched against an associated blank interval rather than against a past single-interval trial.

Feedback to the subject on the correctness of the choice was given only in those fewer series of trials so labelled. In these, a different small green light turned on briefly if the choice was correct or a similar red light if incorrect; this occurred immediately after the subject had acted to press button 1, 2 or 3 for awareness level which he did following his forced choice of $L_1$ or $L_2$.

With randomized actual deliveries of stimulus in $L_1$ or $L_2$, the subject will have a 50% chance of being correct even if he chooses $L_1$ or $L_2$ without any relationship to the stimulus. Detection would then be discernible only if correct responses are achieved in significantly more than 50% of the trials. The full statistical methods for handling and analysis of the observations are given below.

*Awareness of the stimulus by the subject*

The subject was asked to report his awareness or unawareness of a sensory experience (due to the stimulus), whether during $L_1$ or $L_2$. This was indicated in each trial by pressing 1 of 3 buttons *after* having made his forced-choice detection response, as described above. The subject had to press button 1 if a sensation was felt in roughly the same bodily area as that experienced during the preliminary testing trials before each regular series of trials; he was to press button 1 even if the sensation was very weak and/or brief. If he simply felt nothing he was to press button 3. However, if he was uncertain about a sensory experience, or if he felt there was something more than nothing during one interval, he was to press button 2. It is important to note that the instruction to report button 2 extended beyond being uncertain about a sensation with the usual quality and location; it included any feeling of something being different about 1 of the 2 lighted intervals. The latter part of the instruction provided a possibility for responses that did not necessarily signify any sensory awareness, and these became evident in the descriptions by the subjects of what button 2 reponses actually meant to them.

For stimuli consisting of $1-19$ pulses (i.e., train durations of approximately 0.26 s or less with 72 pps) it was initially expected that subjects would never feel any sensation, as stimulus intensities were set near the liminal level required by a $1-2$ s train in the preliminary trials. (A 1 s train is normally distinctly longer than the minimum required with liminal I (defined later), but some subjects seemed to require $> 1$ s at the start of testing.) Therefore, for those series of trials in which all stimulus pulse numbers were in the range of $1-19$, the subject was asked to give a confidence rating, even if he felt no sensation: button 1 was to be pressed if he was 'confident' about his forced-choice answer; button 2 if he had only a 'low confidence' in the forced-choice answer; button 3 if he had 'no confidence at all'. However, on questioning each subject after many such series of trials, it turned out that the basis for the confidence rating was indistinguishable from that used in reporting the corresponding awareness level buttons 1, 2 or 3. Consequently, 1, 2 or 3 were all treated in the analysis as if respectively equivalent, whether the question to the subject was one of awareness or confidence level. The appropriateness of this is further discussed below.

It was emphasized to the subject that there was no 'correct' answer expected from him about his awareness report, and that we wanted and accepted his report of whether and what he felt, if anything.

*General procedure*

The subject sat in a comfortable chair at a table with the panel of responding buttons in an acoustically shielded room. The experimental observer sat alongside. The computer, Grass stimulator and other accessories, with the operator, were located in an adjacent anteroom.

*Liminal I and utilization train durations (TDs).* With 72 pps stimulation running continuously, the subject was initially asked to set the coarse intensity (I) control on the (modified) stimulus box to the level at which he just began to feel a weak sensation (usually a tingling) in the usual body part associated with therapeutic stimuli. The stimulation was then turned off and a series of trials was carried out to determine liminal I; with each stimulus set at 72 pps and a 1 or 2 s TD the observer used the attenuator fine control to achieve the intensity level at which the subject reported the weakest (threshold) sensory awareness. He reported his sensory response by pressing button 1, 2 or 3 (*see above*), as well as verbally when questioned. Holding to this fixed liminal I, the minimum TD (or pulse number) of the stimulus that could still elicit a conscious sensory experience was determined; this value had been named the 'utilization TD' (Libet *et al.*, 1964; Libet, 1973).

As in earlier studies, both liminal I and utilization TD were established by the common method of limits, in a sequence of stepwise changes up or down, with at least 2 consecutive trials in agreement on the threshold and subthreshold values. In the present study, establishing statistically rigorous values for liminal I and utilization TD was not essential; these values were needed only as indicators as to what level of I was to be adopted for the experimental trials in which *TDs* were varied in a statistically rigorous fashion. Consequently, the method of stepwise limits tested minimally practicable step changes in I and TD here, to save time and the subject's energy. However, liminal I or the just supraliminal I to be used in the actual experimental trials were redetermined at the beginning and, if any uncertainty arose, at the end of each study session; this was necessary partly because stability of the actually delivered intensity depends on a constancy in the position of the transmitter coil on the patient's chest.

Utilization TDs were found to average approximately 0.4 – 0.5 s, close to the values previously reported (e.g., Libet, 1973). On this basis, the experimental design was arranged to have two types of series, each series containing blocks or cycles of 20 trials each, allowing the cycles to be repeated to the degree tolerated well by the subject. (1) *'Awareness' series* in which pulse (p) numbers would be randomly varied from 19 to 55 p in different trials, in intervals of 2 pulses (i.e., 19, 21, 23, etc.). This would provide 19 different TDs with 1 trial of the 20 total being a blank (0 p). Such a range of TDs was expected to elicit many responses with awareness (i.e., from button 1) and at least some with no awareness at all (button 3), if the intensity was suitably set at near liminal I (i.e., minimum required TD of about 30 p, but always $>$ 19 p). (2) *'Detection without awareness' series*, in which pulse numbers were randomly varied 1 – 19 p with intervals of 1 pulse, plus 1 blank (0 p) to make 20 total trials. Such a range was expected to elicit no definite awarenesses (button 1) and, as indicated above, subjects were asked instead to make a confidence rating on their forced-choice response. (The actual results did not bear out this expected sharp division between awareness and non-awareness series, as seen below.)

*Study sessions.* The goal was to study each subject in 4 half-day sessions (about 2 h each) on 2 successive days. This schedule was adhered to in most but not all cases; interruptions occurred in some schedules when the patient felt indisposed to continue (because of pain or other discomforts) with a second session either on the same day or on the next day. The patient in such cases returned at a later date to complete the 4 sessions.

Each series of 'awareness' or 'detection' trials consisted of blocks of 20 trials which were repeated, after brief resting intervals of a minute or so, for up to 5 times, giving a desired total of 100 trials. Within each block or cycle of 20 trials, the pulse numbers were varied by the computer in a 'without repeating' random sampling. This was done to achieve comparability of trial numbers at all of the tested pulse numbers. However, the randomization sequence was changed by the computer program for each successive cycle of 20, preventing any possible development of expectation of pulse number. Also varied at random was the appearance of the stimulus during $L_1$ or $L_2$ for each trial. Each individual trial began with a 0.5 s warning auditory bleep, followed after 1.0 s by the sequence of $L_1$ flashing on for 1 s, an interval of 1 s, and $L_2$ on for 1 s. The stimulus began 0.25 s after onset of the 1 s light. When the subject had pressed $L_1$ or $L_2$ as his forced choice for the presence of stimulus, and then also had pressed button 1, 2 or 3 for his subjective response, the computer allowed an additional 15 s before starting the next trial with a warning bleep. When he reported, before pressing any buttons, that he had 'missed' any signal by inadequate attentiveness, etc., the program permitted the operator to repeat the same stimulus test, so that the randomized series of 20 was retained.

A brief series of training trials was administered on a preceding day or at the start of the first session. Each study session of about 2 h was started with determination of near liminal I and rough utilization TD values (*see above*) to establish a fixed intensity for the session. In sessions 1 and 2, 'awareness' and 'detection' series were alternated, with appropriate breaks between series, but the selection of which series came first was randomized among subjects. In sessions 3 and 4, usually on the next day, 'detection with feedback' alternated with 'awareness' series. In those sessions in which the subject (e.g., H.C.) gave relatively few button 1 responses in the 'awareness' series, i.e., reported feeling few stimuli, liminal I was redetermined at the end of that session; this was done to establish that liminal I had *not* risen during the session to a level at which little or no excitation of sensory axons was occurring with the I used in the trials. In such cases, it was consistently found that TDs in the range of 0.5 – 1.0 s were still effective at the end of the session.

*Verbal descriptions of introspective experiences.* These were elicited from the subject in several ways. (1) The quality and referred anatomical location of stimulus-produced sensation was described in the initial testing for liminal I/utilization TD values, and at times thereafter. Most often a localized tingling or twinge-like sensation was reported, although other qualities appeared in some cases especially with brief TDs ($< 0.5$ s). For example, with short TDs J.S. reported 'feathery' or 'breeze wind-like' sensations; similarly, M.P. reported a 'light feathery brush' with both her low confidence and uncertain (both button 2) ratings in some detection and awareness series, respectively. (Both J.S. and M.P. reported tinglings with button 1 replies.) K.K.'s chief report was one of a 'warmth', 'warm air or warm water moving over surface', but she also reported a slight tingle with brief TDs or at the end of a sensation of warmth elicited by a longer TD ($> 0.5$ s). (2) Verbal descriptions of their sensory experiences during the trials were requested from the subjects at the end of some cycles in both 'awareness' and 'detection' series. This included queries as to what they felt in connection with giving button 1, 2 or 3 response ratings of awareness or confidence, respectively. They were also asked, at these times, why they chose $L_1$ or $L_2$ in those trials when they reported feeling nothing or 'no confidence'. (3) Additionally, one or two cycles of 20 trials were devoted to obtaining verbal descriptions after each individual trial, so that the subject's memory at the end of a whole cycle would not be an issue. This was usually done in the fourth study session during a 'detection' series.

### *Statistical treatment*

The results are presented in three different formats: (1) a tabulation of the raw data, that is, percentage correct choices (of $L_1$ vs $L_2$) for different pulse numbers (TDs) of stimulus, but given separately for each of the reported awareness/confidence ratings (buttons 1, 2 or 3); (2) a nonparametric analysis involving a few simple assumptions about the mechanism generating the data; (3) a parametric analysis which assumes a logistic regression model. The basis for formats (2) and (3) is given here; a fuller description of the statistical model for (2) is given in the Appendix.

*Nonparametric analysis.* This assumes (1) that different trials within any series are independent (i.e., that the probability of the subject's responses, as to choice of $L_1$ vs $L_2$ and awareness level report in a given trial is independent to that in other trials); (2) that when *no* pulses are administered there is a 50/50 probability of the subject being right or wrong, regardless of their awareness/confidence; (3) that the administered pulses can only increase the probability of a correct response; (4) that the two parameters being estimated, $\theta$ and $\alpha$, are constants over the course of a single study session.

Assumptions (1) and (2) were basically validated by the experimental design, including the randomization of pulse numbers and of the settings as to whether $L_1$ or $L_2$ was right or wrong. (The feedback series are a slight exception to assumption (1), since a positive dependence was introduced by design; actual results, however, showed only a slight difference from nonfeedback series.) Assumption (2) is in fact validated by the results, in which very close to 50/50 correct was found for the aggregate of blank (0 p) trials. Assumption (3) is a reasonable postulate, and is borne out in the results. Assumption (4) is difficult to assess, but conditions within a session were fairly homogeneous and all sessions were terminated if the subject seemed to show any fatigue, even when he did not spontaneously report this.

Theta ($\theta$) is defined as equal to the *mean number* of pulses required to make an *otherwise incorrect* response to be correct. Note again that even with 0 pulses, the subject was expected to make 50% correct choices; $\theta$ focuses, then, on making any of the otherwise (50%) incorrect responses into a correct choice. The qualitative meaning of this approach may be seen in a first approximation type of look at raw data values. There were a total of 1290 trials in which stimuli consisted of 1 – 10 pulses. In these, 758 produced correct choices (as to whether stimulus was delivered in $L_1$ or $L_2$), with 645 correct expected on sheer chance alone. Thus 758 – 645 = 113 (i.e., 17.5%) of 645 otherwise expectedly incorrect responses were instead correct, when administering stimuli with a range of 1 – 10 p. From moment to moment a different number of pulses, Y, may be required to ensure a correct response. For 113 of the 645 trials Y was 10 or less. Usually, more pulses were needed, and Y would be larger. $\theta$ is defined as the expected value of the number of pulses required to make any such otherwise incorrect response a correct one. Details of the procedure to estimate $\theta$ are deferred to the Appendix.

Alpha ($\alpha$) is defined as equal to the average number of pulses required to move a subject, from the threshold of being correct *but just guessing*, to being correct but also with at least minimal or uncertain *awareness* of a stimulus induced sensation. That is, $\alpha$ = the mean additional pulses required, over and above those for detection, for the subject to achieve some possible awareness/confidence experience noted by him as rating 2 instead of 3 (which is no awareness at all). As noted above, a key assumption in estimating $\alpha$ by the statistical technique employed (*see* Appendix) is that $\alpha$ is a constant and thus independent of Y. However, this assumption did not play a role in our conclusions since recomputation of the estimated values of $\alpha$ under a worst case scenario (of an invalid assumption of constancy) produced only minimally different values.

Awareness or confidence rating 2 was used in defining $\alpha$, rather than level 1. Rating 2 indicated only an uncertain and often debatable actual awareness (*see below*), not a definite feeling or sensation as in rating 1. Consequently, any convincing evidence that $\alpha$ was greater than zero would strongly support the hypothesis that it takes significantly more pulses (greater TD) to achieve *any* awareness of the stimulus than it does for detection.

*Parametric analysis*. In addition to assumptions (1) – (3) in the nonparametric analysis (*above*), the parametric one also assumed that the log of the odds of being correct depends in a linear fashion on both the number of pulses (in a stimulus) and also on the awareness/confidence level or rating by the subject. Here

$$\ln \text{[probability of correct response/probability of incorrect response]}$$

$$= \text{A (no. of pulses)} + B_i$$

where A is the increase in $\log_e$ odds per pulse, B is a value which depends on the awareness/confidence rating i in the trial (i.e., i = 1, 2 or 3).

As the study sessions were differently arranged to carry out the 'awareness' and 'detection' series (*see above*), logistical models that included a term which depended on this difference were also investigated. However, the magnitude of the estimated effect of this difference was almost always very small, and for 27 of the 33 cases its inclusion in the model gave no improvement in terms of goodness-of-fit.

The parametric analysis is the type more commonly encountered in problems such as the present one. However, the nonparametric analysis involves fewer assumptions and should be regarded as the most convincing approach. Fortunately, conclusions derived from all these statistical formats were in basic agreement, relative to the hypothesis being tested.

## RESULTS

### *Actual detection and awareness responses*

In each trial, the subject chose $L_1$ or $L_2$ (as the lighted period in which a stimulus was delivered), and his choice plus the correct answer were recorded by the computer. The subject's indication of his level of awareness of a sensation elicited by the stimulus in that same trial was also recorded.

Table 1 presents a compilation of all the correct responses and total trials, sorted for ranges of stimulus pulse numbers. (Recall that 72 pulses = 1 s stimulus train duration.) These results are also segregated into 1 of the 3 levels of awareness reported by the subjects in those same trials. As described in Methods, level 3 meant no awareness of any sensation at all, level 2 meant uncertain (or low confidence) about any sensation although possibly something there, level 1 meant something was felt (in the appropriate body area) even if very weak or brief (with 'high confidence', when a confidence rating

TABLE 1. PROPORTIONS OF CORRECT CHOICES VS PULSE NO. (TD) AND AWARENESS LEVEL*

| *Subject* | *Awareness level* | *0 pulses* | *1–10 p* | *11–19 p* | *19–37 p* | *39–55 p* | *Total 1–55 p* |
|---|---|---|---|---|---|---|---|
| J.S. | | | | | | | |
| | 3 | 15/29 (0.52) | 102/173 (0.59) | 63/92 (0.68) | 21/34 (0.62) | 17/20 (0.85) | 203/319 (0.64) |
| | 2 | 1/2 | 20/23 (0.87) | 59/61 (0.97) | 43/48 (0.90) | 34/34 (1.0) | 156/166 (0.94) |
| | 1 | 0/1 | 4/4 | 27/27 (1.0) | 38/38 (1.0) | 54/54 (1.0) | 123/123 (1.0) |
| S.D. | | | | | | | |
| | 3 | 11/18 (0.61) | 69/122 (0.57) | 69/98 (0.70) | 11/25 (0.44) | 1/2 | 150/247 (0.61) |
| | 2 | 1/1 | 5/8 | 19/19 (1.0) | 19/19 (1.0) | 26/26 (1.0) | 69/72 (0.96) |
| | 1 | 0/0 | 0/0 | 0/0 | 16/16 (1.0) | 26/26 (1.0) | 42/42 (1.0) |
| H.C. | | | | | | | |
| | 3 | 13/24 (0.54) | 81/148 (0.55) | 102/126 (0.84) | 69/82 (0.84) | 61/66 (0.92) | 313/422 (0.74) |
| | 2 | 0/0 | 2/2 | 6/7 | 3/3 | 5/5 | 16/17 (0.94) |
| | 1 | 0/0 | 0/0 | 2/2 | 5/5 | 10/10 (1.0) | 17/17 (1.0) |
| K.D. | | | | | | | |
| | 3 | 8/17 (0.47) | 43/89 (0.48) | 39/65 (0.55) | 17/23 (0.74) | 3/7 | 99/184 (0.54) |
| | 2 | 2/5 | 46/76 (0.61) | 63/78 (0.81) | 10/15 (0.67) | 14/14 (1.0) | 133/183 (0.73) |
| | 1 | 0/0 | 3/5 | 10/10 (1.0) | 11/12 (0.92) | 23/24 (0.96) | 47/51 (0.92) |
| M.P. | | | | | | | |
| | 3 | 9/14 (0.64) | 49/77 (0.64) | 19/33 (0.58) | 10/15 (0.67) | 2/3 | 80/128 (0.63) |
| | 2 | 1/4 | 23/35 (0.66) | 49/57 (0.86) | 15/19 (0.79) | 12/13 (0.92) | 99/124 (0.80) |
| | 1 | 0/0 | 5/8 | 17/18 (0.94) | 26/26 (1.0) | 38/38 (1.0) | 86/90 (0.96) |
| L.R. | | | | | | | |
| | 3 | 13/22 (0.59) | 90/145 (0.62) | 90/113 (0.80) | 33/50 (0.66) | 32/38 (0.84) | 245/346 (0.71) |
| | 2 | 2/3 | 4/5 | 18/18 (1.0) | 16/18 (0.89) | 14/14 (1.0) | 52/55 (0.95) |
| | 1 | 1/1 | 0/0 | 4/4 | 41/42 (0.98) | 47/47 (1.0) | 92/93 (0.99) |
| K.K. | | | | | | | |
| | 3 | 8/18 (0.44) | 48/96 (0.50) | 50/71 (0.70) | 9/16 (0.56) | 6/8 | 113/191 (0.59) |
| | 2 | 1/2 | 26/39 (0.67) | 24/33 (0.73) | 22/25 (0.88) | 19/20 (0.95) | 91/117 (0.78) |
| | 1 | 0/1 | 9/15 (0.60) | 28/31 (0.90) | 17/19 (0.89) | 25/26 (0.96) | 79/91 (0.87) |
| J.P. | | | | | | | |
| | 3 | 4/7 | 48/70 (0.69) | 55/61 (0.90) | 2/4 | 0/0 | 105/135 (0.78) |
| | 2 | 1/5 | 15/20 (0.75) | 20/20 (1.0) | 23/26 (0.88) | 20/20 (1.0) | 78/86 (0.91) |
| | 1 | 0/0 | 0/0 | 0/0 | 0/0 | 7/7 | 7/7 |
| B.D. | | | | | | | |
| | 3 | 9/20 (0.45) | 66/130 (0.51) | 96/114 (0.84) | 8/21 (0.38) | 0/0 | 170/265 (0.64) |
| | 2 | 0/1 | 0/0 | 3/3 | 30/31 (0.97) | 21/21 (1.0) | 54/55 (0.98) |
| | 1 | 0/0 | 0/0 | 0/0 | 28/28 (1.0) | 51/51 (1.0) | 79/79 (1.0) |

* Ratios: correct choices/no. of trials.

was requested). Table 2 summarizes the totals for these values for all subjects. The following points may be derived from these tables.

1. When the 0 pulse trials are taken together, we have 100 correct answers in 195 trials, or 51% correct. This is close to the 50% correct expected on pure chance from our randomized delivery of a stimulus (or none) during $L_1$ vs $L_2$.

2. Although no awareness would be expected and therefore a level 3 response for all 0 p trials, a surprising 12% of the blank trials reported awareness level 2, though only 1.5% for awareness level 1 (*see* also fig. 5). The significance of these reports in relation to the interpretation of awareness level 2 will be discussed below.

3. The percentage of correct responses when subjects reported level 3, that is *no awareness* of any somatic sensation (during $L_1$ or $L_2$) or *no confidence* at all about their choice of $L_1$ or $L_2$, is of special interest. For all stimuli in the 1 to 10 pulse range, there were 57% correct responses out of 1050 such trials (Table 2), that is, with stimuli of up to 10 p, the estimated probability of eliciting a correct response from among the 50% of expected incorrect responses equals 14%, SE ± 3%. This value is seen to increase with increasing pulse numbers. For the total of all trials (2237) in which subjects reported no awareness at all (level 3) we observed 66% correct answers. In this group, there was an estimated probability of 32 ± 2% for eliciting a correct response with stimuli up to 55 p in duration, from among the otherwise expected 50% of incorrect responses in a randomized situation, even though subjects reported feeling nothing. On the other hand, with reports of some kind of awareness, even when uncertain or unclear as to its nature (i.e., level 2), there was a 71% estimated probability of eliciting a correct response (from among the otherwise expected incorrect ones) with SE ± 1%. For some definite awareness of a sensation (level 1) the otherwise incorrect responses overall are almost all made correct (i.e., 93% ± 1%). (If the 0−10 p range is omitted from this total, the values become 96.4% ± 1.1%.)

TABLE 2. TOTALS OF RESPONSES FOR ALL SUBJECTS IN TABLE 1

| Awareness level | 0 p No. of tr. | | 1−10 p No. | 1−10 p % cor. | 11−19 p No. | 11−19 p % cor. | 19−37 p No. | 19−37 p % cor. | 39−55 p No. | 39−55 p % cor. | 1−55 p No. | 1−55 p % cor. |
|---|---|---|---|---|---|---|---|---|---|---|---|---|
| Level 3 | 169 | 53 | 1050 | 57 | 773 | 75 | 270 | 67 | 144 | 85 | 2237 | 66 |
| Level 2 | 23 | 39 | 208 | 68 | 296 | 88 | 204 | 89 | 167 | 99 | 875 | 85 |
| Level 1 | 3 | | 32 | 66 | 92 | 96 | 186 | 98 | 283 | 99 | 593 | 97 |
| Totals | 195 | 51 | 1290 | 59 | 1161 | 80 | 660 | 84 | 594 | 96 | | |

*Feedback of correct answers to subject*

The correctness of the subject's choice of $L_1$ vs $L_2$ was indicated to him after he made his response, in some sessions in which stimulus pulse numbers were in the range of 1−19 p. In this range there were mostly reports of level 3 of awareness, i.e., no awareness of a sensation to serve as a clue to correctness. There is a slightly higher percentage of correct responses in the second 10 trials of each cycle of 20 (Table 3). This is true rather consistently for most subjects individually, as well for the totals for all subjects. This suggests that they may have been learning to improve slightly their ability to detect the stimuli in this range.

TABLE 3. FEEDBACK SESSIONS ONLY*

| *Subject* | *Period* | *1–10 p* | *11–19 p* | *Total* |
|---|---|---|---|---|
| S.D. | First 10 trials | 14/23 (0.61) | 16/26 (0.62) | 30/49 (0.61) |
| | Second 10 trials | 14/27 (0.52) | 15/19 (0.79) | 29/46 (0.63) |
| H.C. | First 10 trials | 26/54 (0.48) | 34/42 (0.81) | 60/96 (0.625) |
| | Second 10 trials | 27/46 (0.59) | 40/48 (0.83) | 67/94 (0.71) |
| K.D. | First 10 trials | 50/93 (0.54) | 51/72 (0.71) | 101/165 (0.61) |
| | Second 10 trials | 42/77 (0.55) | 58/81 (0.72) | 100/158 (0.63) |
| M.P. | First 10 trials | 19/24 (0.79) | 23/24 (0.96) | 42/48 (0.875) |
| | Second 10 trials | 20/26 (0.77) | 20/21 (0.95) | 40/47 (0.85) |
| J.P. | First 10 trials | 8/13 (0.62) | 13/14 (0.93) | 21/27 (0.78) |
| | Second 10 trials | 13/17 (0.76) | 13/13 (1.0) | 26/30 (0.87) |
| B.D. | First 10 trials | 13/24 (0.54) | 23/24 (0.96) | 36/48 (0.75) |
| | Second 10 trials | 13/26 (0.50) | 21/21 (1.0) | 34/47 (0.72) |
| Total | First 10 trials | 130/231 (0.56) | 160/202 (0.79) | 290/433 (0.67) |
| | Second 10 trials | 129/219 (0.59) | 167/203 (0.82) | 296/422 (0.70) |

* Ratios: correct responses/no. of trials.

The quantitative effect of the feedback procedure, however, was too small to affect the overall values for detection requirements. The percentage of correct responses with feedback, even for the second 10 trials of each 20, was not appreciably different from the comparable values (in the same pulse number range) for all trials, most of which were without feedback (*see* Table 2). Secondly, when estimated separately for these feedback sessions, the nonparametric value of $\theta$ (the mean TD required for detection, *see below*) shows a reduction of only about 1.2 p ($\pm 0.6$) from the overall mean value of about 19.7 p for all sessions. (We should note, however, that the 1 blank test (0 p) in each cycle of 20 trials was treated in the feedback report as if there were a correct answer. We realized belatedly that we should have informed the subject properly after each blank test that this trial had no correct answer, but we did not want to change the procedure in midstudy. Such misinformation on the blank trial could have been a confusing element that reduced the ability of subjects to improve during feedback cycles of trials.)

*Parametric logistical regression analysis of data*

The $\log_e$ of the odds of being correct, i.e., the $\log_e$ of the (probability of correct responses/probability of being incorrect), did in fact increase with the number of pulses in a stimulus. The estimated *increase per pulse*, for this $\log_e$ value, is given in Table 4 for each study session with each subject. To summarize this analysis, a histogram of these values for all trials in all subjects is given in fig. 1. It shows a skewed distribution although most values are near the median of about 0.05. For some of the sessions the increase per pulse in the log odds was not consistent enough to be statistically significant.

With respect to awareness/confidence levels reported by the subjects, the following points appear. (1) With level 3 (i.e., felt nothing) the odds of being correct are substantial

TABLE 4. LOGISTIC REGRESSION (PARAMETRIC) ANALYSIS

| Subject | Date | a.m./p.m. | *ln (probability correct/probability incorrect)* Increase per pulse | Increase. With rise from awareness level 3 to 2 | | Increase. With rise from awareness level 2 to 1 |
|---|---|---|---|---|---|---|
| J.S. | 03/11/87 | p.m. | 0.131 | 0.882 | τ | 1.718 |
| | 05/11/87 | p.m. | 0.035 | 1.463 | | −0.146 |
| | 26/04/88 | a.m. | 0.042 | 2.056 | | 1.656 |
| | 26/04/88 | p.m. | 0.048 | 1.236 | τ | 2.034 |
| S.D. | 01/12/87 | p.m. | 0.027* | 1.854 | | 1.735 |
| | 03/12/87 | a.m. | 0.032* | 1.972 | | 1.683 |
| | 03/12/87 | p.m. | −0.001* | 9.623 | | −3.202 |
| H.C. | 07/12/87 | p.m. | 0.056 | 1.260 | τ | 1.836 |
| | 08/12/87 | a.m. | 0.079 | 5.326 | τ | −2.425 |
| | 08/12/87 | p.m. | 0.054 | 5.363 | τ | −1.752 |
| K.D. | 20/01/88 | a.m. | 0.034 | 1.013 | | −0.305 |
| | 20/01/88 | p.m. | 0.048 | 0.602 | | 0.452 |
| M.P. | 15/02/88 | p.m. | 0.048 | 0.906 | | −0.261 |
| | 16/02/88 | a.m. | 0.183 | −0.439 | τ | 0.260 |
| | 16/02/88 | p.m. | 0.037 | −0.366 | τ | 3.155 |
| L.R. | 05/04/88 | p.m. | 0.004* | 1.236 | | 0.158 |
| | 06/04/88 | a.m. | 0.048 | 4.365 | τ | (no level 1 response) |
| | 06/04/88 | p.m. | 0.043 | 0.751 | | 2.139 |
| K.K. | 07/06/88 | a.m. | 0.040 | −0.031 | τ | 0.233 |
| | 07/06/88 | p.m. | 0.043 | 0.198 | τ | 0.258 |
| | 08/06/88 | a.m. | 0.018* | 1.222 | τ | −0.371 |
| | 28/06/88 | p.m. | 0.053* | 2.438 | | 1.500 |
| | 08/07/88 | a.m. | 0.252 | 6.613 | τ | −3.085 |
| J.P. | 14/07/88 | p.m. | 0.091 | −1.198 | τ | 2.749 |
| | 15/07/88 | a.m. | 0.106* | −0.132 | τ | (no level 1 response) |
| | 15/07/88 | p.m. | 0.175 | 0.656 | τ | 1.0117 |
| B.D. | 21/07/88 | p.m. | 0.046* | 9.051 | | −3.085 |
| | 21/07/88 | p.m. | 0.082 | 7.723 | τ | −2.759 |
| | 22/07/88 | a.m. | 0.123 | 7.048 | | −2.652 |
| | 22/07/88 | p.m. | 0.106 | −0.239 | τ | 2.304 |

* Indicates that including the number of pulses in the model did not significantly improve the fit of the model to the data. τ Indicates that including the awareness level in the model did not significantly improve the fit of the model to the data.

and consistent. For example, the model indicates that typically about 14 pulses would be required to give 2:1 odds for a correct answer, with reports of awareness level 3. This indicates the existence of detection without any awareness of the stimulus. (2) However, there are large increases in the odds of being correct when the reported awareness level is 2 instead of 3. Actual estimated increases are given in Table 4; these values control for the pulse number, thus excluding the factor of higher pulse numbers (which produce a greater incidence of level 2 reports). Note, however, that the odds

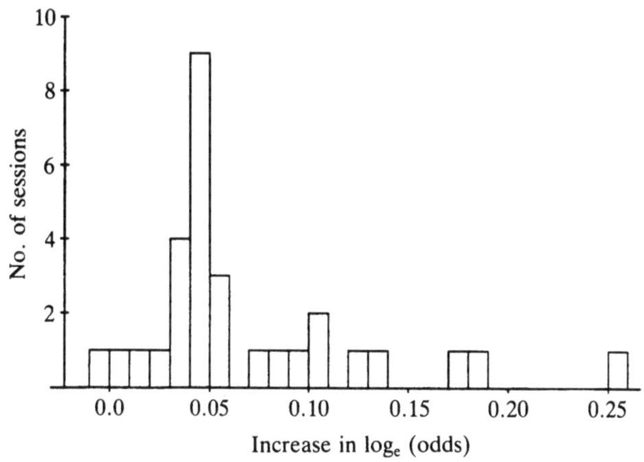


FIG. 1. Histogram summarizing all values for the estimated increase per pulse in the natural logarithm (i.e., $\log_e$) of the odds of being correct, i.e., ln (probability of correct responses/probability of being incorrect), as derived from logistical regression. Each value is an estimate for each session of trials (*see* Table 4). Although the range is large, most values cluster about the mean of 0.069 (median = 0.048), with an average-within-subjects SD of 0.041.

of being correct went down in some cases, as in all 3 sessions with subject J.P. and in 2 of 3 sessions with subject M.P. Similar calculations comparing trials with reports of awareness level 1 vs those with level 2, but at the same pulse numbers, show a less consistent change in odds of being correct, even for the same subject in different sessions. At least 1 subject, B.D., generally performed much better with reports of awareness level 2 than at 1. A histogram in fig. 2 summarizes the changes when going from awareness level 3 to level 1, for all subjects. Note that 29 of these 30 whole session values are positive, indicating a consistently greater chance of being correct with

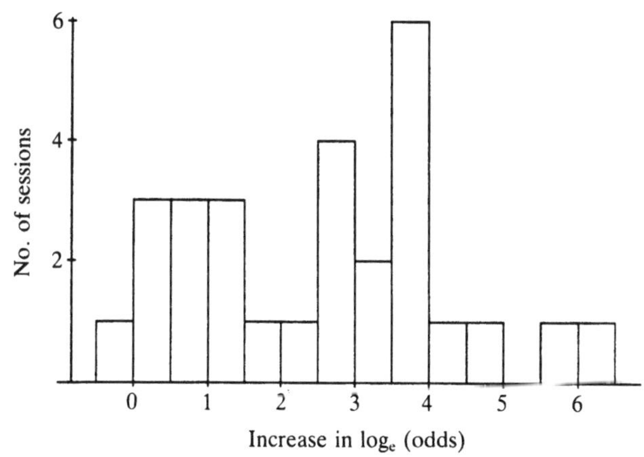


FIG. 2. Histogram summarizing effects of going from awareness level 3 (pure guess) to level 1 (felt it) on the ln of the odds of being correct (*cf* Table 4). The changes are estimated while holding the number of pulses fixed for both awareness levels.

awareness level 1 than with level 3. These quantitative, parametric estimations are all in accord with the differences seen in the raw data (Tables 1, 2), when the percentage of correct responses is compared for the 3 levels of awareness for each range of stimulus pulse numbers.

Variation among subjects was relatively small for the increase in the odds of being correct with each additional pulse. But there was significant person to person variation in how much the odds increase as subjects went from awareness level 3 to level 1.

*Nonparametric analysis of data*

As described in Methods and the Appendix, the parameter $\theta$ = the mean number of pulses required to make an otherwise incorrect response to be correct; while parameter $\alpha$ = the mean number of *additional* pulses which are required to move from mere detection to elicit even the uncertain questionable awareness level 2 (instead of to some definite awareness, level 1).

The estimated values for $\theta$ are given for each subject, in each separate session, in Table 5. A histogram summarizing the distribution of $\theta$ for all sessions in all subjects is given in fig. 3. The mean value for all $\theta$ estimates was 19.71 pulses. Individual subjects showed some variation in $\theta$ for different sessions, with an average-within-subjects SD 7.03 (Table 5).

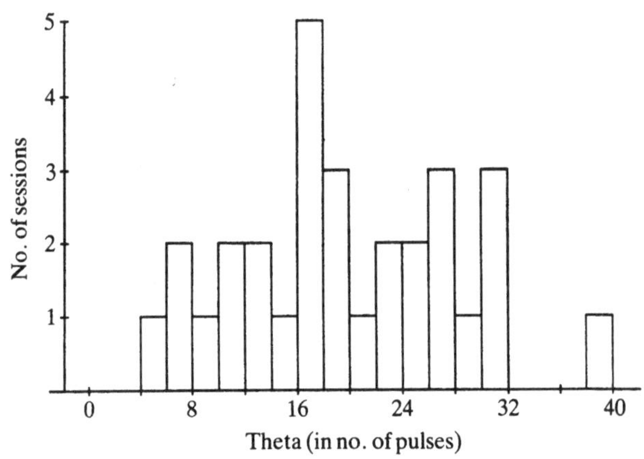


FIG. 3. Histogram summarizing results for $\hat{\theta}$, from nonparametric analysis (*cf* Table 5). $\theta$ = mean number of pulses required to make an otherwise incorrect response to be correct (*see* Methods and Appendix). The mean value of all the $\theta$s here is 19.71 (median 18.42), with an average-within-subjects SD of 7.03.

Estimated values for $\alpha$ are given in Table 6, with a histogram of all values in fig. 4A. This gives a mean value for $\alpha$ of 27.80, average-within-subjects SD = 8.50. Of additional interest is the z value for each $\alpha$; z value is how many SEs above zero it takes to equal the estimated value of $\alpha$. The mean z = 11.98, average-within-subjects SD 2.83 (histogram of z values in fig. 4B).

The following points emerge from these tables of $\hat{\theta}$, $\hat{\alpha}$ and z. (1) Values of $\hat{\alpha}$ are much greater than zero. This means it takes substantially more stimulus pulses to elicit

TABLE 5. NONPARAMETRIC ESTIMATES OF $\theta$ FOR INDIVIDUAL SESSIONS*

| *Subject* | *Date* | *a.m./p.m.* | $\hat{\theta}$ | *Estimated SE* ($\hat{\theta}$) | *Upper limit for SE* ($\hat{\theta}$) |
|---|---|---|---|---|---|
| J.S. | 03/11/87 | p.m. | 5.328 | 1.138 | 2.444 |
| | 05/11/87 | p.m. | 18.100 | 2.356 | 3.560 |
| | 26/04/88 | a.m. | 17.700 | 2.615 | 4.611 |
| | 26/04/88 | p.m. | 26.557 | 3.326 | 5.474 |
| S.D. | 01/12/87 | p.m. | 22.128 | 3.291 | 5.654 |
| | 03/12/87 | a.m. | 25.385 | 2.248 | 4.468 |
| | 03/12/87 | p.m. | 17.700 | 3.030 | 4.531 |
| H.C. | 07/12/87 | p.m. | 18.742 | 2.953 | 4.464 |
| | 08/12/87 | a.m. | 17.967 | 2.889 | 4.891 |
| | 08/12/87 | p.m. | 24.608 | 3.349 | 5.157 |
| K.D. | 20/01/88 | a.m. | 29.142 | 3.522 | 5.157 |
| | 20/01/88 | p.m. | 27.238 | 3.064 | 5.803 |
| M.P. | 15/02/88 | p.m. | 23.728 | 3.617 | 5.654 |
| | 16/02/88 | a.m. | 7.700 | 1.734 | 3.459 |
| | 16/02/88 | p.m. | 30.500 | 3.758 | 5.267 |
| L.R. | 05/04/88 | p.m. | 14.366 | 2.439 | 4.679 |
| | 06/04/88 | a.m. | 30.800 | 3.580 | 5.026 |
| | 06/04/88 | p.m. | 13.566 | 2.476 | 4.111 |
| K.K. | 07/06/88 | a.m. | 38.300 | 3.867 | 6.044 |
| | 07/06/88 | p.m. | 31.166 | 2.196 | 7.914 |
| | 08/06/88 | a.m. | 21.333 | 1.976 | 7.914 |
| | 28/06/88 | p.m. | 6.500 | 1.457 | 2.151 |
| | 08/07/88 | a.m. | 10.000 | 1.172 | 3.002 |
| J.P. | 14/07/88 | p.m. | 19.250 | 1.940 | 5.535 |
| | 15/07/88 | a.m. | 9.917 | 1.511 | 6.217 |
| | 15/07/88 | p.m. | 11.166 | 1.468 | 4.318 |
| B.D. | 21/07/88 | p.m. | 26.366 | 2.841 | 4.066 |
| | 21/07/88 | p.m. | 13.500 | 1.947 | 3.725 |
| | 22/07/88 | a.m. | 16.208 | 2.322 | 3.454 |
| | 22/07/88 | p.m. | 16.433 | 2.244 | 3.540 |

* Combining the series for 1 − 19 p with those for 19 − 55 p, when done within 20 min of each other. $\theta$ = mean number of pulses to make correct an otherwise incorrect response.

even a minimal and questionable level of awareness (level 2) than it does to produce correct detection of the stimulus without any awareness of a difference (level 3). (2) The generally large z values confirm the significance of the conclusion in point (1). If our hypothesis were false, $\hat{\alpha}$ (the extra pulse requirement for any awareness) would be zero. If $\hat{\alpha}$ were actually zero we would rarely see z values greater than 3. However, most z values here are much greater even than 3.

Additionally, subject-to-subject variation in values of $\hat{\theta}$ was on the whole relatively small. The *P* value is 0.81 in a significance test of the null hypothesis that all subjects are equal. However, inter-subject variation in values of $\hat{\alpha}$ was significant, with a *P* value of 0.02 (*see* Discussion section for some analysis of this inter-subject variation).

TABLE 6. NONPARAMETRIC ESTIMATES OF α FOR INDIVIDUAL SESSIONS*

| *Subject* | *Date* | *a.m./p.m.* | $\hat{\alpha}$ | *Upper limit* *for SE* ($\hat{\alpha}$) | *z value* |
|---|---|---|---|---|---|
| J.S. | 03/11/87 | p.m. | 5.871 | 1.211 | 4.85 |
| | 05/11/87 | p.m. | 32.224 | 1.754 | 18.37 |
| | 26/04/88 | a.m. | 30.676 | 2.271 | 13.51 |
| | 26/04/88 | p.m. | 42.095 | 2.696 | 15.61 |
| S.D. | 01/12/87 | p.m. | 45.13 | 2.7828 | 16.22 |
| | 03/12/87 | a.m. | 32.6732 | 2.179 | 14.99 |
| | 03/12/87 | p.m. | 31.585 | 2.2146 | 14.262 |
| H.C. | 07/12/87 | p.m. | 40.649 | 2.196 | 18.51 |
| | 08/12/87 | a.m. | 47.267 | 2.412 | 19.60 |
| | 08/12/87 | p.m. | 51.267 | 2.547 | 20.13 |
| K.D. | 20/01/88 | a.m. | 21.725 | 2.549 | 8.52 |
| | 20/01/88 | p.m. | 23.976 | 2.863 | 8.37 |
| M.P. | 15/02/88 | p.m. | 26.095 | 2.788 | 9.36 |
| | 16/02/88 | a.m. | 1.292 | 1.723 | 0.75 |
| | 16/02/88 | p.m. | 15.267 | 2.619 | 5.83 |
| L.R. | 05/04/88 | p.m. | 44.999 | 2.313 | 19.45 |
| | 06/04/88 | a.m. | 49.267 | 2.481 | 19.86 |
| | 06/04/88 | p.m. | 26.199 | 2.028 | 12.92 |
| K.K. | 07/06/88 | a.m. | 25.033 | 2.980 | 8.40 |
| | 07/06/88 | p.m. | 14.125 | 3.952 | 3.57 |
| | 08/06/88 | a.m. | 43.408 | 3.897 | 11.14 |
| | 28/06/88 | p.m. | 6.000 | 1.491 | 4.02 |
| | 08/07/88 | a.m. | 16.001 | 1.491 | 10.73 |
| J.P. | 14/07/88 | p.m. | 19.000 | 2.756 | 6.89 |
| | 15/07/88 | a.m. | 18.816 | 3.090 | 6.09 |
| | 15/07/88 | p.m. | 24.000 | 2.055 | 11.68 |
| B.D. | 21/07/88 | p.m. | 28.000 | 1.978 | 14.16 |
| | 21/07/88 | p.m. | 19.267 | 1.832 | 10.52 |
| | 22/07/88 | a.m. | 29.267 | 1.679 | 17.43 |
| | 22/07/88 | p.m. | 22.897 | 1.698 | 13.48 |

* Combining the series for 1−19 p with those for 19−55 p when done within 20 min of each other. α = average number of pulses to move from correct but guessing to correct with at least minimal uncertain awareness (level 2).

The possibility of variation between results for morning vs afternoon sessions for each subject was also checked, as there could be a fatigue factor or other difference involved. No consistence or significant difference was found for a.m. vs p.m. results.

*Awareness levels relative to pulse numbers*

With stimulus pulses at near liminal intensity (I), minimum train durations required to elicit a sensory experience were determined for the present subjects at the start of a study session (employing the same method of limits as for determining liminal I, but varying the TD in this case). The resulting 'utilization-TD' (U-TD) values were in the same range as reported previously, average U-TD about 0.5 s (Libet, 1973). (Some

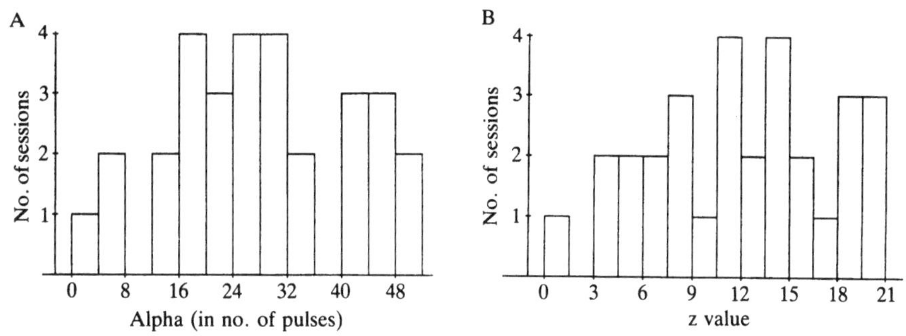


FIG. 4. Histogram summarizing results for $\hat{\alpha}$ and z from the nonparametric analysis (*cf* Table 6). A, $\alpha$ = the average number of (additional) pulses required to move a subject from being correct but just guessing to being at least minimally (even if uncertainly) aware of a stimulus-induced sensation (as well as being correct) (*see* Methods and Appendix). The mean value of all the alphas here is 27.80 (median 26.15), with an average-within-subject SD of 8.50. B, z = number of standard errors, above zero, that it takes to equal the estimated value of $\alpha$. The mean value of all the z values here is 11.98 (median 12.30) with an average-within-subject SD of 2.83.

greater variability of values measured at different times in a given subject here is attributed to the greater difficulty in setting the liminal I value accurately and consistently, under the conditions in this study.) Present subjects were on the average 10–20 yrs younger than subjects in the previous studies, who were mostly patients with motor dyskinesias (although some had intractable pain); yet both groups exhibited a similar average U-TD. For use in the experimental forced-choice series of trials the intensity (I) was often raised somewhat above the apparent liminal I in order to minimize the chance of falling below the threshold for exciting any axons by some possible variations in the effective I actually transmitted to the subdermal receiver coil. However, the maximal accepted level of I was that at which the minimum TD, required for any sensory awareness, was not less than 0.3 s (22 p).

Reports of awareness during the experimental series did not precisely match the relationship to pulse number (TD) established in the initial tests. In the initial test not even an uncertain awareness was reported with TDs less than 0.3–0.5 s (i.e., 22–36 p, according to the I set in each case), but there were such reports with stimuli of < 22 p in the experimental series. Fig. 5 shows a plot of incidence of all awareness report levels 1, 2 and 3 relative to stimulus pulse number. There were significant differences between subjects with respect to the overall summed data plotted here. It may be seen that reports of level 1 (definite if slight sensory awareness) were mostly insubstantial until TDs achieved > 20 p (> 0.28 s TD) and then the incidence continued to rise to about 50% of responses at about 34 p or more (TD > 0.47 s). These TD requirements for awareness level 1 are not far off from those determined initially by the stepwise method of limits, but with two differences. (1) A small but significant number of level 1 reports appeared even with pulse numbers < 20 p, whereas this almost never occurred in the initial stepwise determinations of minimum TD. (2) The average proportion of level 1 responses rose only to a maximum of < 50 to 60% even with pulse numbers of 55 p (0.77 s TD). In the minimum TD determinations, level 1 or level 3 was reported in at least 2 consecutive trials in order for the TD value to be regarded as positive, or negative, respectively.

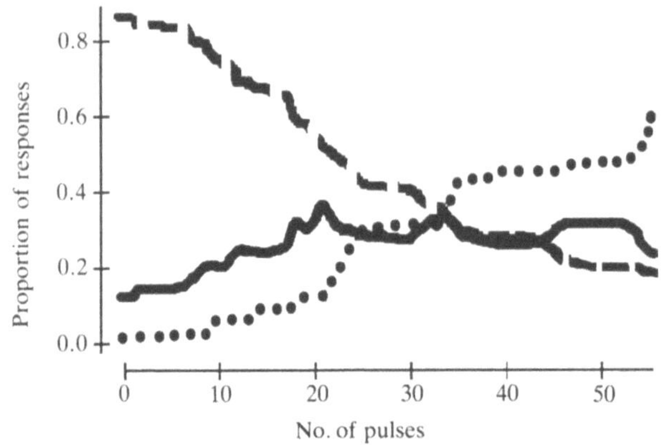


FIG. 5. Incidence of reports of each awareness level relative to number of stimulus pulses (train duration). Proportions of all trials (at each pulse number) which have reports of a given awareness level are plotted against the pulse number. At the standard 72 pps stimulus employed throughout, the 0–55 p range the abscissa represents the 0–760 ms range of train durations (TDs). Dotted line = reports of awareness level 1 (felt sensation, even if very weak); solid line = level 2 (uncertain about any sensation); broken line = level 3 (felt nothing) (*see text*). (These are isotonized line plots; values are computed under the assumption that the expected proportion of level 3 responses decreases with pulse number and the expected proportions of level 1 responses increase with pulse number. This made the curves easier to distinguish visually but it did not appreciably alter the form of the plots of the direct data points here.)

Reports of level 3 (no awareness at all of a sensation or any difference between $L_1$ and $L_2$ intervals) start high but are not 100% of responses even for 0 p and for 1 – 10 p. On the other hand, a surprising average of > 20% of reports were still level 3 at the maximal TD of 55 p.

Reports of level 2 (uncertain sensory awareness, or maybe something different between $L_1$ and $L_2$ intervals) exhibited surprisingly substantial incidences in the range of low pulse numbers (< 20 p). Indeed, even with 0 p trials (when there was no objective difference between the $L_1$ and $L_2$ interval), 12% of the trials (23/197) reported level 2! This levelled off at about 30% in the whole range of about 20 – 50 p. The interpretation of the incidence and meaning of level 2 reports will be dealt with in the Discussion.

*Verbal description of awareness levels*

All descriptions made by each subject for each awareness level were reviewed and summarized. Descriptions given after each individual trial, during 1 or 2 cycles of 20 trials so tested, were uniformly similar to those given by each subject in other situations, that is, at the end of some other cycles of 20 trials and during preliminary testings for liminal I and U-TD. Descriptions for levels 1, 2 or 3 which were given when awareness level was requested (i.e., felt slightly, uncertain, or not felt, respectively) were similar to those given by each subject when a confidence rating (for choosing $L_1$ vs $L_2$) was requested (i.e., confident, low confidence, no confidence or guessing, respectively). The introspective descriptions associated with each level were the same with both kinds of instructed requests.

Awareness level 1 was virtually always associated with the description by subjects that they felt a slight and brief though definite sensation, of the same quality and in the same body location as the sensation produced by a somewhat stronger and longer-lasting (1–2 s) stimulus in the preliminary testings.

For response level 3, subjects consistently said they felt no sensation during either of the light intervals, $L_1$ or $L_2$.

The in-between response, level 2, was associated with more complicated and not uniformly consistent descriptions even by the same subject. However, the introspective descriptions of response 2 could be grouped into one of several types. (a) 'Maybe there was a slight sensation'—similar to the one felt more definitely in a level 1 response and located in the same or a nearby site. This is the 'uncertain' type of awareness that we expected for a level 2 report. Actually, the next description (b) turned out to be the most common type. In (b) the subject reported 'something different, more than nothing', not definable as a real sensation but nevertheless localizable to the general vicinity of (though often less distinctly than) the definite sensation reported for level 1. One subject (K.K.) added that it had an expectancy nature, as if something is about to touch you but does not touch. In type (c) subjects reported something different about one light interval, but this was not localized to the usual body site or (in most cases) to any site, and it was not regarded as a sensation. One subject (K.D.) added that to report this diffuse nebulous feeling was something like finding a black cat in a dark room. Another (B.D.) reported that it was just an intuitive impression of something different, and another (J.S.) reported that one light simply seemed a little different from the other light.

All three types of descriptions were given for level 2 by 2 subjects (S.D., J.S.) in different series of trials. One (H.C.) gave only type a, 1 (L.R.) only type b; 2 subjects (M.P., K.K.) gave a or b; (B.D.) gave b or c; 1 (K.D.) gave only c. Subject B.D. stated, spontaneously in his last session, that retrospectively most or all of his level 2 awarenesses were 'really intuitive, just an impression of something different'.

*Reasons given for choosing $L_1$ or $L_2$.* The subjects were also asked to tell the observer (B.L.) why they chose $L_1$ or $L_2$ as the interval in which a stimulus was delivered, in those trials when they reported an awareness level 3 (i.e., 'feeling nothing' or 'guessing'). H.C. consistently reported he 'had no reason for his choice, he was guessing'; yet his proportions of correct choices (associated with awareness level 3) were as good or better than for most other subjects who did give some reasons (*see* Table 1). Most other subjects gave this report for many but not all such trials. Several reported having some sort of hunch that one of the lights was a 'better choice' or was associated with some undefinable difference (M.P., S.D., J.P., K.D., B.D.). Some revealing psychological projections onto the lights themselves were described: B.D. and J.S. at times said *both* $L_1$ and $L_2$ seemed to be associated with some kind of bodily experience (both sides of the body in case of J.S.), but that it was something more during one light than the other. L.R. stated that 'the light itself tells him more about which one to choose' than any feeling of something in a body part! B.D. additionally volunteered that he went into a kind of trance of concentration, during each cycle of 20 trials, and that he felt he made worse choices when the 'trance' was interrupted; he based the latter conclusion on experiences during 'feedback' series in which he was informed of his correctness after each choice. On the other hand, several subjects (J.S., L.R., B.D.) reported that they often felt frustrated or surprised in a feedback series when they guessed something was different about one light and found their choice was incorrect.

## DISCUSSION

### *Detection with no awareness*

Stimulation of ascending somatosensory fibres at the thalamus can clearly be meaningfully detected even when the subject has no reportable introspective awareness of the signal. The evidence is especially striking for those trials in which subjects reported awareness level 3 (i.e., nothing felt, just guessing). They chose the correct light interval in 66% of all such trials (Table 2). This means the probability of correct detection (in the 50% of trials expected to be incorrect by chance alone) was 32% (SE$\pm$2%). The parametric analysis also showed that the odds of being correct were substantially greater than 1 when they reported level 3. In the nonparametric analysis, incorrect chance choices became correct ones at a mean of 19.7 pulses (TD = 0.27 s). To move from mere detection to elicit even the uncertain and questionable awareness level 2 required an additional mean of 27.8 pulses ($\alpha$).

*Interpretation of reports of awareness level 2.* The subjects were instructed to report level 2 when they were uncertain about feeling a sensation, or if they felt there was something more than nothing. The latter instruction left open the possibility of reporting level 2 when they did not experience any sensory awareness at all, even one of an uncertain nature.

As reported in Results, the subjects descriptions of their associated introspective experiences could be grouped into three types. For two of these (types b and c) there can be serious doubt that an actual sensory awareness of even an uncertain nature occurred. Instead, the reports of 'something different, more than nothing, but not definable', etc., could represent feelings related to intuitive guesses, hunches, or rationalizations about the answer, based on unconscious cerebral effects of the stimulus input. (1) Some of the descriptions by subjects were expressed in just such terms. (2) Level 2 was reported inappropriately (especially by 3 subjects) in a significant proportion (12%) of blank trials (0 p) as compared with 1.5% for level 1 reports (Tables 1, 2). (3) In the series with feedback of correctness, several subjects volunteered that they were surprised and somewhat annoyed to find that they were often incorrect when they had reported a level 2 awareness; this reaction almost never occurred with level 1 reports, as those were associated with almost 100% correct answers (except for the small number of level 1 reports with stimuli below 11 p). (4) Further, more reports of level 2 were made in the trials with lower pulse numbers ($<$ 20 p) than occurred in the preliminary trials for setting of the TD values just able to elicit a threshold sensation. The preliminary trials used only a single lighted period to indicate delivery of a stimulus and did not require a forced choice answer of $L_1$ or $L_2$; there was therefore no motive to look for a hunch or rationalization to justify the answer.

A substantial though unknown proportion of reports of awareness level 2 may thus represent expressions of intuitive hunches not based on any actual sensory awareness. This proposal is not inconsistent with our criterion of awareness, which accepts the subject's introspective report; the instruction permitted the subject to report level 2 if he/she felt there was something different, more than nothing, during one of the lighted intervals, even in the absence of any sensory awareness of any kind. The proposal is also in accordance with the reasons given by subjects for choosing $L_1$ vs $L_2$ (*see* Results). The putative need to rationalize the choice of $L_1$ or $L_2$, by reporting awareness level 2 with this kind of diffuse, undefinable feeling, appeared to vary between the subjects. For trials with brief TDs (1 – 10 p), subject K.D. (an engineer) reported the highest proportion of level 2, 45% of the trials; for M.P., K.K. and J.P., level 2 reports constituted respectively 29%, 26% and 22% in this range of TDs, with much smaller proportions for other subjects (Table 1). (Interestingly, the inter-subject variation in values of $\alpha$ (*see* Table 6) seems to be related inversely to the proportion of level 2 reports by respective subjects. Those with the lowest proportion of level 2 reports (H.C., L.R.) had the highest $\alpha$s, while the converse relation existed for K.D., M.P., K.K. and J.P. (B.D. appears to be an exception, with a low level 2 incidence but relatively low $\alpha$ values). This suggests that subjects who more readily reported awareness level 2 required a smaller addition of TD, over that for mere detection, to produce the level 2 report. Even so, their z values are still high enough to indicate their lower $\alpha$s are highly significant.)

The degree of detection was related to the reported introspective level of awareness, regardless of the mean of these reports (Table 4, fig. 2). Even in the low range of 1 – 10 p the probability of a correct answer was substantially greater with reports of level 2 than with level 3 (Table 2). For example, of the 2451 trials in which only 1 – 19 p were delivered (TDs up to 0.26 s) there were 504 trials in which level 2 was reported, of which 80% were correct in the choice of $L_1$ vs $L_2$. (Probability of correct detection of otherwise incorrect chance reponses was 60% ±3.5%, in those trials.) This is also seen in the parametric analysis giving the ($\log_e$ of the) odds of being correct; the latter shows, in most subjects, a large increase (at the same p number) when awareness levels rises from 3 to 2. This indicates that whatever the basis for reporting level 2, that is, even if many did not represent an actual awareness of a sensation, it did signify an improved detection process compared with that in 'pure guess' reports.

*Relation to reports, by others, of detection without awareness.* The history of such phenomena, often referred to as 'subliminal perception', has been reviewed elsewhere (e.g., Shevrin and Dickman, 1980; Holender, 1986; Weiskrantz, 1986). In normal human subjects, the spread between stimulus intensity for threshold sensory awareness and that for any possible detection without awareness is relatively small, when stimuli are delivered to peripheral sensory receptors or nerve fibres (e.g., Libet *et al.*, 1967; Libet, 1982). It has been found that a human subject can probably *detect* the absolutely minimum possible peripheral sensory input to the CNS, i.e., a single nerve impulse in a single sensory nerve fibre from the skin (Johansson and Vallbo, 1979). That study employed forced-choice methods including one similar to that used in our study. Johansson and Vallbo did not raise or study the question of sensory awareness per se. In any case, the duration of cerebral, not peripheral, repetitive activations is at issue in the 'time-on' theory. Indeed, detection of microstimulation of striate cortex by macaques was achieved with minimum intensities only when 20 – 100 pulses at 50 Hz were applied; threshold intensity for detection increased 6 – 12 fold when only a single pulse was used (Bartlett and Doty, 1980). *If* a single peripheral nerve impulse were to be found to elicit awareness and not merely simple detection, we would predict that the later components of evoked potential would appear; late responses up to 0.5 s or more seem necessary to elicit sensory experience (*see* Libet *et al.*, 1967, 1975), and they would provide durations of cortical activity much longer than the primary cortical response to a single peripheral stimulus.

Much of the recent evidence to test for unconscious detection has employed indirect techniques, such as 'semantic activation without conscious identification' (e.g., Holender, 1986), or masking procedures to block awareness of signals (e.g., Marcel, 1983); interpretations of these experiments have been the target of much argument (*see* Holender, 1986) although the issue has in part been confused by differing definitions of 'subliminal' and 'awareness' (e.g., Dixon, 1986). Holender's proposal, that 'responding discriminately to a stimulus' could be a sufficient definition of conscious identification, is contradicted by our experimental results. It would seem also contradicted by the reported ability to respond to a skin stimulus with a reaction time that is independent of the subject's awareness of that stimulus (Taylor and McCloskey, 1990).

Direct evidence for detection of peripheral sensory inputs without awareness also has been found under other circumstances, as in the demonstration of 'blindsight' in which destruction of primary sensory cortex essentially eliminates the subject's conscious visual experience (Weiskrantz *et al.*, 1974; Weiskrantz, 1986; Stoerig and Cowey, 1989). The significant feature of the present contribution is to demonstrate a role for duration of repetitive input, at cerebral levels, in distinguishing detection-without-awareness from with-awareness. Such a role could not be directly demonstrated with a peripheral stimulus, with which durations of *cerebral* activations can only be minimally and uncertainly controlled. By contrast, stimuli in ventrobasal thalamus provided a potentially large range of durations in which production of awareness was mostly restricted to the longer durations of stimulus.

*Greater stimulus durations required for awareness*

The main hypothesis tested here had two components, one of which was discussed above. We can conclude that the evidence also strongly supports the other component, that the transition from detection-without-awareness to detection-with-awareness requires a significant increase in train duration of repetitive ascending volleys initiated in ventrobasal thalamus.

It should be emphasized that intensity (peak current) remained constant in any given series of trials; only the TD (number of pulses) was varied. There is no doubt that the electrically most excitable neural elements responding to each stimulus pulse were axons, either those arriving from the medial lemniscus and/or those ascending from the thalamic neurons to cerebral cortex (or to other forebrain structures). These axons are said to be predominantly in the small myelinated fibre range (H. J. Ralston III, personal communication). Myelinated axons are relatively insensitive to repetitive excitation with the brief TDs and the pulse frequency (72/s) employed (*see* Ochs, 1965, pp. 45–49). Increases in subliminal excitability persist for $< 1$ ms, so that one would not expect progressive recruitment of more firing axons during each train of stimulus pulses here. That is, each stimulus pulse (at a given intensity in that series of trials) would elicit delivery of essentially the same number of nerve impulses to the forebrain structures; only the number of such identical volleys—the duration of their repetition—would vary between stimuli with different TDs. Dropping out of some firing axons during each stimulus train is a possibility, if sufficient postfiring subnormality develops; but such an eventuality would not significantly affect our conclusions. It should also be noted that the absence of either sensory awareness or detection cannot be attributed to an inability of a stimulus to excite any axons; all stimuli, regardless of TD, must be exciting at least some ascending axons, since awareness of a sensory experience is elicited by a sufficiently long TD in all series.

Evidence that a longer TD is required to elicit awareness, than for simple detection, is seen qualitatively even from the raw data (Tables 1, 2). However, the nonparametric analysis provided rigorous and definitive evidence for the difference in TD requirements of simple detection as opposed to awareness. A relatively large mean number ($\alpha$) of *additional* pulses (27.8 p, or 0.39 s TD) was required to move from mere detection to elicit *any* even questionable awareness (level 2). Furthermore, the values for $\alpha$ were highly significant in almost all sessions for all subjects, as seen from the values of z (Table 6; *see also* fig. 4B). It should be emphasized that $\alpha$ was calculated for uncertain questionable awareness level 2 (*see* discussion above). The greater TD requirement for any awareness is therefore unambiguous; it does not depend upon achieving the more definite report level 1 and is independent of any argument that all reports of level 2 may be a form of actual awareness.

*Signal detection theory and 'criterion' for response.* It might be argued on signal detection theory that the distinction between detection without and with awareness is based on a difference in the 'criterion' adopted by the subject when making each type of response. But signal detection theory was developed specifically in relation to *detection* (Green and Swets, 1966); awareness was not directly studied or involved, and any extension of the theory to awareness would require ad hoc assumptions about the neural mediation of awareness.

The implication in the signal detection argument would be that subjects are really aware of the signal in some fashion at all degrees of detection but that *reportability* of awareness requires a higher criterion. Since subjects report no awareness of such a condition, the proponents would have to assume that they simply will not or cannot report this awareness; such an assumption is nonfalsifiable and without merit when it contradicts observation. When the subject reports that he feels nothing it would be a distortion of the primary evidence to insist, on the basis of a theory, that he really felt something (Libet, 1973, 1985, 1987, 1989). The subject's introspective report is the only valid primary evidence for his conscious subject experience (e.g., Libet, 1987). A possible secondary argument, that to *verbalize* an introspective report requires more stimulus input than the nonverbal response employed for detection, cannot be made for our study. Subjects indicated in every trial both their forced-choice selection ($L_1$ vs $L_2$) and their awareness level by nonverbal acts in pressing different sets of buttons. Actually, the verbal descriptions of awareness level given by subjects, after making their nonverbal responses, matched the latters' intended meanings. It may also be noted that they reported experiencing differing durations of the conscious sensation elicited by the different stimulus durations; namely, a 750 ms train felt longer than a 500 ms train, etc. This reported awareness of variable brevities of sensation would argue against the possibility that reports of no awareness with brief stimulus durations represented dismissal of actual but brief sensations.

It might be suggested that a 'criterion' for awareness was demonstrated operationally by our experimental finding that awareness does require substantially longer stimulus durations (in the thalamus) than does simple detection. But in doing so any implication that such a greater requirement is set consciously by the subject must be avoided (*see above*). Also, we are *not* proposing that there is an absolute minimum duration of repetitive cerebral activation required to elicit awareness; this duration can be altered by increasing intensity of input and, at least conceivably, by the subject's motivation and attention. Any such alterations in required duration, however, would presumably also be achieved by unconscious processes, i.e., without the subject's awareness of a new 'criterion' for duration of cerebral activities. Consequently, the use of the 'criterion' concept would provide, at best, only a semantically different way of describing our findings. Simple detection can remain unconscious, and the greater duration of repetitive ascending activations required to elicit some awareness of the signal would reflect a meaningful physiological difference, not the continuous, qualitatively identical process for both detection and awareness proposed by some signal detection theorists.

*Meaning of 'time-on' theory.* The 'time-on' theory was based on several independent lines of evidence and makes predictions for a number of different though fundamental relationships between subjective timing of events and actual behaviour (e.g., Libet, 1989). One such prediction, which served as the specific hypothesis tested in the present study, is that a discontinuity will be found between duration of cerebral input producing detection-with-no awareness and duration for detection-with-awareness. A substantial discontinuity of great statistical significance (*see* the values of $\alpha$ and z in the nonparametric analysis) was indeed demonstrated. The specific hypothesis tested in this study was potentially refutable, had no difference been found between durations required for simple detection vs sensory awareness. The 'time-on' or duration of cortical activation can thus be a controlling factor in determining whether a mental function such as detection proceeds unconsciously or with conscious awareness.

The conclusion is not intended to mean that cerebral 'time-on' is the exclusive determinant of the transition between conscious and unconscious functions. Additional or alternative factors could include specificities of cerebral areas, or of dynamic activity patterns, as previously discussed (Libet, 1989). Attentional, motivational or psycho-dynamic processes could influence the transition by acting through the 'time-on' feature. Increase in intensity of an input to the cortex can reduce the 'time-on' required for awareness (Libet *et al.*, 1964; Libet, 1973, 1982). The present study was deliberately carried out with near-liminal intensities in input in order to maximize the potential role of the duration of repetitive inputs ('time-on'), in differentiating conscious and unconscious functions. The goal in the present study was to compare the *relative* TD requirements of simple detection and sensory awareness, in the same subject during the same trials, under a given set of psychological conditions. Had we performed the experiments with progressively supraliminal intensities we would expect, within limits, to have progressively briefer ranges of stimulus durations within which to distinguish detections-with from detections-without awareness.

There is, however, evidence against the possible suggestion that any integrative mechanism sensitive simply to intensity and duration produces awareness, instead of some more specific role in this for 'time-on' per se. (1) If stimuli to ventrobasal thalamus or S-I cortex (postcentral gyrus) are just *below* the liminal intensity (for eliciting sensory experience with long stimulus durations of $> 1$ s), then no conscious sensation can be elicited even with durations up to 5 s or longer (Libet *et al.*, 1964; Libet, 1973). Such 'subliminal' intensities are not below threshold for eliciting neuronal responses; substantial electrophysiological responses of large populations of neurons are recordable with each such 'subliminal' stimulus pulse. Were simple integration of intensity and duration the controlling mechanism, a sufficiently long train duration of stimulus pulses would be expected to become effective for awareness. (2) At a liminal intensity which becomes effective with an average 0.5 s of train duration, the neuronal responses recordable electrically at the cortex exhibit no progressive alteration during the train and no unique event at the end of this effective train (Libet, 1973, 1982). Obviously, not all the possible neuronal activities were recordable, but this evidence at least offers no support for a progressive integrative factor. (3) The minimum train duration that can elicit awareness, when intensity is raised as high as possible, has not firmly been established although it would appear to be in the order of 100 ms. However, it was empirically quite definite that a single stimulus pulse localized to the medial lemniscus could not elicit any conscious sensation no matter how strong (Libet *et al.*, 1967); this was true even when the intensity of the single pulse was 20–40 times the strength of the liminal I sufficient with a 0.5 s train of pulses. Although intensity of stimulus may not necessarily correlate linearly with the number of axons excited here, the effectiveness of 10 pulses (at 20/s) at liminal I contrasted with the ineffectiveness of a single pulse at 40 times this liminal I argues against the proposal of a simple integrative mechanism.

Regardless of the potential contribution from intensity of input or of activation to the production of a conscious experience, the present findings demonstrate that 'time-on' of cortical activation can be one controlling factor in this production and in distinguishing an unconscious detection function from detection with conscious awareness. It would seem likely that many cerebral processes mediating mental function proceed at lower and relatively minimal levels of intensity, judging from normal patterns of recordable electrophysiological and metabolic activities relative to levels that are possible in states of hyperactivity such as seizures. If this is so, the quantitative role of the 'time-on' factor would be substantial.

Finally, 'time-on' theory implies (among other things) that conscious sensory awareness can lag behind the real world by as much as 0.5 s (depending on intensity of input). This counter-intuitive implication has been specifically addressed in an experimental fashion (Libet *et al.*, 1979). That earlier work produced evidence for a subjective referral of sensory experiences, from their delayed neural time of production backwards to the time of the initial fast projection signal, so that there would be no appreciable delay in the *subjective* timing of sensory events.

## ACKNOWLEDGEMENTS

We wish to express our deep gratitude to the patients whose cooperation and interested participation as subjects made the investigation possible. We are also indebted to Dr Christoph Schreiner for helpful initial consultations, to Sharon Lamb, RN, for her friendly and cheerful assistance in locating suitable subjects and stimulator devices, and to Mr Jen-fu Maa at Ohio State University for assistance in carrying out computer calculations of the statistical values. The work was supported by USPHS grant NS-24298.

## REFERENCES

Bartlett JR, Doty RW (1980) An exploration of the ability of macaques to detect microstimulation of striate cortex. *Acta Neurobiologiae Experimentalis,* **40,** 713–727.

Dixon NF (1986) On private events and brain events. *Behavioral and Brain Sciences,* **9,** 29–30.

Green DM, Swets JA (1966) *Signal Detection Theory and Psychophysics.* New York: Wiley.

HOLENDER D (1986) Semantic activation without conscious identification in dichotic listening, parafoveal vision, and visual masking: a survey and appraisal. *Behavioral and Brain Sciences*, **9**, 1–66.
HOSOBUCHI Y (1986) Subcortical electrical stimulation for control of intractable pain in humans: report of 122 cases (1970–1984). *Journal of Neurosurgery*, **64**, 543–553.
JOHANSSON RS, VALLBO ÅB (1979) Detection of tactile stimuli. Thresholds of afferent units related to psychophysical thresholds in the human hand. *Journal of Physiology, London*, **297**, 405–422.
LEVY, RM, LAMB S, ADAMS JE (1987) Treatment of chronic pain by deep brain stimulation: long term follow-up and review of the literature. *Neurosurgery*, **21**, 885–893.
LIBET B (1965) Cortical activation in conscious and unconscious experience. *Perspectives in Biology and Medicine*, **9**, 77–86.
LIBET B (1973) Electrical stimulation of cortex in human subjects, and conscious sensory aspects. In: *Somatosensory System. Handbook of Sensory Physiology*, Volume II. Edited by A. Iggo. Berlin: Springer, pp. 743–790.
LIBET B (1981) The experimental evidence of subjective referral of a sensory experience backwards in time. *Philosophy of Science*, **48**, 182–197.
LIBET B (1982) Brain stimulation in the study of neuronal functions for conscious sensory experiences. *Human Neurobiology*, **1**, 235–242.
LIBET B (1985) Unconscious cerebral initiative and the role of conscious will in voluntary action. *Behavioral and Brain Sciences*, **8**, 529–566.
LIBET B (1987) Consciousness: conscious, subjective experience. In: *Encyclopedia of Neuroscience*. Edited by G. Adelman. Boston: Birkhäuser, pp. 271–275.
LIBET B (1989) Conscious subjective experience vs unconscious mental functions: a theory of the cerebral processes involved. In: *Models of Brain Function*. Edited by R. M. J. Cotterill. Cambridge: Cambridge University Press, pp. 35–49.
LIBET B, ALBERTS WW, WRIGHT EW, DELATTRE LD, LEVIN G, FEINSTEIN B (1964) Production of threshold levels of conscious sensation by electrical stimulation of human somatosensory cortex. *Journal of Neurophysiology*, **27**, 546–578.
LIBET B, ALBERTS WW, WRIGHT EW, FEINSTEIN B (1967) Responses of human somatosensory cortex to stimuli below threshold of conscious sensation. *Science*, **158**, 1597–1600.
LIBET B, ALBERTS WW, WRIGHT EW, LEWIS M, FEINSTEIN B (1975) Cortical representation of evoked potentials relative to conscious sensory responses, and of somatosensory qualities—in man. In: *The Somatosensory System*. Edited by H. H. Kornhuber. Stuttgart: Thieme, pp. 291–308.
LIBET B, WRIGHT EW, FEINSTEIN B, PEARL DK (1979) Subjective referral of the timing for a conscious sensory experience: a functional role for the somatosensory specific projection system in man. *Brain*, **102**, 193–224.
MARCEL AJ (1983) Conscious and unconscious perception: an approach to the relation between phenomenal experience and perceptual processes. *Cognitive Psychology*, **15**, 238–300.
OCHS S (1965) *Elements of Neurophysiology*. New York: John Wiley.
SHEVRIN H, DICKMAN S (1980) The psychological unconscious: a necessary assumption for all psychological theory? *American Psychologist*, **35**, 421–434.
STOERIG P, COWEY A (1989) Wavelength sensitivity in blindsight. *Nature, London*, **342**, 916–918.
TAYLOR JL, MCCLOSKEY DI (1990) Triggering of preprogrammed movements as reactions to masked stimuli. *Journal of Neurophysiology*, **63**, 439–446.
WEISKRANTZ L (1986) *Blindsight: A Case Study and Implications*. Oxford: Clarendon Press.
WEISKRANTZ L, WARRINGTON EK, SANDERS MD, MARSHALL J (1974) Visual capacity in the hemianopic field following a restricted occipital ablation. *Brain*, **97**, 709–728.

*(Received June 19, 1990. Revised October 30, 1990. Accepted December 6, 1990)*

## APPENDIX*

*Nonparametric analysis*

In order to eliminate the possibility that the validity of our conclusions would depend on the uncertain assumptions of a statistical model, a nonparametric technique was developed which relies only on assumptions built into the nature of the experiment. For example, when no pulses are administered, a subject has a 50/50 chance of being correct due to the randomization in setting $L_1$ or $L_2$ as correct. Secondly, we may assume that the administered pulses can only increase the chance of a correct response.

Focusing on the possibility of an incorrect choice, define the quantity Y as the number of pulses required to move an otherwise incorrect response just past the threshold into correct. From moment to moment, a different number of pulses, Y, may be required to insure a correct response and we call the expected value of this random variable $\theta$.

In order to estimate the parameter $\theta$, notation is needed for the observable random variables:

$$I_y = \begin{cases} 1 \text{ if subject gives a correct response when given y pulses} \\ 0 \text{ otherwise.} \end{cases}$$

Taking the expected value of this indicator variable, and using the two assumptions above, gives:

$$\begin{aligned} E[I_y] &= P[\text{subject gives a correct response when given y pulses}] \\ &= P[\text{would be correct with 0 pulses}] + \\ &+ P[\text{no. of pulses administered} \geq Y] \cdot P[\text{would be incorrect with 0 pulses}] \\ &= 0.5 + P[Y \leq y](0.5). \end{aligned}$$

Hence $P[Y \leq y] = 2E[I_y] - 1$.

Since Y is a nonnegative valued random variable

$$\theta = B - \int_0^B P[Y \leq y]dy$$

where B is a number for which $P[Y > B] = 0$ (that is, when we administer B pulses we must be *certain* that the subject will give a correct response). In our work we assumed that a correct response is automatic when at least $B = 57$ pulses are administered. Substituting the result above gives

$$\theta = 2\{57 - \int_0^{57} E\,[I_y]dy\}. \tag{1}$$

A natural estimate of $E[I_y]$ is given by the proportion of times a subject is correct when given y pulses. This estimate has variance

$$\sigma_y^2 = E[I_y]\{1 - E[I_y]\}/n_y$$

where $n_y$ is the number of trials with y pulses.

Finally, to compute the estimate of $\theta$ we just substitute the estimates of $E[I_y]$ into formula (1) above, using linear interpolation between the observed y values (that is, use the trapezoidal rule to approximate the integral).

We can also gauge the variability of $\hat{\theta}$ (the estimator of $\theta$). In particular, we use the fact that the estimator is just a linear combination of the $E[I_y]$ estimates and that these estimates are independent random quantities. Mathematically, we put our estimator in the form

$$\hat{\theta} = 114 - \Sigma w_y\, E[\hat{I}_y]$$

which gives us

$$\text{Var}[\hat{\theta}] = \Sigma w_y^2\, \sigma_y^2 \tag{2}$$

($w_y$ depends on $n_y$ and on the incremental number of pulses between administered values). The estimated standard errors which are reported in Table 5 come from substituting the estimates of $E[I_y]$ into the formula for $\sigma_y^2$ in (2) and taking the square root. Also, since $E[I_y]$ is a number between 0 and 1 we know that $\sigma_y^2$ cannot be larger than $1/(4\, n_y)$. Substituting this limit into formula (2) and taking the square root, gives the reported values for the upper limit of the SEs.

The validity of this theory depends only on the assumption that the different trials were administered independently (this is reasonable since, for example, the number of pulses delivered was randomized).

Statistical theory tells us that the sampling distribution of $\hat{\theta}$ will approximately follow the normal distribution. This gives validity to inferential statements that can be made. For example, a conservative 95% confidence interval for $\hat{\theta}$ is given by $\hat{\theta} \pm 2$[upper limit for SE($\hat{\theta}$)].

*Additional theory behind the estimates of $\alpha$*

The random variable Y describes the minimum number of pulses it would take to generate detection. The parameter $\alpha$ is then defined as the number of *additional* pulses it would take for the subject to achieve possible awareness (i.e., just past the threshold into awareness level 2). The key assumption that was needed for the estimation technique described below is that $\alpha$ is constant over the course of a single session and therefore independent of Y.

Since we assume the certainty of a correct response when more than 57 pulses are administered:

$$\alpha = \int_{\alpha}^{57+\alpha} E[I_y]dy - \int_{\alpha}^{57} E[I_y]dy = \int_{0}^{57} E[I_{y+\alpha}]dy - \int_{\alpha}^{57} E[I_y]dy. \quad (3)$$

An interesting consequence of our key assumption is that the chance that a subject gives a correct response when administered $y+\alpha$ pulses is the same as the chance of being correct or just-guessing-and-incorrect when administered only y pulses.

Now we may proceed in a similar fashion as with the estimation of $\theta$. In particular, we can estimate $E[I_{y+\alpha}]$ by the proportion of times a subject is correct or just guessing when given y pulses. The integrals in expression (3) are then estimated by substituting the estimates of the expectations and interpolating between observed values (as before each integral is approximated by a weighted sum using the trapezoidal rule). Finally, $\alpha$ is estimated using an iterative procedure to find the value satisfying (3). The variability in this estimator of $\alpha$ and its approximate normal sampling distribution can be seen by expressing the final answer as a linear combination of independent proportions. The upper limit for the SE of $\alpha$ given in Table 6 again relies on the fact that

$E[I_{y+\alpha}](1-E[I_{y+\alpha}]) \leq 0.25$ since $E[I_{y+\alpha}]$ is between 0 and 1.

* Contributed by Dennis K. Pearl.

# The Neural Time – Factor in Perception, Volition and Free Will

We should all agree that the brain and the mind are intimately linked. One need only drink some wine to remind one of the physicochemical controls of mental function, acting via the brain. And one need merely decide at will to flick a finger, or not to flick it, at any time one wishes to do so, to demonstrate thet a mental event can apparently « order » the brain to produce an action. Yes this kind of prima facie evidence of a mutually causative interaction between mental and cerebral processes has been subjected to an immense amount of philosophical analyses and arguments over its « true » basis. These ranged from materialist views of conscious mental functions as meaningless epiphenomena to dualist, spiritualistic views of a separable mental entity with an ability to control neural function. Althought a causative role for conscious mental processes is readily accommodated within a dualist interactionist theory of mind and brain (e.g. Popper and Eccles, 1978) it has also been argued within the framework of a monist determinist theory (e.g. Sperry, 1980).

Clearly, a knowledge of how the mind and brain are in fact interrelated is fundamental to our views of the nature of man and the human experience. The question of the « how » in this relationship is potentially amenable to scientific experiments investigation. The question of « why » any given relationship exists is a more metaphysical one and presumably not addressable experimentally. But experimental findings about the nature of the mind-brain relationship can at least provide constraints to which any metaphysical theories should be asked to conform.

In our experimental approach to the issue I set for myself two epistemological principles which I believe must be followed.

1. *Introspective report as the operational criterion*

Conscious experience, as an awareness of some thing or some event is directly accessible only to the individual having that experience, not to an external observer. In one version of identity theory, subjective experience (the « mental ») is termed an « inner quality » of the systemn whereas the externally observable neural events (the « physical ») are referred to as an « outer quality » (e.g. Pepper, 1960). Consequently, only an introspective report by the subject can have primary validity as an operational measure of a subjective experience. The « transmission » of an experience by a subject to an external observer can, of course involve distortions and inaccuracies. Indeed there can be no absolute certainty (objectively demonstrable by externally observable events) about the validity of the report (Libet, 1987). (Of course, absolute certainty does not hold even for physical observations). But in all social interactions we accept introspective reports as evidence of a subject's experience, thought we may subject it to some analyses and tests which affect our acceptance of their validity. This same general approach can be applied more rigorously in a scientific investigation. We restricted our study to the simplest kinds of sensory and volitional experiences, not involved with emotional issues and subject to rigorous tests of reliability. An important corollary of this principle is that any behavioral evidence which does not require a convincing introspective report cannot be assumed to be an indicator of conscious, subjective experience ; this is so regardless of the purposeful nature of the action or of the complexity of cognitive and abstract problem-solving processes, since all of these can and often do proceed unconsciously, without awareness by the subject.

2. *No a priori rules for mind-brain relationship*

Mental events and externally observable (physical) events constitue two categorically separate kinds of phenomena. They are mutually irreducible categories in the sense that one cannot, *a priori*, be described in terms of the other. This leads to the principle that there are no *a priori* rules or laws that can describe the relationship between neural-brain events and subjective-mental events. No doubt rules do exist, but they must be discovered. They can only be established by simultaneous observation of both phenomenological categories of events.

It also follows that even a complete knowledge of all the neural-physical events and conditions (in the unlikely case that were possible) would not, in itself, produce a description of any associated subjective experience (Libet, 1966 ; Nagel, 1979). This would constitute a flat rejection of a reductionist view that an adequate knowledge of neuronal functions and strutures would be sufficient for defining and explaining consciousness and mental activities (e.g. Churchland, 1981).

However, one would hope that scientific discoveries of the actual relationships between mental and cerebral-neural processes would lead to theoretical frameworks for the relationships. In such theories, comprehensible patterns of brain function would describe, predict or account for our ability to move at will, perceive consciously, think, feel etc, and vice-versa (i.e. how mental processes of these kinds may provide analogous accounts of brain functions). Such a goal is becoming increasingly realizable with progress in several lines of neuroscientific investigation. Our own direct experimental attack on the issue has yielded evidence of a neural time-factor which may provide a sinificant part of a theoretical framework for the mind-brain relationship.

## THE NEURAL TIME FACTOR IN PERCEPTION

### Cerebral sensory stimulation

We embarked on our long range investigation of the issue by attempting to determine what kinds of activation patterns, in a primary sensory area of cerebral cortex, seemed uniquely required for the production of a conscious sensory experience. This was done by controlling the electrical stimuli applied to somato-sensory cortex in awake human subjects, via electrodes implanted subdurally during therapeutic neurosurgical procedures. It was already well known that electrical stimulation there can elicit somatic sensations which are subjectively referred by the patient to a localized body structure on the side contralateral to the stimulus site. In adopting this experimental approach we were in accord with a suggestion by the great British neuroscientist Lord Adrian (1952). He stated « I think there is reasonable hope that we may sort out the particular activities that coincide with quite simple mental processes like seeing or hearing » (or, in our studies, bodily sensations). « At all events that is the first thing a physiologist must do if he is trving to find out what happens when we think and how the mind is influenced by that goes on in the brain ».

It became evident that duration, of the train of repetitive brief electrical pulses, was a critical factor in producing a conscious sensory response to stimulation of sensory cortex (see Libet, 1973 review). A surprisingly long stimulus duration of about 500 msec was required when the peak intensity of the pulses was at the liminal level for producing any conscious sensation. Changes in frequency of the pulses, electrode contact area or polarity, etc. did not affect the requirement of a long duration. Nor was this requirement unique to locating the stimulus on the surface of cerebral cortex. The same temporal relationship was found with stimulating electrodes placed at any point in the cerebral portion of the direct ascending sensory pathway that leads up to the somatosensory cortex, - i.e. in subcortical white matter, or in the sensory relay nucleus in ventrobasal thalamus or in the medical lemniscus bundle that ascends from nuclei in the medulla to the thalamic nucleus. However, the requirement does not hold for the spinal cord below the medulla, or indeed for peripheral sensory nerves from the skin itself ; there, a single pulse can be effective for sensation even at liminal intensity.

The average 500 msec requirement in cerebral structures is not a rigidly fixed one ; with intensities stronger than the liminal level, briefer stimulus durations can be effective. But a single pulse could not elicit a sensation even with very strong intensities (up 40 times the liminal intensity needed for a 500 msec stimulus train). It appears probable that a minimum duration of about 100 msec is required for even the strongest inputs.

The ineffectiveness of a stimulus duration less than the minimum, (for example, a 400 msec stimulus when 500 msec is necessary at a liminal intensity) is not due to an absence of neural excitation by the ineffective brief stimulus duration; large electrophysiological responses are recordable throughout the stimulus train from the start. This favors, though it does not prove, the hypothesis that the neural code representing conscious sensation involves a minimum duration of appropriate neural activities, rather than the production of some unique neural end-event (e.g. Libet, 1965, 1989). (See further discussion of this below). It also suggests an all-or-nothing character for conscious experience. That is, awareness of an event would not develop in a gradual progressively increasing manner with the series of neural responses to an input; rather, awareness would appear relatively sharply only when an adequate duration of neural activities is achieved (see *Fig. 1*). This could make conscious experiences of events have a unitary quality rather than that inherent in the « stream of consciousness » originally proposed by William James.

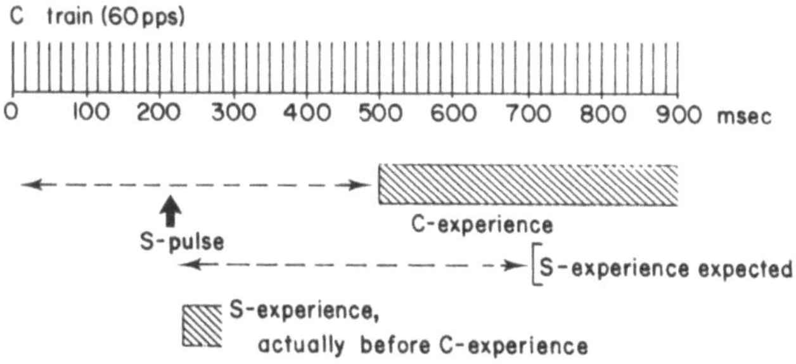


Fig. 1. – Subjective time order of a cerebrally-induced vs. a skin-induced sensation. Diagram of experiment in which a stimulus train at the somatosensory cerebral cortex (C) was paired with a single stimulus pulse to the skin (S). C stimulus consisted of repetitive brief pulses (at 60 per sec), each at the same lowest (liminal) intensity sufficient to elicit any reportable conscious sensory experience. Note that, under these conditions, there was *no sensory experience* at all *until about 500 msec* of the C stimulus train had elapsed; and the sensory experience then proceeds without change in its weak subjective intensity as the stimulus continues.

At 200 msec after the start of the C stimulus train a single near-threshold stimulus pulse (S) was given to the skin on the hand opposite to that to which the cortically-induced sensation was referred. Previous evidence had indicated that there should be a delay of up about 500 msec before cerebral « neuronal adequacy » for eliciting the skin-induced sensation is achieved. The subject *reported* that this skin-induced sensation actually appeared before the C-induced experience.

To explain this it was postulated that the skin-induced sensation is subjectively referred back to the time of its primary cortical response, i.e. about 15 msec after the S pulse (see Fig. 2). This hypothesis was tested by changing the location of the C stimulus, to the fast subcortical pathway (medial lemniscus) that delivers the sensory message from skin to cerebral cortex. Unlike the cortical stimulus, each stimulus pulse in medial lemniscus elicits the same primary cortical response that is evoked by the skin pulse. But, unlike the skin, medial lemniscus requires 500 msec of repetitive pulses (at liminal intensity) to elicit a conscious sensation. Nevertheless, the subject reported that the medial lemniscus-induced sensation appeared subjectively before the skin-sensation, with the same relative pairing of the stimuli shown in the figure ; this confirmed the hypothesis.

## Peripheral sensory inputs

Although a single stimulus pulse to the skin can elicit a conscious sensory experience there are several lines of evidence for the view that there is actually a cerebral neural delay of up to about 500 msec before the experience can appear. We must distinguish between the required duration of the peripheral sensory activation and that of the required cerebral responses to that peripheral input. A single peripheral nerve volley leads to an early « primary » evoked response in sensory cortex, after a latency of about 10-20 msec. But this is followed by a series of endogenous cortical neural responses that persist for 500 msec or more. The primary evoked response is not sifficient for a conscious sensory experience; the later components are required (see Libet, 1973, 1982). Similar persisting endogenous cortical activity is not generated by a single pulse in the cerebral sensory pathway; this would explain the requirement for the longer duration of repetitive stimulus pulses there.

There are other lines of evidence that a substantial neural delay is required before cerebral activity achieves adequacy for the appearance of a conscious sensation that is induced even by a single peripheral sensory stimulus (Libet 1981, 1982), but they will not be detailed here. Such a neural delay does, however, raise the question of whether there is also a corresponding *subjective* delay for the sensory experience.

## Subjective timing of a sensory experience

Intuitively, we do not experience sensory events as having delays relative to their occurrence. In a direct test of this issue we paired a skin stimulus (single pulse) with a cortical stimulus train of pulses (at liminal intensity, 500 msec duration required), and asked the subject to report which of these stimulus – induced sensations came first. Subjects in fact reported that the skin – induced sensation came first even when the skin pulse was applied a few hundred msec *after* the onset of the cortical train (*Fig. 1*). That is, there appeared to be no subjective delay for the skin-induced sensation relative to the delayed cortically-induced sensation. To reconcile this finding with the evidence that there is a similar neural delay in both cases (see above), we proposed that there is a subjective referral of the timing of the skin-induced experience, from its actually delayed appearance back to the time of the initial fast response of the cortex (which has only a 10-20 msec, indiscernable delay) (*Fig. 2*). The experience would thus be subjectively « antedated » and would seem to the subject to have occurred without any delay. No such antedating would occur with the cortical stimulus since the normal initial or primary response, to a sensory volley ascending from below, is not generated with our sur-

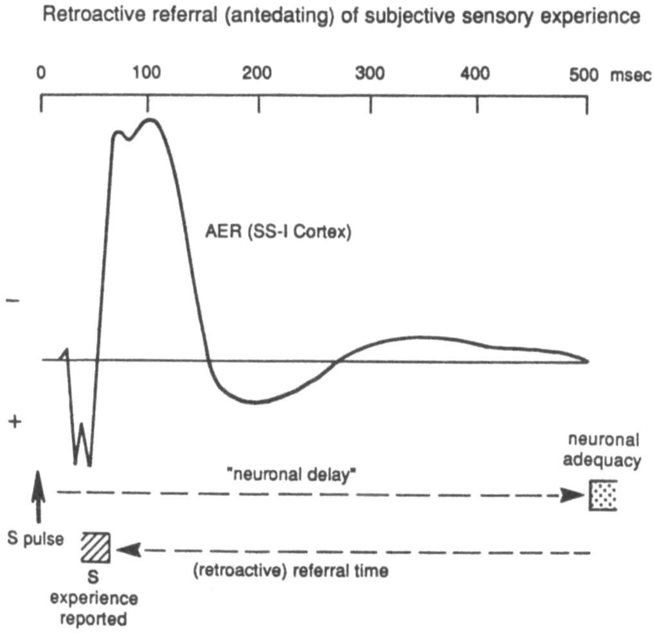


Fig. 2. – Diagram of subjective referral of a sensory experience backward in time. The averaged evoked response (AER) recorded at SI (somatosensory) cortex was evoked by single pulses just above threshold for sensation (at about 1/sec., 256 averaged responses) delivered to skin of contralateral hand. Below the AER, the first line shows the approximate delay in achieving the state of neuronal adequacy that appears (on the basis of other evidence) to be necessary for eliciting the sensory experience. The second line shows the apparent retroactive referral of the subjective timing of the experience (as reported by the subject) from the time of neuronal adequacy backward to some time associated with the primary surface-positive component of the evoked potential. This would explain the subject's reporting the sensations as simultaneous, whether elicited by a single pulse to the skin or by a 0.2 to 0.5 sec train of pulses in medial lemniscus, while a sensation elicited by a similar train of pulses toSI cortex is reported as delayed.

The primary component of AER is relatively highly localized to an area on the contralateral postcentral gyrus in these awake human subjects. The secondary or later components, especially those following the surface negative component after the initial 100 to 150 msec of the AER, are wider in distribution over the cortex and more variable in form even when recorded subdurally (see, for example, Libet *et al.*, 1979). It should be clear, therefore, that the present diagram is not meant to indicate that the state of neuronal adequacy for eliciting conscious sensation is restricted to neurons in primary SI cortex of postcentral gyrus ; on the other hand, the primary component or « timing signal » for retroactive referral of the sensory experience would be a function more strictly of this SI cortical area. (The later components of the AER shown here are small compared to what could be obtained if the stimulus repetition rate were lower than 1/sec and if the subjects had been asked to perform some discriminatory task related to the stimuli.) (From Libet *et al*, 1979).

face-cortical stimuli. This hypothesis was tested and confirmed as follows: A skin timulus was paired with one in the direct subcortical pathway that leads to sensory cortex (Libet et *al.,* 1979). Unlike the cortical stimulus, the subcortical one does elicit the initial primary response in the cortex with each pulse; but, unlike the skin stimulus, it resembles the cortical one in its requirement of a long (up to 500 msec) train duration of pulses. As predicted by the hypothesis the subcortically-induced sensation was reported to have no delay relative to the skin – induced sensation, even though it was empirically established that the subcortically-induced experience could not have appeared before the end of the stimulus train (whether 200 or 500 msec, depending on intensity); see (*Fig. 1*).

## Initiation of a voluntary act

Does a substantial neural delay apply to conscious experiences other than perceptual sensory ones ? It became experimentally feasible to test this for the initiation or appearance of the conscious intention to perform a voluntary act. The results not only supported the existence of a comparable neural delay in this mental function but also led to some far-reaching implications for the nature of volition and of free will.

### *Neural delay for conscious intention*

The possibility to investigate this issue had its roots in the discovery by Kornhuber and Deecke (1965; Deecke *et al.*, 1976) that an electrical change (the « readiness-potential » or RP) is recordable on the head starting up to a second or more prior to the actual muscle activity in a simple « self-paced » movement. Accepting this electrical change as an indicator of brain activity that is involved in the onset of a volitional act, we asked the question: Does the conscious wish or intention to perform that act precede or coincide with the onset of the preparatory brain processes, or does the conscious intention follow that cerebral onset ?

We first established that an RP is recordable even in a fully spontaneous voluntary act, with an average onset of about –550 msec, before the first indication of muscle action (« O time »). (The presence of a component of « preplanning » when to act makes RP onset even earlier; this probably accounts for the longer RP values given by others). We then devised and tested a method whereby clock-time indicators of the subject's first awareness of his/her wish to move

(« W ») could be obtained reliably. RPs (neural processes) and Ws (the times of conscious intention) were then measured simultaneously in large numbers of these simple voluntary acts (Libet *et al.*, 1983).

The results clearly showed that onset of RP precedes W by about 350 msec (Fig. 3). This means that the brain has begun the specific preparatory processes for the voluntary act well before the subject is even aware of any wish or intention to act; i.e. the volitional process must have been initiated unconsciously (or non-consciously). (Discussions of such a conclusion may be seen in Commentaires by others at the end of my article, in Libet, 1985). This sequence is also in accord with the principle of a substantial neural delay in the production of a conscious experience generally. The issue of conscious control of the act will be considered below.

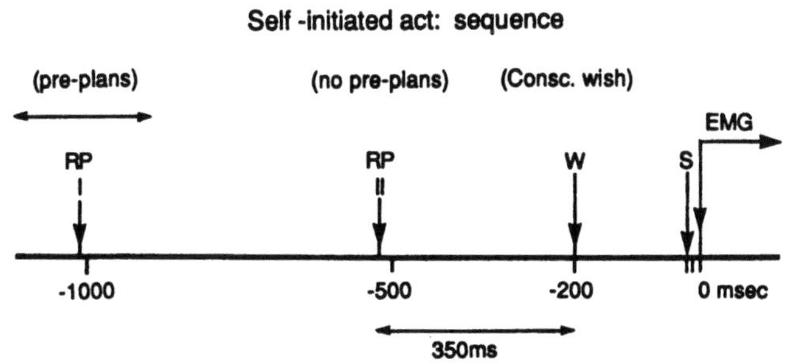


Fig. 3. – Diagram of sequence of events, cerebral and subjective, that precede a fully self-initiated voluntary act. Relative to 0 time, signalled by the onset of EMG (electromyogram) of the suddenly activated muscle, the RP (readiness-potential, a recordable indicator of related cerebral neuronal activities) begins first, at about – 1050 msec when some preplanning is reported (RP I) or about – 550 msec with spontaneous acts lacking immediate preplanning (RP II). Subjective awareness of the wish to move (W) appears at about – 200 msec, some 350 msec after onset even of RP II but well before the act (EMG). Subjective timings reported for awareness of the randomly delivered S (skin) stimulus averaged about – 50 msec relative to actual delivery time.

## CONSCIOUS AND UNCONSCIOUS MENTAL FUNCTIONS

It is generally agreed that we do not have conscious access to all (or perhaps even most) of our mental activities. Unconscious (or non-conscious) mental functions include all varieties of cognitive, evaluative and decision-making operations which can be shown to go on without awareness, i.e. they are not reportable as introspective experiences. They would even include learning and recalling of learned responses, and thought processes even of the most complex, abstract and creative nature. (Recall the anecdotes by the great French mathematician Poincare about his unconscious solving of major problems, the solution of which then sprang fully formed into his consciensness). This broad view of unconscious mental functions extends far beyond a more limited one of Freudian « repression » of emotionally difficult thoughts. However, it would not include those cerebral operations which are « non-mental », in the sense that they do not ever achieve a potentiality to rise into awareness; these would include control processes for blood pressure, heart rate, automatic postural adjustments, etc.

How, then, does the brain distinguish between conscious and unconscious mental events ? An obvious possibility would be that different areas or structures of brain are involved in representing the two types of events. Indeed, the neurosurgeon Wilder Penfield (1958) (to whom we owe much of the mapping of human cerebral functions) had proposed that the centrencephalic system of the brain is responsible for the integration and expression of conscious experience. The centrencephalic system constitutes the central, gray core of the upper brain stem, including that in the thalamus but below the cerebral hemispheres. There is no doubt that normal function of this system is essential for the maintenance of the conscious state and probably for mediating shifts in attention and vigilance. But being a necessary condition does not make it a sufficient one. That is, given all the background of brain structure and function required to be able to elicit a conscious experience, are there either specific structures or kinds of neural activities which can then elicit or represent the conscious mental even? On the basis of our experimental findings I have proposed a relatively simple theory for this issue.

## Cerebral « time-on » theory

The theory states that the transition, from an unconscious mental event to one that reaches awareness and is consciously experienced, can be a function of a sufficient increase in the duration (or « time-on ») of appropriate neural activities (Libet, 1989). That is, appropriate neural activities whose duration is below some minimum substantial duration could mediate a mental function that remains unconscious; but when such activities persist for longer than a minimum time of up to about 500 msec, subjective awareness of the mental function can appear. The changes in durations of the appropriate neural activities may be affected, for example, by changes in the attention process.

It should be clear that the theory does not exclude other important or even controlling distinctions between processes mediating conscious vs. unconscious mental functions. It is only proposed that the « time-on » factor is superimposed, as a critical requirement, upon any other differentiating factors. Indeed, one wants at present to avoid designing any specificity or mechanism into the theory other than the « time-on » factor *per se*. On the other hand, the theory does imply that both conscious and unconscious mental events can be elaborated or represented in the same cerebral areas or structures. Such a potentiality would be in good accord with the fact that the psychological processes (cognition, evaluation, goal-directed decision, complexity, etc.) can be basically similar, in both conscious and unconscious functions. The chief difference is simply in the existence of awareness of the mental event.

There are two basic neural-mental propositions inherent in the theory, each of which is experimentally testable and falsifiable. (1) A minimum duration of appropriate neural activity, of up to about 500 msec depending on conditions, is required in order to elicit a conscious experience or awareness of an event. The available evidence already supports this proposition, as indicated above. (2) When appropriate neural activity has a duration briefer than that required for awareness, it may still mediate an unconscious mental function, without any subjective awareness of it. A direct experimental test of one prediction of the theory has been carried out (Libet *et al.*, 1991). This involved human subjects in whom stimulating electrodes had been chronically implanted in the somatosensory (ventrobasal) nucleus of the thalamus, for the therapeutic control of certain kinds of intractable pain. At a given fixed liminal intensity, each brief electrical pulse of a stimulus train excited essentially the same nerve fibers and delivered the same amount of ascending sensory input to the somatosensory cortex. Only the duration of the train of repetitive pulses was varied among different trials. The results clearly demonstrated that subjects could correctly detect the presence or absence of a stimulus in a forced choice paradigm, even when durations of repetitive inputs were too brief to elicit any conscious sensory awareness of the stimulus, i.e. even when they were « guessing ». The mean stimulus duration for a correct response, of what should otherwise have been incorrect on a randon chance basis, was about 250 msec. But, in order to obtain correct answers *and* also sensory awareness of the stimulus, a substantially greater average stimulus duration of about 500 msec was required. The results confirmed the theoretical expectations for both of the neural-mental propositions stated above. In other words, the transition between psychological detection of a sensory signal without awareness and the detection with awareness can be controlled simply by differences in duration of repetitive ascending activations of sensory cortex.

## CONSCIOUS SENSORY EXPERIENCE AND PERCEPTION

How do our experimental findings of the neural time-factor and the resulting « cerebral time-on » theory additionally affect the concepts of sensory experience and perception? Perception is usually formally regarded as a conscious sensation related to some particular input. If the neural « time-on » requirement can be generalized for all modalities of sensation, we would be constrained to adopt a rather remarkable view of sensory perceptions: All of our sensory experiences are delayed with respect to the time of the sensory stimulus; i.e., we do not experience the world in real time. The delays would be variable, depending on the intensity of the input and other factors. But the delays would be substantial, ranging from a probable minimum of about 100 msec up to about 500 msec (and even 1000 msec in some individuals).

However, the neural delays imposed on the generation of a sensory experience need not be reflected in the subjective content of that experience. As noted above, the experimental evidence supports the view that the neurally delayed sensory experience is subjectively antedated, i.e. referred backwards in time to the early initial cortical response to the fast, direct sensory pathway. Thus, the experience would subjectively appear to have occurred without any appreciable delay after the input signal. For a group of different stimili there are likely to be varying neural delays for generating the different experiences in the group. That could lead to a temporal jitter (asynchrony) in this group of sensory experiences when the various sensory stimuli are in fact applied synchronously. However, if all these asynchronously produced experiences are subjectively antedated to the initial fast cortical response for each, they would subjectively be experienced as synchronous, without jitter.

Such subjective referral thus serves to « correct » the temporal distortion of the real sensory event, a distortion imposed by the cerebral requirements of a neural delay for the experience. An analogous subjective correction occurs for the spatial distortion of the real sensory image, imposed by the spatial representation of the image the responding neurones of the sensory cortex. Nowhere in the brain is there a response configuration that matches the sensory image as perceived subjectively. Both the spatial and the temporal subjective referrals are probably learned processes. There is direct evidence for the learning in the case of spatial referral of a visual image; when the visual image is altered by glasses that rotate the field or displace its horizontal or vertical position, the subjective image corrects back to a more « real » one after some days. One would expect that in some period of infant life there are indeed spatial and temporal distortions in the subjective experiences of the real world.

Subjective referrals of the timing or the spatial configuration of an experience are clear examples of an event in the mental sphere that was not evident in or predictable by a knowledge of the associated neural events. This illustrates the second epistemological principle stated above. On the other hand, subjective referral should not be regarded as an illusion. It provides the normal way in which we perceive temporal and spatial features in a more accurate relationship to the real world; but an illusion constitutes a perception distorted from the real stimulus configuration by inputs that lead to abnormal or unusual neural configurations.

« *Unconscious perceptions* »

Behavioral responses to sensory signals or images can be made within as little as 100 msec. As the responses may involve recognition, evaluation and decision in a meaningful way they are not infrequently said to follow a perception of the signal. Since the usual definition of perception includes conscious experience, this loose usage of the word is often accompanied by the assumption that quick, meaningful behavioral responses are also made after consious recognition of the signal.

The « time-on » theory implies that such quick responses are occurring before conscious awareness of the signal could have developed. Such an implication has in fact been experimentally confirmed (e.g. Taylor and McCloskey, 1990). Unconscious responsiveness even to complicated sensory signals is in accord with our common experiences in daily activities and in sports. Most of us have at times driven an automobile for some distance, responding properly to various traffic signals and requirements with no reportable awareness of all that for some interval.

*Unconscious mental functions in general*

In a cerebral « time-on » theory for conscious awareness in general, unconscious mental processes could proceed quickly. This would be so since they would not require the fuller « time-on » needed for a conscious one. Such a feature is obviously advantageous not only for quick meaningful reactions to signals but also for the ability to carry on creative, complex and intuitive thinking. Given a neural delay of some hundreds of msec for developing conscious awareness of a mental event, it is clearly of great adaptive value that much or most of our thought processes proceed unconsciously, without the lenghtier neural requirements for awareness. Note, however, that when an endogenous unconscious mental process reaches conscious awareness, as when a solution to an unconsciously processed problem springs into consciousness, there would not be any subjective antedating of the experience ; subjective temporal referral is postulated only for sensory experiences and then only when the early « timing signal » is present.

Unconscious modulation of the content of a conscious experience is also accommodated by time-on theory. It is well known that the content of the introspectively reportable experience of an event may be modified considerably in relation to the content of the actual signal, whether this be an emotionally laden sensory or endogenous mental event (which may be fully repressed, in Freud's terms). In order for this to happen, some delay between the initiating event and the appearrance of the conscious experience of it is essential. The « time-on » theory provides a built-in basis for the appropriate delays. We have produced some direct experimental evidence for such modulatory actions on the awareness of a simple sensory signal from the skin; in this, an appropriate cortical stimulus was begun up to 200 msec or more after the skin pulse, but could still either inhibit or enhance the sensory experience elicited by the skin stimulus. (Libet, 1978, 1982).

## CONTROL OF VOLUNTARY ACTION; FREE WILL

The experimental evidence indicates that a voluntary act is initiated in the brain unconsciously, before the appearance of the *conscious* intention (see above). The question then arises, what role if any does the conscious process itself have in volitional actions? (In this we are considering only the processes immediately involved in the performance of a voluntary movement. The issue of conscious deliberation or planning of what, whether and when to act is a separate one; if and when such forethought leads to a voluntary act, then we have the case under consideration in our investigation). First, the immediate initiation of the voluntary act appears to be an unconscious cerebral process. Clearly, free will or free choice of whether « to act now » could not be the initiating agent, contrary to one widely held view. This is of course also contrary to each individual's own introspective feeling that he/she consciously initiates such voluntary acts, this provides an important empirical example of the possibility that the subjective experience of a mental causality need not necessarily reflect the actual causative relationship between mental and brain events.

However, we must distinguish the initiation of a process leading to a voluntary action from *control of the outcome* of that process. The experimental results showed that W (conscious wish to act) did appear at about minus 200 msec, i.e. before the motor act, even though it followed the onset of cerebral process (RP) by about 350 msec. This permits at least a potential role for the conscious function, in determining whether the volitional process will go on to completion. That could come about by a conscious choice either to promote the culmination of the process into an action (whether passively or by a conscious « trigger »), or to prevent the progress to action by a conscious blockade or « veto ». The potentiality for such conscious veto power, within the last 100-200 msec before an anticipated action, was experimentally demonstrated by us. It is also in accord with a common subjective experience, that one can veto or stop oneself from performing an act after a conscious urge or wish to perform it has appeared (even when the latter is sudden and spontaneous).

Assuming that one can extrapolate these results to volitional acts generally, they do not exclude a possible role for free will, even though the volitional process starts with unconscious cerebral activity. However, the potential role of free will would be constrained; it would be changed from being an initiator of the voluntary act to one only of controlling the outcome of the volitional process, after the individual becomes aware of an intention aware of an intention or wish to act now. In a general sense, free will could only select from among the brain activities that are a part of a given individual's makeup.

Such a « self-control » role for free will is actually in accord with much of ethical strictures propounded in many religious and philosophical systems (e.g. in existentialism). On the other hand some systems hold the individual responsible for having merely a conscious desire or wish to act, even without actually performing the act. The individual is thus held morally guilty of being wicked if the urge in question is a socially or theologically unacceptable one. Such a view would appear to impose an insoluble burden on people, if the intentions are initiated unconsciously with no conscious control being possible (except possibly in an indirect general way by procedures that might affect such unconscious initiatives). In any case, it is only the actual performance of a voluntary movement that can have a practical impact on others, and so a potentiality for conscious self-control of performance would still serve the purposes of functional ethical systems.

## Time-on theory, conscious control and free will

If we generalize the « time-on » theory to apply to all mental functions, a serious potential difficulty arises if it should also apply to the initiation of the *conscious control* of a volitional outcome. As discussed above, the evidence indicates that the volitional process, leading to « action now », is initiated in the brain well before the appearance of the conscious intention (W) to act; this is in accordance with « time-on » theory. But, once W is present, the subject can consciously control the outcome of that volitional process, for example by vetoing, i.e. blocking its final expression as a motor act. If the conscious control function itself is initiated by unconscious cerebral processes, one might argue there is no role at all for conscious free will even as a controlling agent.

However, conscious control of an event is not the same as becoming aware of the volitional intent. Control implies the imposing of a change, in this case *after* the appearance of W (the conscious awareness of the wish to act now). In this sense conscious control may not necessarily require the same neural « time-on » feature that may precede the appearance of awareness *per se*. We should recognize that there is presently no specific experimental test of the possibility that conscious control may require an unconscious cerebral process to produce it. Given the difference between a control and an awareness phenomenon, a proposed absence of the requirement for conscious control qould not be in conflict with a general « time-on » theory for awareness. Thus, a potential role for free would remain viable in the conscious control, though not in the initiation, of a voluntary act.

## General conclusion

It should be evident, from the studies described in this and in other papers of this volume, that the question of how neural and mental events are inter-related is amenable to rigourous experimental investigation. Our findings have shown that the production of a conscious experience involves some unique requirements of neuronal activities and that much cerebral neural activity proceeds without being able to elicit any conscious experience. The discovery of a substantial neural time factor, in the production of conscious experiences (as opposed to unconscious mental functions), imposes both restrictions and opportunities for philosophical theories of the mind-brain relationship. This refers especially to those theories attempting to deal with conscious, subjective experiences, and affects the ways in which concepts of perception, conscious volition and free will may be acceptably developed.

Benjamin LIBET

RÉFÉRENCES

ADRIAN, E.D. (1952), « What happens when we think », *in The Physical Basis of Mind*, ed. Laslett, P., Oxford, Basil Blackwell.

CHURCHLAND, P.S. (1981), « On the alleged backwards referral of experiences and its relevance to the mind-body problem », *Philosophy of Science*, 48, p. 165-181.

DEECKE, L., GROZINGER, B. et KORNHUBER, H.H. (1976), « Voluntary finger movement in man / Cerebral potentials and theory », *Biological Cybernetics*, 23, p. 99-119.

KORNHUBER, H.H. et DEECKE, L. (1965), « Hirnpotentialänderungen bei Willkürbewegungen und passiven Bewegungen des Menschen : Bereitschaftspotential und reafferente Potentiale », *Pflugers Archiv fur Gesamte Physiologie*, 284, p. 1-17.

LIBET, B. (1965), « Cortical activation in conscious and unconscious experience », *Perspectives in Biology and Medicine*, 9, p. 77-86.

LIBET, B. (1966), « Brain stimulation and the threshold of conscious experience », *in Brain and conscious experience*, J.C. Eccles (ed.), Springer, p. 168-181.

LIBET, B. (1973), « Electrical stimulation of cortex in human subjects, and conscious sensory aspects », *in Handbook of sensory physiology*, A. Iggo (ed.), Springer, p. 743-790.

LIBET, B. (1978), « Neuronal vs. subjective timing for a conscious sensory experience », *in Cerebral correlates of conscious experience*, P.A. Buser et A. Rougeul-Buser (ed.), Elsevier/North Holland Biomedical Press, p. 69-82.

LIBET, B. (1981), « The experimental evidence for subjective referral of a sensory experience backwards in time », *Philosophy of Science*, 48, p. 182-197.

LIBET, B. (1982), « Brain stimulation in the study of neuronal functions for conscious sensory experiences », *Human Neurobiology*, 1, 235-242.

LIBET, B. (1985), « Unconscious cerebral initiative and the role of conscious will in voluntary action », *Behavioral and Brain Sciences*, 8, p. 529-566.

LIBET, B. (1987), « Consciousness : Conscious, Subjective Experience », *in Encyclopedia of Neuroscience*, G. Adelman (ed.), Birkhauser, Boston, p. 271-275.

LIBET, B. (1989), « Conscious subjective experience vs. unconscious mental functions : A theory of the cerebral processes involved », *in Models of brain function*, R.M.J. Cotterill (ed.), Cambridge U. Press, p. 35-49.

LIBET, B., GLEASON, C.A., WRIGHT, E.W. et PEARL, D.K. (1983), « Time of conscious intention to act in relation to onset of cerebral activities (readiness-potential) ; the unconscious initiation of a freely voluntary act », *Brain*, 106, p. 623-642.

LIBET, B., PEARL, D.K., MORLEDGE, D.M., GLEASON, C.A., HOSOBUCHI, Y. et BARBARO, N.M. (1991), « Control of the transition from sensory detection to sensory awareness in man by duration of a thalamic stimulus », *Brain*, 114, p. 1731-1757.

LIBET, B., WRIGHT, E.W., JR., FEINSTEIN, B. et PEARL, D.K. (1979), « Subjective referral of the timing for a conscious sensory experience : A functional role for the somatosensory specific projection system in man », *Brain*, 102, p. 193-224.

NAGEL, T. (1979), *Mortal questions*, Cambridge University Press.

PENFIELD, W. (1958), *The excitable cortex in conscious man*, Liverpool University Press.

PEPPER, S.C. (1960), « A neural - identity theory of mind », *in Dimensions of Mind*, S. Hook (ed.), New York University Press, Washington Square, p. 37-56.

POPPER, K.R. et ECCLES, J.C. (1977), *The self and its brain*, Springer.

SPERRY, R.W. (1980), « Mind-Brain interaction : Mentalism, yes ; dualism, no », *Neuroscience*, 5, p. 195-206.

TAYLOR, J.L. et MCCLOSKEY, D.I. (1990), « Triggering of preprogrammed movements as reactions to masked stimuli », *J. Neurophysiology*, 63, p. 439-446.

# Epilogue
# I. Some Implications of "Time-on" Theory

The "time-on" theory (Libet 1989; Libet et al. 1991) proposes a controlling neural factor in the production and coding of a conscious mental event (with awareness), as distinguished from that for an unconscious/non-conscious mental event (with no subjective awareness of it). Some implications of the theory were listed but they merit elaboration, in view of their broad fundamental nature.

## 1. Our Sensory World is Delayed

The subjective experiences elicited by sensory signals would be delayed by substantial times, if they require periods of neuronal acitivites of up to about 0.5 sec before cerebral "adequacy" is achieved for producing a conscious experience. However, the actual delays are subjectively "corrected" for by subjectively antedating the experience back to the time of the cortical response to the initial fast projection afferents impulses (with small delays of 20 msec ±). The *content* of the sensory experience is thus normally altered, in relation to the fast initial timing signal, in a way that makes the actually delayed experience appear to occur without significant delay.

## 2. All Mental Events may Begin Unconsciously

We have found that a substantial duration of cerebral activity is required not only to produce the experience or awareness of somatosensory signal but also for appearance of an endogenous mental event, the conscious intention to act. We could suggest extrapolating these findings to apply to the production of any kind of awareness (a principle requiring further experimental investigation). In such a case all mental events or processes would begin unconsciously. That would includc thc initiation of thoughts of which one would become conscious only if the requirements of cerebral duration are met. One should note that the awareness of any endogenous mental event would not be subjectively antedated, as in the absence of antedating for the intention to act. Subjective referral backwards in time has been found only for a sensory input and then only when the fast ascending sensory projection is present and functions to elicit a timing signal in the form of a primary cortical evoked response.

## 3. Conscious Events Have an "all-or-nothing," Discontinuous Character

In the experiments that provided evidence for the cerebral "time-on" requirement for a conscious experience two important features are noteworthy: (i) Durations of cortical activations that were less than the minimun requirement (about 0.5 sec at liminal levels) elicited considerable neuronal responses which were clearly insufficient to produce a conscious experience. (ii) But with such subminimal durations (of < 0.5 sec) the subject reported feeling or aware of absolutely nothing. That is, there were no reports of any experience building up gradually, even of anything vague developing into something definite. A sensory experience either appeared after a sufficiently long activation even if very weak, or it was not present at all with a briefer activation. (Of course, detection of the signal without any awareness of it was demonstrable with durations insufficient for awareness, see Libet et al., 1991).

Such evidence implies that conscious events must arise discontinuously and that the onset (though not the intensity or persistence) of a conscious experience has an all-or-nothing character. This would mean that the common notion of a "stream of consciousness" (discussed by William James and many others) is not strictly compatible with the evidence and requires some re-evaluation. In our terms, any experience of a stream of consciousness would have to be based on some temporal overlapping in a sequence of actually discontinuous conscious events.

## 4. Cerebral Structures for Awareness

The question of which dynamic, physiological neural activities may be unique mediators in eliciting a conscious, subjective experience was the chief focus of our experimental studies. However, there is another fundamental question about which specific cerebral structures are uniquely involved in this conscious process, and whether unconscious mental functions may be mediated by structures that are the same or different from those for conscious functions. The question of the locus of conscious experience fell into some disrepute when speculative answers like that of Descartes (the pineal body) were found experimentally untenable. But the question is nevertheless a legitimate and fascinating one when approached more scientifically.

Clearly, many cerebral structures are *necessary* for conscious experience. Among these are the centrencephalic system, essentially located in the core of the brain stem and thalamus (and the outer reticular nucleus of the thalamus) which provides the reticular activating system etc. Lesions or abnormal activity in this system can lead to loss of all conscious functions; this, and the retention of consciousness with large losses of cerebral cortex tissue, led the neurosurgeon Penfield to propose the centrencephalic system as the site of consciousness. But being necessary to a function does not necessarily make a site *sufficient* for or the actual locus of that function. Normal cardiac function is also necessary for maintenance of conscious function, but that does not place the conscious function in the heart (although this has been mistakenly proposed in earlier times).

However, some cerebral areas appear to be more uniquely involved than others for subjective awareness at least of sensory inputs. For example, lesions of the primary visual area of cortex have long been known to produce subjective blindness; it has more recently become clear that it is visual awareness that is specifically lost since objects in the affected visual field can still be correctly detected and responded to without any reportable awareness (the "blindsight" phenomenon described by Weiskrantz, 1986). This phenomenon distinctly separates structural mediation of the awareness function from the function of simple sensory detection (with no awareness) and the ability to respond to such detection. But even in this case one cannot conclude that conscious visual experience is necessarily located in or directly generated by the calcarine, visual cortex; this cortical area may only be uniquely necessary to such conscious functions and the latter may actually be generated elsewhere or in a cerebrally more global fashion.

In any case, our proposed "time-on" factor could be an operative and controlling requirement in any specially involved cerebral area. On the other hand, the "time-on" theory raises the interesting possibility that there is not any exclusively sufficient site where a conscious experience develops. That is, the controlling factor or coding for the appearance of a subjective experience could simply lie in the "time-on" or duration of suitable neural activities, regardless of the specific cortical or other area. Conscious and unconscious mental processes, if distinguished from each other by the "time-on" factor, could potentially both be developed by the same cerebral areas. This could obtain whether the immediate and sufficient production of a conscious experience (i.e., awareness of an event) resides in certain special structures that mediate all conscious events or in any one of many areas or structures for various events, or in a cerebrally more global process required for all awareness.

## 5. "Filter" Function

Most sensory inputs, even if they reach cerebral cortical levels, do not enter conscious awareness. If one of the controlling requirements for awareness is the achievement of a sufficiently long duration of neural activities ("time-on"), this requirement would provide a kind of filter mechanism which can prevent achieving awareness by all sensory inputs that do not achieve the required cerebral duration. The "time-on" requirement can thus be part of the mechanism which permits conscious awareness to be uncluttered and to focus on one or a few specific issues at a time.

Mechanisms that affect the duration of cortical responses to a sensory input would provide the critical controls in this. Selective attention mechanisms are thought to provide at least one major influence on what reaches awareness. The "time-on" theory would propose that attentional and other such controlling mechanisms may act by changing the durations of cortical responses to given inputs; selective increase in durations of some responses could convert them from unconscious to conscious status.

## 6. Unconscious Mental Functions Proceed Speedily

The "time-on" theory proposes that unconscious mental functions may be mediated by relatively brief durations of neural activity, even 100 msec or less in duration. In contrast to conscious, subjectively experienced mental events, the unconscious ones could proceed in rapid sequences. Such a distinction would seem to be in accord with how we view the apparent speed of unconscious vs. conscious mental functions. Unconscious functions would include problem solving, intuitive thoughts, creative thinking, etc. It is commonly believed that such mental functions proceed usually and more successfully at unconscious levels. Brief time-on requirements would clearly be advantageous in such functions.

## 7. Quick Behavioral Responses are Initially Unconscious

Meaningful behavioral responses to selective signals can be made within as little as 100 msec or less, depending on the complexity of the signal and of the selective processing required. Reaction-time tests have provided the chief quantitative approach in studying quick responses to a signal. However, quick meaningful responses are made frequently in daily routine activities, like avoiding obstacles in walking or running, in responding to sudden signals while driving an automobile, etc. They are essential and predominant in most sports, such as tennis, baseball, football, boxing, etc.; in these it is commonly accepted that the immediate fast responses to sensory signals are made unconsciously and that the intrusion of conscious awareness fatally delays the responses. It should be emphasized that all such quick behavioral responses, including reaction times, involve detection and recognition of the sensory signal, appropriate processing of the signal's significance and making a decision on whether to respond or not.

The "time-on" theory proposes that such quick meaningful responses would be developed unconsciously, before awareness of the sensory signal has been neurally achieved. That durations of cerebral activity too brief for awareness can nevertheless subserve meaningful detection of and response to the signal, was demonstrated in paper no. 21 (Libet et al. 1991). There have also been direct confirmations of the proposition that reaction times to a sensory signal can be the same whether the subject is or is not aware of the signal (Taylor and McCloskey, 1990). Furthermore, when subjects are asked to slow their reaction time deliberately, a procedure that presumably requires conscious awareness of the signal, it was found that the reaction times could only be slowed discontinuously, to values suddenly more than 300–400 msec greater than those obtained in the usual manner with no deliberateness involved (Jensen, 1979). A favorite thought ("Gedanken") example of mine would be that of the racer who may start running within perhaps 100 msec or less after the starting gun-shot signal; such a subject would putatively start running before becoming consciously aware of the shot but would later report having heard it before take-off (because of subjective referral that antedates the experience—see above).

## 8. Modulation of a Conscious Experience

It is well known that the *content* of an introspectively reportable experience of an event may be modified considerably in relation to the actual content of the signal or originating mental events. A repression or distorted awareness of certain emotionally laden sensory images or endogenous mental events had of course been postulated by Freud, who backed this up with psychoanalytical analyses. But modulations at a simpler level have also been demonstrated under controlled experimental conditions. These include retroactive masking of a visual signal when followed by another appropriate stimulus at intervals up to 200 msec, and the effect on the meaning or nature of a signal when it is preceded by a "priming" signal for which the subject has no awareness (e.g., Holender, 1986). We had ourselves demonstrated that awareness of a single stimulus pulse to the skin could be retroactively either depressed or enhanced, by a suitably placed cortical stimulus train that began more than 200 msec after the skin stimulus (Libet et al., 1972; Libet, 1978; Libet et al., 1992a – paper no. 10).

In order for any modulations of the experimental content to take place there clearly must be some delay between the initiating cerebral event and the development and actual appearance of the conscious experience. Without a substantial delay, there would be no possibility for neural activities (that are "aroused" by the signal, or imposed belatedly by external inputs) to intervene to affect the *content* of the experience as introspectively reported. (Any *ad hoc* assumption, that subjects are concealing some immediate conscious experience and reporting only a modified version, is unwarranted on the available evidence). The "time-on" theory, provides a built-in physiological mechanism for appropriate neural delays needed to accommodate modulatory influences brought to bear after arrival of a signal.

## 9. Conscious Control of Voluntary Action

On the basis of our experimental studies (Libet et al. 1982b, 1983a,b; Libet 1985) I concluded (a) that the performance of even a freely voluntary act is initiated unconsciously, some 350 msec before the individual is consciously aware of wanting to move, but also (b) that conscious control of whether the act will actually be performed is still possible during the remaining 150 to 200 msec before activating the muscles. This would appear to preserve the possibility for at least a controlling role for conscious free choice or will, and thereby also a basis for individual conscious responsibility for the act. However, the argument has been made that the conscious control process, whether a vetoing or permissive/triggering one, may itself require unconscious cerebral processes to precede its own development and appearance (e.g., McKay, Doty, Wood, 1985). If this were the case, the awareness of which control to apply would have followed an unconscious initiation of the control process; conscious control would then be no different, in principle, from the initiating volitional process, in that both could not be regarded as consciously free in origin.

In making such an argument it is assumed that awareness of any kind requires a substantial prior period of activity, as indeed proposed on a general basis by the "time-on" theory (Libet, 1965; 1989b). However, there may be an important distinction between awareness (of the wish to act) and conscious control (of a volitional outcome). Conscious control appears only after awareness of the wish to move has developed; the control process depends upon prior awareness of the volitional direction but it is not an awareness in itself. On this basis, the "time-on" theory would not necessarily apply to the phenomenon of conscious control. Obviously, suitably functional brain processes must precede as well as mediate any conscious (or unconscious) mental function. However, sudden appearance of conscious control and of the set of mediating brain processes involved in such control, without a *specific* sequence of preceding processes, can be shown to be compatible with either monist or dualist theories of mind-brain interaction. There is no logical imperative in any mind-brain theory that requires specific neural activity to precede the appearance of a conscious control function. Even identity theory would be compatible with the occurrence of sudden spontaneous neural patterns that were immediately associated with conscious events. In an issue as fundamentally important to our view of human conduct and identity, we should not rely on *a priori* judgements of how the process must function, without further evidence.

The argument about conscious control in relation to preceding processes may take another form, as set forth particularly by Doty (1985). In this, the question is raised as to how the conscious control function would "know" what will ensue (if it vetoes or permits the process to go on), "if the preparatory movement is wholly outside (a concomitantly developing) conscious control" (Doty, 1985). A reply to this reasonable argument may take two related forms: (1) Even after the delayed appearance of conscious awareness of intention to act, there remain 100–150 msec in which the conscious function could evaluate the options and decide on whether to veto, etc. We are still far from being able to say with any confidence how much time a conscious function needs to perform these tasks. (2) Another option, that would be compatible with Doty's position, would be that awareness of the potential consequences of the developing voluntary act may also be developing and arising unconsciously, within the same matrix of neuronal activities. In such a case, when awareness of the whole process does appear it could include being conscious of both the intention/wish to act and the options of the consequences of the act. A conscious control of these options, following on and distinguished from the awareness of them, could then appear and fulfill its role without other specific preceding activities.

*Free will and responsibility.* The foregoing discussion is clearly of fundamental importance to views of free will and of individual responsibility for one's actions. Since these issues are crucial to our views of the human condition my position has been to avoid falling into apparently "scientifically-based" conclusions on one side or the other when, in fact, this involves *a priori* judgements (biased by one's metaphysical views) rather than experimental analysis and evidence. It is essential to note, therefore, that the present evidence does not exclude the existence of free will or a basis for individual responsibility, although it puts constraints on the permissible nature of these propositions.

Free will in the usual and traditional sense is taken to include a *conscious* intention to act and a *conscious* ability to control the act, i.e., whether and when to act or not to act. With this view, one would conclude from the present evidence that free will (if it exists) does *not initiate* a voluntary act, but the possibility of free will controlling the actual outcome of the unconscious cerebral initiative (by veto, or permissiveness etc.) is not excluded by the evidence and remains viable. This conclusion refers to the volitional process that leads immediately and directly to actual motor action, not to deliberative processes planning a course of future possible voluntary action (see Libet, 1985).

The intuitive feelings about these phenomena are among the most difficult to deal with in the scientific study of mind-brain interaction, and great care should be taken not to make allegedly scientific conclusions which actually depend upon hidden ad hoc assumptions. The subjective introspective experience of free will, to initiate and control voluntary actions, would appear to provide *prima facie* evidence that the conscious mind can affect neuronal functions. However, there is the alternative possibility that the content of the introspective experience has itself been formed by neuronal activities and is thus an effect rather than a causal agent in this process. Such an alternative would remain an ad hoc construction which cannot take precedence or "over-rule" the indications from the primary experiential report, unless it is somehow experimentally tested. It is just such a test, of the nature of initiation of a freely voluntary act, that we carried out (Libet et al. 1983b; Libet, 1985). As already discussed, that test led to the startling conclusion that the act is initiated by the brain unconsciously, in spite of the experimental reports, although a control function for conscious will remained possible. This illustrates how assumed relationships between conscious activity and brain processes may be subjected to experimental analysis. It is even possible to conceive of experiments to test whether conscious mental activity can in fact causally modulate or control neuronal functions, a cardinal issue in all such theories (see the following essay on a "testable theory of mind-brain interaction").

In conclusion, I would re-emphasize the principle that, for any scientific investigation of the relation of conscious experience to brain to be valid, introspective reports must be part of in the experimental design. I allude to the principles made in the Preface and elsewhere in the book: 1) Subjective experience is in a phenomenologically different category from the physical brain processes; the experience is an "inner quality," not directly accessible to an external observer and only indirectly via the introspective report. The problem of possible distortion of the experiential contents during the process of translating it into a report must be recognized but can be dealt with. When the experience under study is a relatively simple and unemotional one, and when the introspective reports exhibit sufficient reliability in the face of manipulative challenges by the experimenter, one is entitled to have confidence in the credibility of the measurements even though they cannot achieve an objectively absolute status of validity. 2) There are no *a priori* rules that can describe the relationship between the mental and the physical. Any such rules, and many certainly exist, must be experimentally discovered. Such experimental discovery was the objective of the work reported in this book.

# Epilogue
# II. A Testable Theory of Mind–Brain Interaction

One of the most mysterious and seemingly intractable problems in the mind-brain relationship is that of the unitary and integrated nature of conscious experience. We have a brain with an estimated 100 billion neurons, each of which may have thousands of interconnections with other neurons. It is increasingly evident that many functions of cerebral cortex are localized. This is not merely true of the primary sensory areas for each sensory modality of the motor areas which command movement, and of the speech and language areas — all of which have been known for some time. Many other functions now find other localized representations, including visual interpretations of color shape and velocity of images; recognition of human faces; preparation for motor actions, etc. Localized function appears to extend even to the microscopic level within any given area. The cortex appears to be organized into functional and anatomical vertical columns of cells, with discrete interconnections within the column and with other columns near and far, as well as with selective subcortical structures. This columnar view began with findings by Mountcastle and co-workers in the 1950s and has been greatly extended by him and others; for example, there are the columnar localizations of visual shapes and motions and of binocular vision as discovered by Hubel and Wiesel.

In spite of the enormously complex array of localized functions and representations, the conscious experiences related to or elicited by these neuronal features have an integrated and unified nature. Whatever does reach awareness is not experienced as an infinitely detailed array of widely individual events. It may be argued that this amazing discrepancy between particularized neuronal representations and unitary integrated conscious experiences, should simply be accepted as part of a general lack of isomorphism for the relationship between mental and neural events. But that would not exclude the possibility that some unifying process or phenomenon may mediate the profound transformation in question.

The general problem had been recognized by many others, going back at least to Sherrington (1940) and probably earlier. Eccles (in, Popper and Eccles, 1977, p. 362) specifically proposed that "the experienced unity comes not from a neurophysiological synthesis but from the proposed integrating character of the self-conscious mind." This was proposed in conjunction with a dualist–interactionist view in which a separate non-material mind could detect and integrate the neuronal activities. Some more monistically inclined neuroscientists have also been arriving at related views (e.g., Sperry, 1952, 1980; Doty, 1984) i.e., that integration seems to be best accountable for in the mental sphere even if one views subjective experience as an inner quality of the brain "substrate" (as in "identity theory") or as an emergent property of it. There has been a growing consensus that no single cell or group of cells is likely to be the site of a conscious experience, but rather that conscious experience is an attribute of a more global or distributed function of the brain. Recent discovery of a widespread synchronization of oscillatory neuronal responses to certain visual configurations (Gray and Singer, 1989; Singer, 1991) has led to some speculation that a "correlation" model might represent the neural coding for recognizing a unified image in an otherwise chaotic background. This speculation is still to be tested.

A second apparently intractable problem in the mind-brain relationship involves the reverse direction. There is no doubt that cerebral events or processes can influence, control and presumably "produce" mental events, including conscious ones. The reverse of this, that mental processes can influence or control neuronal ones, has been generally unacceptable to many scientists on (often unexpressed) philosophical grounds. Yet, our own feelings of conscious control of at least some of our behavioral actions and mental operations would seem to provide *prima facie* evidence for such a reverse interaction, unless one assumes that these feelings are illusory. Eccles (1990; Popper and Eccles, 1977) proposed a dualistic solution, in which separable mental units (called psychons) can affect the probability of presynaptic release transmitters. Sperry (1952, 1969, 1980) proposed a monistic solution, in which mental activity is an emergent property of cerebral function; although the mental is restrained within a macro-deterministic framework it can "supervene" though not "intervene" in neuronal activity. However, both views remain philosophical theories, with explanatory power but without experimentally testable formats.

As one possible experimentally testable solution to both features of the mind-brain relationship, I would propose that we may view conscious subjective experience as if it were a field, produced by appropriate though multifarious neuronal activities of the brain.

## The Conscious Mental Field (CMF)

A chief quality or attribute of the CMF would be that of a unified or unitary subjective experience. A second attribute would be a causal ability to affect or alter neuronal function. The additional meaning or explanatory power of describing subjective experience in terms of a CMF will become more evident with the proposed experimental testing of the theory. That is, the CMF is proposed as more than just another term for referring to "unified subjective experience."

The putative CMF would *not* be in any category of known physical fields, such as electromagnetic, gravitational, etc. The conscious mental field would be in a phenomenologically independent category; it is not describable in terms of any externally observable physical events or of any known physical theory as presently constituted. In the same sense as for subjective experience, the CMF would be detectable only in terms of subjective experience, accessible only to the individual who has the experience. An external observer could only gain valid direct evidence about the conscious mental field from an introspective report by the individual subject. In this respect the conscious mental field would differ from all known physical fields whose existence and characteristics are derived from physical observations.

On the other hand, the proposed CMF should be viewed as an operational phenomenon, i.e., as a working and testable feature of brain function. It is not proposed as a view of the metaphysical origin and nature of the mind; indeed, it could be shown to be potentially compatible with virtually any philosophical mind-brain theory. The CMF may be viewed as somewhat analagous to known physical fields. For example, a magnetic field is produced by electric current flowing in a conductor, but it can in turn influence the flow of the current. However, as indicated, the CMF cannot be observed directly by external physical means.

The proposed interaction between brain and CMF differs from the "unitary hypothesis of mind-brain interaction" proposed by Eccles (1990). Eccles postulates that each putative unit of mental function (a "psychon") is associated with a specific neural aggregate (a "dendron"); the present theory does not postulate such a specific and fixed relationship. In Eccles' theory the question of how neural activation is translated into a mental event is dealt with by hypothesizing a specific synaptic-psychon interaction. He proposes that when the synaptic input to a dendrite makes it possible for a "psychon (to) successfully select a (presynaptic) vesicle for exocytosis (that is for release), the 'micro-success' is registered in the psychon for transmission through the mental world." In the present theory the appropriate (presently unspecified) neural activity directly contributes some alteration in the overall CMF; the contribution does *not* depend upon an action by the mental phase (the psychon, in Eccles' theory) on synaptic function.

How is the CMF attribute of unified subjective experience related to its production by contributions from local neuronal areas? Local alterations in the CMF would be reflected in a changed overall field, but there would not be a separately required mechanism for transmission and integration of such local contributions. To think in terms of a transmission and integrative process would be to continue thinking in terms of the externally observable neural events. To do so would be a misunderstanding of the nature of the proposed CMF, which is in a phenomenological category not reducible to (although intimately related with) neuronal processes. There are no doubt rules for (at least much of) the relationship between the CMF and the physically (externally) observable neural processes. But the rules are not describable *a priori*, i.e., before they are discovered by studying both phenomena simultaneously (e.g., Libet 1987, 1989).

It seems evident, from the "split-brain" studies of Sperry et al. (1969; Sperry, 1985), that transection of the main communicating commissures between the two cerebral hemispheres can result in simultaneously different contents of experience for the two sides. I shall avoid here the argument about whether the isolated non-speaking right hemisphere does or does not actually "have" conscious experience. What is clear, however, is that the contents of conscious mental events in the right hemisphere, are not available to the left hemisphere in this condition. This would imply that any contributions of right hemisphere activity to a mental field cannot directly alter the CMF of the left hemisphere. That is, unity of the CMF would, in these circumstances, be restricted to a given hemisphere. It would also imply that contributions of local neural areas to the overall CMF of a hemisphere are effective only when contiguous with those of other areas; i.e., the contributions would not be effective across substantial gaps of space or of non-neural barriers.

## Experimental Design to Test Theory

The theory of a CMF makes some crucial predictions that can, at least in principle, be tested experimentally. If local areas of cerebral cortex could independently contribute to or alter the larger, unitary CMF, it should be possible to demonstrate such contributions when (a) that cortical area is completely isolated or cut off from neuronal communication with the rest of the brain, but (b) the area remains *in situ*, alive and kept functioning in some suitable manner that sufficiently resembles its normal behavior. The experimental prediction to be tested would be as follows: Suitable electrical and/or chemical activation of the isolated tissue should produce or affect a conscious experience, even though the tissue has no neural connections to the rest of the brain. Possibilities of spread of influences from the isolated slab via physical non-neural paths (e.g., electrical current flow) would have to be controlled for. If a subjective experience is induced and reportable within a second or so, that would tend to exclude spread by chemical diffusion or by changes in vascular circulation or in contents of circulating blood as a cause (see Ingvar, 1954).

Suitable neuronal isolation could be achieved either (a) by surgically cutting all connections to the rest of the brain, but leaving sufficient vascular connections and circulation intact, or (b) by temporarily blocking all nerve conduction into and out of an area. Surgical isolation (a) will be discussed further below. Functional isolation (b) might be achievable by injecting blocking agents in small amounts so as to form a ring of blockade around and under a selected block of cerebral cortex. A local anesthetic agent might be used, such as procaine suitably buffered to pH 7.4 in Ringer's solution. Or, the selective blocker of sodium-conducted action potentials, tetrodotoxin, could be combined with a calcium channel blocker like verapamil (to insure that calcium-mediated action potentials would not escape blockade; see Garcia Ramos and Ibarra, 1973). The advantage of pharmacological method (b) for isolation is its reversibility; this would permit its use on areas of cortex not scheduled for surgical excision, thus greatly enlarging the potential pool of subjects (if risk factors are suitably met). The disadvantages of method (b) are (i) the difficulty of limiting the blockade to a narrow band around the slab, because of diffusibility, (ii) the need to prove that complete blockade has been achieved; (iii) a reduced ability to introduce neural inputs into the isolated slab by the excitation of ascending nerve fibers within the slab but near its lower borders.

## Surgically Isolated Slab of Cortex, *in situ*

A slab of cerebral cortex can be neurally isolated surgically, remaining in place but viable by retaining its blood supply as the only connection with the rest of the brain. This is accomplished by making all of the cuts *subpially.* Studies of the electrophysiological activity of such isolated cortex in situ have been reported (Kristiansen and Courtois, 1949; Burns, 1951, 1954; Echlin et al. 1952; Ingvar, 1954, 1955; Goldring et al. 1961). The basic method involved introducing a narrow curved blade through an opening in an avascular area of the pia-arachnoid membrane. This could undercut a block or slab or cortex and, by bringing its tip up to meet the pia at some distance away, also cut the connections to adjacent cortex. In an earlier study (of how vertical cuts in cortico-cortical connections might affect the integrated, organized function of the sensorimotor cortex in monkeys) Sperry (1947) had used a somewhat different technique. The cutting instrument was an extremely thin double-edged blade made from a fine wire or sewing needle. The sharpened end of this wire was bent to a right angle; this terminal portion of the blade could be sunk vertically into the cortex so that its horizontal arm lay just below the pia. When the vertical knife is pushed forward it cuts through the cortex while its horizontal carrying arm slides just below the pia. This technique could easily be arranged to produce undercutting of the cortex as well. The potential advantage to Sperry's method lies in the very thin line of tissue damage created by this knife, capable of producing chronic scars less than $100\mu m$ thick. That would be particularly desirable if the isolated slab were to remain in situ for therapeutic reasons.

Isolation of a cortical slab has also been performed in *human* subjects, by Echlin et al. (1952), with both general and local anesthesia (patient awake). They reported an immediate reduction but not complete abolition of rhythmic electrical activity (EEG) in the area. After 20 min., paroxysmal bursts of high velocity activity appeared. This kind of seizure pattern in normal brain is usually associated with disruption or distortion of normal functions and, in the motor area, convulsive motor actions. There was no spread of activity from the isolated block to surrounding areas. With only undercutting of a cortical area in human subjects under local anesthesia, Henry and Scoville (1952) also reported autogenous spontaneous activity but of a markedly decreased amount. Also, bursts of fast and slow waves alternated with quiet periods; these were confined to the undercut areas even though there was superficial neural continuity with adjacent areas. In one case a prolonged period of high voltage electrical activity followed probing for the sphenoid ridge below frontal cortex; this indicates that stimulation (in this case mechanical) of already cut input fibers can induce further activity in acutely isolated cortex.

The physiological properties of the isolated slab are obviously immediately altered because of the sudden loss of all inputs. For example, it is well known that destruction of the reticular activating system in the brain stem results in a coma; this afferent input would have to be properly excited so as to "wake up" the isolated slab of cortex. Some procedures to restore some level of activity would be necessary. These could involve local electrical stimulation (e.g., Libet et al. 1964) or the application of exciting chemical agents. Chemical stimulation of isolated cortex has already been studied (Kristiansen and Courtois, 1949; Echlin et al. 1952; Rech and Domino, 1960). With longer term chronic isolation, the nerve fibre inputs and their synaptic contacts with cells in the block would degenerate and no longer provide these normal structural contacts. The studies in question here should be better carried out in the acute phase, during the initial period after isolation. Indeed, with the afferent cut axons still viable and potentially functional, they could be utilized to restore some degree of neural inputs by electrically stimulating them within the slab in a highly localized and controlled fashion.

With surgical isolation the irretrievable loss of normal neural function for a cortical block would limit studies to cases in which a block of tissue has been designated for therapeutic surgical removal from the brain. The study would then be carried out in the operating room before the actual excision of the tissue, if other conditions are also met. These include — the patient being awake and responsive; using local rather than general anesthesia; informed consent and ready cooperativeness by the patient; approval of any risk assessments by all concerned, particularly the hospital/university committee for protection of human subjects. Actually, many patients have been found to tolerate brain surgery under local anesthesia and to participate fruitfully in many past studies (e.g., Penfield, 1958; Libet et al., 1964; Libet, 1973).

A further special requirement of the experiment is that the cortical slab to be isolated should be one for which, when that cortical tissue is still intact before isolation, local electrical stimulation can elicit a conscious subjective experience that is introspectively reportable. The obvious candidates are any of the primary sensory areas — somatic, auditory or visual, for which suitable surface electrical stimulation is known to elicit a primitive sensory experience (e.g., Libet 1973, 1982). However, stimulation of some other cortical areas particularly in temporal lobe have been reported to elicit more complex conscious experiences (e.g., Penfield, 1958). In every case, it would be desirable that a bit of fairly normal responding tissue be included within the block scheduled for excision; but neurosurgeons almost always include such normal bits in order to achieve an adequately therapeutic removal of pathological tissue.

The test of the existence of a CMF that can unify subjective experience would be to see whether electrical or chemical stimulation of a suitably "normalized" isolated cortical slab can elicit an introspective report of an experience. The cortical site of the slab would have to be one at which suitable electrical stimulation does elicit a reportable experience when the brain is intact. There is the possibility that such a cortical site must secondarily activate certain other additional areas in order to produce the conscious experience. In that case these other areas may have to be identified, and multiple isolated slabs that include such areas be included in the experimental stimulation test. In the event of a positive result, possible sources of physical spread of the stimulus to the rest of the brain would have to be excluded, as noted above.

A test of the causal ability of the putative CMF to affect neuronal functions is already implicit in the test just described for the existence of the CMF. If stimulation of the isolated cortical slab can elicit an introspective report by the subject, that could only come about if the CMF could activate the appropriate cerebral areas required to produce the verbal report. However, other specific tests are also possible with cortical areas that have been found specifically to increase their activity when a subject with an intact brain imagines making some movements or imagines some sensory experience. For example, neural activity (as indicate by measurements of regional blood flow or metabolic rate) has been shown to increase selectively in the supplementary motor area (SMA) when the subject is asked to imagine moving his fingers without actually moving them (Ingvar et al.; Roland et al.). Eccles has taken this to be a demonstration of a mental action on neural activities. But there are difficulties with such a conclusion from that experiment: a) There is the technical limitation of temporal resolution by the blood flow - metabolism measurement; this is not fine enough to permit a definite conclusion about which came first — the mental imaging or the increase in SMA activity. b) If it were shown that the mental event did come first here, that would certainly be suggestive of Eccles' kind of interpretation; but there is always the possibility that the whole process was initiated by some neural events elsewhere in the brain, too small or so oriented as to be not recorded by a given recording method. Unless the mental event (of imagining or command) could be shown to precede *any* possible neural event specifically related to the process studied, there could always be doubt about the nature of the causal interaction. With the neurally isolated cortical slab, there are no such difficulties of interpretation. On the other hand, any indirect "extraneuronal" influences from elsewhere in the brain would have to be evaluated and excluded. For example, Ingvar (1954) had reported that stimulation of the reticular activating system in the brain stem could influence electrophysiological activity in completely isolated cortex. This effect appeared to be mediated by a change in blood circulation, but had a long latency of 10 to 70 sec.

If, for example, one had available a neurally isolated cortical slab in the SMA, one could repeat the above described experiment, of asking the subject to imagine moving his fingers. Recordings of electrophysiological responses could be added to those of blood flow and metabolism. Isolation of an SMA block is one of the more impractical possibilities, but an experiment similar in principle could be designed for certain other cortical areas more accessible to therapeutic isolation.

## General Conclusions

Suppose that the experimental results prove to be positive, i.e., suitable stimulation of the neurally isolated cortex elicits some reportable subjective response that is not attributable to stimulation of adjacent non-isolated cortex or of other cerebral structures. That would mean that activation of a cortical area can contribute to over-all unified conscious experience by some mode other than by neural messages delivered via nerve conduction, etc. This would provide crucial support of the proposed field theory, in which a cortical area can contribute to or affect the larger conscious field. It would provide an experimental basis for a unified field of subjective experience and for mental intervention in neuronal functions.

With such a finding one may ask, what would be the role for all the massive and complex neural interconnections, cortico-cortical, cortical-subcortical and hemisphere to hemisphere? An answer might be — to subserve all the cerebral functions other than that directly related to the appearance of the conscious subjective experience and its role in conscious will. It should be noted that all cognitive functions (receipt, analysis, recognition of signals etc.), information storage, learning and memory, processes of arousal and attention and of states of affect and mood, etc. are *not* proposed as functions to be organized or mediated by the postulated CMF (conscious mental field). In short, it is only the phenomenon of conscious subjective experience, associated with all the complex cerebral functions, that is modelled in the CMF.

It may be easy to dismiss the prospect of obtaining "positive" results in the proposed experimental tests, since such results would be completely unexpected from prevalent views of brain functions based on physical connectivities and interactions. But the improbability of positive results is strictly a function of existing views which do not deal sucessfully with the problems of unity of subjective experience and of apparent mental controls of brain processes. The potential implications of the CMF theory and of the positive results it predicts are clearly profound in nature. On those grounds, and because the proposed experiments are in principle workable although difficult, the proposed experimental design should merit a serious place in investigations of the mind-brain problem.

## References

Burns BD (1951): Some properties of isolated cerebral cortex in the unanesthetized cat. *J Physiol (Lond.)* 112:156–175

Burns BD (1954): The production of after-bursts in isolated unanesthetized cerebral cortex. *J Physiol (Lond.)* 125:427–446

Doty RW (1984): Some thoughts and some experiments on memory. In: *Neuropsychology of Memory*, LR Squire and N Butters,eds., New York: Guilford, pp. 330–339

Doty RW (1985): The time course of conscious processing: Vetoes by the uninformed? *Behavioral and Brain Sci* 8:541–542

Eccles JC (1990): A unitary hypothesis of mind-brain interaction in cerebral cortex. *Proc Roy Soc Lond B* 240:433–451

Echlin FA, Arnett V and Zoll J (1952): Paroxysmal high voltage discharges from isolated and partially isolated human and animal cerebral cortex. *Electroenceph & Clin Neurophysiol* 4:147–164

Garcia Ramos J and Ibarra BH (1973): Studies on the mechanisms of learning, II, on the ionic nature of the dendritic action potentials and mescaline spikes. *Acta Physiol Latino Amer* 23:202–212

Goldring S, O'Leary JL, Holmes TG and Jerva MJ (1961): Direct response of isolated cerebral cortex of cat. *J Neurophysiol* 24:633–650

Gray CM and Singer W (1989): Stimulus-specific neuronal oscillations in orientation columns of cat visual cortex. *Proc Natl Acad Sci USA* 86:1698–1702

Henry CE and Scoville WB (1952): Suppression-burst activity from isolated cerebral cortex in man. *Electroenceph & Clin Neurophysiol* 4:1–22

Holender D (1986): Semantic activation without conscious identification in dichotic listening, parafoveal vision, and visual masking: A survey and appraisal. *Behavioral and Brain Sci* 9:1–66

Ingvar D (1955a): Electrical activity of isolated cortex in the unanesthetized cat with intact brain stem. *Acta Physiol Scand* 33:151–168

Ingvar D (1955b): Extraneuronal influences upon the electrical activity of isolated cortex following stimulation of the rettricular activating system. *Acta Physiol Scand* 33:169–193

Ingvar D and Phillipson L (1977): Distibutions of cerebral blood flow in the dominant hemisphere during motor ideation and motor performance. *Ann Neurol* 2:230–237

Jensen AR (1979): "g": Outmoded theory or unconquered frontier? *Creative Science and Technology* 11:16–29

Kristiansen K and Courtois G (1949): Rhythmic electrical activity from isolated cerebral cortex. *Electroenceph & Clin Neurophysiol* 1:265–272

Libet B (1965) [see this book, no. 2]

Libet B et al. (1972) [see this book, no. 7]

Libet B (1973) [see this book, no. 5]

Libet B (1978) [see this book, no. 8]

Libet B (1982) [see this book, no. 12]

Libet B (1985) [see this book, no. 16]

Libet B (1987) [see this book, no. 18]

Libet B (1989) [see this book, no. 20]

Libet B, Alberts WW, Wright EW, De Lattre LD, Levin G and Feinstein B (1964) [see this book, no. 1]

Libet B, Wright EW and Gleason C (1982) [see this book, no. 13]

Libet B, Wright EW and Gleason C (1983a) [see this book, no. 14]

Libet B, Wright EW, Gleason CA and Pearl DK (1983b) [see this book, no. 15]

Libet B, Wright EW, Feinstein B and Pearl DK (1979) [see this book, no. 9]

Libet B, Pearl DK, Morledge DM, Gleason CA, Hosobuchi Y and Barbaro NM (1991) [see this book, no. 21]

Libet B, Wright EW, Feinstein B and Pearl DK (1992) [see this book, no. 10]

McKay DM (1985): Do we control our brains? *Behavioral and Brain Sci* 8:546

Penfield W (1958): *The Excitable Cortex in Conscious Man.* Liverpool: Liverpool University Press

Popper KR and Eccles JC (1977): *The Self and its Brain.* Heidelberg: Springer

Rech RH and Domino EF (1960): Effects of various drugs on activity of the neuronally isolated cerebral cortex. *Exper Neurol* 2:364–378

Roland PE and Friberg L (1985): Localization of cortical areas activated by thinking. *J Neurophysiol* 53:1219–1243

Sherrington CS (1940): *Man in His Nature.* Cambridge University Press

Singer W (1991): Response synchronization of cortical neurons: an epiphenomenon or a solution to the binding problem. *IBRO News* 19:6–7 (New York: Pergamon)

Sperry RW (1947): Cerebral regulation of motor coordination in monkeys following multiple transections of sensorimotor cortex. *J Neurophysiol* 10:275–294

Sperry RW (1952): Neurology and the mind-brain problem. *American Scientist* 40:291–312

Sperry RW (1980): Mind-brain interaction: Mentalism yes; dualism no. *Neuroscience* 5:195–206

Sperry RW (1985): *Science and Moral Priority.* Westport, Conn.: Praeger Publ

Sperry RW, Gazzaniga MS and Bogen JE (1969): Interhemispheric relationships: The neocortical commissures: Syndromes of hemisphere disconnection. In: *Handbook of Clinical Neurology* vol. 4, PJ Vinken and GW Bruyn (eds), Amsterdam: North Holland Publ

Taylor JL and McCloskey DI (1990): Triggering of preprogrammed movements as reactions to masked stimuli. *J Neurophysiol* 63:439–446

Weiskrantz L (1986): *Blindsight: A Case Study and Implications.* Oxford: Clarendon Press

Wood CC (1985): Pardon, your dualism is showing. *Behavioral and Brain Sci* 8:557–558

# Permissions

Birkhäuser Boston would like to thank the original publishers of the papers of Benjamin Libet for granting permission to reprint specific papers in this collection.

[1] Reprinted from *J. Neurophysiol* **27**, ©1964 by The American Physiological Society.
[2] Reprinted from *Perspectives in Biology and Medicine* **9** ©1965 by the University of Chicago.
[3] Reprinted from *Brain and Conscious Experience*, edited by J.C. Eccles, ©1966 by Springer-Verlag New York.
[4] Reprinted from *Science* **158**, 1597–1600, ©1964 by the AAAS.
[5] Reprinted from *Handbook of Sensory Physiology, Vol. II: Somatosensory System*, edited by A. Iggo, ©1973 by Springer-Verlag New York.
[6] Reprinted from *The Somatosensory System*, edited by H.H. Kornhuber, ©1975 by Georg Thieme. Verlag Stuttgart.
[7] Reprinted from *Neurophysiology Studies in Man*, edited by G.G. Somjen, ©1972 by Elsevier Science Publishers.
[8] Reprinted from *Brain: A Journal of Neurology* **102**, ©1979 by Oxford University Press.
[9] Reprinted from *Cerebral Correlates of Conscious Experience*, edited by P.A. Buser and A. Rougeul-Buser, ©1978 by Elsevier Science Publishers.
[10] Reprinted from *Consciousness and Cognition* **1**, ©1992 by Academic Press.
[11] Reprinted from *Philosophy of Science* **48**, ©1981 by the Philosophy of Science Association.
[12] Reprinted from *Human Neurobiology* **1**, ©1982 by Springer-Verlag New York.
[13] Reprinted from *Electroencephalography and Clinical Neurophysiology* **54**, ©1982 by Elsevier Science Publishers.
[14] Reprinted from *Electroencephalography and Clinical Neurophysiology* **56**, ©1983 by Elsevier Science Publishers.
[15] Reprinted from *Brain: A Journal of Neurology* **106**, ©1983 by Oxford University Press.
[16] Reprinted from *The Behavioral and Brain Sciences* **8** ©1985 by Cambridge University Press.
[17] Reprinted from *The Behavioral and Brain Sciences* **10**, ©1987 by Cambridge University Press.

MIX
Papier aus verantwortungsvollen Quellen
Paper from responsible sources
FSC® C105338

If you have any concerns about our products,
you can contact us on
**ProductSafety@springernature.com**

In case Publisher is established outside the EU,
the EU authorized representative is:
**Springer Nature Customer Service Center GmbH**
**Europaplatz 3, 69115 Heidelberg, Germany**

Printed by Libri Plureos GmbH
in Hamburg, Germany